Biofuel from Microbes and Plants

Biofuel from Microbes and Plants

Green Energy Alternative

Edited by

Nitish Kumar

CRC Press is an imprint of the
Taylor & Francis Group, an **informa** business

First edition published 2022
by CRC Press
6000 Broken Sound Parkway NW, Suite 300, Boca Raton, FL 33487–2742

and by CRC Press
2 Park Square, Milton Park, Abingdon, Oxon, OX14 4RN

© 2022 Taylor & Francis Group, LLC

CRC Press is an imprint of Taylor & Francis Group, LLC

Reasonable efforts have been made to publish reliable data and information, but the author and publisher cannot assume responsibility for the validity of all materials or the consequences of their use. The authors and publishers have attempted to trace the copyright holders of all material reproduced in this publication and apologize to copyright holders if permission to publish in this form has not been obtained. If any copyright material has not been acknowledged please write and let us know so we may rectify in any future reprint.

Except as permitted under U.S. Copyright Law, no part of this book may be reprinted, reproduced, transmitted, or utilized in any form by any electronic, mechanical, or other means, now known or hereafter invented, including photocopying, microfilming, and recording, or in any information storage or retrieval system, without written permission from the publishers.

For permission to photocopy or use material electronically from this work, access www.copyright.com or contact the Copyright Clearance Center, Inc. (CCC), 222 Rosewood Drive, Danvers, MA 01923, 978–750–8400. For works that are not available on CCC please contact mpkbookspermissions@tandf.co.uk

Trademark notice: Product or corporate names may be trademarks or registered trademarks and are used only for identification and explanation without intent to infringe.

Library of Congress Cataloging-in-Publication Data
Names: Kumar, Nitish, editor.
Title: Biofuel from microbes and plants : green energy alternative / edited by Nitish Kumar.
Description: First edition. | Boca Raton : CRC Press, 2022. | Includes bibliographical references.
Identifiers: LCCN 2021026749 (print) | LCCN 2021026750 (ebook) | ISBN 9780367207052 (hbk.) |
    ISBN 9781032106816 (pbk.) | ISBN 9780429262975 (ebk.)
Subjects: LCSH: Biomass energy.
Classification: LCC TP339 .B5378 2022 (print) | LCC TP339 (ebook) | DDC 662/.88—dc23
LC record available at https://lccn.loc.gov/2021026749
LC ebook record available at https://lccn.loc.gov/2021026750

ISBN: 978-0-367-20705-2 (hbk)
ISBN: 978-1-032-10681-6 (pbk)
ISBN: 978-0-429-26297-5 (ebk)

DOI: 10.1201/9780429262975

Typeset in Times New Roman
by Apex CoVantage, LLC

Contents

Preface vii
Acknowledgments ix
Editor xi
Contributors xiii

1 **Biofuel: The Green Alternative** 1
  *Nikita Patel, Swetal Patel, A.A. AbdulRahaman,
  and R. Krishnamurthy*

2 **Algal-Based Biofuel: Challenges and Future Perspectives** 23
  *Waleed M.M. El-Sayed, Hassan A.H. Ibrahim,
  Mohamed A.A. Abdrabo, and Usama M. Abdul-Raouf*

3 **Waste to Energy: A Means of Sustainable Development through Bioethanol Production** 43
  Girish Venkatachalapathy and Girisha Shiringala Thimappa

4 **The Role of Microbes in Biofuel Production** 65
  *Arpan Kumar Basak, Sayantani M. Basak, Kazimierz Strzałka,
  and Pradip Kumar Chatterjee*

5 **A Comparative Account on Biodiesel Production from Forest Seeds** 129
  Jigna G. Tank and Rohan V. Pandya

6 **Development of Low-Cost Production Medium and Cultivation Techniques of Cyanobacteria *Arthrospira platensis* (*Spirulina*) for Biofuel Production** 165
  *R. Dineshkumar, N. Sharmila Devi, M. Duraimurugan,
  A. Ahamed Rasheeq, and P. Sampathkumar*

7 **Biofuel and Halophytes** 179
  Aneesha Singh and Krupali Dipakbhai Vyas

8 **Eco-Friendly Applications of Natural Secondary Metabolites and Status of Siderophores** 189
  *Pratika Singh, Azmi Khan, Micky Anand, Hemant Kumar,
  Shivpujan Kumar, and Amrita Srivastava*

9 **Genetic Diversity Analysis of *Jatropha curcas*: A Biofuel Plant** 211
  Nitish Kumar

10 **Cellulase Immobilization on Magnetic Nanoparticles for Bioconversion of Lignocellulosic Biomass to Ethanol** 225
  Prabhpreet Kaur and Monica Sachdeva Taggar

11 Biogas Production in a Biorefinery Context: Analysis of the Scale Based on Different Raw Materials 245
J.A. Poveda Giraldo, M. Ortiz-Sánchez, S. Piedrahita-Rodríguez, J.C. Solarte-Toro, A.M. Zetty-Arenas, and C.A. Cardona Alzate

12 Plant-Based Biofuels and Sustainability Issues 269
Lakshmi Gopakumar

13 Microbial-Based Biofuel Production: A Green Sustainable Approach 285
J. Ranjitha, Shreya Subedi, M. Anand, R. Shobana, S. Vijayalakshmi, and Bhaskar Das

14 Biodiesel Production from *Mimusops elengi* Seed Oil through Means of Co-Solvent–Based Transesterification Using an Ionic Liquid Catalyst 299
Gokul Raghavendra Srinivasan, Shalini Palani, Mamoona Munir, Muhammed Saeed and Ranjitha Jambulingam

15 Conventional Breeding Methods for the Genetic Improvement of *Jatropha curcas* L.: A Biodiesel Plant 319
Nitish Kumar

Preface

The depletion of petroleum-derived fuel and environmental concerns has prompted many millennials to consider biofuels as alternative fuel sources. But completely replacing petroleum-derived fuels with biofuels is currently impossible in terms of production capacity and engine compatibility. Nevertheless, the marginal replacement of diesel with biofuel could delay the depletion of petroleum resources and abate the radical climate change caused by automotive pollutants. Energy security and climate change are the two major driving forces for worldwide biofuel development and also have the potential to stimulate the agroindustry. The development of biofuels as alternative and renewable sources of energy has become critical in national efforts toward maximum self-reliance, the cornerstone of our energy security strategy. At the same time, the production of biofuels from various types of biomass such as plants, microbes, algae, and fungi is now an ecologically viable and sustainable option. This book describes the biotechnological advances in biofuel production from various sources, while also providing essential information on the genetic improvement of biofuel sources at both the conventional and genomic levels. These innovations and the corresponding methodologies are explained in detail. *Biofuel from Microbes and Plants: Green Energy Alternative* contains 15 chapters, which cover the latest developments in the research on a promising biofuel crop *Jatropha* and address the role of microorganisms in biofuel production.

Acknowledgments

Thanks to all the authors of the various chapters for their contributions. It was a bit of a long process from the initial outlines to developing the full chapters and then revising them in light of reviewers' comments. We sincerely acknowledge the authors' willingness to go through this process. I also commend the work and knowledge of the members of our review panels, many of which had to be done at short notice. Thanks to all the people at CRC Press/Taylor & Francis, India, especially Ms. Renu Upadhyay and Ms. Jyotsna Jangra, with whom we corresponded for their advice and facilitation in the production of this book. I am grateful to my family members Mrs. Kiran (wife), Miss Kartika Sharma (daughter), and my parents for their continuous incredible and selfless support.

Editor

Dr. Nitish Kumar is Senior Assistant Professor at the Department of Biotechnology, Central University of South Bihar, Gaya, Bihar, India. Dr. Kumar completed his doctoral research at the Council of Scientific & Industrial Research–Central Salt & Marine Chemicals Research Institute, Bhavnagar, Gujarat, India. He has published more than 60 research articles and book chapters in leading international and national journals and books. He has a wide area of research experience in the field of renewable source of energy. He has received many awards/fellowships/projects from various organizations, for example, the CSIR, DBT, ICAR and SERB-DST, BRNS-BARC, among others. He is an active reviewer for journals, including *Biotechnology Reports, Aquatic Botany, Industrial Crops and Products, PLoS One, Plant Biochemistry and Biotechnology*, and *3Biotech*. He also serves as an associate editor of the journal *Gene*.

Contributors

Mohamed A.A. Abdrabo
Marine Microbiology Department
National Institute of Oceanography and Fisheries
Egypt

A.A. AbdulRahaman
Department of Plant Biology
Faculty of Life Sciences
University of Ilorin
Kwara State, Nigeria

Usama M. Abdul-Raouf
Botany and Microbiology Department
Faculty of Science
Al-Azhar University
Egypt

C.A. Cardona Alzate
Universidad Nacional de Colombia Sede Manizales
Instituto de Biotecnología y Agroindustria, Laboratorio de Equilibrios Químicos y Cinética Enzimática
Departamento de Ingeniería Química
Manizales, Colombia

M. Anand
Department of Science and Humanities
Kingston Engineering College
Vellore, India

Micky Anand
Department of Life Science
School of Earth, Biological and Environmental Sciences
Central University of South Bihar
Gaya, Bihar, India

Arpan Kumar Basak
Malopolska Centre of Biotechnology
Jagiellonian University
Krakow, Poland

Sayantani M. Basak
Faculty of Biology
Jagiellonian University
Krakow, Poland

Pradip Kumar Chatterjee
Central Mechanical Engineering Research Institute (CSIR)
Durgapur, India

Bhaskar Das
Department of Environment and Water Resources Engineering
School of Civil and Chemical Engineering
Vellore Institute of Technology
Vellore, India

N. Sharmila Devi
Department of Microbiology
Karpagam Academy for Higher Education
Coimbatore, Tamil Nadu, India

R. Dineshkumar
Department of Microbiology
Karpagam Academy for Higher Education
Coimbatore, Tamil Nadu, India

M. Duraimurugan
Department of Microbiology
Karpagam Academy for Higher Education
Coimbatore, Tamil Nadu, India

Waleed M.M. El-Sayed
Marine Microbiology Department
National Institute of Oceanography and Fisheries
Egypt

J.A. Poveda Giraldo
Universidad Nacional de Colombia Sede Manizales
Instituto de Biotecnología y Agroindustria, Laboratorio de Equilibrios Químicos y Cinética Enzimática
Departamento de Ingeniería Química
Manizales, Colombia

Lakshmi Gopakumar
School of Environmental Studies
Cochin University of Science and Technology
Kerala, India

Hassan A.H. Ibrahim
Marine Microbiology Department
National Institute of Oceanography and Fisheries
Egypt

Prabhpreet Kaur
Department of Biochemistry
Punjab Agricultural University
Ludhiana, Punjab, India

Azmi Khan
Department of Life Science
School of Earth, Biological and Environmental Sciences
Central University of South Bihar
Gaya, Bihar, India

R. Krishnamurthy
C.G. Bhakta Institute of Biotechnology
Uka Tarsadia University
Surat, Gujarat, India

Hemant Kumar
Department of Life Science
School of Earth, Biological and Environmental Sciences
Central University of South Bihar
Gaya, Bihar, India

Nitish Kumar
Department of Biotechnology
Central University of Bihar
Gaya, Bihar, India

Shivpujan Kumar
Department of Life Science
School of Earth, Biological and Environmental Sciences
Central University of South Bihar
Gaya, Bihar, India

Mamoona Munir
Department of Plant Sciences
Quaid-i-Azam University
Islamabad, Pakistan

M. Ortiz-Sánchez
Universidad Nacional de Colombia Sede Manizales
Instituto de Biotecnología y Agroindustria, Laboratorio de Equilibrios Químicos y Cinética Enzimática
Departamento de Ingeniería Química
Manizales, Colombia

Shalini Palani
CO_2 Research and Green Technologies Centre
Vellore Institute of Technology
Vellore, India

Rohan V. Pandya
Department of Microbiology
Faculty of Science
Atmiya University
Rajkot, Gujarat, India

Nikita Patel
C.G. Bhakta Institute of Biotechnology
Uka Tarsadia University
Surat, Gujarat, India

Swetal Patel
C.G. Bhakta Institute of Biotechnology
Uka Tarsadia University
Surat, Gujarat, India

S. Piedrahita-Rodríguez
Universidad Nacional de Colombia Sede Manizales
Instituto de Biotecnología y Agroindustria, Laboratorio de Equilibrios Químicos y Cinética Enzimática
Departamento de Ingeniería Química
Manizales, Colombia

Pratika Singh
Department of Life Science
School of Earth, Biological and Environmental Sciences
Central University of South Bihar
Gaya, Bihar, India

J. Ranjitha
CO_2 Research and Green Technologies Centre
Vellore Institute of Technology
Vellore, India

A. Ahamed Rasheeq
Department of Fisheries Resources and Management
College of Fisheries
Karnataka Veterinary, Animal and Fisheries Sciences University
Mangalore, Karnataka, India

Mohammed Saeed
Department of Chemistry
Quaid-i-Azam University
Islamabad, Pakistan

P. Sampathkumar
Faculty of Marine Sciences
Centre of Advanced Study in Marine Biology
Annamalai University
Parangipettai, Tamil Nadu, India

Contributors

R. Shobana
CO_2 Research and Green Technologies Centre
Vellore Institute of Technology
Vellore, India

Aneesha Singh
Applied Phycology and Biotechnology Division
Central Salt and Marine Chemicals Research Institute (CSIR)
Gijubhai Badheka Marg
Bhavnagar, Gujarat, India

J.C. Solarte-Toro
Universidad Nacional de Colombia Sede Manizales
Instituto de Biotecnología y Agroindustria, Laboratorio de Equilibrios Químicos y Cinética Enzimática
Departamento de Ingeniería Química
Manizales, Colombia

Gokul Raghavendra Srinivasan
CO_2 Research and Green Technologies Centre
Vellore Institute of Technology
Vellore, India

Amrita Srivastava
Department of Life Science
School of Earth, Biological and Environmental Sciences
Central University of South Bihar
Gaya, Bihar, India

Kazimierz Strzałka
Faculty of Biochemistry, Biophysics and Biotechnology
Department of Plant Physiology and Biochemistry
Jagiellonian University
Krakow, Poland

Shreya Subedi
Department of Mechanical Engineering
Vellore Institute of Technology
Vellore, India

Monica Sachdeva Taggar
Department of Renewable Energy Engineering
Punjab Agricultural University
Ludhiana, Punjab, India

Jigna G. Tank
UGC-CAS Department of Biosciences
Saurashtra University
Rajkot, Gujarat, India

Girisha Shiringala Thimappa
Department of Microbiology and Biotechnology
JB Campus, Bangalore University
Bengaluru, Karnataka, India

Girish Venkatachalapathy
Department of Microbiology and Biotechnology,
JB Campus, Bangalore University
Bengaluru, Karnataka, India

S. Vijayalakshmi
CO_2 Research and Green Technologies Centre
Vellore Institute of Technology
Vellore, India

Krupali Dipakbhai Vyas
Academy of Scientific & Innovative Research (AcSIR)
Ghaziabad, India
and
Applied Phycology and Biotechnology Division
Central Salt and Marine Chemicals Research Institute (CSIR)
Bhavnagar, Gujarat, India

A.M. Zetty-Arenas
Universidad Nacional de Colombia Sede Manizales
Instituto de Biotecnología y Agroindustria, Laboratorio de Equilibrios Químicos y Cinética Enzimática
Departamento de Ingeniería Química
Manizales, Colombia

1 Biofuel: The Green Alternative

Nikita Patel, Swetal Patel, A.A. AbdulRahaman, and R. Krishnamurthy

CONTENTS

1.1	Introduction	1
1.2	Types of Biofuels	3
	1.2.1 Bioalcohol (Ethanol, Propanol, Butanol)	3
	1.2.2 Biodiesel (Fischer–Tropsch Biodiesel, Green Diesel)	3
	1.2.3 Bioethers	4
	1.2.4 Biogas	4
	1.2.5 Solid Biofuels	4
	1.2.6 Cellulosic Ethanol	4
	1.2.7 Algae-Based Biofuels	4
	1.2.8 Biohydrogen	4
	1.2.9 Dimethylfuran	5
	1.2.10 Aviation Biofuels	5
1.3	Algae as a Prominent Source of Biofuel	5
1.4	Agronomic Practices in Algal Biomass Production	6
1.5	Biodiesel: Source Materials and Production Technology	7
	1.5.1 Biodiesel Production Source Materials	8
	1.5.1.1 *Jatropha curcas* L.	8
	1.5.1.2 Soapnut Seeds (*Sapindus mukorossi*)	9
	1.5.1.3 Karanj (*Pongomia pinnata* [L.] Pierre)	9
	1.5.1.4 Neem (*Azadirachta indica* A.Juss.)	9
	1.5.1.5 Mahua (*Madhuca indica* J.Konig, J.F.Macbr.)	9
	1.5.1.6 Date Palm Seeds (*Phoenix dactylifera* L.)	10
	1.5.1.7 *Moringa oleifera* Lam. Seeds (Drumstick Tree)	10
	1.5.1.8 *Ricinus communis* L. (Castor Beans)	10
	1.5.1.9 *Thevetia peruviana* (Pers.) K. Schum. (Yellow Oleander)	11
	1.5.1.10 Sapota (*Manilkara zapota* [L.] P.Royen)	11
1.6	Biodiesel Production Process and Technology	11
1.7	Agrowaste as a Source of Biofuel	13
1.8	Biofuels from Fungi	14
References		15

1.1 Introduction

Biofuels are the renewable source of energy generally produced from organic or waste materials which plays an important role in environmental reduction of carbon dioxide (CO_2). In 2018, energy usage grew twice the average growth rate since 2010, due to the global economy and higher heating and cooling needs in some parts of the world, according to the International Energy Agency's *Global Energy & CO_2 Status Report* (2019), and as a result, there was an increase in fuel demand. As a result, that year, CO_2

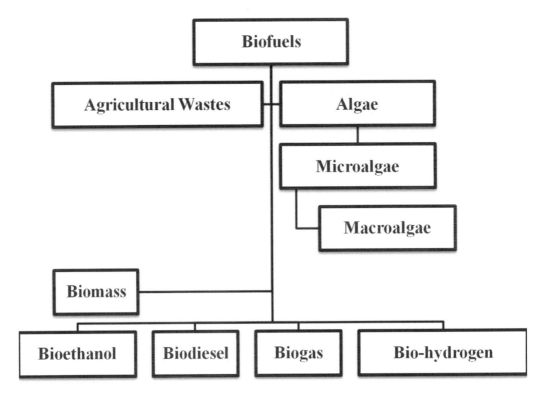

FIGURE 1.1 Biofuel sources.

emission rose 1.7% and reached a new record (International Energy Agency, 2019). Recently, biofuels are gaining much importance because of their eco-friendly nature. As they are eco-friendly in nature, they could replace gasoline and diesel, respectively. Biofuels can be produced from various sources, such as microbes, algae, plants, and agricultural or industrial waste (Figure 1.1). The worldwide energy requirement is produced from 9 billion tons of oil, which is sourced from fossil fuels and which emit more than 6.3 billion tons of carbon dioxide and is equivalent to a 45% increase in worldwide annual release of carbon dioxide by humans since 1990 (Fernando et al., 2006). According to the United Nations (2018), urbanization increased from 27% to 59% in Asia and it is estimated that it will increase upto 74% by 2050. More than half of the world population depends on fossil fuels such thatthey consume 76% anthracite or coal, 62% oil, and 82% natural gases (Hirani et al., 2018; Sullivan, 2010) which are recognized as non-renewable resources of energy which creates global warming—and as a result, the ozone layer is depleting day by day. Biofuels can serve as great environmentally friendly alternatives in mitigating climate change (Kothari & Gujral, 2013).

Biogas is also an important biofuel which is generated by anaerobic digestion of organic waste, which produces methane gas and carbon dioxide as byproducts. Biogas production is an underground process which has been carried out in rural areas throughout the ages. Apart from biogas production, anaerobic digestion is also useful in wastewater treatment and for household purposes (Beschkov, 2017).

According to their processing technologies and feedstocks, biofuels are often divided into four generations (Table 1.1).

First-generation biofuels are produced from starch, sugars, or vegetable oil through fermentation, distillation, or transesterification, and the products are generally ethanol, methane, and biodiesel. Feedstocks play a significant role in production-to-cost ratio of first-generation biofuel, as food crops are the major sources of production. They are also classified into: (a) sugar-based feedstocks and

TABLE 1.1

Classification of Biofuels

Biofuel Generation	Production Sources	Production Technologies	Products	References
First-generation biofuels	Starch, sugar or vegetable oil	Fermentation, distillation, transesterification	Ethanol, biodiesel, methane	(Lü *et al.*, 2011)
Second-generation biofuels	Wood, wheat straw, food waste	Thermochemical or biochemical process to break lignins and sugars	Syngas (hydrocarbon, carbon monoxide and hydrogen mixture)	(Lü *et al.*, 2011; Patel *et al.*, 2016)
Third-generation biofuels	Algae	Extraction of oils from algae	Biodiesel, glycerin and protein residues	(Patel *et al.*, 2016)
Fourth-generation biofuels	Genetically modified algae	Genetic engineering, pyrolysis, gasification	Hydrocarbons and Biogasoline	(Abdullah *et al.*, 2019)

(b) starch-based feedstocks (Hirani *et al.*, 2018). Biofuels of the second generation are produced from wood, wheat straw, and food waste by thermochemical or biochemical process, with syngas as their byproduct (Lü *et al.*, 2011; Patel *et al.*, 2016). Third-generation biofuels are potentially produced from algae/microalgae through an extraction process which results in production of biodiesel, glycerin, and important protein residues. As they are smaller in size with exponential growth rate within 24 hours, they are good sources of biofuel production. Fourth-generation biofuels are produced through genetically modified algae, which results in enhanced biomass production. These biofuels are produced by modifying or targeting specific markers or genes in the selected algal species for biomass production. However, they may impose great risk to environment and human health if they are not processed properly (Tandon & Jin, 2017; Abdullah *et al.*, 2019).

1.2 Types of Biofuels

On the basis of their processing and production technologies, biofuels are categorized into the following.

1.2.1 Bioalcohol (Ethanol, Propanol, Butanol)

Bioalcohols are the alcohols that are produced from biological resources through enzymatic action or fermentation by microorganism (Melikoglu *et al.*, 2016). The most common biofuel produced worldwide is ethanol and the less common biofuels are biobutanol and biopropanol. It is claimed that biobutanol (also known as biogasoline) can replace gasoline, which can be used directly to run engines (Liu *et al.*, 2013). Their energy density, on the other hand, ranges from 30–36.6 megajoules/kg.

1.2.2 Biodiesel (Fischer–Tropsch Biodiesel, Green Diesel)

According to the United States National Biodiesel Board, biodiesels are mono alkyl esters produced from vegetable oil/animal fats (Demirbas, 2008). In general, biodiesel is produced from triglycerides by reacting them with alcohol to form fatty acid alkyl/methyl esters. Fischer–Tropsch biodiesel is generated by gasification of biomass to release a mixture of carbon monoxide and hydrogen gas. The diesel produced by this method is same as the diesel derived from fossil fuels in terms of density, viscosity, and energy (Sauciuc *et al.*, 2011). Contrary to biodiesel, which is produced by a chemical process, green diesel is produced by a refining process. It is produced by hydrogenation of oils. It is identical to fossil fuel diesel with no sulfur content. It can be used directly without modifying the engines, whereas biodiesel requires a slight modification in the engines (Morone *et al.*, 2016).

1.2.3 Bioethers

Bioethers are produced upon dehydration of biomass. Sugar beets or wheat are mostly used in production of bioethers. Apart from sugar beet or wheat, they can also be produced from glycerol. They possess high combustion and emission rates when compared to other biofuels. Dimethyl ether and diethyl ether are mostly used in the fuel industry. Dimethyl ether possesses high volatility and a high cetane number, whereas diethyl ether possesses a very high cetane number with reasonable energy density (Shukla et al., 2007).

1.2.4 Biogas

Biogas is a type of biofuel produced by anaerobic decomposition/breakdown of organic waste or animal waste which results in formation of a mixture of gas, generally methane and carbon dioxide also known as syngas. Wood gas is also a type of biogas which is produced by gasification of wood and other biomass materials. This type of gas basically possesses traces containing methane, nitrogen, and carbon monoxide. Biogas is generally used as household fuel to cater easily to daily requirements. It also plays an important role in solid waste management in this modern era. Further, compressed natural biogas can be used as renewable fuel to run motor engines, replacing large amounts of vehicle fuel (Bhatia & Gupta, 2019; Bhatia, 2014; Kristoferson & Bokalders, 1986).

1.2.5 Solid Biofuels

Solid biofuels include sawdust or wood pellets/chips which can be used directly to generate energy. More than 2 billion people all around the world use animal dung, most commonly cow dung, to generate energy where limited resources are available to meet their daily requirements. Cow dung can generate 67.9% methane (Ukpai & Nnabuchi, 2012) and 50% carbon dioxide. Municipal waste can also generate large amounts of energy by proper processing. It mainly possesses starchy materials, and hardly digestible lignocellulosic compounds which proved to be potential biofuel production materials (Mahmoodi et al., 2018).

1.2.6 Cellulosic Ethanol

Cellulosic ethanol is produced from lignocellulosic material, e.g. it can be produced by proper processing of sawdust, wheat straw, or jatropa. Field residual materials such as leaves, stems, or other parts of the plant can be used to produce ethanol efficiently (Ziolkowska, 2020). Corncob contains 40–70% hemicellulose and 72–90% cellulose which could be processed to alcohol/ethanol via specific microorganisms (Chen et al., 2010). Recently, it has been discovered that pineapple waste, paper waste, several fibrous wastes, and coffee residue proved to be efficient in ethanol production on a large scale (Chen et al., 2010; Dutta et al., 2014; Choi et al., 2012; Ziolkowska, 2020).

1.2.7 Algae-Based Biofuels

Biofuels produced by extracting oil from macroalgae or microalgae are known as algal-based biofuels. Recently, they have gained importance, as they have short doubling rates. They could be easily cultivated in open ponds or photobioreactors, and result in low emission of greenhouse gases (Patel et al., 2016).

1.2.8 Biohydrogen

Biohydrogen is a biofuel produced biological conversion of biomass. It is also produced from algae on a larger scale by thermochemical conversion (thermochemical process refers to combustion, gasification, and pyrolysis of biomass) (Sadhwani et al., 2013; Zhang & Zhang, 2019). Several production technologies are incorporated to produce biohydrogen, such as steam reforming of methane gas produced in biogas which further reacts with natural gas to produce hydrogen and by fermentation technology using specific microbial strain. Several algal species under anaerobic conditions also produce hydrogen in bulk quantity (Bioenergy, 2019; Wang & Yin, 2018).

1.2.9 Dimethylfuran

Dimethylfuran is an important biofuel that could replace low carbon biofuels such as ethanol and butanol. It could be produced on a large scale by converting cellulosic materials. It possesses high density and a high octane number with a suitable boiling point, water insolubility, and miscibility with gasoline and diesel oil at any ratio (Qian *et al.*, 2015).

1.2.10 Aviation Biofuels

To reduce carbon emissions, aviation biofuels were designed for aircraft. Ethanol and methanol were not used as aviation biofuels because of their low energy levels (biofuel.org.uk). Aviation biofuels are also known as biojet fuels, and were initially generated from first-generation feedstocks. Later, third-generation feedstocks were used in production of biojet fuels because they do not compete with major food crops (Hajilary *et al.*, 2019). Jet A and Jet B are the two types of petroleum-based jet fuels. They are non-renewable in nature. Hence, an attempt was made by Baroutian and his co-workers (2013), blending methyl esters (ME) of waste vegetable oil and *Jatropha* oils with Jet A-1 aviation fuel, and it was found that 10% and 20% ester contents have similar characteristics with the petroleum-based aviation fuel (Jet A-1) and it could replace the commercial fuel.

1.3 Algae as a Prominent Source of Biofuel

Algae are simple aquatic plants which possess chlorophyll but lack true roots, stems, and leaves. They are unicellular or multicellular in nature. There are more than 300,000 algal species distributed all over the world in diverse habitats including seawater, wastewater and freshwater habitats (Oilgae, 2010; Bhateria & Dhaka, 2014). Algae are generally classified into: (a) macroalgae or (b) microalgae, depending on their size and growth pattern. Macroalgae are generally multicellular in nature, whereas microalgae are the most primitive forms of plants which undergo photosynthesis, same as do higher plants. They possess a simple cellular structure and transform solar energy into chemical energy very efficiently and rapidly. Generally, they grow in suspension in water bodies (Chang, 2007). Their doubling period varies from species to species. Some microalgae grows exponentially within 24 hours and some within 3–4 hours (Metting, 1996). Microalgae are efficient CO_2 fixers and are able to remove CO_2 from flue gases of power plants, which thus reduce greenhouse gases.

Microalgae are cultivated through open-pond system such as natural ponds, raceways, lakes or inclined systems which comprise a paddle wheel and a CO_2 supply. They are also cultivated in closed systems such as photobioreactors under controlled environments. Both of the methods are reliable and useful in cultivation of microalgae on a larger scale for biofuel production. Algal biomass possesses carbohydrates, proteins, and lipids in form of triacylglycerol, which is in great consideration for biodiesel production. There are four main groups of microalgae such as: (a) green algae; (b) diatoms; (c) blue-green algae; and (d) golden algae (Khan *et al.*, 2009). The lipid content varies from species to species; for example, *Schizochytrium* sp. contains 55–70% lipid while *Ankistrodesmus* sp., *Botryococcus braunii, Chaetoceros muelleri* and *Ellipsoidion* sp. contains 12–37% lipid (Patel *et al.*, 2016).

Algae are also classified on the basis of their metabolic activity as: (a) photoautotrophic algae; (b) heterotrophic algae; and (c) mixotrophic algae. Photoautotrophic algae are those that use light as a sole energy source for photosynthesis, whereas heterotrophic algae are those that use organic compounds as a sole source of energy. Mixotrophic algae use both light energy and organic compounds as sources of energy to carry out its metabolic activity (Harun *et al.*, 2014; Bhateria & Dhaka, 2014). The flowchart of bioethanol production from an algal source is depicted in (Figure 1.2).

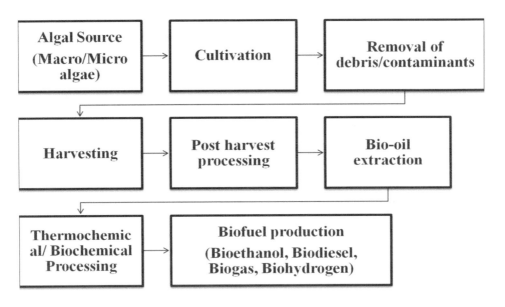

FIGURE 1.2 Flow chart of biofuel production from algae.

1.4 Agronomic Practices in Algal Biomass Production

Macroalgae are mulicellular organisms which possess plant-like structures but lack roots, stems, and leaves. They are distributed all over the world where suitable substratum is available for attachment. Generally, they are distributed in free-floating form as seaweed. They serve as a good source of natural compounds on the sea surface (Kingsford, 1992; Vandendriessche *et al.*, 2006). Macroalgae contains good amounts of structural polysaccharides that are useful for biodiesel production. A bulk amount of biomass is required for energy production that can be achieved from varied species of macroalgae such as *Macrocystis pyrifera*, *Sargassum* sp., *Carpophyllum maschalocarpum*, *Macrocystis pyrifera*, *Laminaria*, *Gracilaria*, and *Ulva* (Safran & Omori, 1990; Kingsford, 1992, 1995; Edgar, 1987; Helmuth *et al.*, 1994; Kokita & Omori, 1998; Hobday, 2000a, 2000b, 2000c; Oilgae, 2010). For biomass production and processing, the seaweeds are harvested from a proper source or collection site. After the seaweeds are collected, they are generally washed with water to remove debris or unwanted material, which is followed by dewatering to maintain 25–30% water for easy transportation and fermentation process for fuel production (Burton *et al.*, 2009). Large-scale production also includes removal of phenols, various salts, and sulfated compounds to enhance the fermentation process. After proper pre-processing, the raw materials are transported to large-scale companies or industries for fuel production.

Algal cultivation and processing includes site selection, cultivation, and harvesting. Recently, enormous research work has been carried out in biofuel production units, as this represents the economic viability of production process. According to Slade and Bauen (2013), several criteria need to be fulfilled for site selection of cultivation unit, i.e. the climatic conditions, temperature, evaporation, and precipitation should be well maintained. There should be adequate water supply with easily accessible nutrients. The site selected should also have proper geographical location. Apart from site selection, the sampling conditions should also be well maintained considering various factors such as:

1. The growth rate and doubling time should be well maintained and monitored regularly.
2. Availability of nutrients to be monitored.
3. Lipid content and free fatty acid content to be known properly.
4. Biomass collected per unit time and volume must be measured properly.
5. Biomass must be separated easily.
6. Toxins, if any, secreted from algae must also be monitored.

Cultivation of macroalgae or microalgae is generally conducted in open ponds (raceway ponds), closed ponds (photobioreactors), closed fermentors (heterotrophic bioreactors), hybrid algae production systems (HAPS), batch and continuous systems of algae cultivation (Williams, 2002), desert-based algae cultivation, and cultivation next to power stations (Bhateria & Dhaka, 2014). There are certain species that can acclimatize in desert areas, such as *Microcoleus vaginatus, Chlamydomonas perigranulata, Synechocystis* and *Haematococcus pluvialis*; such species proved to be potential in biofuel production (Friedmann et al., 1967). For proper growth and cultivation of algae, optimum conditions such as abiotic and biotic conditions including sunlight, air, water, temperature, nutrient concentration, and harvesting density must be maintained (Juneja et al., 2013). After proper processing of microalgae, harvesting is carried out through several techniques. It is harvested through flocculation and sedimentation by altering the pH and by adding additives such as alum, lime, chitosan, etc. It was also evident that electroflocculation and electrocoagulation enhances harvesting processes (Lee et al., 1998; Knuckey et al., 2006; Poelman et al., 1997; Chen, 2004; Mollah et al., 2004).

On the other hand, macroalgae is initially harvested manually from suitable sources and then it is pre-processed to remove debris and unwanted materials. Recently, mechanical harvesters have been employed in the harvesting process of seaweeds (Burton et al., 2009). Later, oil is extracted from algae through various procedures, and subsequently is further used as biofuel.

1.5 Biodiesel: Source Materials and Production Technology

Biodiesel, an alternative fuel, is produced by the chemical method of reacting vegetable oil or animal fat with an alcohol, such as methanol. The biodiesel development reaction process includes catalysts as a robust base such as sodium or caustic potash producing methyl esters, which are referred to as biodiesel (Patel & Krishnamurthy, 2013; Van Gerpen, 2005).

In the future, biofuels will be seriously taken into account as potential energy sources. The demand for petroleum diesel is increasing day by day due to the recent petroleum crisis and the unavailability of petroleum diesel, so there is an incentive to look for an appropriate solution. In order to restructure the national energy economies, the use of renewable resources for energy production could be a strategic focus of government institutions, and many efforts are being made to increase the share of renewable energies within the nation; biodiesel is currently the most widely accepted alternative fuel for diesel engines because of its technical, environmental, and strategic nature. As the carbon present in the exhaust is originally fixed within the atmosphere, biodiesel is considered to be a carbon-neutral fuel. In addition, various vegetable oils—such as cotton seed oil, *Jatropha* seed oil, palm seed oil, sunflower seed oil, maize oil, and corn oil—could be used for biodiesel production (Shalaby, 2015). Their physiochemical properties are depicted in Table 1.2.

TABLE 1.2

Physiochemical Properties of Biodiesel Produced from a Variety of Sources

Biodiesel	Density	Kinematic Viscosity	Flash Point	Pour Point	Cetane Number	References
Jatropha Biodiesel	920 kg/m^3	4.82 mm^2/s	128°C	2°C	50	(Chhetri et al., 2008; Raja, 2011)
Neem Biodiesel	820 kg/m^3	3.5 mm^2/s	147°C	16°C	–	(Banu et al., 2018)
Soapnut Biodiesel	870 kg/m^3	4.86 mm^2/s	175°C	5.1°C	58	(Tiwari and Singh, 2017; Chen et al., 2012)
Karanj Biodiesel	877 kg/m^3	5.02 mm^2/s	152°C	5.4°C	55.84	(Tiwari and Singh, 2017; Scott et al., 2008)
Mahua Biodiesel	955 kg/m^3	6.04 mm^2/s	170°C	–	46	(Solaimuthu et al., 2013)
Moringa Biodiesel	877.5 kg/m^3	4.91 mm^2/s	206°C	3°C	62.12	(Kafuku and Mbarawa, 2010)
Rubber oil Biodiesel	910 kg/m^3	66.2 mm^2/s	198°C	–	–	(Guharaja et al., 2016)
ASTM Standard	845 kg/m^3	0.9–5 mm^2/s	130°C	18°C	–	(Banu et al., 2018)

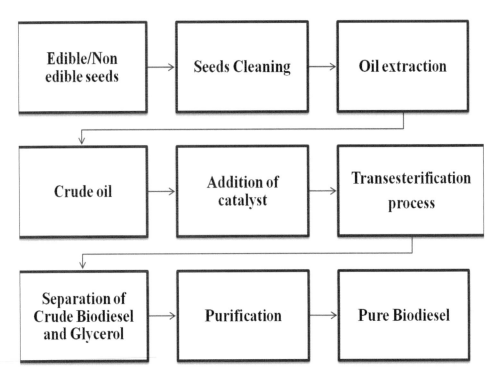

FIGURE 1.3 Biodiesel from edible/non-edible seed oil.

1.5.1 Biodiesel Production Source Materials

A biomass or source raw material plays vital role in biodiesel production. To overcome cost-per-production ratio, there is a need to use unused, non-coasted biomass material like non-edible oils or waste materials rather than edible oils. The schematic diagram of biodiesel production from edible/non-edible feedstock is depicted in Figure 1.3.

1.5.1.1 Jatropha curcas L.

Jatropha curcas L. of the Euphorbiaceae family is native to Central America. It is a tropical plant and can withstand high to low rainfall easily. It is also cultivated widely in various parts of the globe, including India, Thailand, Mexico, and Africa. Due to its potential in producing large amounts of oil, it has also been promoted for cultivation in Nepal, southern Africa, and Brazil (Openshaw, 2000; Chand, 2002). This plant is generally cultivated for seed production, as it is the main reservoir of oil which can be used as biodiesel. Apart from seeds, leaves and stems are not used by animals or humans because of the presence of toxic compounds, or toxalbumin, called curcin, ricin, and cyanic acid (Singh et al., 2010). After proper pretreatment, the seed cake can be used as cattle feed (Raja, 2011). The proximate analysis of *Jatropha curcas* seeds revealed the presence of 4% moisture, 6% ash, 40% lipid, 8% fiber, 27.65% protein and 18.35% carbohydrate (Magu et al., 2018). *Jatropha* oil contains 21% saturated and 79% unsaturated fatty acid. The oil can be extracted from seeds through a mechanical process, a chemical process or an enzymatic process. It possesses about 37%–38% oil, which can be used for combustion as fuel without any refining process, and it burns with clear smoke-free flame. The byproducts formed during processing of *Jatropha* seeds can be used as organic fertilizer as it is rich in nitrogen, potassium and phosphorous. The fatty acid composition plays an important role in physico-chemical properties of biodiesel production, and *Jatropha* oil proved to be a potent source of biodiesel in terms of fuel from non-edible feedstocks (Raja, 2011).

1.5.1.2 **Soapnut Seeds (*Sapindus mukorossi*)**

Soapnut, belonging to family Sapindaceae, is found mainly in India and Nepal. It is also found in America, Europe, Bangladesh, and other countries (Chhetri *et al.*, 2008). The non-edible oil from soapnuts was found to be a possible and potential source of biomass for biodiesel production. A study on different parts of soapnuts (Chhetri *et al.*, 2008) stated that soapnuts have medicinal properties along with surfactant and cleaning property. The proximate analysis of soapnut seeds revealed the presence of 5.36% moisture content, 2.81% ash content, 74.33% volatiles, and 17.5% fixed carbon (Mathiarasu & Pugazhvadivu, 2019). An experiment was carried out for production of biodiesel from soapnut by the pyrolysis method which exhibited that soapnuts could be potential source of biodiesel production on a large scale. The viscosity of the produced bio-oil from soapnuts was estimated to be 10 times higher than that of diesel fuel, with a flash point of 54°C, which is much comparable to diesel fuel. The fire point was found to be 74°C with acidic pH, and hence it was concluded that soapnut oil could be used as biodiesel with minor upgradation and modifications (Solikhah *et al.*, 2018).

1.5.1.3 **Karanj (*Pongomia pinnata* [L.] Pierre)**

Pongomia pinnata belonging to Leguminosae (Fabaceae) family is native to tropical regions of Asia, Australia, China, and New Zealand. *Pongomia pinnata* is also known as Karanj or *Millettia pinnata* (Archana, 2012). *Pongomia pinnata* is a tree with multifarious properties and can survive in extreme conditions, such as they can easily withstand saline conditions (Ismail *et al.*, 2014). They are often planted along roadsides to reduce soil erosion, and this property is due to the presence of lateral root system (Dwivedi *et al.*, 2011; Sangwan *et al.*, 2010; Arpiwi, 2013). The proximate analysis of Karanj seeds revealed the presence of 3.8% ash, 7–8% protein, 23–24% oil, 10% free amino acids, and 0.24% free fatty acids. Karanj seeds also have erucic acid and toxic flavonoids, for instance karanjin, pongapin, and pongaglabrin, which render the oil its inedible property. However, low levels of saturated and polyunsaturated fatty acids with favorable cetane number and iodine value suggested its application as a biodiesel fuel (Bala *et al.*, 2011). Consistent with the findings of Harreh and his colleagues (2018), Karanj seed oil proved to be a sustainable fuel feedstock from Malaysia for biodiesel production. Biodiesel was produced by transesterification reaction using NaOH and MeOH. Physiochemical properties of crude Karanj oil and Karanj biodiesel were investigated per biodiesel standards (EN 14214). The biodiesel blends fueled engine test results like BHP (brake horsepower), BTE (brake thermal efficiency), and BSFC (brake-specific fuel consumption) that were used to describe the suitability of Karanj methyl esters (Karanj ME) as a replacement fuel for diesel engines, and the performance was found to be optimum.

1.5.1.4 **Neem (*Azadirachta indica* A.Juss.)**

The neem tree, which is scientifically known as *Azadirachta indica*, belongs to the family Meliaceae. It is indigenous to tropical and subtropical regions of India, and is also found in several regions of Iran and Bangladesh. The neem tree seeds contain 40% oil, which is a potential source for biodiesel production. The color of neem oil is basically light to dark brown, with a strong odor which may be due to the presence of bioactive component azadirachtin. Several reports showed the presence of large amounts of fatty acids such as oleic acid, linoleic acid, palmitic acid, and stearic acid in neem oil (Ashrafuzzaman *et al.*, 2008; Muthu *et al.*, 2010; Fazal *et al.*, 2011; Ali *et al.*, 2013). An experiment on biodiesel production from neem oil was carried out using calcium oxide (Banu *et al.*, 2018) and NaOH as catalyst (Ali *et al.*, 2013), and it was found that the biodiesel produced through both the catalysts were relevant with the values prescribed by ASTM (American Standard Testing Method), and neem oil could be a potential biodiesel source in areas where neem seeds are available in abundance.

1.5.1.5 **Mahua (*Madhuca indica* J.Konig, J.F.Macbr.)**

Madhuca indica belonging to family Sapotaceae is native to tropical regions of India. Seeds are the important source of biofuel. It is generally grown in humid regions for seed production. Mahua trees possess high socioeconomic value. The proximate analysis of Mahua fruit seeds was carried out and it revealed the presence of 3.4% ash, 3.2% fibers, 22% carbohydrates, and 51.5% oil (Singh & Singh, 1991; Pradhan & Singh, 2013). *Madhuca indica* can adapt and grow in extreme conditions including acidic to alkaline soils. A study on production of biodiesel from mahua seeds was carried out by following two-step transesterification process, and it was noted that this process can yield upto 90% biodiesel with nominal cost at high production rate (Sabariswaran *et al.*, 2014). In another report by Vijay and his co-workers (2019), two-step transesterification was carried out which yielded 85% Mahua methyl esters with less amounts of free fatty acid content (1%) and proved *Madhuca indica* seeds to be a potential and sustainable source of biodiesel.

1.5.1.6 Date Palm Seeds *(Phoenix dactylifera* L.*)*

Phoenix dactylifera belonging to family Arecaceae is a flowering species native to semi-arid and arid regions all over the world. It is also found in Saudi Arabia, the United Arab Emirates, Egypt, Iran, and Pakistan. Date palm is known for its sweet and edible fruits (Habib & Ibrahim, 2011). There are more than 2,000 date palm cultivars cultivated nationally and internationally (Al-Hooti *et al.*, 1997; Mrabet *et al.*, 2008). Date seeds collected from date palm fruit are often used as fodder for animals (animal feed) and are also used in cosmetic and pharmaceutical industries (Vandepopuliere *et al.*, 1995; Al-Farsi & Lee, 2011; Devshony *et al.*, 1992). In a study conducted by Kamil and his coworkers (2019), four different biodiesel blends were prepared and evaluated. Their biodiesel properties were correlated and confirmed with standards; the values for brake power, brake thermal efficiency, and brake-specific fuel consumption showed analogous results with petrol-diesel, which demonstrated that date palm seeds could be a potent and suitable source of biomass for the production of biodiesel. Biodiesel was also produced by transesterification reaction of four different cultivars of date palm seeds viz. Ramly, Haiany, Sewy, and Amhat, which exhibited that highest biodiesel was produced from Ramly (2.43%) cultivar, followed by Amhat (2.05%), Sewy (1.90%), and Haiany (2.84%) cultivars (Shanab *et al.*, 2014).

1.5.1.7 *Moringa oleifera* Lam. Seeds (Drumstick Tree)

Moringa oleifera belonging to family *Moringaceae* is the most widely cultivated species from all the 13 species of *Moringa*. This plant is native to India but is also found in several other parts of the world, including Malaysia, the Philippines, Singapore, Thailand, Mexico, Caribbean Islands, Paraguay, and Brazil (Coppin, 2008). Being a fast-growing tree species, *Moringa oleifera* can tolerate drought conditions, as well as high rainfall (Beaulah, 2001; Palada & Chang, 2003). *Moringa oleifera* dried seeds can produce about 33%–45% w/w oil (Sengupta & Gupta, 1970). The fatty acid profile revealed that *Moringa oleifera* oil is rich in oleic acid (Somali *et al.*, 1984; Anwar *et al.*, 2005; Anwar *et al.*, 2007; Lalas & Tsaknis, 2002). To produce biodiesel in bulk quantity, a high oil-yielding variety designated as MOMAX3 was developed by Indian bioenergy crop research and development company Advanced Biofuel Center (ABC) which could show real promise to transform wasteland into a productive farm for biomass production for biofuel generation, as well as supporting food security to meet basic nutritional requirements (Advanced Biofuel Center, 2016). A recent investigation on biodiesel production from *Moringa* seeds was conducted by following transesterification reaction using MgO as nanocatalyst, and it was revealed that the biodiesel produced was in comparable range with the international standards of ASTM D6751 and ASTM D14214, and proved *Moringa* oil to be a favorable source for biodiesel production (Esmaeili *et al.*, 2019; Kafuku & Mbarawa, 2010).

1.5.1.8 *Ricinus communis* L. (Castor Beans)

Ricinus communis is a flowering plant native to India and Iran that belongs to the Euphorbiaceae family. It can withstand drought conditions easily (Chen *et al.*, 2004; Scholz & Silva, 2008). Castor bio-oil produced from castor seeds has wide array of importance in different sectors, including biodiesel production. Castor seeds possess 50–60% oil, and ricinolic acid is the important fatty acid component of castor oil. The proximate analysis of castor seeds revealed the presence of 51.20% fat, 20.78% protein, 7.96% carbohydrates and 9.40% ash (Annongu & Joseph, 2008). An investigation was carried out in producing ethyl esters fatty acid by transesterification process using NaOH as catalyst which revealed that density, viscosity, cloud point, and flash point of castor oil based biodiesel were 0.93 g/cm³, 14.79 cSt, −23°C, and 120°C, respectively, and it was also evident that highest rate of linoleic acid (41.38%) was observed through GC analysis in comparison to other acids (Hajlari *et al.*, 2019).

1.5.1.9 Thevetia peruviana (Pers.) K. Schum. (Yellow Oleander)

Thevetia peruviana belonging to family Apocynaceae is a flowering plant native to Central America and Mexico. It is perennial and evergreen plant which is cultivated for ornamental purposes (Usman *et al.*, 2009). This plant provides a steady supply of seeds, as fruiting is seen throughout the year. *Thevetia peruviana* is still an underused plant with lesser known importance (Adebowale *et al.*, 2012). This plant is also known to produce important metabolites with diverse properties such as insecticidal, rodenticidal, antifungal, and antibacterial properties (Gata-Gonçalves, 2003; Hassan *et al.*, 2011). The proximate analysis of *Thevetia peruviana* seeds revealed the presence of 2% moisture, 3.33% ash, 60%–65% oil, 30% crude protein, 4.79% crude fiber, and 58.3% fat (Akintelu & Amoo, 2016; Deka & Basumatary, 2011; Oluwaniyi & Ibiyemi, 2007; Usman *et al.*, 2009). The seeds remain non-edible because of the presence of cardiac glycosides, but they may be a great protein diet for animals if properly processed (Usman *et al.*, 2009). Oleic acid, myristic acid, palmitic acid, stearic acid, linoleic acid, linolenic acid, arachidic acid, and arachidonic acid were found in the fatty acid profile (Usman *et al.*, 2009; Deka & Basumatary, 2011; Sanjay, 2015).

1.5.1.10 Sapota (Manilkara zapota [L.] P.Royen)

Manilkara zapota is an evergreen tree belonging to Sapotaceae family, and is commonly known as Sapota or Chikoo. This plant is indigenous to Central America and southern Mexico (Anjaria *et al.*, 2002). It is also widely cultivated in bulk quantities in India, Malaysia, Thailand, Pakistan, and Bangladesh (Bano & Ahmed, 2017). Sapota is mostly cultivated for its fruit value because of its sweet taste and high nutritional value. It possesses a good amount of protein, fiber, carbohydrates, and important minerals (Chadha, 2001). The seed kernels possess varying amounts of oil from 25%–30%. High amounts of oleic acid (64.15%) and linoleic acid (17.92%) were observed in the fatty acid profile of Sapota seed oil, A study was conducted on biodiesel production from Sapota seeds by adopting acid-catalyzed reaction and alkali-catalyzed reaction using a mixture of H_2SO_4 and KOH, respectively, in methanol, which yielded 80% biodiesel through acid-catalyzed reaction and 52% through alkali-catalyzed reaction (Ware *et al.*, 2013). Kumar and colleagues (2015) adopted the Taguchi method for optimizing four important parameters for biodiesel production which yielded 94.83% biodiesel. The flash point (174°C), pour point (−6°C) and cetane number (51–52) (Kumar *et al.*, 2019) was also reported and compared with standard reference which suggested Sapota-based biodiesel to be a suitable substitute for petroleum-based diesel.

1.6 Biodiesel Production Process and Technology

Due to environmental consequences and diminishing petrofuels, much emphasis is made on the biodiesel production process and technology to meet diesel engine necessities and fuel efficiency. Hence, processes such as direct use/blending, pyrolysis, microemulsions, and transesterification have been investigated to produce the bulk quantity of biodiesel.

1. Direct Use/Blending

The invention of running a diesel engine by peanut oil by Dr. Rudolf Diesel made a way of using vegetable oil in replacing petroleum-based fuels. However, intensive use of vegetable oils have been considered impractical due to theirhigh viscosity and free fatty acid contents. To overcome thisproblem, it was indicated that diluting highly viscous oil by specific solvents could be a solution (Bilgin *et al.*, 2002; Patel & Krishnamurthy, 2013). Ma and Hanna (1999) compared the

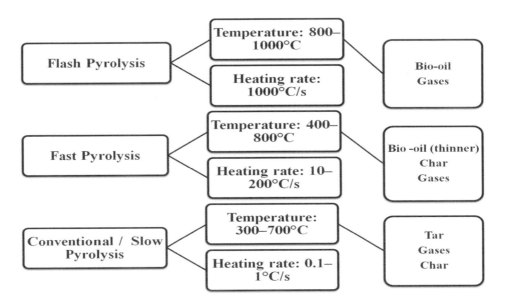

FIGURE 1.4 Pyrolysis classification.

viscosities of different blends prepared from *Brassica napus* seed oil in the ratio of 50:50 and 50:70 (*Brassica napus* seed oil:diesel) and it was found that 50:70 blend was successful in running a single-cylinder diesel engine. Another researcher worked on the applicability of using different blends (1:1 and 1:2 ratios) of degummed soybean oil and No. 2 diesel fuel, which suggested that 1:2 fuel blend was found to be efficient in running a diesel engine (Adams et al., 1983).

2. Pyrolysis

Pyrolysis is a thermochemical process applied at high temperature to organic materials in an inert environment in presence of a catalyst, resulting in separation of bonds and formation of molecules. The end products are generally oil, gases and biochar. Pyrolysis is classified into: (a) flash pyrolysis; (b) fast pyrolysis; and (c) conventional pyrolysis (Figure 1.4) (Czernik & Bridgwater, 2004; Zhang et al., 2019). Several researchers have stated that biodiesel generated through pyrolysis process can be used for engines and thermal pyrolysis or thermal cracking could reduce production costs. However, low heating rate, instability, and separate distillation units for separating various fractions impose a great disadvantage to this process (Ranganathan et al., 2008; French & Czernik, 2010; Singh & Singh, 2010; Abbaszaadeh et al., 2012).

3. Microemulsions

Microemulsions refer to thermodynamically stable immiscible liquids forming a stable and homogeneous end product. Basically, a biodiesel microemulsion includes alcohol (methanol/ethanol) as viscosity balancing additives, diesel fuel, vegetable oil, cetane improver (alkyl nitrates), and a surfactant (higher alcohols) (Chiaramonti et al., 2003; Slomkowski et al., 2011). The problems associated with highly viscous oils could be resolved by using microemulsions (Agarwal, 2007). However, microemulsion-based biodiesels have certain disadvantages such as partial or incomplete combustion, problems associated with needle (sticking of injection needle) and deposition of carbon (Parawira, 2010).

4. Transesterification

Transesterification is an exothermic reaction in which fat/oil is reacted with an alcohol, which results in formation of fatty acid esters or methyl esters (Parawira, 2010; Patel & Krishnamurthy, 2013).

Alcoholysis is also referred to as transesterification and the end product produced (fatty acid/methyl esters) is known as biodiesel (Fukuda et al., 2001). In general, transesterification is a reversible reaction which also results in formation of glycerol. It is the most widely accepted and preferred method for biodiesel production. Basically, the transesterification reaction is classified into: (a) acid catalyzed; (b) alkali catalyzed; (c) lipase catalyzed; (d) heterogeneous catalysis; (e) supercritical transesterification; (f) ionic liquid catalysis; and (g) nanocatalysis, on the basis of their reaction process; acid-catalyzed transesterification and alkali-catalyzed transesterification are most widely accepted.

a. Acid-catalyzed reaction

Acid-catalyzed reaction involves mixing of oil and acidified alcohol, which is a single-step process whereby alcohol acts as both and esterification reagent and solvent (Abbaszaadeh et al., 2012). This method was adopted to produce biodiesel from palm oil using H_2SO_4 and methanol. However, it requires high temperature with longer reaction time (Gerhard et al., 2005). Different catalysts are used in acid-catalyzed transesterification; for example, hydrochloric acid, sulphonic acid, and sulfuric acid. However, several researchers suggested that $AlCl_3$ could be an alternative catalyst for biodiesel production from vegetable oils (Soriano et al., 2009).

b. Alkali-catalyzed reaction

In alkali-catalyzed reactions, alkaline catalysts such as sodium or potassium carbonates or alkaline metal alkoxides/hydroxides are used in reacting fats or oils with an alcohol to produce fatty acid esters with formation of glycerol. These types of reactions are more efficient, faster, and less corrosive in comparison to acid-catalyzed reactions (Ranganathan et al., 2008; Marchetti et al., 2007; Grebemariam & Marchetti, 2017). There are chances of saponification of oil due to presence of high amounts of fatty acid content, which leads to improper separation of glycerol (Parawira, 2010). Owing to the cost ratio and catalytic effect, sodium hydroxide is the most widely used catalyst in alkali or base-catalyzed reactions, but several researchers demonstrated that sodium methoxide was also found useful in catalysis process (Nye et al., 1983; Freedman et al., 1984; Atadashi et al., 2013).

1.7 Agrowaste as a Source of Biofuel

Agrowastes or agricultural wastes are defined as residues of crops or plants that are lost during transportation, handling, and storage of agricultural plants. Plant residues like stems, leaves, husks, seeds, seed pods, and straws are included in agricultural wastes. These crop/plant residues could easily reduce the emission of greenhouse gases, and they could also minimize the effect of global warming (Champagne, 2007; Panpatte & Jhala, 2019; Soccol et al., 2019). For biofuel production, these agrowastes need to be pretreated by following specific processes or methods just to convert lignocellulosic materials into biofuel. The pretreatment process includes enzymatic pretreatment, mechanical/physical pretreatment, and biological pretreatment (Balat et al., 2008; Panpatte & Jhala, 2019). Generally, banana plant residues, rice straws, and sugarcane waste or bagasse are the major agrowastes produced in India, and disposal of such materials is not carried out easily (Xie et al., 2010). These agrowaste materials can be used for large-proportion biofuel production; for example, a one-hectare banana plantation can produce 220 tons of residues which can be used for producing bioethanol. Ingale and colleagues (2014) produced 17.1g/L (1.71%) of ethanol by using *A. fumigates*, *A. ellipticus* (fungal strains), and *S. cerevisiae* (yeast strain) under co-culture fermentation of banana plant residues. Similarly, Krishnamurthy et al. (2014) attempted bioethanol production through yeast fermentation process using *Ficus carica* (fig) and *Phoenix dactylifera* (date) fruit mash, which yielded 8.76% of ethanol from *F. carica* (93.3% purity) and 10.64% ethanol from *P. dactylifera* (92.6% purity). About 600–900 million tons of rice straw and 280 kg of bagasse from one metric ton (MT) of sugarcane produced worldwide can also be used in production of biofuel (Canilha *et al.*, 2012; Karimi *et al.*, 2006). Cocoa pod husk (*Theobroma cacao* L.) from cocoa plants, a waste material or residue produced during processing of cocoa beans, is also useful in production of biodiesel. Cocoa pod is abundantly found in Nigeria with a varied range of sugar (Igbinadolor, 2012; Rachmat *et al.*, 2018). Around 1 million tons of cocoa pod husk is generated in Nigeria as agricultural waste, and an attempt was made by (Akinola *et al.*, 2018) to produce biodiesel by pyrolysis at varying temperatures which resulted in 36.23 MJ/kg heating energy with a negligible amount of sulfur. Beside this, cocoa pods husk and palm kernel fronds were used as catalysts in production of biodiesel from *Jatropha*. It was reported that biodiesel produced using cocoa pod husks and palm kernel fronds showed comparable results in terms of specific gravity, density, flash point, and refractive index. However, cocoa pod husks yielded 85.33% biodiesel, compared to palm kernel fronds biodiesel (66%) (Belewu *et al.*, 2018).

1.8 Biofuels from Fungi

Fungi are eukaryotic organisms that includeyeast, molds, and mushrooms. Some of these species are oleaginous (more than 20–25% lipid content/dry weight) in nature (Ratledge, 1991; Ma, 2006). These oleaginous species have been explored for biofuel production because of their minimal requirement and their ability to use different carbon source that could assist in higher biomass and lipid production (Ageitos *et al.*, 2011). Fungi belonging to Zygomorphic phylum are considered as a great source of lipid content. *Thamnidium elegan, Cunninghamella echinulata, Mortierella isabellina, Cunninghamella bainieri, Zygorhynhus moelleri, Mortierella alpine, Aspergillus niger, Mortierella ramanniana, Epicoccum purpurascens, Chaetomium globosum, Colletotrichum* sp., and *Alternaria* sp. are potent lipid-accumulating species of fungi that are assisted in production of biofuel (Athenaki *et al.*, 2018). Generally, the production process of biofuels from fungi includes: (a) direct method and (b) indirect method.

In the direct method for biofuel production, the biomass obtained from a fungal strain is directly converted into biofuel without extracting lipids from the fungal strain, whereas in the indirect method, extraction of lipids from biomass is carried out which is then converted to biofuel (Raven *et al.*, 2019). A study was carried out on a strain of *Aspergillus*—namely, *Aspergillus terreus*—by irradiating it with UV light, and it was observed that there was an increase in biomass yield, which was considered as an alternative source for biofuel production (Shafiq, 2017).

Owing to the production of biofuel from fungi, researchers have recently identified endophytic fungi that produce volatile organic compounds (VOCs). These VOCs have potential similar to petroleum diesel, and hence they are also known as mycodiesel (Strobel, 2014; Raven *et al.*, 2019). Certain microorganisms are able to produce bioethanol, biohydrogen, and biodiesel, which is depicted in Table 1.3. In terms of environmental safety, biofuels are the need of the hour, and fungal sources proved to be better alternative.

TABLE 1.3

Fungal Strains Used in Biofuel Production

Biofuel Type	Microbial Species	References
Bioethanol	*Aspergillus niger, Aspergillus fumigates, Aspergillus ellipticus, Trichoderma reesei*	(Xin *et al.*, 2013; Ingale *et al.*, 2014; Ezeonu *et al.*, 2014)
Biodiesel	*Mucor fragilis* sp., *Aspergillus* sp., *Cunninghamella* sp., *Mortierella* sp.	(Vicente *et al.*, 2009; Shafiq, 2017; Huang *et al.*, 2016; Meng *et al.*, 2009)
Biohydrogen	*Gymnopus contrarius* J2	(Sheng *et al.*, 2018)

REFERENCES

Abbaszaadeh, A., Ghobadian, B., Omidkhah, M. R., & Najafi, G. (2012). Current biodiesel production technologies: A comparative review. *Energy Conversion and Management*, *63*, 138–148.

Abdullah, B., Muhammad, S. A. F. A. S., Shokravi, Z., Ismail, S., Kassim, K. A., Mahmood, A. N., & Aziz, M. M. A. (2019). Fourth generation biofuel: A review on risks and mitigation strategies. *Renewable and Sustainable Energy Reviews*, *107*, 37–50.

Adams, C., Peters, J. F., Rand, M. C., Schroer, B. J., & Ziemke, M. C. (1983). Investigation of soybean oil as a diesel fuel extender: Endurance tests. *Journal of the American Oil Chemists Society*, *60*(8), 1574–1579.

Adebowale, K. O., Adewuyi, A., & Ajulo, K. D. (2012). Examination of fuel properties of the methyl esters of *Thevetia peruviana* seed oil. *International Journal of Green Energy*, *9*(3), 297–307.

Advanced Biofuel Center. (2016). MOMAX3 *Moringa* variety: The *Moringa* industry's most productive seeds search for higher seed oil yield. http://www.jatrophaworld.org/momax3-moringa-perennial-seed-variety.html.

Agarwal, A. K. (2007). Biofuels (alcohols and biodiesel) applications as fuels for internal combustion engines. *Progress in Energy and Combustion Science*, *33*(3), 233–271.

Ageitos, J. M., Vallejo, J. A., Veiga-Crespo, P., & Villa, T. G. (2011). Oily yeasts as oleaginous cell factories. *Applied Microbiology and Biotechnology*, *90*(4), 1219–1227.

Akinola, A. O., Eiche, J. F., Owolabi, P. O., & Elegbeleye, A. P. (2018). Pyrolytic analysis of cocoa pod for biofuel production. *Nigerian Journal of Technology*, *37*(4), 1026–1031.

Akintelu, M. T., & Amoo, I. A. (2016). Proximate characterisation and physicochemical properties of raw and boiled milk bush *(Thevetia peruviana)* seed. *International Journal of Science*, *5*(3), 16–21.

Al-Farsi, M. A., & Lee, C. Y. (2011). Usage of date (*Phoenix dactylifera* L.) seeds in human health and animal feed. In *Nuts and seeds in health and disease prevention* (pp. 447–452). New York, NY: Academic Press, Elsevier.

Al-Hooti, S., Sidhu, J. S., & Qabazard, H. (1997). Physicochemical characteristics of five date fruit cultivars grown in the United Arab Emirates. *Plant Foods for Human Nutrition*, *50*(2), 101–113.

Ali, M. H., Mashud, M., Rubel, M. R., & Ahmad, R. H. (2013). Biodiesel from neem oil as an alternative fuel for diesel engine. *Procedia Engineering*, *56*, 625–630.

Anjaria, J., Parabia, M., & Dwivedi, S. (2002). *Indian ethnoveterinary medicine—An overview*. Ahmedabad, Gujarat: Pathi Enterprise.

Annongu, A. A., & Joseph, J. K. (2008). Proximate analysis of castor seeds and cake. *Journal of Applied Sciences and Environmental Management*, *12*(1), 39–41.

Anwar, F., Ashraf, M., & Bhanger, M. I. (2005). Interprovenance variation in the composition of *Moringa oleifera* oilseeds from Pakistan. *Journal of the American Oil Chemists' Society*, *82*(1), 45–51.

Anwar, F., Latif, S., Ashraf, M., & Gilani, A. H. (2007). *Moringa oleifera*: A food plant with multiple medicinal uses. *Phytotherapy Research: An International Journal Devoted to Pharmacological and Toxicological Evaluation of Natural Product Derivatives*, *21*(1), 17–25.

Archana, C. (2012). Chemical examination of the seeds of *Pongamia pinnata* (L.) Pierre. *Journal of Pharmaceutical Research*, *2*(146).

Arpiwi, N. L. (2013). *Millettia pinnata (L.) Panigrahi a biodiesel tree, characterization of traits for production on marginal land* (Doctoral dissertation). University of Western Australia.

Ashrafuzzaman, M., Haque, M. S., & Luna, L. N. (2008). In vitro clonal propagation of the neem tree (*Azadirachta indica* A. Juss). *African Journal of Biotechnology*, *7*(4).

Atadashi, I. M., Aroua, M. K., Aziz, A. A., & Sulaiman, N. M. N. (2013). The effects of catalysts in biodiesel production: A review. *Journal of Industrial and Engineering Chemistry*, *19*(1), 14–26.

Athenaki, M., Gardeli, C., Diamantopoulou, P., Tchakouteu, S. S., Sarris, D., Philippoussis, A., & Papanikolaou, S. (2018). Lipids from yeasts and fungi: Physiology, production and analytical considerations. *Journal of Applied Microbiology*, *124*(2), 336–367.

Bala, M., Nag, T. N., Kumar, S., Vyas, M., Kumar, A., & Bhogal, N. S. (2011). Proximate composition and fatty acid profile of *Pongamia pinnata*, a potential biodiesel crop. *Journal of the American Oil Chemists' Society*, *88*(4), 559–562.

Balat, M., Balat, H., & Öz, C. (2008). Progress in bioethanol processing. *Progress in Energy and Combustion Science*, *34*(5), 551–573.

Bano, M., & Ahmed, B. (2017). *Manilkara zapota* (L.) P. Royen (*Sapodilla*): A review. *International Journal of Advance Research, Ideas and Innovations in Technology*, *3*, 1364–1371.

Banu, H. D., Shallangwa, T. B., Innocent, J., Thomas, O. M., Hitler, L., & Sadia, A. (2018). Biodiesel production from neem seed (*Azadirachta indica*) oil using calcium oxide as heterogeneous catalyst. *Journal of Physical Chemistry & Biophysics, 8*(2), 266.

Baroutian, S., Aroua, M. K., Raman, A. A. A., Shafie, A., Ismail, R. A., & Hamdan, H. (2013). Blended aviation biofuel from esterified *Jatropha curcas* and waste vegetable oils. *Journal of the Taiwan Institute of Chemical Engineers, 44*(6), 911–916.

Beaulah, A. (2001). *Growth and development of Moringa (Moringa oleifera Lam) under organic and inorganic systems of culture* (Doctoral dissertation). Tamil Nadu Agricultural University, Coimbatore.

Belewu, M., Ameen, M., & Krishnamurthy, R. (2018). Green chemistry for low carbon emission and healthy environment. *Chemistry Research Journal, 3*(1), 28–33.

Beschkov, V. (2017). Biogas, biodiesel and bioethanol as multifunctional renewable fuels and raw materials. *Frontiers in Bioenergy and Biofuels*, 185–205.

Bhateria, R., & Dhaka, R. (2014). Algae as biofuel. *Biofuels, 5*(6), 607–631.

Bhatia, S. C. (Ed.). (2014). *Advanced renewable energy systems (part 1 and 2)*. New York: CRC Press.

Bhatia, S. C., & Gupta, R. K. (2019). *Textbook of renewable energy*. New Delhi, India: Woodhead Publishing India PVT. Limited.

Bilgin, A., Durgun, O., & Sahin, Z. (2002). The effects of diesel-ethanol blends on diesel engine performance. *Energy Sources, 24*(5), 431–440.

Bioenergy. (2019). *Value chains applying advanced conversion technologies. E. T. I. P.* European Technology and Innovation Platform.

Biofuel.org.uk

Burton, T., Lyons, H., Lerat, Y., Stanley, M., & Rasmussen, M. B. (2009). *A review of the potential of marine algae as a source of biofuel in Ireland*. Dublin: Sustainable Energy Ireland—SEI.

Canilha, L., Chandel, A. K., Suzane dos Santos Milessi, T., Antunes, F. A. F., Luiz da Costa Freitas, W., . . . & da Silva, S. S. (2012). Bioconversion of sugarcane biomass into ethanol: An overview about composition, pretreatment methods, detoxification of hydrolysates, enzymatic saccharification, and ethanol fermentation. *Journal of Biomedicine and Biotechnology*, 1–15.

Chadha, K. L. (2001). *Hand book of horticulture*. New Delhi: ICAR.

Champagne, P. (2007). Feasibility of producing bio-ethanol from waste residues: A Canadian perspective: Feasibility of producing bio-ethanol from waste residues in Canada. *Resources, Conservation and Recycling, 50*(3), 211–230.

Chand, N. (2002). Plant oils: Fuel of the future. *Journal of Scientific and Industrial Research, 61*, 7–16.

Chang, H. (2007). *Marine biodiversity and systematics*. Research Programmes. nap.edu/read/4923/chapter/8. Retrieved April, 14, 2009.

Chen, G. (2004). Electrochemical technologies in wastewater treatment. *Separation and Purification Technology, 38*(1), 11–41.

Chen, G. Q., He, X., Liao, L. P., & McKeon, T. A. (2004). 2S albumin gene expression in castor plant (*Ricinus communis* L.). *Journal of the American Oil Chemists' Society, 81*(9), 867–872.

Chen, L., Liu, T., Zhang, W., Chen, X., & Wang, J. (2012). Biodiesel production from algae oil high in free fatty acids by two-step catalytic conversion. *Bioresource Technology, 111*, 208–214.

Chen, Y., Dong, B., Qin, W., & Xiao, D. (2010). Xylose and cellulose fractionation from corncob with three different strategies and separate fermentation of them to bioethanol. *Bioresource Technology, 101*(18), 6994–6999.

Chhetri, A. B., Tango, M. S., Budge, S. M., Watts, K. C., & Islam, M. R. (2008). Non-edible plant oils as new sources for biodiesel production. *International Journal of Molecular Sciences, 9*(2), 169–180.

Chiaramonti, D., Bonini, M., Fratini, E., Tondi, G., Gartner, K., Bridgwater, A. V., & Baglioni, P. (2003). Development of emulsions from biomass pyrolysis liquid and diesel and their use in engines—part 1: Emulsion production. *Biomass and Bioenergy, 25*(1), 85–99.

Choi, I. S., Wi, S. G., Kim, S. B., & Bae, H. J. (2012). Conversion of coffee residue waste into bioethanol with using popping pretreatment. *Bioresource Technology, 125*, 132–137.

Coppin, J. (2008). *A study of the nutritional and medicinal values of Moringa oleifera leaves from sub-Saharan Africa: Ghana, Rwanda, Senegal and Zambia* (Doctoral dissertation). Rutgers University-Graduate School, New Brunswick.

Czernik, S., & Bridgwater, A. V. (2004). Overview of applications of biomass fast pyrolysis oil. *Energy & Fuels, 18*(2), 590–598.

Deka, D. C., & Basumatary, S. (2011). High quality biodiesel from Yellow oleander (*Thevetia peruviana*) seed oil. *Biomass and Bioenergy, 35*(5), 1797–1803.

Demirbas, A. (2008). *Biodiesel* (pp. 111–119). London: Springer.

Devshony, S., Eteshola, E., & Shani, A. (1992). Characteristics and some potential applications of date palm (*Phoenix dactylifera* L.) seeds and seed oil. *Journal of the American Oil Chemists' Society, 69*(6), 595–597.

Dutta, K., Daverey, A., & Lin, J. G. (2014). Evolution retrospective for alternative fuels: First to fourth generation. *Renewable Energy, 69*, 114–122.

Dwivedi, G., Jain, S., & Sharma, M. P. (2011). *Pongamia* as a source of biodiesel in India. *Smart Grid and Renewable Energy, 2*(3), 184–189.

Edgar, G. J. (1987). Dispersal of faunal and floral propagules associated with drifting *Macrocystis pyrifera* plants. *Marine Biology, 95*(4), 599–610.

Esmaeili, H., Yeganeh, G., & Esmaeilzadeh, F. (2019). Optimization of biodiesel production from *Moringa oleifera* seeds oil in the presence of nano-MgO using Taguchi method. *International Nano Letters, 9*(3), 257–263.

Ezeonu, C. S., Otitoju, O., Onwurah, E. I., Ejikeme, C., Ugbogu, O. C., & Anike, E. (2014). Enhanced availability of biofuel and biomass components in *Aspergillus niger* and *Aspergillus fumigatus* treated rice husk. *European Scientific Journal, 10*(18), 96–117.

Fazal, M. A., Haseeb, A. S. M. A., & Masjuki, H. H. (2011). Biodiesel feasibility study: An evaluation of material compatibility; performance; emission and engine durability. *Renewable and Sustainable Energy Reviews, 15*(2), 1314–1324.

Fernando, S., Hall, C., & Jha, S. (2006). NOx reduction from biodiesel fuels. *Energy & Fuels, 20*(1), 376–382.

Freedman, B., Pryde, E. H., & Mounts, T. L (1984). Variables affecting the yields of fatty esters from transesteritied vegetable oils. *Journal of the American Oil Chemists Society, 61*, 1638–1643.

French, R., & Czernik, S. (2010). Catalytic pyrolysis of biomass for biofuels production. *Fuel Processing Technology, 91*(1), 25–32.

Friedmann, I., Lipkin, Y., & Ocampo-Paus, R. (1967). Desert algae of the Negev (Israel). *Phycologia, 6*(4), 185–200.

Fukuda, H., Kondo, A., & Noda, H. (2001). Biodiesel fuel production by transesterification of oils. *Journal of Bioscience and Bioengineering, 92*(5), 405–416.

Gata-Gonçalves, L., Nogueira, J. M. F., Matos, O., & de Sousa, R. B. (2003). Photoactive extracts from *Thevetia peruviana* with antifungal properties against *Cladosporium cucumerinum*. *Journal of Photochemistry and Photobiology B: Biology, 70*(1), 51–54.

Gerhard, K., Jon, V. G., & Jürgen, K. (2005). *The biodiesel handbook*. Champaign, IL: AOCS Press.

Grebemariam, S., & Marchetti, J. M. (2017). Biodiesel production technologies. *AIMS Energy, 5*(3), 425–457.

Guharaja, S., DhakshinaMoorthy, S., Hasan, Z. I., Arun, B., Ahamed, J. I., & Azarudheen, J. (2016). Biodiesel production from Mahua (*Madhuca indica*). *International Journal of Nano Corrosion Science and Engineering, 3*(1), 34–47.

Habib, H. M., & Ibrahim, W. H. (2011). Effect of date seeds on oxidative damage and antioxidant status in vivo. *Journal of the Science of Food and Agriculture, 91*(9), 1674–1679.

Hajilary, N., Rezakazemi, M., & Shirazian, S. (2019). Biofuel types and membrane separation. *Environmental Chemistry Letters, 17*(1), 1–18.

Hajlari, S. A., Najafi, B., & Ardabili, S. F. (2019). Castor oil, a source for biodiesel production and its impact on the diesel engine performance. *Renewable Energy Focus, 28*, 1–10.

Harreh, D., Saleh, A. A., Reddy, A. N. R., & Hamdan, S. (2018). An experimental investigation of *Karanja* biodiesel production in Sarawak, Malaysia. *Journal of Engineering*, 1–8.

Harun, R., Yip, J. W., Thiruvenkadam, S., Ghani, W. A., Cherrington, T., & Danquah, M. K. (2014). Algal biomass conversion to bioethanol—A step-by-step assessment. *Biotechnology Journal, 9*(1), 73–86.

Hassan, M. M., Saha, A. K., Khan, S. A., Islam, A., Mahabub-Uz-Zaman, M., & Ahmed, S. S. U. (2011). Studies on the antidiarrhoeal, antimicrobial and cytotoxic activities of ethanol-extracted leaves of yellow oleander (*Thevetia peruviana*). *Open Veterinary Journal, 1*(1), 28–31.

Helmuth, B., Veit, R. R., & Holberton, R. (1994). Long-distance dispersal of a subantarctic brooding bivalve (*Gaimardia trapesina*) by kelp-rafting. *Marine Biology, 120*(3), 421–426.

Hirani, A. H., Javed, N., Asif, M., Basu, S. K., & Kumar, A. (2018). A review on first-and second-generation biofuel productions. In *Biofuels: Greenhouse gas mitigation and global warming* (pp. 141–154). New Delhi: Springer.

Hobday, A. J. (2000a). Age of drifting *Macrocystis pyrifera* L.C. Agardh rafts in the Southern California Bight. *Journal of Experimental Marine Biology and Ecology, 253*, 97–114.

Hobday, A. J. (2000b). Persistence and transport of fauna on drifting kelp (*Macrocystis pyrifera* L.C. Agardh) rafts in the Southern California blight. *Journal of Experimental Marine Biology and Ecology, 253*, 75–96.

Hobday, A. J. (2000c). Abundance and dispersal of drifting kelp *Macrocystis pyrifera* rafts in the Southern California Bight. *Marine Ecology Progress Series, 195*, 101–116.

Huang, G., Zhou, H., Tang, Z., Liu, H., Cao, Y., Qiao, D., & Cao, Y. (2016). Novel fungal lipids for the production of biodiesel resources by *Mucor fragilis* AFT7-4. *Environmental Progress & Sustainable Energy, 35*(6), 1784–1792.

Igbinadolor, R. (2012). *Fermentation of cocoa pod husk (Theobrama cacao) and its hydrolysate for ethanol production using improved starter culture* (PhD thesis). University of Ibadan, Ibadan, Nigeria.

Ingale, S., Joshi, S. J., & Gupte, A. (2014). Production of bioethanol using agricultural waste: Banana pseudo stem. *Brazilian Journal of Microbiology, 45*(3), 885–892.

International Energy Agency. (2019). *Global Energy & CO_2 status report*. Paris, France: IEA.

Ismail, S. A. A. S., Abu, S. A., Rezaur, R., & Sinin, H. (2014). Biodiesel production from castor oil and its application in diesel engine. *ASEAN Journal on Science and Technology for Development, 31*(2), 90–100.

Juneja, A., Ceballos, R. M., & Murthy, G. S. (2013). Effects of environmental factors and nutrient availability on the biochemical composition of algae for biofuels production: A review. *Energies, 6*(9), 4607–4638.

Kafuku, G., & Mbarawa, M. (2010). Alkaline catalyzed biodiesel production from *Moringa oleifera* oil with optimized production parameters. *Applied Energy, 87*(8), 2561–2565.

Kamil, M., Ramadan, K., Olabi, A. G., Ghenai, C., Inayat, A., & Rajab, M. H. (2019). Desert palm date seeds as a biodiesel feedstock: Extraction, characterization, and engine testing. *Energies, 12*(16), 3147.

Karimi, K., Kheradmandinia, S., & Taherzadeh, M. J. (2006). Conversion of rice straw to sugars by dilute-acid hydrolysis. *Biomass and Bioenergy, 30*(3), 247–253.

Khan, S. A., Hussain, M. Z., Prasad, S., & Banerjee, U. C. (2009). Prospects of biodiesel production from microalgae in India. *Renewable and Sustainable Energy Reviews, 13*(9), 2361–2372.

Kingsford, M. J. (1992). Drift algae and small fish in coastal waters of northeastern New Zealand. *Marine Ecology Progress Series*, 41–55.

Kingsford, M. J. (1995). Drift algae: A contribution to near-shore habitat complexity in the pelagic environment and an attractant for fish. *Marine Ecology Progress Series. Oldendorf, 116*(1), 297–301.

Knuckey, R. M., Brown, M. R., Robert, R., & Frampton, D. M. (2006). Production of microalgal concentrates by flocculation and their assessment as aquaculture feeds. *Aquacultural Engineering, 35*(3), 300–313.

Kokita, T., & Omori, M. (1998). Early life history traits of the gold-eye rockfish, *Sebastes thompsoni*, in relation to successful utilization of drifting seaweed. *Marine Biology, 132*(4), 579–589.

Kothari, A., & Gujral, S. S. (2013). Introduction to bio-fuel and its production from Algae: An overview. *International Journal of Pharma and Biosciences, 3*(1), 269–280.

Krishnamurthy, R., Animasaun, D. A., Ingalhalli, R. S., & Ramani, N. D. (2014). A preliminary attempt of ethanol production from fig *(Ficus carica)* and date *(Phoenix dactylifera)* fruits using *Saccharomyces cerevisiae*. *International Journal of Pure & Applied Bioscience, 2*(2), 174–180.

Kristoferson, L. A., & Bokalders, V. (1986). Biogas. *Renewable Energy Technologies*, 100–113. https://doi.org/10.1016/b978-0-08-034061-6.50015-9

Kumar, R. S., Sivakumar, S., Joshuva, A., Deenadayalan, G., & Vishnuvardhan, R. (2019). Data set on optimization of ethyl ester production from sapota seed oil. *Data in Brief, 25*, 104388.

Kumar, R. S., Sureshkumar, K., & Velraj, R. (2015). Optimization of biodiesel production from *Manilkara zapota* (L.) seed oil using Taguchi method. *Fuel, 140*, 90–96.

Lalas, S., & Tsaknis, J. (2002). Extraction and identification of natural antioxidant from the seeds of the *Moringa oleifera* tree variety of Malawi. *Journal of the American Oil Chemists' Society, 79*(7), 677–683.

Lee, S. J., Kim, S. B., Kim, J. E., Kwon, G. S., Yoon, B. D., & Oh, H. M. (1998). Effects of harvesting method and growth stage on the flocculation of the green alga *Botryococcus braunii*. *Letters in Applied Microbiology, 27*(1), 14–18.

Liu, H., Wang, G., & Zhang, J. (2013). The promising fuel-biobutanol. *Liquid, Gaseous and Solid Biofuels-Conversion Techniques*, 175–198.

Lü, J., Sheahan, C., & Fu, P. (2011). Metabolic engineering of algae for fourth generation biofuels production. *Energy & Environmental Science, 4*(7), 2451–2466.

Ma, F., & Hanna, M. A. (1999). Biodiesel production: A review. *Bioresource Technology, 70*(1), 1–15.

Ma, Y. L. (2006). Microbial oils and its research advance. *Chinese Journal of Bioprocess Engineering, 4*(4), 7–11.

Magu, T. O., Louis, H., Nzeata-Ibe, N., Sunday, E. A., & Udowo, V. M. (2018). Proximate analysis and mineral composition of *Jatropha curcas* seeds obtained from Pankshin local government area of Plateau State of Nigeria. *Journal of Physical Chemistry & Biophysics, 8*(1). https://doi.org/10.4172/2161-0398.1000265

Mahmoodi, P., Karimi, K., & Taherzadeh, M. J. (2018). Efficient conversion of municipal solid waste to biofuel by simultaneous dilute-acid hydrolysis of starch and pretreatment of lignocelluloses. *Energy Conversion and Management, 166*, 569–578.

Marchetti, J. M., Miguel, V. U., & Errazu, A. F. (2007). Possible methods for biodiesel production. *Renewable and Sustainable Energy Reviews, 11*(6), 1300–1311.

Mathiarasu, A., & Pugazhvadivu, M. (2019, September). Production of bio-oil from soapnut seed by microwave pyrolysis. In *IOP conference series: Earth and environmental science* (Vol. 312, No. 1, pp. 12–22). Bristol: IOP Publishing.

Melikoglu, M., Singh, V., Leu, S. Y., Webb, C., & Lin, C. S. K. (2016). Biochemical production of bioalcohols. In *Handbook of biofuels production* (pp. 237–258). Duxford: Woodhead Publishing, Elsevier.

Meng, X., Yang, J., Xu, X., Zhang, L., Nie, Q., & Xian, M. (2009). Biodiesel production from oleaginous microorganisms. *Renewable Energy, 34*(1), 1–5.

Metting, F. B. (1996). Biodiversity and application of microalgae. *Journal of Industrial Microbiology, 17*(5–6), 477–489.

Mollah, M. Y., Morkovsky, P., Gomes, J. A., Kesmez, M., Parga, J., & Cocke, D. L. (2004). Fundamentals, present and future perspectives of electrocoagulation. *Journal of Hazardous Materials, 114*(1–3), 199–210.

Morone, P., Cottoni, L., Luque, R., Lin, C., Wilson, K., & Clark, J. (2016). Biofuels: Technology, economics, and policy issues. In *Handbook of biofuels production* (pp. 61–83). Duxford: Elsevier.

Mrabet, A., Ferchichi, A., Chaira, N., & Mohamed, B. S. (2008). Physico-chemical characteristics and total quality of date palm varieties grown in Southern Tunisia. *Pakistan Journal of Biological Sciences, 11*(7), 1003–1008.

Muthu, H., SathyaSelvabala, V., Varathachary, T. K., Kirupha Selvaraj, D., Nandagopal, J., & Subramanian, S. (2010). Synthesis of biodiesel from neem oil using sulfated zirconia via tranesterification. *Brazilian Journal of Chemical Engineering, 27*(4), 601–608.

Nye, M. J., Williamson, T. W., Deshpande, W., Schrader, J. H., Snively, W. H., Yurkewich, T. P., & French, C. L. (1983). Conversion of used frying oil to diesel fuel by transesterification: Preliminary tests. *Journal of the American Oil Chemists' Society, 60*(8), 1598–1601.

Oilgae. (2010). *Oilgae guide to fuels from macroalgae*. Tamilnadu, India: Oilgae.

Oluwaniyi, O. O., & Ibiyemi, S. A. (2007). Extractability of *Thevetia peruviana* glycosides with alcohol mixture. *African Journal of Biotechnology, 6*(18).

Openshaw, K. (2000). A review of *Jatropha curcas*: An oil plant of unfulfilled promise. *Biomass and Bioenergy, 19*(1), 1–15.

Palada, M. C., & Chang, L. C. (2003). Suggested cultural practices for *Moringa*. In *International cooperators' guide AVRDC*. Taiwan: AVRDC Pub. 03–545.

Panpatte, D. G., & Jhala, Y. K. (2019). Agricultural waste: A suitable source for biofuel production. In *Prospects of renewable bioprocessing in future energy systems* (pp. 337–355). Cham: Springer.

Parawira, W. (2010). Biodiesel production from *Jatropha curcas*: A review. *Scientific Research and Essays, 5*(14), 1796–1808.

Patel, P., Patel, S., & Krishnamurthy, R. (2016). Microalgae: Future biofuel. *Indian Journal of Geo-Marine Sciences, 45*, 823–829.

Patel, Z., & Krishnamurthy, R. (2013). Biodiesel: Source materials and future prospects. *International Journal of Geology, Earth and Enviromental Sciences, 3*(2), 10–20.

Poelman, E., De Pauw, N., & Jeurissen, B. (1997). Potential of electrolytic flocculation for recovery of microalgae. *Resources, Conservation and Recycling, 19*(1), 1–10.

Pradhan, D., & Singh, R. K. (2013). Bio-oil from biomass: Thermal pyrolysis of mahua seed. In *2013 International conference on energy efficient technologies for sustainability* (pp. 487–490). Nagercoil, India: IEEE.

Qian, Y., Zhu, L., Wang, Y., & Lu, X. (2015). Recent progress in the development of biofuel 2, 5-dimethylfuran. *Renewable and Sustainable Energy Reviews, 41*, 633–646.

Rachmat, D., Mawarani, L. J., & Risanti, D. D. (2018). Utilization of cacao pod husk (*Theobroma cacao* L.) as activated carbon and catalyst in biodiesel production process from waste cooking oil. In *IOP conference series: Materials science and engineering, Malang* (Vol. 299, pp. 2–9). Bristol: IOP Publishing.

Raja, S. A. (2011). Biodiesel production from *Jatropha* oil and its characterization. *Research Journal of Chemical Sciences, 1*, 81–87.

Ranganathan, S. V., Narasimhan, S. L., & Muthukumar, K. (2008). An overview of enzymatic production of biodiesel. *Bioresource Technology, 99*(10), 3975–3981.

Ratledge, C. (1991). Microorganisms for lipids. *Acta Biotechnologica, 11*(5), 429–438.

Raven, S., Francis, A., Srivastava, C., Kezo, S., & Tiwari, A. (2019). Fungal biofuels: Innovative approaches. In *Recent advancement in white biotechnology through fungi* (pp. 385–405). Cham: Springer.

Sabariswaran, K., Selvakumar, S., & Kathirselvi, A. (2014). Biodiesel production from Mahua oil by using two-step transesterification process. *Chemical*, 52–57.

Sadhwani, N., Liu, Z., Eden, M. R., & Adhikari, S. (2013). Simulation, analysis, and assessment of CO_2 enhanced biomass gasification. In *Computer aided chemical engineering* (Vol. 32, pp. 421–426). Netherlands: Elsevier.

Safran, P., & Omori, M. (1990). Some ecological observations on fishes associated with drifting seaweed off Tohoku coast, Japan. *Marine Biology, 105*(3), 395–402.

Sangwan, S., Rao, D. V., & Sharma, R. A. (2010). A review on *Pongamia pinnata* (L.) Pierre: A great versatile leguminous plant. *Nature and Science, 8*(11), 130–139.

Sanjay, B. (2015). Yellow oleander *(Thevetia peruviana)* seed oil biodiesel as an alternative and renewable fuel for diesel engines: A review. *International Journal of ChemTech Research, 7*(6), 2823–2840.

Sauciuc, A., Potetz, A., Weber, G., Rauch, R., Hofbauer, H., & Dumitrescu, L. (2011). Synthetic diesel from biomass by Fischer-Tropsch synthesis. *Renewable Energies and Power Quality Journal, 1*(8), 1–6.

Scholz, V., & Da Silva, J. N. (2008). Prospects and risks of the use of castor oil as a fuel. *Biomass and Bioenergy, 32*(2), 95–100.

Scott, P. T., Pregelj, L., Chen, N., Hadler, J. S., Djordjevic, M. A., & Gresshoff, P. M. (2008). *Pongamia pinnata*: An untapped resource for the biofuels industry of the future. *Bioenergy Research, 1*(1), 2–11.

Sengupta, A., & Gupta, M. P. (1970). Studies on the seed fat composition of *Moringaceae* family. *Fette, Seifen, Anstrichmittel, 72*(1), 6–10.

Shafiq, S. A. (2017). Biodiesel production by *Oleaginous fungi* before and after exposing of UV light. *International Journal of ChemTech Research, 10*(12), 357–363.

Shalaby, E. A. (2015). A review of selected non-edible biomass sources as feedstock for biodiesel production. *Biofuels-Status and Perspective*, 3–20.

Shanab, S. M. M., Hanafy, E. A., & Shalaby, E. A. (2014). Biodiesel production and antioxidant activity of different Egyptian date palm seed cultivars. *Asian Journal of Biochemistry, 9*(3), 119–130.

Sheng, T., Zhao, L., Gao, L., Liu, W., Wu, G., Wu, J., & Wang, A. (2018). Enhanced biohydrogen production from nutrient-free anaerobic fermentation medium with edible fungal pretreated rice straw. *RSC Advances, 8*(41), 22924–22930.

Shukla, M. K., Bhaskar, T., Jain, A. K., Singal, S. K., & Garg, M. O. (2007). *Bio-ethers as transportation on fuel: A review*. Dehradun: Indian Institute of Petroleum.

Singh, A., & Singh, I. S. (1991). Chemical evaluation of Mahua (*Madhuca indica*) seed. *Food Chemistry, 40*(2), 221–228.

Singh, R. K., Singh, D., & Mahendrakar, A. G. (2010). *Jatropha* poisoning in children. *Medical Journal, Armed Forces India, 66*(1), 80.

Singh, S. P., & Singh, D. (2010). Biodiesel production through the use of different sources and characterization of oils and their esters as the substitute of diesel: A review. *Renewable and Sustainable Energy Reviews, 14*(1), 200–216.

Slade, R., & Bauen, A. (2013). Micro-algae cultivation for biofuels: Cost, energy balance, environmental impacts and future prospects. *Biomass and Bioenergy, 53*, 29–38.

Slomkowski, S., Alemán, J. V., Gilbert, R. G., Hess, M., Horie, K., Jones, R. G., & Stepto, R. F. (2011). Terminology of polymers and polymerization processes in dispersed systems (IUPAC Recommendations 2011). *Pure and Applied Chemistry, 83*(12), 2229–2259.

Soccol, C. R., Faraco, V., Karp, S. G., Vandenberghe, L. P., Thomaz-Soccol, V., Woiciechowski, A. L., & Pandey, A. (2019). Lignocellulosic bioethanol: Current status and future perspectives. In *Biofuels: Alternative feedstocks and conversion processes for the production of liquid and gaseous biofuels* (pp. 331–354). New York, NY: Academic Press.

Solaimuthu, C., Senthilkumar, D., & Ramasamy, K. K. (2013). An experimental investigation of performance, combustion and emission characteristics of mahua (*Madhuca indica*) oil methyl ester on four-stroke direct injection diesel engine. *Indian Journal of Engineering and Materials Sciences, 20*(1), 42–50.

Solikhah, M. D., Pratiwi, F. T., Heryana, Y., Wimada, A. R., Karuana, F., Raksodewanto, A. A., & Kismanto, A. (2018, April). Characterization of bio-oil from fast pyrolysis of palm frond and empty fruit bunch. In *IOP conference series: Materials science and engineering* (Vol. 349, No. 1). Bristol: IOP Publishing.

Somali, M. A., Bajneid, M. A., & Al-Fhaimani, S. S. (1984). Chemical composition and characteristics of *Moringa peregrina* seeds and seeds oil. *Journal of the American Oil Chemists' Society*, *61*(1), 85–86.

Soriano Jr, N. U., Venditti, R., & Argyropoulos, D. S. (2009). Biodiesel synthesis via homogeneous Lewis acid-catalyzed transesterification. *Fuel*, *88*(3), 560–565.

Strobel, G. (2014). The story of mycodiesel. *Current Opinion in Microbiology*, *19*, 52–58.

Sullivan, P. (2010). ESSAY: Energetic cities: Energy, environment and strategic thinking. *World Policy Journal*, *27*(4), 11–13.

Tandon, P., & Jin, Q. (2017). Microalgae culture enhancement through key microbial approaches. *Renewable and Sustainable Energy Reviews*, *80*, 1089–1099.

Tiwari, M., & Singh, A. P. (2017). A comparative evaluation of biodiesel blends of Soapnut, Palm, and Karanja for usage in CI engine. *International Journal of Applied Engineering Research*, *12*(17), 6440–6446.

Ukpai, P. A., & Nnabuchi, M. N. (2012). Comparative study of biogas production from cow dung, cow pea and cassava peeling using 45 litres biogas digester. *Advances in Applied Science Research*, *3*(3), 1864–1869.

United Nations. (2018). *World urbanization prospects: The 2018 revision, key facts*. Technical Report.

Usman, L. A., Oluwaniyi, O. O., Ibiyemi, S. A., Muhammad, N. O., & Ameen, O. M. (2009). The potential of Oleander (*Thevetia peruviana*) in African agricultural and industrial development: A case study of Nigeria. *Journal of Applied Biosciences*, *24*, 1477–1487.

Van Gerpen, J. (2005). Biodiesel processing and production. *Fuel Processing Technology*, *86*(10), 1097–1107.

Vandendriessche, S., Vincx, M., & Degraer, S. (2006). Floating seaweed in the neustonic environment: A case study from Belgian coastal waters. *Journal of Sea Research*, *55*(2), 103–112.

Vandepopuliere, J. M., Al-Yousef, Y., & Lyons, J. J. (1995). Dates and date pits as ingredients in broiler starting and Coturnix quail breeder diets. *Poultry Science*, *74*(7), 1134–1142.

Vicente, G., Bautista, L. F., Rodríguez, R., Gutiérrez, F. J., Sádaba, I., Ruiz-Vázquez, R. M., . . .& Garre, V. (2009). Biodiesel production from biomass of an oleaginous fungus. *Biochemical Engineering Journal*, *48*(1), 22–27.

Vijay Kumar, M., Veeresh Babu, A., & Ravi Kumar, P. (2019). Producing biodiesel from crude Mahua oil by two steps of transesterification process. *Australian Journal of Mechanical Engineering*, *17*(1), 2–7.

Wang, J., & Yin, Y. (2018). Fermentative hydrogen production using pretreated microalgal biomass as feedstock. *Microbial Cell Factories*, *17*(1), 22.

Ware, P., Patel, D., Patel, J., & Krishnamurthy, R. (2013). Synthesis and characterization of biofuel from non-edible seed oil. *Journal of Energy, Environment & Carbon Credits*, *3*(2), 25–33.

Williams, J. A. (2002). Keys to bioreactor selections. *CEP Magazine*, *98*(3), 34–41.

Xie, G., Wang, X., & Ren, L. (2010). China's crop residues resources evaluation. *Sheng wu gong cheng xue bao Chinese Journal of Biotechnology*, *26*(7), 855–863.

Xin, F., Zhang, H., & Wong, W. (2013). Bioethanol production from horticultural waste using crude fungal enzyme mixtures produced by solid state fermentation. *BioEnergy Research*, *6*(3), 1030–1037.

Zhang, J., & Zhang, X. (2019). The thermochemical conversion of biomass into biofuels. In *Biomass, biopolymer-based materials, and bioenergy* (pp. 327–368). Duxford: Woodhead Publishing.

Zhang, Y., Cui, Y., Chen, P., Liu, S., Zhou, N., Ding, K., & Wang, Y. (2019). Gasification technologies and their energy potentials. In *Sustainable resource recovery and zero waste approaches* (pp. 193–206). Amsterdam: Elsevier.

Ziolkowska, J. R. (2020). Biofuels technologies: An overview of feedstocks, processes, and technologies. In *Biofuels for a more sustainable future* (pp. 1–19). Amsterdam: Elsevier.

2

Algal-Based Biofuel: Challenges and Future Perspectives

Waleed M.M. El-Sayed, Hassan A.H. Ibrahim, Mohamed A.A. Abdrabo, and Usama M. Abdul-Raouf

CONTENTS

2.1 Introduction 23
2.2 Energy from Microalgae 24
2.3 Potential Challenges to Biofuel Production from Microalgae 24
    2.3.1 Algal Crop Protection 26
        2.3.1.1 Rotenone: A Potential Protective Agent for Microalgal Cultivation and Biomass Production 27
        2.3.1.2 The Pesticide TFM: A Potential Protective Agent for Microalgal Cultivation and Biomass Production 29
        2.3.1.3 Niclosamide: A Potential Protective Agent for Microalgal Cultivation and Biomass Production 30
    2.3.2 Infochemicals as Microalgal Self-Defending 31
        2.3.2.1 β-Cyclocitral ($C_{10}H_{16}O$) 31
        2.3.2.2 Trans,Trans-2,4-Decadienal (DDE) 32
        2.3.2.3 Hydrogen Peroxide (H_2O_2) 32
        2.3.2.4 Norharmane 32
        2.3.2.5 Tryptamine 32
        2.3.2.6 Kairomone 33
        2.3.2.7 2-Butoxyethanol 33
        2.3.2.8 Organic Sulfonates 33
2.4 Microbial Community Associated with Algal Ponds 33
2.5 Optimization of Algal Cultivation Systems 34
2.6 The Technology of Harvesting Microalgae 35
2.7 Conclusion 36
References 36

2.1 Introduction

The industrial revolution has raised the demand for energy from sources dependent on fossil fuels such as coal, oil, and natural gas. Moreover, fossil fuels are running out due to continuous consumption and increased development in all sectors of human life. Depletion of fossil fuels, increases in environmental impacts, and climate change have led more investigators to find sustainable and natural sources of energy to face the limitations on energy resources and find alternative fuel sources (Sambusiti et al., 2015).

New energy resources can fulfill energy needs and will partially replace fossil fuels. Thus, several nations and countries are concentrating their attention on the creation of new, renewable, and sustainable energy resources. Biofuels are of greatest interest among the numerous potential sources of renewable energy, and currently play a critical role in the global energy infrastructure (Chisti, 2007).

Biofuels are biomass-based liquid, gas, and solid fuels such as bioethanol, biogas, biodiesel, and biohydrogen (Demirbas, 2008). Given the large amount of atmospheric carbon and industrial gases they consume, algae are a very promising source of biomass for biofuel production and algae, are also very successful in using nutrients from industrial byproducts and municipal wastewater. Algal biomass cultivation therefore offers dual benefits, supplying biomass for biofuel production and protecting our atmosphere from air and water pollution (El-Naggar et al., 2014; Singh & Olsen, 2011).

Because of their high growth rate, lipid content, carbon dioxide (CO_2) absorption and uptake rate, and the relatively limited amount of land on which algae can be grown, algal biofuels are an attractive option. In order to explore the use of microalgae as an energy feedstock, comprehensive research has been carried out, with applications being developed for biodiesel, bioethanol, and biohydrogen development (Singh & Olsen, 2011; Van Ginkel et al., 2020).

In this way, algae may play a major role in wastewater treatment/use and reduce the associated environmental impacts and critical problems. It is possible to grow algae in saline/coastal sea water and on non-agricultural lands (desert, arid, and semi-arid land), and the production of algae will not create competition for food/fuel. Compared to knowledge about other advanced feedstocks based on cellulose and considered for biofuel production, algal genomics and basic research are more advanced and gaining momentum (Subhadra & Edwards, 2010).

To generate several biofuels, such as biodiesel, biogas, biohydrogen, and syngas, microalgal biomass can be used. The residual algal biomass generated in the biodiesel production process of lipid extraction can be appropriately used to produce bioethanol or biomethane. However, in order to produce commercially viable biofuels, substantial improvements must be made in performance, cost structure, and ability to scale up microalgal growth, lipid extraction, and biofuel production. A given collection of technological breakthroughs will be needed for this purpose, in order to maximize the use of algal biomass for commercial biofuel production. Biofuels based on algal biomass will play a role in potential energy systems as these technological breakthroughs occur (El-Sayed et al., 2016). It is still too early to comment on any favored routes of biofuel production from algal biomass at the current development level. Finally, a detailed lifecycle evaluation of algal biofuels that highlights environmental benefits and impacts can and should be an instrument for driving technical progress and policy decisions (Singh & Olsen, 2011).

2.2 Energy from Microalgae

Microalgae are unicellular photosynthetic microorganisms living in marine or freshwater environments that generate algal biomass using sunlight, water, and carbon dioxide (Demirbas, 2010). Microalgae have been classified into a number of groups by biologists, primarily characterized by their pigmentation, lifecycle, and basic cell structure (Demirbas, 2010).

Algae are divided into four major categories: diatoms, green algae, golden algae, and blue-green algae. Algae are photosynthetic and aquatic organisms, and the features of ideal algal biofuel production strains are rapid growth rate, high density, high oil content, and ease of harvesting. While there are more than 500,000 recognized algae species, only a few species have been assessed as potential sources of biofuels to date (Table 2.1), and none have been found to be suitable. Thus, while current and future prospecting is important, algae breeding and genetic engineering are likely to be needed to grow or acquire commercially viable strains of algae (Borowitzka & Moheimani, 2013).

2.3 Potential Challenges to Biofuel Production from Microalgae

Microalgae biofuel production is feasible on a small scale, but large-scale production is still not economically viable. Producing biofuel from microalgae requires a large-scale system for cultivation and economically smart harvesting systems to meet the challenge of reducing the cost per unit area.

TABLE 2.1

Lipid Content (% Dry Weight) and a Class of Some Promising Microalgae for Biofuel Production.

Algal Species	Class	Lipid Dry Weight (%)
Botryococcus braunii	Trebouxiophyceae	80
Chlorella protothecoides		57.9
Nannochloris sp.		30–50
Chlorella pyrenoidosa		46.7
Chlorella vulgaris		14–22
Pleurochrysis carterae	Prymnesiophyceae	30–50
Prymnesium parvum		22–38
Scenedesmus dimorphus	Chlorophyceae	16–40
Dunaliella tertiolecta		35.6
Hormidium sp.		38
Scenedesmus obliquus		12–14
Tetraselmis suecica	Chlorodendrophyceae	20
Euglena gracilis	Euglenoidea	14–20

Source: Borowitzka & Moheimani (2013).

Therefore, algal growth conditions must be controlled at a large scale, and algae must be provided with the optimum growth environment (Brennan & Owende, 2010; Greenwell et al., 2009; Mata et al., 2010).

Much like an agricultural crop, large-scale algal cultures are attacked by predatory and pathogenic microorganisms; therefore, the protection of algal crops is a major challenge in producing microalgae biofuel. Several strategies for algae crop protection must be employed. For instance, algal cultures are initiated with high-density inoculation. Algae have a rapid population growth rate, and the loss of algal biomass yield may be reduced by minimizing contamination. Unfortunately, this strategy may not be feasible to meet the large-scale demand for the production of biofuels, especially if a continuous system of harvesting is implemented. Another strategy to eliminate contamination is to cultivate microalgae that can grow under unique conditions that are not suitable for most potential predators. Microalgae have developed defenses against predators and pathogens (Hannon et al., 2010).

The infochemical strategy is commonly present in the "algae group" against several attackers, including viruses, bacteria, fungi, protozoans, and other algae. Fundamentally, the challenges mentioned can be overcome in large-scale algal cultivation systems, but this might require several years of investigation and research to understand the interactions of pathogens and algae (Fierer, 2008).

Optimizing the algal cultivation system is necessary to achieve economic viability. Algae can be cultivated at a range of temperatures, with growth being restricted mainly by the availability of nutrients and light. Light stores energy for carbon fixation and is transformed by photosynthesis into chemical energy, providing the building blocks for the production of biofuels. Algal growth rates are mostly restricted by light penetration into ponds from both self-shading and light absorption by the water, and these constraints are major factors in determination of pond depth (Amin, 2009; El-Sayed et al., 2016).

Algal biomass harvesting is another vital challenge due to the small size of the algal cells (ranging from 3–30 μm in diameter), the large volumes of water, and their similar density to water that must be addressed to recover the biomass yield. Algae harvesting requires solid and/or liquid separation steps, including the collection and dewatering process. Generally, the harvesting technologies used are flocculation, coagulation, filtration (membrane and screen), flotation, centrifugation, and gravity sedimentation (Carmichael et al., 2000; Heasman et al., 2000; Munoz & Guieysse, 2006; Wang et al., 2008).

2.3.1 Algal Crop Protection

Protecting algae crops is the single most significant obstacle facing the production of algal biofuel. The defense of algae crops involves combat against pathogens (viruses, bacteria, and fungi), predators (protozoa, planktonic crustaceans, rotifers, and helminths), weeds (undesired algae and aquatic plants) and high-density growth inhibitors (bioactive chemical compounds) developed by algae or associated bacteria (Muradov et al., 2015; Neethirajan et al., 2018).

There are many advantages of algal biofuel. The actual productivity of algal biodiesel is almost 10 times higher than that of biodiesel from agricultural oil crops, such as canola and rapeseed, and therefore, algae cultivation is not expected to compete with conventional agriculture for available land. However, because algae are a food source for higher organisms such as amoebas, ciliates, and rotifers, predation, especially in open-pond systems, must be controlled (Figure 2.1). The contamination of the open algal pond by chytrids has been reported as one of the most significant obstacles to the producing of astaxanthin from the green algae *Haematococcus pluvialis* (Han et al., 2013).

The marine rotifer *Brachionus plicatilis*, another of these higher species, has the capacity to consume up to 12,000 algal cells per hour and can cause pond crashes in a few days, and other rotifers can eat thousands of algal cells per hour and cause a pond crash in a couple of days (Lubzens, 1987). Consequently, to realize the benefits of algal biofuel production, predators grazing on algae must be controlled (Van Ginkel et al., 2015). Since the removal of rotifers by mechanical methods is difficult, chemical treatment has been proposed as an alternative solution (Lubzens, 1987). The addition of a chemical to an algal culture at concentrations at which the predator is more susceptible than the algae is one method of controlling predation. Ideally, predation could be reduced or prevented without a substantial effect on algal growth (Van Ginkel et al., 2015).

Unlike rotifers, which have a true mouth and consume water while feeding on algae, ciliates engulf their prey and thus may be less susceptible to chemical agents. In highly eutrophic environments, rotifers grow faster than ciliates and can reach broad population densities of 1,000–500,000 individuals per liter (Montemezzani et al., 2015). Rotifers belong to the animal kingdom, live in both freshwater and saltwater environments, and are ideal test organisms because of their global distribution, sensitivity, ease of cultivation, and short generation time (Dahms et al., 2011; Yang et al., 2011). Because of their microscopic size and ease of use, genetic uniformity and high resistance to toxic chemicals, they are ideal model algal predators (Van Ginkel et al., 2015).

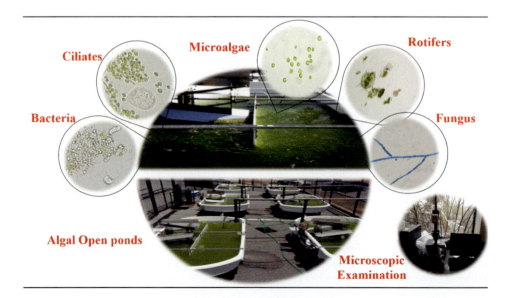

FIGURE 2.1 Potential contamination of microorganisms associated with algal open ponds.

Another aspect of seeking effective control agents for algae crops is that it is difficult to correctly dose such conventional compounds such as copper and hypochlorite to achieve selective grazer toxicity because they are rapidly degraded (hypochlorite) or algae (copper) adsorbed, decreasing their bioavailability to grazers (Wang et al., 2012). The task is to find compounds that are selectively toxic to grazers, that are easy to apply, that are not detoxified by algae, that are not environmentally persistent upon discharge, and that are cost-effective at the appropriate treatment doses. It is understood that rotifers are sensitive to a wide variety of chemicals, including hypochlorite, mercury, cadmium, zinc, aluminum, and copper (Snell & Persoone, 1989).

Chemical protective agents that inhibit the electron transport chain (ETC), which is important for all breathing organisms, may be the most effective and promising. For instance, rotenone is a traditional Complex I inhibitor of the ETC system. Algae are considered to be insensitive to ETC disruptors because algae have alternate pathways for the oxidation of NADH2 in addition to Complex I (Rasmusson et al., 2004). This differential ETC disruptor sensitivity could be exploited to control several types of predators while not affecting the crop of algae.

One potential chemical agent for algae crop protection is rotenone. Rotenone is a broad-spectrum pesticide that exists naturally in many members of the bean family. It is a close structural homologue to the electron carrier ubiquinone found in Complex I of the respiratory mitochondrial ETC. One of the first steps in the oxidation of NADH2 used to pump protons and produce ATP is ubiquinone reduction. Rotenone binds to Complex I's ubiquinone binding site and is an inhibitor of ETC. Rotenone is better both environmentally and toxicologically than many of the pesticides currently in use. In the environment, the persistence of rotenone is well known and depends on pH, temperature and light (El-Sayed et al., 2018; Rasmusson et al., 2004). Rotenone can persist for as long as two weeks or longer, which is ample time for biodiesel production to control predation in algae ponds.

The piscicide 3-trifluoromethyl-4-nitrophenol (TFM) is another potential chemical agent for algal crop protection. The invasive sea lamprey in the Great Lakes has been widely used to control TFM, and its toxicity against algal predators is less known (Smith & Crews, 2014). Not only is TFM toxic to lamprey larvae, but it also affects different teleost fish (Niblett & McKeown, 1980). TFM is one of several phenolic piscicides. Earlier researchers have studied its toxicity in both target and non-target organisms, especially fish larvae (Lech & Statham, 1975; Smith & Crews, 2014). It is well known to inhibit mitochondrial ATP production due to its similarity to other phenolic compounds, and can target all organisms that breathe. Generally, previous studies found that ponds containing algae and treated with TFM at concentrations typically used to control lampreys demonstrated small decreases in dissolved oxygen (DO), which indicates that TFM has a slight effect on algae (Smith & Crews, 2014).

TFM is very soluble in water with a solubility limit of 5 g/L. According to (Dawson et al., 1986), at the pH in most natural waters, TFM is almost entirely ionized and is not deeply divided into sediments where anaerobic conditions occur. The persistence of TFM in the environment is well known and is a function of DO and biochemical oxygen demand (BOD) (Eganhouse et al., 1983). In *in vitro* aqueous, sediment-free environments, TFM is persistent and resists biodegradation during a period of 2.5 months under aerobic conditions (Thingvold & Lee, 1981).

Niclosamide (NC) can be considered a protective agent for algae crops. In commercially operated fish ponds, it is used to remove unwanted fish before restocking. Previous literature mentioned that open ponds containing microalgae and treated with NC to control fish demonstrated a small effect on the algae present, which indicates that NC has a slight effect on microalgae (Karuppasamy et al., 2018).

2.3.1.1 Rotenone: A Potential Protective Agent for Microalgal Cultivation and Biomass Production

Rotenone is a broad-spectrum pesticide that occurs naturally in many members of the Fabaceae bean family. Rotenone is an especially effective agent in preventing pond crashes because it interferes with their ETC by inhibiting all non-photosynthetic species. Cellular oxygen is unusable and reduced to a reactive species of oxygen that can damage DNA and mitochondria (Stowe & Camara, 2009). Rotenone is safer both environmentally and toxicologically than many pesticides commonly used. Rotenone "is moderately harmful to humans and other mammals, but highly toxic, including fish, to insects and aquatic life." This greater toxicity in fish and insects is because lipophilic rotenone is readily absorbed through the gills or trachea, but not through the skin or gastrointestinal tract as easily. In the environment, the persistence of rotenone is well known and is a function of pH, temperature, and light (Perrone, 2011).

In the presence of light, rotenone readily breaks down into at least 20 components, of which only one is toxic, i.e. 6 alphaβ, 12 alphaβ-rotenolone. None of the other degradation products are hazardous, which means that they are considered safe to use. Rotenone can persist for up to two weeks, which is ample time to control predation in algae ponds used in the production of biodiesel (Turner et al., 2007; Van Ginkel et al., 2015).

While rotenone is well known as a piscicide and an insecticide, little research is available on the toxic impact of rotenone on oxygen-producing aquatic algae predators or algal cultures, and can therefore generate reactive oxygen species (Stowe & Camara, 2009). Salk (1975) concluded that rotenone inhibited the isolates of *Stichococcus chodati*, *Chlorococcum*, and *Bracteacoccus* algal, whereas it did not affect the isolates of *Chlorella* and *Klebsormidium flaccidum* algal. Salk (1975), however, tested rotenone concentrations as high as 50 ppm, which could be too cost-prohibitive for biodiesel production to be used in algal ponds. Rotenone is a pesticide and a close structural homologue of ubiquinone, an electron carrier found in Complex I of the all-life respiratory mitochondrial ETC (Figure 2.2) (Van Ginkel et al., 2015). Algae are considered to be rotenone-insensitive because they have alternative pathways for the oxidation of NADH2 in addition to Complex I. To minimize several kinds of predators while not affecting biodiesel-producing algae, this differential sensitivity to rotenone can be exploited. To decrease predation in open ponds, a universal tool such as rotenone is needed, since predation is well known to restrict cultivation to a few weeks in the warmer months (El-Sayed et al., 2018).

French Guiana's indigenous people used rotenone-containing plants to kill fish that are subsequently consumed. To protect algae ponds from predation, bleach, copper, and quinine have all been used, and rotenone appears to be more protective at lower concentrations. Rotenone can persist for a period of two weeks or longer, which is ample time to control predation in ponds of algae used for the production of biodiesel. While rotenone is a well-known pesticide, little research on the toxic effect of rotenone on predators of aquatic algae is available (Van Ginkel et al., 2015).

The effect of rotenone on the reproduction and mortality of *C. kessleri* and *B. calyciflorus* was examined (Van Ginkel et al., 2015). The inhibition of rotifer populations with minor effects on the algae was

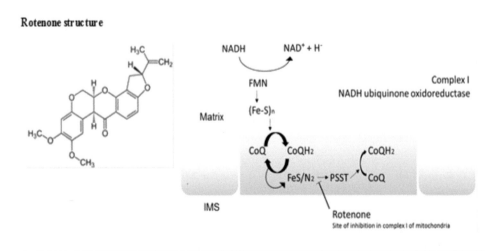

FIGURE 2.2 Rotenone is a classic inhibitor of Complex I of the electron transport chain (ETC). (From Van Ginkel et al. 2015.)

determined by Rotenone. For rotifers, the rotenone 24-h LC50 ranged from 0.074–0.35 µM (0.029–0.14 mg/L), while *C. kessleri* growth was uninhibited at 0.8 µM (0.32 mg/L) (Van Ginkel et al., 2015). The rotenone concentration of 0.1 µM (0.04 mg/L) caused near-complete *B. calyciflorus* mortalities in co-cultivation, while leaving *C. kessleri* growth unaffected (Van Ginkel et al., 2015).

Although rotenone is a well-known pesticide, there is little research available on its toxic effect on aquatic algae predators. It was concluded (Salk, 1975) that rotenone inhibited algal isolates of *Stichococcus chodati, Chlorococcum*, and *Bracteacoccus*, whereas algal isolates of *Chlorella* and *Klebsormidium flaccidum* were unaffected. However, rotenone concentrations as high as 50 mg/L were examined (Salk, 1975), which may be too cost-prohibitive for use in biodiesel algal ponds. We thus tested concentrations of rotenone as low as 1.1 µg/L (2.5 µM).

Overall, previous studies have shown more evidence that rotenone provides an effective and inexpensive method for algae farmers to control grazing predators in mass algal cultures in order to avoid pond crashes.

2.3.1.2 The Pesticide TFM: A Potential Protective Agent for Microalgal Cultivation and Biomass Production

Since 1958, the pesticide 3-trifluoromethyl-4-nitrophenol (TFM) has been targeted to control sea lamprey (*Petromyzon marinus*) and other larval lampreys in the Great Lakes region as part of an integrated pest management program (Smith & Tibbles, 1980). TFM is one of a number of pesticides which are phenolic. Previous studies have investigated its toxicity in both target and non-target organisms, in particular in fish and larvae (Applegate et al., 1961; Kanayama, 1963; Lech & Statham, 1975; Smith & Tibbles, 1980). Because of the similarity to the construction of phenolic compounds, it is well known to impair mitochondrial ATP production. TFM exerts its toxicity by disrupting ATP production via the uncoupling of oxidative phosphorylation by acting as a protonophore (Figure 2.3).

In general, previous studies described that ponds containing algae treated with a TFM dose typical of those used to monitor sea lampreys in streams have shown only slight changes in pH levels of less than 0.1 and a modest decrease in the DO ratio of approximately 8% in underexposed light and 11% in dark conditions. The pH levels and DO of open ponds may not modify the effectiveness of TFM, especially

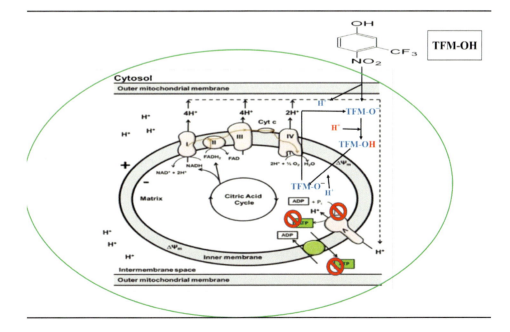

FIGURE 2.3 TFM uncoupling oxidative phosphorylation of mitochondria.

in open algal ponds. Moreover, TFM has no harmful effects on algal biomass production; aquatic plants and algae absorb TFM more readily than do other aquatic photosynthetic organisms (Dawson et al., 1992; Gilderhus et al., 1975; Maki et al., 1975; Maki & Johnson, 1976), and microorganisms degrade TFM in aquatic systems (Kempe, 1973).

TFM is better both environmentally and toxicologically than many pesticides commonly used. In addition to being toxic to lamprey larvae, TFM, the key toxicant used to control ammocoete populations, also affects various teleost fish (Niblett & McKeown, 1980). According to Dawson et al. (1986), TFM is almost completely ionized at the pH level of most natural waters and is not strongly partitioned to the sediment were anaerobic conditions exist. The solubility limit of TFM is known to be 5 g/L, which makes it a very soluble pesticide compared to others. However, for safety reasons, proper care should be taken when handling more concentrated stock solutions (Dawson et al., 1986).

TFM's persistence in the environment is well known and is a DO and BOD feature (Eganhouse et al., 1983).

Thingvold and Lee (1981) presented experimental results showing that in *in vitro* aqueous, sediment-free environments, TFM is persistent and resists biodegradation (Thingvold & Lee, 1981). In aquatic systems, TFM may be degraded by microorganisms. Thingvold and Lee (1981) found that TFM was integrated into the cell composition, and enzymes involved in the degradation of substrates similar to TFM may decay TFM incidentally. TFM may also be deactivated by high electron localization. Some studies reported no degradation of TFM during a period of 2.5 months under aerobic conditions. Some sorption of TFM to sediment material was also reported, and was dependent on the sediment source (Science et al., 1967; Thingvold & Lee, 1981).

Although TFM is well known as a pesticide for parasitic and invasive organisms, little research is available on the toxic effects of TFM on oxygen-producing aquatic algae predators or algal cultures, which can therefore generate reactive oxygen species (Conover et al., 2007).

The effect of TFM on the reproduction and mortality of *C. kessleri* and *B. calyciflorus* was evaluated (Van Ginkel et al., 2020). The TFM EC50 for *C. kessleri* was 84.5 µM, and the IC50 for *B. calyciflorus* was 28.67 µM. Mortality (≥95%) was noticed at TFM concentrations ≥68 µM. The co-culture of *B. calyciflorus* with *C. kessleri* at TFM concentrations above ~68 µM showed significant rotifer mortality by 48 hours of treatment. While the relative growth rate of *C. kessleri* was not affected at concentrations below 68 µM, Therefore, we suggest that microalgae farmers use approximately 75 µM TFM as a protective dose during cultivation in open ponds.

2.3.1.3 Niclosamide: A Potential Protective Agent for Microalgal Cultivation and Biomass Production

In 1958, NC was discovered. The most effective and stable medicines needed in a health system are on the World Health Organization's List of Essential Medicines. NC kills a wide variety of snails, cestodes, and flatworms. It is applied in public health and hygiene programs to control snails such as *Biomphalaria glabrata*, which are intermediate hosts for *Schistosoma* sp., the infectious agents of schistosomiasis in open waters in tropical Africa (Cioli et al., 2014).

NC inhibits glucose uptake, anaerobic metabolism, and oxidative phosphorylation in tapeworms (Weinbach & Garbus, 1969). The metabolic effects of NC are relevant to a wide range of organisms, and accordingly, it has been applied as a control measure to organisms other than tapeworms. For example, it is an active ingredient in some formulations for killing lamprey larvae, as a molluscicide, and as a general-purpose pesticide in aquaculture. NC has a short half-life in water in field conditions, thus making it valuable in ridding commercial fish ponds of unwanted fish; it loses its activity soon enough to permit re-stocking within a few days of eradicating the previous population (Cirkovic et al., 2015).

NC was found in a 2015 study to display "strong *in vivo* and *in vitro* activity against methicillin-resistant *Staphylococcus aureus* (MRSA)" (Rajamuthiah et al., 2015). In addition, NC was found to inhibit the gastric pathogen *Helicobacter pylori*, both *in vitro* and *in vivo* (Tharmalingam et al., 2018). A 2016 drug repurposing screening study suggested that NC may inhibit Zika virus replication *in vitro* (Xu et al., 2016).

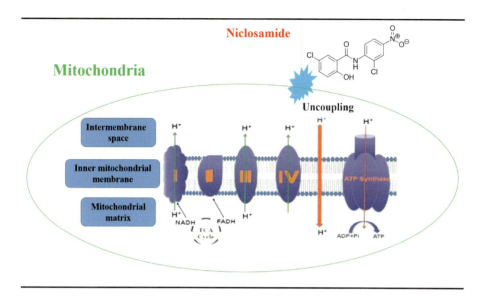

FIGURE 2.4 Niclosamide inhibits glucose uptake and uncoupling oxidative phosphorylation of mitochondria.

NC seems to inhibit the STAT3 signaling pathway, which is active in some cancers. The first use of NC was approved by the U.S. Food and Drug Administration (FDA) to treat worm infections in both humans and domestic livestock (Li et al., 2017).

NC is highly toxic to fish, yet it has a short half-life in water due to attenuation by UV light. NC acts by uncoupling oxidative phosphorylation (Figure 2.4). The proposed mechanism works by reducing the inner mitochondrial membrane's ability to inhibit oxidative phosphorylation (Wilson, 2012).

The solubility of NC in water is 1.6 mg/L at 20°C (Buchmann et al., 1990). Earlier studies found that in ponds containing algae and treated with NC to control undesired fish larvae, NC had little effect against algae (Oliveira-Filho & Paumgartten, 2000). Therefore, we expect NC to act as a potential protective agent for algae crops against various types of contamination.

The effect of NC on the reproduction and mortality of *C. kessleri* and *B. calyciflorus* was investigated (Van Ginkel et al., 2020). Moreover, The NC EC50 was 26.54 µM for *C. kessleri*, and the IC50 for the rotifer *B. calyciflorus* was 13.24 µM. In a previous study (Karuppasamy et al., 2018), the sensitivity of *C. vulgaris* to NC was tested. The researchers found that NC was effective against grazers such as *Oxyrrhis* sp. and *Euplotes* sp., and it is quite toxic to *C. vulgaris*. Thus, NC was compatible and a very good candidate for crop protection of this strain (Karuppasamy et al., 2018).

2.3.2 Infochemicals as Microalgal Self-Defending

Infochemicals may manage the aquatic organism's physiology and behavior, and as such, change the structure, mechanism, and regulation of nutrition maps. Therefore, we need more insight into how and to what extent infochemical maps are capable of controlling interactions among aquatic life (Ianora & Miralto, 2010; Poulson et al., 2009; Taylor et al., 2012; Vidoudez et al., 2011). Six known infochemicals (β-cyclocitral; trans,trans-2,4-decadienal; hydrogen peroxide; norharmane; kairomone; and tryptamine) and two chemical agents, 2-butoxyethanol and organic sulfonates, will be described as the infochemicals and agents affecting algal crop protection.

2.3.2.1 β-Cyclocitral ($C_{10}H_{16}O$)

β-Cyclocitral is a volatile compound produced by cyanobacteria and some *Microcystis* species and used as a flavoring ingredient. β-Cyclocitral belongs to the class of chemical entities known as organic oxides

(Robinson et al., 2009). Generally, β-cyclocitral plays a crucial protective role in photosynthetic organisms. β-Cyclocitral elicits changes in the expression of oxygen-responsive genes and leads to an enhancement of photo-oxidative stress tolerance (Ramel et al., 2013; Triantaphylidès & Havaux, 2009). β-Cyclocitral, as an infochemical, was assayed in previous studies for impacts on *Nannochloropsis oculata*. The lipid neutral content of *N. oculata* increased significantly; however, high concentrations of β-cyclocitral resulted in a significant decrease in *N. oculata* growth (Taylor et al., 2012). β-Cyclocitral is produced by cyanobacteria during periods of high biomass growth and triggers behavioral changes in grazing invertebrates, although direct the impacts on microalgae have yet to be fully assessed (Jüttner et al., 2010; Taylor et al., 2012).

2.3.2.2 Trans,Trans-2,4-Decadienal (DDE)

DDE is a member of the organic compound class known as medium-chain aldehydes. DDE is an aromatic substance produced from lipid peroxidation degradation. DDE, a polyunsaturated aldehyde produced mainly by diatom species from the oxidation of lipid precursors, promotes biochemical and cytological changes in microalgal cells (Vardi et al., 2008; Vardi et al., 2006) and can affect the reproductive success of grazing invertebrates (Caldwell, 2009). Moreover, DDE is cytotoxic to mammalian cells and is well known to be implicated in DNA damage. DDE was investigated in previous studies for effectiveness against *N. oculata*. The oil content of *N. oculata* increased significantly; however, high concentrations of DDE resulted in a significant reduction in *N. oculata* growth (Taylor et al., 2012). The increase in cell lipid levels in microalgae such as *Nannochloropsis* sp. is well known with exposure to a source of stress; for example, nitrogen restriction, high light intensity, or photo-oxidation (Borowitzka & Moheimani, 2013; Rodolfi et al., 2009; Taylor et al., 2012). This stress-related increase in lipid production is the result of the breakdown of the photosynthetic membrane and accumulation of lipid bodies in the cell cytosol (Hu et al., 2008).

2.3.2.3 Hydrogen Peroxide (H_2O_2)

H_2O_2 plays vital roles in host defense and oxidative biosynthetic reactions. H_2O_2 is naturally produced in organisms as a byproduct of oxidative metabolism, and belongs to the class of chemical entities known as homogeneous non-metal compounds. Samuilov and colleagues (2004) reported that H_2O_2 inhibits the evolution of photosynthetic O_2 because hydroxyl and hydroperoxyl radicals can lead to photosystem II inactivation, disrupt pigment synthesis and membrane integrity, and trigger cyanobacterial cell death. A selective effect of H_2O_2 on the phytoplankton population has been reported by Drábková and colleagues (2007). H_2O_2 affected cyanobacteria at concentrations 10 times lower than green algae and diatoms, and high light-dependent toxicity increased these differences (Drábková et al., 2007). H_2O_2 toxicity is due to the formation of hydroxyl radicals and the subsequent oxidation of biomolecules in general (Russell, 2003). Moreover, H_2O_2 was reported to be highly mobile in membranes and to diffuse to a vulnerable area inside of the cell (Florence, 1984).

2.3.2.4 Norharmane

Norharmane is a nitrogen-containing heterocycle. Norharmane is a naturally occurring β-carboline alkaloid exhibiting a wide range of biological, psychopharmacological, and toxicological actions. Norharmane belongs to the harmala alkaloids (HAlks) family of compounds (Arribam et al., 1994). Norharmane was tested in previous studies for impacts on some green algae. Norharmane demonstrated a strong effect on neutral lipid accumulation in the tested algae; however, the high concentrations of norharmane resulted in some decrease in the algal growth rate (Taylor et al., 2012). Norharmane can influence grazing invertebrates (Caldwell, 2009). Norharmane is released by cyanobacteria and acts in microbial communities as an allelopathic compound and helps to express mutagenic activity (Arribam et al., 1994; Volk & Furkert, 2006).

2.3.2.5 Tryptamine

Tryptamine is an alkaloid called monoamine. It contains the structure of an indole ring and is structurally identical to the tryptophan amino acid. Tryptamine is present in trace quantities in mammalian brains and is expected to play a role as a neuromodulator or neurotransmitter. Tryptamine, the bacillamide precursor monoamine alkaloid (Churro et al., 2009), can induce the formation of reactive oxygen species in microalgae (Churro et al., 2010). Tryptamine was effective in inhibiting the growth of some problematic blooms of cyanobacteria—such as *Microcystis aeruginosa*, *Anabaena circinalis*, *Anabaenopsis circularis*, *Leptolyngbya* sp., *Aphanizomenon gracile*, and *Nodularia spumigena*—at low concentrations (IC50 = 4.15 g/L) (Churro et al., 2010). However, cyanobacteria were less resistant to tryptamine than were most eukaryotic algae. A eukaryotic algae seemed to respond quite effectively to tryptamine-induced oxidative stress (Churro et al., 2010).

2.3.2.6 Kairomone

Kairomone is a trans-specific chemical messenger. When kairomone is released by an organism that is subsequently detected by another species, a response is evoked that adaptively favors the latter. Kairomone is simultaneously a repellent against aggressors, is similar to pheromones (communication chemicals used within a species), and can be used as a pesticide (Ansari et al., 2012).

2.3.2.7 2-Butoxyethanol

2-Butoxyethanol (2-BE) has the chemical formula $CH_3CH_2CH_2CH_2OH$ and is an organic compound. This colorless liquid is a butyl ether of ethylene glycol derived from the glycol ether family. It is used in many domestic and industrial products because of its properties as a surfactant, in addition to being a relatively non-volatile compound and an inexpensive solvent of low toxicity. In many products and industries, 2-BE is widely used as an additive, including agricultural herbicides and pesticides, and food processing additives (Kroes et al., 2007; Manz & Carter, 2016). It exhibits surfactant properties and forms micelles at concentrations above 120 g/L (Manz & Carter, 2016; Quirion et al., 1990).

2.3.2.8 Organic Sulfonates

Sulfonates are organic compounds. They are colorless salts with useful properties as surfactants (Ullmann, 1985). Sulfonates are currently used as chemical detergents and absorbents in pesticides and other agricultural chemicals. In nature, sulfonates are widespread and make up over 95% of the sulfur content of most aerobic soils. Sulfonates can be used as a source of sulfur for growth by many microalgae; sulfonates are often used as an antimicrobial processing aid in washing water development. The intended function of these products is to reduce the number of microorganisms in water. The mode of action of sulfonates commonly includes protein denaturing, inactivation of essential enzymes or membrane disruption, and alteration of cell permeability (Bickerstaff, 1997).

2.4 Microbial Community Associated with Algal Ponds

One of the major barriers to realizing consistently high rates of production in algal biomass cultivation is the ability to maintain culture stability. Daily decreases in production due to outdoor culture crashes can decrease overall production rates. Outdoor algal cultures—such as ciliates, amoebae, and other microorganisms such as bacteria and fungi—are regularly exposed to possible contaminating organisms, which can be transferred via the wind or the water source and act as the main cause of outdoor cultures crashes. Thus, these pests of outdoor culture systems must be controlled (Brennan & Owende, 2010; Hossain et al., 2008; Van Ginkel et al., 2015). Microorganisms are ubiquitous in the environment. Environmental variables play a vital role in the microbial community associated with algal ponds. Moreover, unlike laboratory cultures, the very large quantities of water needed for commercial-scale production of algal biomass cannot be sterilized. Therefore, the control required over a long time while maintaining axenic cultures of the desired microalgae is a significant obstacle. For maintaining long-standing cultures and to reduce contamination issues, it is necessary that the culture conditions be optimized. If the microalgae grow favorably in highly selective conditions, such as under high salinity

(*Dunaliella salina*), high nutrient availability (*Chlorella*), or high alkalinity (*Arthrospira*), then predators are controlled as long as these conditions can be maintained. Some algae also control the growth conditions, inhibiting the growth of predators (Moheimani & Borowitzka, 2006). For instance, cultures of *Pleurochrysis carterae* can increase the pond pH throughout the day to levels as high as pH 11 (Moheimani & Borowitzka, 2006).

Another performative aspect that influences algae growth in aquatic environments is salinity. Species-specific variations in the ability of microorganisms to adapt to salinity changes are always observed. *Nitzschia*, a genus of algae, has been observed growing at high salinities, whereas *Chaetoceros* sp. is abundant at low salinities (Zhang & Dickman, 1999). The content of nutrients is thus a significant factor shaping the microbial community. Depending on the involvement of other microbial species, predation, and contamination by viruses, the microbial community composition often varies. Blasco (1965) performed experiments on the effect of bacterial contamination in open ponds growing *Chlorella pyrenoidosa* in mass culture. This study indicates that bacterial population increased with algal mass increases and found symbiotic bacteria enclosed in the algal cell surface, demonstrating that there was no link between the exchange of gas in highly contaminated ponds and algal proliferation (Blasco, 1965).

In addition, chemical crop management is another potential method for the prevention of algae contaminants and pollutants. Hypochlorite has been used to control protozoans in *Nannochloropsis* mass cultures (Park et al., 2016; Weissman & Radaelli, 2010). Ammonia and acidic pH shock have been used successfully to control rotifers in open ponds (Lincoln et al., 1983; Zmora, 2004).

Nannochloropsis cultures have been shown to be cultivated outdoors, aided by the use of glyphosate and ozone (Weissman et al., 2012). Control of zooplankton in laboratory cultures of *Chlorella* has been achieved through the use of pesticides such as dipterex, parathion, and dichlorodiphenyltrichloroethane (DDT) (Loosanoff et al., 1957). None of these chemical techniques, however, have been effective for prolonged periods in outdoor systems on a commercial scale or for algae cultivation.

2.5 Optimization of Algal Cultivation Systems

In order to achieve effective growth, algae require nutrients, light, water, and a source of carbon, most often CO_2. Phosphorous, iron, and sulfur are the main nutrients that most algae require. Nutrient requirements for algal production are often overlooked because algae are very successful in sequestering these nutrients from their environment (Marchetti et al., 2009; Ruiz-Marin et al., 2010). In terms of eutrophication of lakes and coastal regions, shifts in nutrient load and algal growth have been extensively studied, but not as high in terms of productivity in large-scale aquaculture (Conley et al., 2009). If terrestrial agriculture is a model for some of the algae aquaculture challenges, then it is a major challenge to provide adequate nutrients for large-scale algae growth. In the current terrestrial agricultural industry, micro- and macronutrient supplements, or fertilizers, account for substantial costs, and biofuel production is not expected to vary in this regard. Globally, fertilizer use has been growing. Unfortunately, many fertilizer components are generated or mined from fossil fuels and, as such, are not renewable (Vaccari, 2009; Vance, 2001). Similar to plants, algae need sources of phosphorus, nitrogen, and potassium, which are the major components of agricultural fertilizers, and these already small supplies will be impacted by large-scale aquaculture. Chelated iron and sulfur are also needed for the optimal growth of many algal species. Phosphorous makes up slightly less than 1% of the total algal biomass and is needed to support algal growth in the medium at approximately 0.03–0.06%. Currently, fertilizers used for agriculture in the United States contain less than the optimum phosphate concentration due to insufficient supplies. Currently, less than 40 million tons of phosphate are extracted annually from the United States, and the overall production of phosphate from this mining peaked in the late 1980s. An additional 53 million tons of phosphate must be acquired annually if algal biofuels are to fully replace petroleum in the United States. This is a big challenge, considering that the estimated total volume of phosphate in the United States is approximately 2.8 billion tons. This leaves few choices other than effective phosphate recycling back into the algae ponds or dramatically increasing mining production, a possibility that would at best seem to provide a temporary solution. Nitrogen is not limited in supply, unlike phosphorous, but is also a limiting macronutrient in terms of plant and algal growth. To be nutritionally useful, algae need nitrogen to be fixed into ammonia, nitrates, and similar molecules (Ryther & Dunstan, 1971). Some bacteria, such as rhizobia, are able to fix their own nitrogen and some form symbiotic relationships with terrestrial plants, providing this important nutrient to the plants to promote the synthesis of protein and nucleic acid (Long, 1989; Vance, 2001). Some cyanobacteria are also capable of fixing nitrogen, whereas nearly all algal species known to date need an exogenous source of fixed nitrogen, and most prefer ammonia because it is less energy intensive than nitrate or nitrite (Inokuchi et al., 2002). It will be necessary for algae-based biofuel production to provide an inexpensive source of fixed nitrogen, and the possibility of using nitrogen-fixing cyanobacteria to supply this nitrogen will help reduce these costs (Berman-Frank et al., 2003). Iron is a significant limiting nutrient for algal growth in the open oceans, as seen by the induction of algal blooms by the addition of exogenous iron to the open oceans (Berman-Frank et al., 2003). Interestingly, the addition of iron to induce an algal bloom has been considered and tested as a strategy to sequester CO_2 (Berman-Frank et al., 2003; Boyd et al., 2004; Coale et al., 1996). Biologically, in all known photosynthetic species, iron is needed for electron transport, including *Chlamydomonas reinhardtii*, in a number of photosynthetic proteins, and is usually found in iron-sulfur clusters (Godman & Balk, 2008). Iron is not suitable for uptake in its oxidized state, and most algae prefer chelated iron. Fortunately, iron can be easily obtained and is more accessible than many of the other nutrients needed. Sulfur is also necessary for protein synthesis and lipid metabolism, in addition to its key role in the electron transport chain. Sulfur deficiency was shown to reduce the density of algal and stunt growth (Yildiz et al., 1994).

2.6 The Technology of Harvesting Microalgae

The major obstacle that affects the development of microalgae oil production is the harvesting cost. Use successful harvesting methods to minimize the cost of harvesting, which contributes to 20–30% of the overall cost of growing biomass. With respect to this approach, different methods of harvesting microalgae have been published (Prajapati et al., 2016; Wan et al., 2015). For biomass concentration variations (typically in the range of 0.3–5 g/L), the ideal harvesting method should be productive and scalable and require low operational, energy, and maintenance costs (Olaizola, 2003). Microalgae are commonly harvested in a two-step thickening and dewatering phase for separation. Thickening methods, for example, include coagulation and/or flocculation, gravity sedimentation, flotation, and electrical methods; centrifugal sedimentation and filtration are dewatering methods. Although these techniques are fast and effective, they are either energy intensive or susceptible to microalgal biomass contamination with detrimental chemicals added (Chen et al., 2013; Grima et al., 2003; Schlesinger et al., 2012).

Bioflocculation, due to the use of biological agents, is a form of chemical-free flocculation. The use of bioflocculants is non-toxic to biomass microalgae and enables recycling of the culture medium, which can further reduce the overall cost (Ummalyma et al., 2017). In addition, bioflocculation will increase the efficiency of total biomass production, lipid yield, and wastewater bioremediation (Muradov et al., 2015).

Algae bioflocculation has previously been carried out by the use of acceptable microbial partners (Table 2.2), such as algae–algae, algae–bacteria and algae–fungi (Alam et al., 2016). The fungal spore–assisted (FSA) method of harvesting microalgae could harvest microalgae by co-culturing microalgae with filamentous fungal spores. It has recently been recorded as a successful method of algal harvesting. Filamentous fungi—such as *Aspergillus* sp., *Mucor* sp., and *Penicillium* sp.—can serve as bioflocculating agents because of their self-pelletization and high efficiency on harvesting microalgae (Gultom & Hu, 2013; Xie et al., 2013).

In many industries, dealing with the processing of microalgal biomass, extracting microalgal biomass from growth media, is a major challenge. Cultivation broths have biomass densities below 0.5 kg/m³ in some industrial production systems, which means that large volumes need to be treated before algal oil extraction (Chen et al., 2014).

One of the most difficult areas of algal biofuel research is the development of a cost-effective harvesting system (Greenwell et al., 2009). A major factor restricting the industrial use of microalgae is that 20–30% of the overall cost of production has been stated to be involved in biomass harvesting (Grima et al., 2003; Mata et al., 2010). Other researchers estimated that the cost of the recovery process contributed about 50% to the final cost of oil production in their analysis (Chisti, 2007; Greenwell et al., 2009).

TABLE 2.2
Microorganisms Used for Microalgal Harvesting via Bioflocculation and Their Sources

Microorganism	Sources	Reference
Klebsiella pneumonia	Sediments of the wastewater treatment plant, soil sample	(Wan et al., 2013)
Bacillus mucilaginosus	Soil samples, kaolin suspension, activated sludge	(Zhang et al., 2002)
Citrobacter sp.	Domestic (kitchen) drainages	(Oswald, 2003)
Sorangium cellulosum sp.	Salt soil samples	(Nurdogan & Oswald, 1995)
Enterobacter aerogenes sp.	Soil samples	(Muñoz et al., 2004)
Cobetia sp.	Sediment samples of Algoa Bay	(Robinson, 1998)
Chryseomonas Luteola sp.	Palm oil mill effluent	(Wilde & Benemann, 1993)
Halomonas sp.	West Pacific Ocean (deep-sea)	(Rahman et al., 2016)
Staphylococcus cohnii	Palm oil mill effluent	(Abd-El-Haleem et al., 2008)
Proteus mirabilis sp.	Activated sludge	(Wan et al., 2013)

Many of the microalgal biofuel production studies have focused on lipid yield and biomass composition rather than the method of harvesting.

2.7 Conclusion

It has been demonstrated that microalgae are promising in the development of several forms of biofuels. This chapter provides algal farmers with a practical and feasible strategy to overcome the existing challenges of algal biofuel production and to contribute to the sustainability of the algal-based fuel industry. In ensuring that this technology becomes viable, sustainable production of biomass from algae grown in open ponds by the use of an efficient grazer and pollution management strategy will be crucial. The chemical agents (rotenone, TFM, and NC) are promising to eliminate contaminating organisms. In the same context, Applying the Plackett–Burman design is useful for finding factors for the growth optimization of algae and provide newly formulated media that will cost-efficient for algal-based biofuel. In addition, fungi as *A. fumigatus* exhibits a strong ability to capture microalgae, with a bioflocculation efficiency reached to 98%. Finally, minimizing of the costs of biofuel production from microalgal biomass is an effective way to attract commercial application.

REFERENCES

Abd-El-Haleem, D.A., Al-Thani, R.F., Al-Mokemy, T., Al-Marii, S., Hassan, F. 2008. Isolation and characterization of extracellular bioflocculants produced by bacteria isolated from Qatari ecosystems. *Polish Journal of Microbiology*, **57**(3), 231–239.

Alam, M.A., Vandamme, D., Chun, W., Zhao, X., Foubert, I., Wang, Z., Muylaert, K., Yuan, Z. 2016. Bioflocculation as an innovative harvesting strategy for microalgae. *Reviews in Environmental Science and Bio/Technology*, **15**(4), 573–583.

Amin, S. 2009. Review on biofuel oil and gas production processes from microalgae. *Energy Conversion and Management*, **50**(7), 1834–1840.

Ansari, M.S., Ahmad, N., Hasan, F. 2012. Potential of biopesticides in sustainable agriculture. In: *Environmental Protection Strategies for Sustainable Development*. Springer, pp. 529–595.

Applegate, V.C., Howell, J.H., Moffett, J.W., Johnson, B., Smith, M.A. 1961. *Use of 3-Trifluormethyl-4-Nitrophenol as a Selective Sea Lamprey Larvicide*. Great Lakes Fishery Commission.

Arribam, A.F., Lizcano, J., Balsa, M., Unzeta, M. 1994. Inhibition of monoamine oxidase from bovine retina by β-carbolines. *Journal of Pharmacy and Pharmacology*, **46**(10), 809–813.

Berman-Frank, I., Lundgren, P., Falkowski, P. 2003. Nitrogen fixation and photosynthetic oxygen evolution in cyanobacteria. *Research in Microbiology*, **154**(3), 157–164.

Bickerstaff, G.F. 1997. Immobilization of enzymes and cells. In: *Immobilization of Enzymes and Cells*. Springer, pp. 1–11.

Blasco, R.J. 1965. Nature and role of bacterial contaminants in mass cultures of thermophilic *Chlorella pyrenoidosa*. *Applied and Environmental Microbiology*, **13**(3), 473–477.

Borowitzka, M.A., Moheimani, N.R. 2013. *Algae for Biofuels and Energy*. Springer.

Boyd, P.W., Law, C.S., Wong, C., Nojiri, Y., Tsuda, A., Levasseur, M., Takeda, S., Rivkin, R., Harrison, P.J., Strzepek, R. 2004. The decline and fate of an iron-induced subarctic phytoplankton bloom. *Nature*, **428**(6982), 549.

Brennan, L., Owende, P. 2010. Biofuels from microalgae—a review of technologies for production, processing, and extractions of biofuels and co-products. *Renewable and Sustainable Energy Reviews*, **14**(2), 557–577.

Buchmann, K., Székely, C., Bjerregaard, J. 1990. Treatment of Pseudodactylogyrus infestations of *Anguilla anguilla* I.: trials with niclosamide, toltrazuril, phenolsulfonphtalein and rafoxanide. *Bulletin of European Association of Fish Pathologists*, **10**(1), 14–17.

Caldwell, G.S. 2009. The influence of bioactive oxylipins from marine diatoms on invertebrate reproduction and development. *Marine Drugs*, **7**(3), 367–400.

Carmichael, W.W., Drapeau, C., Anderson, D.M. 2000. Harvesting of *Aphanizomenon flos-aquae* Ralfs ex Born. & Flah. var. flos-aquae (Cyanobacteria) from Klamath Lake for human dietary use. *Journal of Applied Phycology*, **12**(6), 585–595.

Chen, G., Zhao, L., Qi, Y., Cui, Y.-L. 2014. Chitosan and its derivatives applied in harvesting microalgae for biodiesel production: an outlook. *Journal of Nanomaterials*, **2014**, 3.

Chen, L., Wang, C., Wang, W., Wei, J. 2013. Optimal conditions of different flocculation methods for harvesting *Scenedesmus* sp. cultivated in an open-pond system. *Bioresource Technology*, **133**, 9–15.

Chisti, Y. 2007. Biodiesel from microalgae. *Biotechnology Advances*, **25**(3), 294–306.

Churro, C., Alverca, E., Sam-Bento, F., Paulino, S., Figueira, V., Bento, A., PrabhaKar, S., Lobo, A., Calado, A., Pereira, P. 2009. Effects of bacillamide and newly synthesized derivatives on the growth of cyanobacteria and microalgae cultures. *Journal of Applied Phycology*, **21**(4), 429–442.

Churro, C., Fernandes, A., Alverca, E., Sam-Bento, F., Paulino, S., Figueira, V., Bento, A., Prabhakar, S., Lobo, A., Martins, L. 2010. Effects of tryptamine on growth, ultrastructure, and oxidative stress of cyanobacteria and microalgae cultures. *Hydrobiologia*, **649**(1), 195–206.

Cioli, D., Pica-Mattoccia, L., Basso, A., Guidi, A. 2014. Schistosomiasis control: praziquantel forever? *Molecular and Biochemical Parasitology*, **195**(1), 23–29.

Cirkovic, M., Kartalovic, B., Novakov, N., Pelic, M., Djordjevic, V., Radosavljevic, V., Aleksic, N. 2015. Distribution of niclosamide residues in meat and internal organs of common carp. *Procedia Food Science*, **5**, 54–56.

Coale, K.H., Johnson, K.S., Fitzwater, S.E., Gordon, R.M., Tanner, S., Chavez, F.P., Ferioli, L., Sakamoto, C., Rogers, P., Millero, F. 1996. A massive phytoplankton bloom induced by an ecosystem-scale iron fertilization experiment in the equatorial Pacific Ocean. *Nature*, **383**(6600), 495.

Conley, D.J., Paerl, H.W., Howarth, R.W., Boesch, D.F., Seitzinger, S.P., Havens, K.E., Lancelot, C., Likens, G.E. 2009. *Controlling Eutrophication: Nitrogen and Phosphorus*. American Association for the Advancement of Science.

Conover, G., Simmonds, R., Whalen, M. 2007. *Management and Control Plan for Bighead, Black, Grass, and Silver Carps in the United States*. Aquatic Nuisance Species Task Force, Asian Carp Working Group, Washington, DC. Available at: www. asiancarp.org/Documents/Carps_Management_Plan.pdf (accessed April 2010).

Dahms, H.-U., Hagiwara, A., Lee, J.-S. 2011. Ecotoxicology, ecophysiology, and mechanistic studies with rotifers. *Aquatic Toxicology*, **101**(1), 1–12.

Dawson, V., Johnson, D., Allen, J. 1986. Loss of lampricides by adsorption on bottom sediments. *Canadian Journal of Fisheries and Aquatic Sciences*, **43**(8), 1515–1520.

Dawson, V., Johnson, D., Sullivan, J. 1992. *Effects of the Lampricide 3-Trifluoromethyl-4-Nitrophenol on Dissolved Oxygen in Aquatic Systems*. Great Lakes Fishery Commission, US Fish and Wildlife Service.

Demirbas, A. 2008. Comparison of transesterification methods for production of biodiesel from vegetable oils and fats. *Energy Conversion and Management*, **49**(1), 125–130.

Demirbas, A. 2010. Use of algae as biofuel sources. *Energy Conversion and Management*, **51**(12), 2738–2749.

Drábková, M., Admiraal, W., Maršálek, B. 2007. Combined exposure to hydrogen peroxide and light selective effects on cyanobacteria, green algae, and diatoms. *Environmental Science & Technology*, **41**(1), 309–314.

Eganhouse, R.P., Blumfield, D.L., Kaplan, I.R. 1983. Long-chain alkylbenzenes as molecular tracers of domestic wastes in the marine environment. *Environmental Science & Technology*, **17**(9), 523–530.

El-Naggar, M., Abdul-Raouf, U., Ibrahim, H., El-Sayed, W.M. 2014. Saccharification of *Ulva lactuca* via *Pseudoalteromonas piscicida* for biofuel production. *Journal of Energy and Natural Resources*, **3**(6), 77–84.

El-Sayed, W.M., Ibrahim, H., Abdul-Raouf, U., El-Nagar, M. 2016. Evaluation of bioethanol production from *Ulva lactuca* by *Saccharomyces cerevisiae*. *Journal of Biotechnology & Biomaterials*, **6**(226), 2.

El-Sayed, W.M., Van Ginkel, S.W., Igou, T., Ibrahim, H.A., Abdul-Raouf, U.M., Chen, Y. 2018. Environmental influence on rotenone performance as an algal crop protective agent to prevent pond crashes for biofuel production. *Algal Research*, **33**, 277–283.

Fierer, N. 2008. Microbial biogeography: patterns in microbial diversity across space and time. In: *Accessing Uncultivated Microorganisms*. American Society of Microbiology, pp. 95–115.

Florence, T. 1984. The production of hydroxyl radical from hydrogen peroxide. *Journal of Inorganic Biochemistry*, **22**(4), 221–230.

Gilderhus, P.A., Sills, J.B., Allen, J.L. 1975. *Residues of 3-Trifluoromethyl-4-Nitrophenol (TFM) in a Stream Ecosystem after Treatment for Control of Sea Lampreys*. US Fish and Wildlife Service.

Godman, J., Balk, J. 2008. Genome analysis of *Chlamydomonas reinhardtii* reveals the existence of multiple, compartmentalized iron–sulfur protein assembly machineries of different evolutionary origins. *Genetics*, **179**(1), 59–68.

Greenwell, H., Laurens, L., Shields, R., Lovitt, R., Flynn, K. 2009. Placing microalgae on the biofuels priority list: a review of the technological challenges. *Journal of the Royal Society Interface*, rsif20090322.

Grima, E.M., Belarbi, E.-H., Fernández, F.A., Medina, A.R., Chisti, Y. 2003. Recovery of microalgal biomass and metabolites: process options and economics. *Biotechnology Advances*, **20**(7–8), 491–515.

Gultom, S., Hu, B. 2013. Review of microalgae harvesting via co-pelletization with filamentous fungus. *Energies*, **6**(11), 5921–5939.

Han, D., Li, Y., Hu, Q. 2013. Biology and commercial aspects of *Haematococcus pluvialis*. In *Handbook of Microalgal Culture: Applied Phycology and Biotechnology*, Vol. 2, pp. 388–405. Hoboken, NJ: John Wiley & Sons.

Hannon, M., Gimpel, J., Tran, M., Rasala, B., Mayfield, S. 2010. Biofuels from algae: challenges and potential. *Biofuels*, **1**(5), 763–784.

Heasman, M., Diemar, J., O'connor, W., Sushames, T., Foulkes, L. 2000. Development of extended shelf-life microalgae concentrate diets harvested by centrifugation for bivalve molluscs—a summary. *Aquaculture Research*, **31**(8–9), 637–659.

Hossain, A., Salleh, A., Boyce, A.N., Chowdhury, P., Naqiuddin, M. 2008. Biodiesel fuel production from algae as renewable energy. *American Journal of Biochemistry and Biotechnology*, **4**(3), 250–254.

Hu, Q., Sommerfeld, M., Jarvis, E., Ghirardi, M., Posewitz, M., Seibert, M., Darzins, A. 2008. Microalgal triacylglycerols as feedstocks for biofuel production: perspectives and advances. *The Plant Journal*, **54**(4), 621–639.

Ianora, A., Miralto, A. 2010. Toxigenic effects of diatoms on grazers, phytoplankton and other microbes: a review. *Ecotoxicology*, **19**(3), 493–511.

Inokuchi, R., Kuma, K.i., Miyata, T., Okada, M. 2002. Nitrogen-assimilating enzymes in land plants and algae: phylogenic and physiological perspectives. *Physiologia Plantarum*, **116**(1), 1–11.

Jüttner, F., Watson, S.B., Von Elert, E., Köster, O. 2010. β-Cyclocitral, a grazer defence signal unique to the cyanobacterium *Microcystis*. *Journal of Chemical Ecology*, **36**(12), 1387–1397.

Kanayama, R.K. 1963. *The Use of Alkalinity and Conductivity Measurements to Estimate Concentrations of 3-Trifluormethyl-4-Nitrophenol Required for Treating Lamprey Streams*. Great Lakes Fishery Commission.

Karuppasamy, S., Musale, A.S., Soni, B., Bhadra, B., Gujarathi, N., Sundaram, M., Sapre, A., Dasgupta, S., Kumar, C. 2018. Integrated grazer management mediated by chemicals for sustainable cultivation of algae in open ponds. *Algal Research*, **35**, 439–448.

Kempe, L.L. 1973. *Microbial Degradation of the Lamprey Larvicide 3-Trifluoromethyl-4-Nitrophenol in Sediment-Water Systems*. Great Lakes Fishery Commission.

Kroes, R., Renwick, A., Feron, V., Galli, C., Gibney, M., Greim, H., Guy, R., Lhuguenot, J., Van de Sandt, J. 2007. Application of the threshold of toxicological concern (TTC) to the safety evaluation of cosmetic ingredients. *Food and Chemical Toxicology*, **45**(12), 2533–2562.

Lech, J.J., Statham, C.N. 1975. Role of glucuronide formation in the selective toxicity of 3-trifluoromethyl-4-nitrophenol (TFM) for the sea lamprey: comparative aspects of TFM uptake and conjugation in sea lamprey and rainbow trout. *Toxicology and Applied Pharmacology*, **31**(1), 150–158.

Li, X., Ding, R., Han, Z., Ma, Z., Wang, Y. 2017. Targeting of cell cycle and let-7a/STAT3 pathway by niclosamide inhibits proliferation, migration and invasion in oral squamous cell carcinoma cells. *Biomedicine & Pharmacotherapy*, **96**, 434–442.

Lincoln, E., Hall, T., Koopman, B. 1983. Zooplankton control in mass algal cultures. *Aquaculture*, **32**(3–4), 331–337.

Long, S.R. 1989. Rhizobium-legume nodulation: life together in the underground. *Cell*, **56**(2), 203–214.

Loosanoff, V., Hanks, J., Ganaros, A. 1957. Control of certain forms of zooplankton in mass algal cultures. *Science*, **125**(3257), 1092–1093.

Lubzens, E. 1987. Raising rotifers for use in aquaculture. In *Rotifer Symposium IV*. Springer, pp. 245–255.

Maki, A.W., Geissel, L.D., Johnson, H.E. 1975. *Toxicity of the Lampricide 3-Trifluoromethyl-4-Nitrophenol (TFM) to 10 Species of Algae*. US Fish and Wildlife Service.

Maki, A.W., Johnson, H.E. 1976. Evaluation of a toxicant on the metabolism of model stream communities. *Journal of the Fisheries Board of Canada*, **33**(12), 2740–2746.

Manz, K.E., Carter, K.E. 2016. Extraction and recovery of 2-butoxyethanol from aqueous phases containing high saline concentration. *Analytical Chemistry Research*, **9**, 1–7.

Marchetti, A., Parker, M.S., Moccia, L.P., Lin, E.O., Arrieta, A.L., Ribalet, F., Murphy, M.E., Maldonado, M.T., Armbrust, E.V. 2009. Ferritin is used for iron storage in bloom-forming marine pennate diatoms. *Nature*, **457**(7228), 467.

Mata, T.M., Martins, A.A., Caetano, N.S. 2010. Microalgae for biodiesel production and other applications: a review. *Renewable and Sustainable Energy Reviews*, **14**(1), 217–232.

Moheimani, N.R., Borowitzka, M.A. 2006. The long-term culture of the coccolithophore *Pleurochrysis carterae* (Haptophyta) in outdoor raceway ponds. *Journal of Applied Phycology*, **18**(6), 703–712.

Montemezzani, V., Duggan, I.C., Hogg, I.D., Craggs, R.J. 2015. A review of potential methods for zooplankton control in wastewater treatment high rate algal ponds and algal production raceways. *Algal Research*, **11**, 211–226.

Munoz, R., Guieysse, B. 2006. Algal—bacterial processes for the treatment of hazardous contaminants: a review. *Water Research*, **40**(15), 2799–2815.

Muñoz, R., Köllner, C., Guieysse, B., Mattiasson, B. 2004. Photosynthetically oxygenated salicylate biodegradation in a continuous stirred tank photobioreactor. *Biotechnology and Bioengineering*, **87**(6), 797–803.

Muradov, N., Taha, M., Miranda, A.F., Wrede, D., Kadali, K., Gujar, A., Stevenson, T., Ball, A.S., Mouradov, A. 2015. Fungal-assisted algal flocculation: application in wastewater treatment and biofuel production. *Biotechnology for Biofuels*, **8**(1), 24.

Neethirajan, S., Ragavan, V., Weng, X., Chand, R. 2018. Biosensors for sustainable food engineering: challenges and perspectives. *Biosensors*, **8**(1), 23.

Niblett, P., McKeown, B. 1980. Effect of the lamprey larvicide TFM (3-trifluoromethyl-4-nitrophenol) on embryonic development of the rainbow trout (*Salmo gairdneri*, Richardson). *Water Research*, **14**(5), 515–519.

Nurdogan, Y., Oswald, W.J. 1995. Enhanced nutrient removal in high-rate ponds. *Water Science and Technology*, **31**(12), 33–43.

Olaizola, M. 2003. Commercial development of microalgal biotechnology: from the test tube to the marketplace. *Biomolecular Engineering*, **20**(4–6), 459–466.

Oliveira-Filho, E.C., Paumgartten, F.J. 2000. Toxicity of *Euphorbia milii* latex and niclosamide to snails and nontarget aquatic species. *Ecotoxicology and Environmental Safety*, **46**(3), 342–350.

Oswald, W.J. 2003. My sixty years in applied algology. *Journal of Applied Phycology*, **15**(2), 99–106.

Park, S., Van Ginkel, S.W., Pradeep, P., Igou, T., Yi, C., Snell, T., Chen, Y. 2016. The selective use of hypochlorite to prevent pond crashes for algae-biofuel production. *Water Environment Research*, **88**(1), 70–78.

Perrone, P. 2011. *Rotenone Detection in Surface and Ground Waters*. Microbac Laboratories, Inc. Available at: https://www.microbac.com/rotenone-detection-in-waters/

Poulson, K.L., Sieg, R.D., Kubanek, J. 2009. Chemical ecology of the marine plankton. *Natural Product Reports*, **26**(6), 729–745.

Prajapati, S.K., Bhattacharya, A., Kumar, P., Malik, A., Vijay, V.K. 2016. A method for simultaneous bioflocculation and pretreatment of algal biomass targeting improved methane production. *Green Chemistry*, **18**(19), 5230–5238.

Quirion, F., Magid, L.J., Drifford, M. 1990. Aggregation and critical behavior of 2-butoxyethanol in water. *Langmuir*, **6**(1), 244–249.

Rahman, N.N.N.A., Shahadat, M., Omar, F.M., Chew, A.W., Kadir, M.O.A. 2016. Dry *Trichoderma* biomass: biosorption behavior for the treatment of toxic heavy metal ions. *Desalination and Water Treatment*, **57**(28), 13106–13112.

Rajamuthiah, R., Fuchs, B.B., Conery, A.L., Kim, W., Jayamani, E., Kwon, B., Ausubel, F.M., Mylonakis, E. 2015. Repurposing salicylanilide anthelmintic drugs to combat drug resistant *Staphylococcus aureus*. *PLoS One*, **10**(4), e0124595.

Ramel, F., Mialoundama, A.S., Havaux, M. 2013. Nonenzymic carotenoid oxidation and photooxidative stress signalling in plants. *Journal of Experimental Botany*, **64**(3), 799–805.

Rasmusson, A.G., Soole, K.L., Elthon, T.E. 2004. Alternative NAD (P) H dehydrogenases of plant mitochondria. *Annual Review of Plant Biology*, **55**, 23–39.

Robinson, A.L., Ebeler, S.E., Heymann, H., Boss, P.K., Solomon, P.S., Trengove, R.D. 2009. Interactions between wine volatile compounds and grape and wine matrix components influence aroma compound headspace partitioning. *Journal of Agricultural and Food Chemistry*, **57**(21), 10313–10322.

Robinson, P.K. 1998. Immobilized algal technology for wastewater treatment purposes. In: *Wastewater Treatment with Algae*. Springer, pp. 1–16.

Rodolfi, L., Chini Zittelli, G., Bassi, N., Padovani, G., Biondi, N., Bonini, G., Tredici, M.R. 2009. Microalgae for oil: strain selection, induction of lipid synthesis and outdoor mass cultivation in a low-cost photobioreactor. *Biotechnology and Bioengineering*, **102**(1), 100–112.

Ruiz-Marin, A., Mendoza-Espinosa, L.G., Stephenson, T. 2010. Growth and nutrient removal in free and immobilized green algae in batch and semi-continuous cultures treating real wastewater. *Bioresource Technology*, **101**(1), 58–64.

Russell, A. 2003. Similarities and differences in the responses of microorganisms to biocides. *Journal of Antimicrobial Chemotherapy*, **52**(5), 750–763.

Ryther, J.H., Dunstan, W.M. 1971. Nitrogen, phosphorus, and eutrophication in the coastal marine environment. *Science*, **171**(3975), 1008–1013.

Salk, M.S. 1975. *The Effects of Selected Pesticides on Ten Isolates of Terrestrial Algae*. Electronic Theses and Dissertations. Paper 1253.

Sambusiti, C., Bellucci, M., Zabaniotou, A., Beneduce, L., Monlau, F. 2015. Algae as promising feedstocks for fermentative biohydrogen production according to a biorefinery approach: a comprehensive review. *Renewable and Sustainable Energy Reviews*, **44**, 20–36.

Samuilov, V., Timofeev, K., Sinitsyn, S., Bezryadnov, D. 2004. H2O2-induced inhibition of photosynthetic O2 evolution by *Anabaena variabilis* cells. *Biochemistry (Moscow)*, **69**(8), 926–933.

Schlesinger, A., Eisenstadt, D., Bar-Gil, A., Carmely, H., Einbinder, S., Gressel, J. 2012. Inexpensive non-toxic flocculation of microalgae contradicts theories; overcoming a major hurdle to bulk algal production. *Biotechnology Advances*, **30**(5), 1023–1030.

Science, A.A.f.t.A.o., Brady, N.C., Agronomy, A.S.o. 1967. *Agriculture and the Quality of Our Environment*. American Association for the Advancement of Science Washington.

Singh, A., Olsen, S.I. 2011. A critical review of biochemical conversion, sustainability and life cycle assessment of algal biofuels. *Applied Energy*, **88**(10), 3548–3555.

Smith, B., Tibbles, J. 1980. Sea lamprey (*Petromyzon marinus*) in Lakes Huron, Michigan, and Superior: history of invasion and control, 1936–78. *Canadian Journal of Fisheries and Aquatic Sciences*, **37**(11), 1780–1801.

Smith, V.H., Crews, T. 2014. Applying ecological principles of crop cultivation in large-scale algal biomass production. *Algal Research*, **4**, 23–34.

Snell, T., Persoone, G. 1989. Acute toxicity bioassay using rotifers. II. A freshwater test with *Brachionus rubens*. *Aquatic Toxicology*, **14**, 81–92.

Stowe, D.F., Camara, A.K. 2009. Mitochondrial reactive oxygen species production in excitable cells: modulators of mitochondrial and cell function. *Antioxidants & Redox Signaling*, **11**(6), 1373–1414.

Subhadra, B., Edwards, M. 2010. An integrated renewable energy park approach for algal biofuel production in United States. *Energy Policy*, **38**(9), 4897–4902.

Taylor, R.L., Rand, J.D., Caldwell, G.S. 2012. Treatment with algae extracts promotes flocculation, and enhances growth and neutral lipid content in *Nannochloropsis oculata*—a candidate for biofuel production. *Marine Biotechnology*, **14**(6), 774–781.

Tharmalingam, N., Port, J., Castillo, D., Mylonakis, E. 2018. Repurposing the anthelmintic drug niclosamide to combat *Helicobacter pylori*. *Scientific Reports*, **8**(1), 3701.

Thingvold, D.A., Lee, G.F. 1981. Persistence of 3-(trifluoromethyl)-4-nitrophenol in aquatic environments. *Environmental Science & Technology*, **15**(11), 1335–1340.

Triantaphylidès, C., Havaux, M. 2009. Singlet oxygen in plants: production, detoxification and signaling. *Trends in Plant Science*, **14**(4), 219–228.

Turner, L., Jacobson, S., Shoemaker, L. 2007. *Risk Assessment for Piscicidal Formulations of Rotenone*. Compliance Services International, Lakewood, 25.

Ullmann, F. 1985. *Ullmann's Encyclopedia of Industrial Chemistry*. VCH Verlagsgesellschaft.

Ummalyma, S.B., Gnansounou, E., Sukumaran, R.K., Sindhu, R., Pandey, A., Sahoo, D. 2017. Bioflocculation: an alternative strategy for harvesting of microalgae—an overview. *Bioresource Technology*, **242**, 227–235.

Vaccari, D.A. 2009. Phosphorus: a looming crisis. *Scientific American*, **300**(6), 54–59.

Van Ginkel, S.W., El-Sayed, W.M., Johnston, R., Narode, A., Lee, H.J., Bhargava, A., Snell, T., Chen, Y. 2020. Prevention of algaculture contamination using pesticides for biofuel production. *Algal Research*, **50**, 101975.

Van Ginkel, S.W., Igou, T., Hu, Z., Narode, A., Cheruvu, S., Doi, S., Johnston, R., Snell, T., Chen, Y. 2015. Taking advantage of rotifer sensitivity to rotenone to prevent pond crashes for algal-biofuel production. *Algal Research*, **10**, 100–103.

Vance, C.P. 2001. Symbiotic nitrogen fixation and phosphorus acquisition. Plant nutrition in a world of declining renewable resources. *Plant Physiology*, **127**(2), 390–397.

Vardi, A., Bidle, K.D., Kwityn, C., Hirsh, D.J., Thompson, S.M., Callow, J.A., Falkowski, P., Bowler, C. 2008. A diatom gene regulating nitric-oxide signaling and susceptibility to diatom-derived aldehydes. *Current Biology*, **18**(12), 895–899.

Vardi, A., Formiggini, F., Casotti, R., De Martino, A., Ribalet, F., Miralto, A., Bowler, C. 2006. A stress surveillance system based on calcium and nitric oxide in marine diatoms. *PLoS Biology*, **4**(3), e60.

Vidoudez, C., Nejstgaard, J.C., Jakobsen, H.H., Pohnert, G. 2011. Dynamics of dissolved and particulate polyunsaturated aldehydes in mesocosms inoculated with different densities of the diatom *Skeletonema marinoi*. *Marine Drugs*, **9**(3), 345–358.

Volk, R.-B., Furkert, F.H. 2006. Antialgal, antibacterial and antifungal activity of two metabolites produced and excreted by cyanobacteria during growth. *Microbiological Research*, **161**(2), 180–186.

Wan, C., Alam, M.A., Zhao, X.-Q., Zhang, X.-Y., Guo, S.-L., Ho, S.-H., Chang, J.-S., Bai, F.-W. 2015. Current progress and future prospect of microalgal biomass harvest using various flocculation technologies. *Bioresource Technology*, **184**, 251–257.

Wan, C., Zhao, X.-Q., Guo, S.-L., Alam, M.A., Bai, F.-W. 2013. Bioflocculant production from *Solibacillus silvestris* W01 and its application in cost-effective harvest of marine microalga *Nannochloropsis oceanica* by flocculation. *Bioresource Technology*, **135**, 207–212.

Wang, B., Li, Y., Wu, N., Lan, C.Q. 2008. CO_2 bio-mitigation using microalgae. *Applied Microbiology and Biotechnology*, **79**(5), 707–718.

Wang, Z.H., Nie, X.P., Yue, W.J., Li, X. 2012. Physiological responses of three marine microalgae exposed to cypermethrin. *Environmental Toxicology*, **27**(10), 563–572.

Weinbach, E., Garbus, J. 1969. Mechanism of action of reagents that uncouple oxidative phosphorylation. *Nature*, **221**(5185), 1016–1018.

Weissman, J., Radaelli, G. (Inventors) 2010. Aurora Biofuels, assignee. Systems and methods for maintaining the dominance of *Nannochloropsis* in an algae cultivation system. *US Patent No. 20100183744*.

Weissman, J., Radaelli, G., Rice, D. 2012. Systems and methods for maintaining the dominance and increasing the biomass production of *Nannochloropsis* in an algae cultivation system, *Google Patents*.

Wilde, E.W., Benemann, J.R. 1993. Bioremoval of heavy metals by the use of microalgae. *Biotechnology Advances*, **11**(4), 781–812.

Wilson, C.M. 2012. Antiparasitic agents. In: *Principles and Practice of Pediatric Infectious Diseases*. Elsevier, pp. 1518–1545. e3.

Xie, S., Sun, S., Dai, S.Y., Yuan, J.S. 2013. Efficient coagulation of microalgae in cultures with filamentous fungi. *Algal Research*, **2**(1), 28–33.

Xu, M., Lee, E.M., Wen, Z., Cheng, Y., Huang, W.-K., Qian, X., Julia, T., Kouznetsova, J., Ogden, S.C., Hammack, C. 2016. Identification of small-molecule inhibitors of Zika virus infection and induced neural cell death via a drug repurposing screen. *Nature Medicine*, **22**(10), 1101.

Yang, J., Xu, M., Zhang, X., Hu, Q., Sommerfeld, M., Chen, Y. 2011. Life-cycle analysis on biodiesel production from microalgae: water footprint and nutrients balance. *Bioresource Technology*, **102**(1), 159–165.

Yildiz, F.H., Davies, J.P., Grossman, A.R. 1994. Characterization of sulfate transport in *Chlamydomonas reinhardtii* during sulfur-limited and sulfur-sufficient growth. *Plant Physiology*, **104**(3), 981–987.

Zhang, F., Dickman, M. 1999. Mid-ocean exchange of container vessel ballast water. 1: seasonal factors affecting the transport of harmful diatoms and dinoflagellates. *Marine Ecology Progress Series*, **176**, 243–251.

Zhang, J., Liu, Z., Wang, S., Jiang, P. 2002. Characterization of a bioflocculant produced by the marine myxobacterium *Nannocystis* sp. NU-2. *Applied Microbiology and Biotechnology*, **59**(4–5), 517–522.

Zmora, O. 2004. Microalgae for aquaculture: Microalgae production for aquaculture. In: Richmond, A. (Ed.), *Handbook of Microalgal Cultures: Biotechnology and Applied Phycology*. Blackwell Science.

3 Waste to Energy: A Means of Sustainable Development through Bioethanol Production

Girish Venkatachalapathy and Girisha Shiringala Thimappa

CONTENTS

3.1 Introduction 44
    3.1.1 Non-Renewable Energy Resources 45
    3.1.2 Renewable Energy Resources 47
3.2 Bioenergy: Biomass to Bioethanol for Sustainable Development 48
    3.2.1 Worldwide Status of Bioethanol Production 48
        3.2.1.1 India's Status of Bioethanol Production 48
        3.2.1.2 Bioethanol-Producing Techniques 49
    3.2.2 Classification of Biofuels 49
    3.2.3 Generation of Biofuels 49
        3.2.3.1 First-Generation Biofuels 49
        3.2.3.2 Second-Generation Biofuels 49
        3.2.3.3 Third-Generation Biofuels 50
        3.2.3.4 Fourth-Generation Biofuels 51
3.3 Production of Bioethanol 51
    3.3.1 Raw Materials with Available Fermentable Sugars for First-Generation Biofuels 51
        3.3.1.1 Citrus Fruit 52
        3.3.1.2 Papaya Fruit 52
        3.3.1.3 Grape Fruit 52
        3.3.1.4 Chikku Fruit 52
        3.3.1.5 Guava Fruit 52
        3.3.1.6 Banana Fruit 52
        3.3.1.7 Pineapple Fruit 53
        3.3.1.8 Tropical Fruit 53
    3.3.2 Microorganism: Yeast 53
    3.3.3 Pretreatments 54
    3.3.4 Production of Bioethanol at a Large Scale 55
    3.3.5 Bioethanol Production by Rapid Fermentation with High Cell Density 55
        3.3.5.1 Batch Fermentation 55
3.4 Conclusion 56
Acknowledgments 57
Note 57
References 57

3.1 Introduction

Energy in the form of biomass for the past several years has served as a source of foodstuffs for human and animal consumption, of building materials, and for heating and cooking. In Third World countries, use of biomass has not changed since preindustrial times, whereas industrial societies have adapted and added to this list of necessities, specifically to the energy category. In industrialized countries, biomass has become a minor source of energy and fuels; it has been replaced by coal, petroleum crude oil, and natural gas, which have become the raw materials of alternative for the manufacture and production of a host of derived products and energy as heat, steam, and electric power, as well as solid, liquid, and gaseous fuels. The fossil fuel era has indeed had a large impact on civilization and industrial progress. But since the use of fossil fuels increases, its availability decreases, and environmental issues raise due to fossil fuel use. Hence, biomass is one of the few renewable, indigenous, widely dispersed natural resources that can be used to reduce both the amount of fossil fuels burned and the several greenhouse gases emitted by or formed during the fossil fuel combustion processes (Figure 3.1) (Klass, 1998).

Energy is a key issue in debate of environmental, social, and economic magnitude of sustainable development (Dincer, 1999). Energy is essential to record diverse social aspects such as inequality in income, inequality in raw substance and energy wealth and sources, the scientific progress privileged by standards of living, educational attitudes, climatic circumstances, demography, and dissimilarity between developed industrialized society, where the agricultural sectors represent simply a little proportion of the effective residents, and the public among an significant rural sector collectively through service sectors like transportation, industries, etc., of several urban areas due to social constraints (Akella et al., 2009). All through the earlier period of two decades, the threat and authenticity of environmental deprivation have become more evident. Rising proof of environmental evils is owing to an amalgamation of numerous factors because the environmental impact of individual activities has developed significantly since the increase of world population, energy use, industrial action, etc. It is important to facilitate energy as one of the main factors that should be considered in deliberations of sustainable development (Midilli et al., 2006). The majority of our everyday

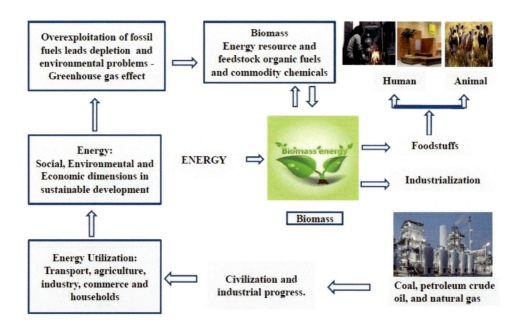

FIGURE 3.1 Energy consumption patterns.

activities engage energy shift and energy change (Dincer & Rosen, 2007), growth of rural areas depends on the interconnections of the people across the country for the mobility of the goods for the social, regional, and economic unity powered by the transport system. The world relies on agriculture to feed humanity; energy is an indispensable constituent of agricultural creation (Clark & Fo, 2009). Commerce energy is a necessary component for socioeconomic progress and economic development. Energy is the lifeblood of the universal economy—a key contribution to almost all of the goods and services of the contemporary world (Yergin & Gross, 2012). The purpose of energy organization is to give energy services for illumination, air-conditioning, refrigeration, transportation, cooking, industrial processes such as conversion of raw materials to final products, etc. (Oyedepo, 2012).

Energy efficiency is a significant constituent of the power wealth. Energy takes an essential role in the progress of nations. To accept the challenges outstanding to growing populations, it is significant to optimize the use of energy, looking for alternative energy resources (Datta et al., 2011). The world around us has altered considerably in the past 20 years. Technology has grown to be one of the major drivers of economic and social progress. It is unnecessary to pronounce that almost all technologies run on electricity, and as a result, allocation of electricity is rising more quickly than total primary energy supply (TPES). Population expansion has forever been and will remain one of the key drivers of energy requirements along with economic and social development. Whereas worldwide population has increased by over 1.5 billion during the past two decades, the overall pace of population enlargement has been slowing down. Global primary energy use amplified by just 1% in 2014, with subsequent increases of 0.9% in 2015 and 1% in 2016. This compares with the ten-year average of 1.8% a year. As was the case in 2015, development was below normal in all regions apart from Europe and Eurasia. All fuels except oil and nuclear power grew at below-average rates. Energy use in China grew by just 1.3% in 2016. Growth during 2015 and 2016 was the lowest over a two-year period since 1997–1998 (Table 3.1). World primary energy use grew by 1.0% in 2016, well below the ten-year average of 1.8% and the third successive year at or below 1%. Oil provided the leading growth to energy use at 77 million tons of oil equivalent (mtoe), followed by natural gas (57 mtoe) and renewable power (53 mtoe) (W.E. Council, 2013).

3.1.1 Non-Renewable Energy Resources

Non-renewable energy comes from sources that will run out or will not be replenished in our life period. The majority of non-renewable energy sources are fossil fuels like coal, petroleum, and natural gas. Coal is a vital supplier to the energy supply in many countries, although it is deprived of environmental recommendation. The ongoing reputation of coal becomes above all obvious when compared to the present generation figures through those from 20 years ago. Whereas the global treasury of coal decreased by 14% between 1993 and 2011, production has gone up by 68% over the similar time period. Compared to the 2010 assessment, the relatively current data shows that the proven coal treasuries have amplified by 1% and production by 16%. Coal resources survive in many developing countries. Coal will therefore take part in a main role in sustaining the development of bottom freight electricity where it is mainly desired (Carlisle & Parker, 2014).

The oil disaster in the 1970s and 1980s resulted in lengthy queues outside petrol stations, as well as the skyrocketing cost of oil. In subsequent years, heated debate about "peak oil" was based on the hope of the world running out of oil within not many decades. Now, the peak oil issue is not creation headlines any longer; however, since oil is a finite resource, this issue will return in the future. Global oil treasuries are approximately 60% larger today than they were 20 years before, and generation of oil has gone up by 25%. The cleanest of the complete fossil-based fuels, natural gas is plentiful and flexible. It is gradually used in the bulk able authority age group technologies, such as combined cycle gas turbines (CCGTs), by means of replacement efficiencies of about 60%. The treasuries of conventional natural gas have developed by 36% over more than the past two decades, and its manufacture by 61%. Compared to the 2010 survey, the proven natural gas treasuries have developed by 3% and manufacturing by 15% (W.E.Council, 2013).

The nuclear industry has a relatively short history: the first nuclear reactor was specially commissioned in 1954. Uranium is the main reserve of fuel for nuclear reactors. Worldwide manufacture of

TABLE 3.1
Different Fruit Waste as Substrates to Produce Bioethanol

Substrate	Sample Preparation	Microorganism	Fermentation Method	Yield of Alcohol	References
Dates	Pulverization	*Hanseniaspora uvarum*, *H. opuntiae* and *Pichia kudriavzevii*	Bath fermentation	80 g/L	(Hashem et al., 2017)
Cashew apple	Pulverization	*Saccharomyces cerevisiae*	Bath fermentation	61.34 g/L	(Srinivasarao et al., 2013)
Mango fruits	Pulverization	*Saccharomyces cerevisiae*	Bath fermentation	15%	(Boyce, 2014)
Carica papaya	Pulverization	Baker's and brewer's yeast	Bath fermentation	5.19% (v/v)	(Akin-Osanaiye et al., 2005)
Pineapple fruit peel	Pulverization	*Saccharomyces cerevisiae*	Bath fermentation	5.98 g/L	(Casabar et al., 2019)
Watermelon	Pulverization	*Saccharomyces cerevisiae*	Bath fermentation	35.5 g/L	(Jahanbakhshi & Salehi, 2019)
Grape waste	Pulverization	*Saccharomyces cerevisiae*	Bath fermentation	6.85%	(Raikar, 2012)
Mango peel	Pulverization	*Saccharomyces cerevisiae* (W6) & Baker's yeast (B1)	Bath fermentation	13 and 10.1 g/L^{-1}	(Somda et al., 2011)
Pineapple wastes	Pulverization	*Saccharomyces cerevisiae*	Batch fermentation	3.9% (v/v)	(Tropea et al., 2014)
Banana waste	Pulverization	*Saccharomyces cerevisiae*	Batch fermentation	7.1%	(Alshammari et al., 2011)

uranium in current times is used in large scale succeeding to an extensive phase of waning manufacture caused by in excess of supply following nuclear disarmament. Whole nuclear electricity generation has been rising throughout the history of two decades and reached an annual output of about 2,600 TWh by the mid-2000s. The nuclear share of total worldwide electricity generation reached its peak of 17% by the late 1980s; ever since then, it has been dropped to 13.5% in 2012 (Bodansky, 2004).

3.1.2 Renewable Energy Resources

Nature offers a selection of freely accessible options for producing energy. Their use is mostly a query of how to alter sunlight, wind, biomass, or water into electricity, heat, or power as competently, sustainably, and cost-effectively as feasible (Edenhofer et al., 2011).

Hydropower provides a considerable quantity of energy all over the world and is present in more than 100 countries, contributing approximately 15% of global electricity generation. In numerous cases, the development in hydropower was facilitated by the abundant renewable energy bear policies and CO_2 penalties. Over the past two decades, the total worldwide installed hydropower ability has improved by 55%, while the authentic production by 21% (i.e., the production decreases though installation increases) (Dincer et al., 2014).

Wind is obtainable practically all over on earth, even though there are broad variations in wind strengths. The total reserve is huge, predicted to be roughly 1 million GW for total land coverage. If only 1% of this area is used, and allowance made for the lower load factors of wind plants (15–40%, compared with 75–90% for thermal plants) that would still match, approximately, the total worldwide ability of all electricity generating plants in process today. World wind energy ability (i.e., installation of wind mills and production of electricity) has been repetition about every three and a half years since 1990 (Hasager et al., 2005). Solar energy is the majority plentiful energy source and it is obtainable for use in its direct (solar radiation) and indirect (wind, biomass, hydro, ocean, etc.) forms. About 60% of the total energy emitted by the sun reaches the earth's plane. Even if only 0.1% of this energy might be transformed at a competence of 10%, it would be four times better than the total world's electricity generating ability of about 5,000 GW. The data regarding solar photovoltaic (PV) installations are erratic and incompatible (Lysen & Ouwens, 2002).

Bioenergy includes a wide group of energy fuels manufactured from a range of feedstocks of biological derivation and by many conversion technologies to produce heat, power, liquid biofuels, and gaseous biofuels. The term "traditional biomass" mostly refers to fuel wood, charcoal, and agricultural residues used for domestic cooking, lighting, and space heating in rising countries. The share of bioenergy in TPES was predictable at about 10% in 1990. Between 1990 and 2010, bioenergy supply has amplified from 38 to 52 EJ as a result of increasing energy demand. New policies to amplify the share of renewable energy and indigenous energy wealth are also driving demand (Kavanagh, 2006).

Demand for energy will persist to rise for decades to come. Population increases and a rising pace of electrification will position enormous necessities on energy provisions. Global primary energy command might amplify by 50% by the center of the century. At least 80% of this increase is predictable to move toward from rising countries. The total primary energy requirements of China alone are likely to double by 2035, and that of India to rise by almost 150% during the identical period. Both countries with enormous populations and high economic growth are predicted to take over the global use of energy capital in the upcoming years. A secure delivery of energy resources is usually decided to be a required but not sufficient prerequisite for progress in the world. Sustainable progress in a society demands a sustainable deliver of energy resources and a valuable and competent consumption of energy resources. The close relationship among renewable energy sources and sustainable development comes out (Midilli et al., 2006).

India is swiftly shifting from an agricultural-based nation to industrial- and services-oriented country. About 31.2% of the population is currently living in urban areas. Over 377 million urban people are living in 7,935 towns/cities. India has diverse geographic and climatic regions and four seasons; accordingly, inhabitants living in these zones have diverse use and waste generation patterns. Municipal solid waste management (MSWM), a serious constituent toward sustainable urban progress, comprises segregation, storage, collection, relocation, processing, and disposal of solid waste to reduce its horrible impact on the environment. Unmanaged municipal solid waste (MSW) becomes an issue for the transmission of countless ailments (Joshi & Ahmed, 2016); hence, technology is applied to convert these wastes to biofuel.

3.2 Bioenergy: Biomass to Bioethanol for Sustainable Development

3.2.1 Worldwide Status of Bioethanol Production

The total global requirement for oil is expected to increase by 1% annually generally due to rising demand in energy market of rising countries, especially India (3.9%/year) and China (3.5%/year). Bioenergy ranks second in renewable U.S. primary energy production and accounts for 3% of the U.S. primary energy production (W.E. Council, 2016). The United States is the world's major producer of bioethanol fuel, accounting for almost 47% of worldwide bioethanol production in 2005 and 2006 (Balat & Balat, 2009). Brazil is the world's major exporter of bioethanol and second biggest producer following the United States. With regard to bioethanol, the allocation of the United States in the worldwide production is 50% and Brazil constitutes 39% of the sum worldwide supply, whereas the share of the Organization for Economic Cooperation and Development (OECD) of Europe is 5% (Gnansounou et al., 2005). European Union (EU) countries are in view of changing various mass flows of sugar beets into ethanol production, and then to use it as energy subsidy. Determination of the yield, economically feasible, technical necessities and optimal circumstances of the overall ethanol production from dissimilar intermediate and byproducts of the sugar beet processing are extremely significant (Grahovac et al., 2012).

3.2.1.1 India's Status of Bioethanol Production

India is a country with a positive outlook toward renewable energy technologies and dedicated to the use of renewable sources to supplement its energy requirements. The country is one among the few nations to have a separate ministry for renewable energy, which speaks to the expansion of biofuels along with other renewable energy sources. The Planning Commission of the Government of India in 2003 brought out a wide report on the progress of biofuels, and bioethanol and biodiesel were recognized as the main biofuels to be developed for the nation. The National Biofuel Policy released in December 2009 by the Ministry of New and Renewable Energy envisioned an aim of whole blending of 5% ethanol by 2011–2012 and then gradually raising it to 10% by 2016–2017 and to 20% after 2017 (Kumar et al., 2010).

Worldwide, biofuels are mainly used to power vehicles, but can also be used for additional purposes. Sustainable biofuels can decrease the dependency on petroleum and thereby improve energy security (Nigam & Singh, 2011). Energy use has amplified progressively more than the last century as the global population has developed and extra countries have spun out to be industrialized; crude oil has been the chief reserve to meet the amplified energy demand (Sun & Cheng, 2002). In recent years, bioethanol manufacturing from renewable resources includes agricultural waste and has acknowledged significant official support in order to contend cost-effectively with petroleum fuels (Thomas & Ingledew, 1990). Global attention to "sustainable energy for the prospect" motivates the investigation for additional resources of energy generation as fossil fuels are increasingly inadequate (Ibeto et al., 2011). Concerns more than the harmful, ecological possessions, related with the burning of fossil fuel, united with forecasts predicting potential future cost may amplify, have confident governments to believe the need for substitute sources for transport fuels. At present the majority activity in this area paying attention on the manufacture of transport fuels from biomass (Hayes & Hayes, 2009).

Biomass is organic material resulting from living organisms. As an energy basis, biomass can also be used straightforwardly via burning to manufacture heat, or ultimately after converting it to a range of forms of biofuel. Alteration of biomass to biofuel can be achieved by diverse methods which are generally classified as thermal, chemical, and biochemical methods. Biomass can be transformed into additional working forms of energy similar to methane gas or transportation fuels similar to ethanol and biodiesel. Bioethanol is another transportation fuel which can be formed from leftover food products like agricultural residues. Also, conversion of biomass to liquids and cellulosic ethanol are still under research (Turcotte & Schubert, 2002). Converting crops into ethanol and biodiesel from rapeseed methyl ester (RME) is happening largely in Brazil, the United States, and Europe. Theoretically, biofuels can be formed from any organic carbon resource, while the most common is photosynthetic plants. Diverse plants and plant-derived resources are used for biofuel production.

Worldwide, 80% of whole bioethanol is formed from eco-friendly renewable biomass such as sugar, starch, corn, sugarcane, bagasse, sugar beet, sorghum, switch grass, barley, hemp, potatoes, sunflower, wheat, wood, paper, straw, cotton, and lignocellulosic materials (Mojović et al., 2009). They can also be produced from some polysaccharides which can be hydrolyzed for obtaining sugars that can be transformed to ethanol (Cardona & Sánchez, 2007). Sergio Capareda and colleagues reported that the ethanol use anticipated reaching 11.2 billion gallons from waste by the year 2012 (Anex et al., 2010).

3.2.1.2 Bioethanol-Producing Techniques

Ethanol is chiefly produced from two main categories, catalytic hydration of ethylene and biofermentation of agricultural feedstocks such as fruits, vegetables, and cereals, which to explore favorable microorganisms (i.e., microorganism that survive in all the parameters to produce more ethanol) (Balasubramanian et al., 2011). Bioethanol from biomass sources is the principal fuel used as petrol alternate for road transport vehicles. The soaring cost of crude oil makes biofuels eye-catching (Scott & Bryner, 2006).

3.2.2 Classification of Biofuels

Biofuels are generally classified as primary and secondary biofuels. Primary biofuels are used in an unrefined form, primarily for heating, cooking, or electricity generation such as fuel wood, wood chips and pellets, etc. Secondary biofuels are formed by processing of biomass that can be used in vehicles and a range of industrial processes. Secondary biofuels are further classified in to first-, second-, third- and fourth-generation biofuels on the basis of raw material and technology used for their manufacture. Biofuels are also classified according to their source and type. They may be resultant from forest, agricultural, or fishery goods or municipal wastes, also including byproducts and wastes originated from agroindustry, food industry and food services. Biofuels can be solid, such as fuel wood, charcoal, and wood pellets; liquid, such as ethanol, biodiesel, and pyrolysis oils; or gaseous, such as biogas (Nigam & Singh, 2011).

3.2.3 Generation of Biofuels

Biofuels are being researched largely to substitute conventional liquid fuels (diesel and petrol). A newly popularized categorization for liquid biofuels includes first-, second-, third- and fourth-generation biofuels. The main difference among them is in the feedstock used (Figure 3.2) (Patil et al., 2008; Larson, 2007).

3.2.3.1 First-Generation Biofuels

The first-generation liquid biofuels are the kind of fuels usually formed from sugars, grains, or seeds, and require a comparatively easy procedure to generate the complete fuel manufactured goods. The majority renowned first-generation biofuel is ethanol prepared by fermenting sugar extracted from crop plants and starch enclosed in maize kernels or other starchy crops (Larsen et al., 2008).

3.2.3.2 Second-Generation Biofuels

Second-generation biofuels are generally produced by two fundamentally different approaches, i.e. biological or thermochemical processing, from agricultural lignocellulosic biomass, which are either non-edible residues of food crop production or non-edible whole plant biomass (Naik et al., 2010). Brazilian ethanol production units shown the amount of residue generated by an industry and the same

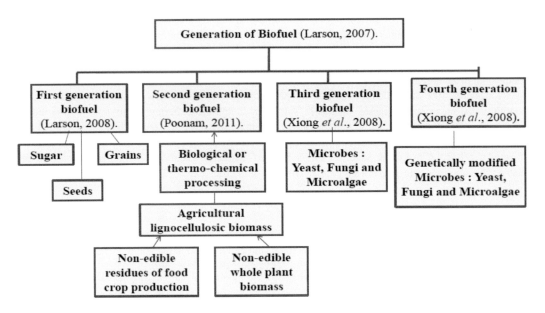

FIGURE 3.2 Schematic diagram of the generation of biofuels.

can be used to manufacture second-generation bioethanol, which processed more than 63,812 tons of sugarcane and produced 2,616 L of ethanol. Canilha and colleagues (2012) demonstrated the high potential of residual biomass each ton of sugarcane processed generates 270–280 kg of bagasse, which used in production of bioethanol. In distilleries to cogeneration heat and electricity, these residues are the source through thermochemical processes, such as direct combustion, pyrolysis, or gasification (Perez et al., 2014). These residues of biochemical transformation have certain technical limitations in producing bioethanol due to the chemical nature of the lignocellulosic residues. Basically, the lignocellulosic biomass is comprised of cellulose, hemicellulose, and lignin (Hayn et al., 1993). Cellulose is a linear, crystalline homopolymer with a repeating unit of glucose strung together with beta-glycosidic linkages. The structure is rigid and harsh treatment is required to break it down (Gray et al., 2006; Saha et al., 2005). Hence, heterogeneous and different pretreatments, or modifications to the raw material, must be applied (Huron et al., 2016).

3.2.3.3 Third-Generation Biofuels

Rapid generation of biomass with high energy density raw materials for ethanol production are found; these crops are perennial grasses, micro and macroalgae, and cyanobacteria (Brennan & Owende, 2010; Chisti, 2008; Hu et al., 2008; Gouveia & Oliveira, 2009; Lardon et al., 2009). Several species of algae have been investigated, including fast-growing species *Chlamydomonas reinhardtii*, *Dunaliella salina*, and various *Chlorella* species (Ahorsu et al., 2018). Starches can be obtained by converting biomass from macroalgae (Adams et al., 2009; Khambhaty et al., 2013; Scholz et al., 2013; Sudhakar et al., 2016; Dürre, 2008). Macroalgae are raw material for bioethanol production because they do not have a high content of lipids, as do microalgae, but they do have a high content of sugars and other carbohydrates that can be fermented. According to George and colleagues (2014), a photosynthetic light intensity of 60 mmol m^{-1} s^{-1} and 12:12 h (light:dark) cycles achieved a biomass production of 7.9 mg L^{-1} d^{-1} with *Ankistrodesmus falcatus*. Microalgae have a high content of lipids, but also of carbohydrates, like *Scenedesmus* sp. (Sivarama krishnan & Incharoensakdi, 2018). Another high-density crop in lakes and lagoons is water hyacinth (*Eichhornia crassipe*), which is composed mainly of cellulose, which can be converted to glucose, the fermentable substrate for ethanol production (Zhang et al., 2018).

3.2.3.4 Fourth-Generation Biofuels

The main idea in fourth-generation (4G) is just trying to translate the available lipid content in algae in real time from technical and economical viewpoints. Sustainable developments in both the production and application of the future fuels and high-value products strongly depend on the collaboration of scientists and engineers in the related fields with significant advances, one of which may be the exploitation of synthetic biology to reengineer photosynthetic microorganisms with an ability to directly fabricate some desired products from CO_2 contributing approximately 60% of global warming (Parry et al., 2007; Resch et al., 2008). Eukaryotic algae are divided into macro and microalgae. Unlike macroalgae, microalgae are considered as efficient converters of solar energy mainly because of their efficient access to water, CO_2, and nutrients, and their simple cellular morphology, which in turn results in an enormous ratio of surface area to mass in comparison with land-based plants. All of these mentioned advantages make microalgae a leading candidate for the latest generation of biofuels (Basile & Dalena, 2019). In fourth-generation biofuels, raw materials that are inexhaustible, cheap, and widely available are used to convert solar energy to solar biofuels. The production of photobiological solar biofuel or electrofuel exploits the synthetic biology of algae and cyanobacteria (Hays & Ducat, 2015; Berla et al., 2013; Scaife et al., 2015). Synthetic biology involves designing and creating new biological parts, and redesigning existing ones, devices, and natural biological systems for useful purposes. Fourth-generation biofuels are produced: (a) by designing photosynthetic microorganisms to produce photobiological solar fuels; (b) by combining photovoltaics and microbial fuel production; and (c) by synthetic cell factories or synthetic organelles tailored explicitly to produce the desired high-value chemicals and biofuels (Aro, 2016). The recombinant DNA technique for algae has been verified to be capable of creating constructs for both prokaryotes and eukaryotes, which may replicate and possess novel functions. In addition to direct metabolic pathway modification, algal metabolic engineering also involves targeted improvement of cellular activities by manipulation of enzymatic, transport, and regulatory functions of the photosynthetic cells using other biological and engineering approaches. Algae are subjected to a wide range of environmental stresses (Mendez-Alvarez et al., 1999; Ahorsu et al., 2018).

3.3 Production of Bioethanol

3.3.1 Raw Materials with Available Fermentable Sugars for First-Generation Biofuels

Global demand for ethanol has increased due to its wide industrial applications. Ethanol is mainly used as a chemical feedstock to produce ethylene with a market demand of more than 140 million tons per year (Lundgren & Hjertberg, 2010; IEA-ETSAP & & I. R. E. N. A., 2013). Bioethanol can be obtained from a variety of feedstocks using cellulosic, starchy, and sugar sources, corn, sugarcane, bagasse, sugar beet, sorghum, switch grass, barley, hemp, maize, potatoes, sunflower, willow, poplar trees, wheat, Jerusalem artichoke, miscanthus and sorghum plants, wood, paper, straw, cotton, and other biomass materials (Ibeto et al., 2011; Coppola et al., 2009). Bioethanol can also be produced from some polysaccharides can be hydolyzed for obtaining sugars that can be converted to ethanol (Cardona & Sánchez, 2007). To avoid conflicts between human food use and industrial use of crops, only the wasted crop, which is defined as crop lost in distribution, is considered as efficient feedstock (Janani et al., 2013; Mojović et al., 2009).

Agricultural wastes (fruit and vegetable wastes) can be explored as possible substrates for ethanol manufacture. The universal production of fruits is approximately 370 million metric tons (MT) per annum. India with its present production of approximately 32 million MT of fruit, accounts for reference to 8% of the world's varied total fruit production. The agroclimatic zones of the country make it feasible to produce roughly all varieties of fresh fruits and vegetables. India is the second major producer of fruits. The main fruits grown are bananas, mangoes, citrus, guava, grapes, apples, and pineapple, which constituted almost 80% of the total fruit production in the country. Banana has the largest share of 31.7% in total fruit production, followed by mango with 28% (Verma & Joshi, 2000). Although India is one of the major producers of fruit and vegetables, around 25%–40% of total produce is wasted, owing to postharvest losses. The majority of fruits are consumed as either fresh foodstuffs or processed fruits; however, studies explain that merely extremely high class fruits are chosen for processing and consignment;

low superiority fruits are therefore left missing to rot on the farm, as well a huge amount of unused segment is also generated from processing industries, which can be used as a possible resource for ethanol manufacture (Tanaka et al., 1999; Janani et al., 2013; Braddock, 1999).

3.3.1.1 Citrus Fruit

Citrus processing industries especially in the developed countries, in 2010, the orange and lemon production is estimated to reach 66.4 million MT and 4.2 million MT, about 30.1 MT of the orange production will be processed to yield juice, essential oils, and other byproducts (Grohmann & Baldwin, 1992; Grohmann et al., 1994; Tripodo et al., 2004; Talebnia et al., 2008) and the rest is a potential feedstock for production of bioethanol due to its high carbohydrate substance (Boluda-Aguilar et al., 2010).

3.3.1.2 Papaya Fruit

Papaya is one of the fruits commonly used as food and medicine in Nigeria. It is eaten as fresh fruit or processed into desserts (Desmond, 1995). The unripe matured papaya fruit is used for the production of papain. Ethanol manufacture by fermentation faces struggle with ethanol production from petroleum-based products as feedstocks (Ahmeh et al., 1988). While the huge quantities of papaya agrowaste are obtainable from plantations cultivated for papain production, their dumping can be a problem—so use of renewable resources would be less expensive, because they are cheaper and effortlessly obtainable, so they attempt to process the fruit waste into alcohol, which will have industrial applications (Akin-Osanaiye et al., 2008).

3.3.1.3 Grape Fruit

Grape cultivation is one of the remunerative farming enterprises in India. India is also the largest grape producer in terms of productivity per hectare with 80,000 hectares of land under cultivation and 1,878,000 MT of grapes produced annually which is nearly 23.5 MT/hectare. Fuel ethanol production from grapes is already a reality in Europe. Although bioethanol production from grapes has negative energy yield, it is possible to improve the performance of grapes as an energy crop, principally by raising yield the suitability of incineration for dried grapes and exhausted grape stalks. This has been tested with good results (Lakhawat et al., 2011).[1]

3.3.1.4 Chikku Fruit

Chikku peels as substrates can serve as a reasonable and renewable less price unprocessed substance for bioethanol production. The benefit of simultaneous saccharification and fermentation (SSF) is that a multistage process for the production of bioethanol is carried out in one reactor and glucose produced during saccharification is simultaneously fermented to bioethanol by yeast cells (Chaudhary et al., 2014).

3.3.1.5 Guava Fruit

Guava is one of the significant profitable fruit crops of India, is obtainable twice a year, and constitutes 6.2% of the total fruit production. Due to seasonal differences, lack of procedure technology and packing methods, huge-scale fruit is extremely prone to spoilage; a substitute to this difficulty is to use surplus guava fruits for the production of ethanol (Srivastava et al., 1997).

3.3.1.6 Banana Fruit

Banana constitutes one of the major principal food resources in the world (Molina & Kudagamage, 2002). Most of the fruit peels/residues are dried, ground, pelletized, and sold to the feed manufacturers at a low price, which is not considered a highly viable proposition (Mamma et al., 2008). As per UN Food and Agriculture Organization (FAO) statistics, India is the largest producer of banana in the world and accounts for nearly 30% of the total world production of banana (Emaga et al., 2008), resulting in readily available agricultural waste that is underused as potential growth medium for yeast strain, despite rich carbohydrate content and other basic nutrients that can support yeast growth (Brooks, 2008). Since banana peels contain lignin in low quantities (Hammond et al., 1996), it could serve as a good substrate for production of value-added products like ethanol. Graefe and colleagues (2011) analyzed ethanol production from bananas. Bananas have an advantage over tubers because of their ripening and, because they are climacteric, they naturally hydrolyze starch and thus no enzyme treatment is required to reach the fermentable sugars (Astedu, 1987; Bugaud et al., 2009). Production of ethanol from bananas at different stages of ripeness showed that the highest ethanol yield was from green bananas, immediately after harvesting (Hammond et al., 1996).

Green bananas discarded owing to quick post-harvesting ripening might attain 40%–50% (Mascarenhas, 1999), but these residues may be able to potentially be used for industrial purposes (Zhang et al., 2005). In Brazil, there are many fruits that can be used for the preparation of fermented distilled beverages (spirits). The banana stands out among these fruits because of its abundance and relatively high concentration of fermentable sugars (Silva, 2004). The use of bananas for the production of spirits, in addition to using the surplus fruit and the stills during the periods between sugarcane harvesting seasons, may provide a new product to the market (Alvarenga et al., 2011). Banana waste has been used to produce bioethanol with the aim of rising an inexpensive, sustainable manufacture process (Taherzadeh & Karimi, 2007). The ethanol production process from banana cultivation waste is diverse from the production process from sugarcane and is still not well known, making it necessary to evaluate the operational and energetic incorporation of the stages, such as hydrolysis (Palacios-Bereche et al., 2011).

3.3.1.7 Pineapple Fruit

Pineapple waste has the potential for recycling in order to obtain precious raw material, transfer into helpful and higher price products, food or feed after biological treatment, and still as raw material for other industries (Kroyer, 1991). One example of raw material is pineapple waste that is converted to bioethanol production (Hossain et al., 2008). The wastes contain valuable components such as sucrose, glucose, fructose, and other nutrients (Hossain & Fazliny, 2010). The usefulness of the juice of rotten or discarded pineapples and the waste material generated throughout production of pineapple juice by *Z. mobilis* as inexpensive substrates for ethanol production was explored by Tanaka and colleagues (1999).

3.3.1.8 Tropical Fruit

Tropical fruits are probable sources for wine production (Ifeanyi, 2004). The ultimate goal is to spread new ethanol industries in countries from Japan to Nigeria (Lynch, 2006). According to Berg (2001), fuel ethanol, besides its environmental value, is and will remain first and a leading instrument to sustain farmers, as they will profit from fuel ethanol programs. The preceding provide helpful intention such as wine production, ethanol making among others, and as a means to salvage fruit wastage, reducing the bulk of solid waste and cleaning the environment (Obire et al., 2015). Pomegranate wine is the product of anaerobic fermentation by yeast in which the sugars are converted into alcohol and carbon dioxide (Adsule et al., 1995; Sevda & Rodrigues, 2011).

3.3.2 Microorganism: Yeast

The fermentation of sugar to ethanol by yeast has important advantages among the numerous processes that are carried out in industry. *S. cerevisiae*, *S. uvarum*, *Sch. pombe*, and *Kluyveromyces* species are the majority significant yeast species as a main interest to industrial operations (Noor et al., 2003). The fermentation process is initiated by a variety of species of *Candida, Debaryomyces, Hanseniaspora, Pichia, Kloeckera, Metschnikowia, Schizosaccharomyces, Torulospora,* and *Zygosaccharomyces*; the majority sturdily fermenting and more ethanol tolerant species of *Saccharomyces* dominates the fermentation (Lee et al., 2011). *Saccharomyces cerevisiae* is a facultative anaerobe that beneath anaerobic conditions can ferment glucose to ethanol extremely competently, and thus is the majority usually used microorganism in industrial ethanol production (Osho, 2005).

Yeast has been established to be more efficient for ethanol production than bacteria due to its good fermentative capability and its ability to tolerate high concentrations of ethanol and the byproducts formed during pretreatment and fermentation. *S. cerevisiae* is capable of fermenting different types of sugars—such as glucose, fructose, and sucrose—to ethanol via the glycolysis pathway under anaerobic conditions (Blieck et al., 2007). Between the attractive character of strains necessary for competent bioethanol production, tolerance to high temperature, acid tolerance, and high ethanol production are vital for decreasing cooling and ethanol recovery expenses, and for minimizing the danger of infectivity (Osho, 2005). Under suitable environmental and nutritional circumstances, *Saccharomyces cerevisiae* can manufacture and tolerate high ethanol concentrations. The very-high–gravity (VHG) fermentation process exploits the observation that the development of S. *cerevisiae* is promoted and prolonged when extremely small but sufficient levels of oxygen are present and assimilable nitrogen levels are not limiting (Chamnipa et al., 2017).

3.3.3 Pretreatments

Fruit waste (FW) to ethanol may not require harsh pretreatment (Tang et al., 2008; Kumar et al., 1998). Instead, autoclave of FW before fermentation is often required for improving product yield and purity, but at the cost of energy and water consumption. It should be noted that thermal treatment may lead to partial degradation of sugars and other nutritional components, as well as side reactions through which the amounts of useful sugars and amino acids are reduced (Sakai & Ezaki, 2006). Moreover, fresh FW crushed juice appears to be more effective (Kim et al., 2005). The progression of ethanol manufacture depends on the types of feedstocks used. Commonly, there are three major steps in ethanol manufacture: obtaining a solution that contains fermentable sugars, converting sugars to ethanol by fermentation, and ethanol separation and purification (Singh & Cheryan, 1998). Yeasts are precise to ferment fermentable sugars into ethanol, and separation technologies are used to improve ethanol ahead of its use as fuel (Azhar et al., 2017). Pretreatment has an important result on the whole process which makes the hydrolysis easier and produces an elevated sum of fermentable sugars. It influences the sum of ethanol yield and manufacture cost (Mata et al., 2010). Methods that are at present used for pretreatments are physical, chemical, biological, and physico-chemical. Physical pretreatment uses mechanical milling to ground the substrate. The ordinary compound pretreatment includes ozonolysis, acid hydrolysis, alkaline hydrolysis, and organosolv-based procedures. Fruit wastes were soaked in water and then autoclaved in steam, then the pretreated wastes were filtered and the filtrate obtained was analyzed for the total reducing sugars and pentose sugars. The obtained solid fruit waste from this treatment was dried and used for the subsequent hydrolysis studies using commercial cellulase or xylanase enzymes separately, as well as for the acidic hydrolysis (Jahid et al., 2018). Acidic pretreatment 1N concentrated sulfuric acid and 1N nitric acid with waste in the ratio for acid pretreatment was maintained as 2:2:1. The pretreated substrate was then filtered, and the filtered substrate further subjected to steam pretreatment (Chahande et al., 2018). The fresh pulp puree and milled peels were slurried with distilled water using a solid-to-liquid ratio of 10% (w/w) and autoclaved. The thermal pretreated samples were cooled to room temperature and used for further analysis. The dilute-acid pretreatment (DAP) was performed using dilute H_2SO_4 (0.05N). The hydrolysates obtained from the pretreated pulps and peels were subjected to a two-step enzymatic saccharification (Arumugam & Manikandan, 2011). Diverse fungal species are occupied in biological pretreatment though physico-chemical pretreatment includes ammonia fiber outburst and steam. Dehydration of hexose and pentoses all through pretreatment liberate furan compounds like 5-hydroxymethyl-2-furaldehyde (HMF) and 2-furaldehyde. These furan derivatives stimulate the inhibition of cell growth and lessen ethanol yield (Hsueh et al., 2007). Yeast fermentation is inhibited by the weak acid pressure induced from lignocellulosic wealth, but the small concentration of weak acids can amplify ethanol manufacture by cellular division. It was reported that the existence of weak acids can progress glucose consumption, ethanol manufacture, and tolerance to HMF and furfural in *S. cerevisiae* (Azhar et al., 2017).

3.3.4 Production of Bioethanol at a Large Scale

The capability to reasonably choose the most excellent type of biological reactor system to congregate the plan criteria and specifications of its working circumstances is possibly the majority significant function of biochemical engineers. The rate of separation and purification of the fermentation products often depends on the competence of the reactor and the economics of the total procedure; it hinges on the right option of the reactor (Tyagi & Ghose, 1980). The majority of fermentors used in industry are of the submerged type, because the submerged fermentor saves gap and is more agreeable to engineering control and design (Okafor et al., 2007). The major common form is the simple stirred tank bioreactor. This is a well-stirred tank designed for ideal mixing; beneath these circumstances, the fluid is homogeneous.

3.3.5 Bioethanol Production by Rapid Fermentation with High Cell Density

Fuel ethanol is an economical bulk chemical, and therefore, the yield and manufactured goods yield based on total sugars added to the reactor are the significant factors for the general economy of the process. The simplest means to get better volumetric efficiency is to amplify the quantity of biocatalyst, i.e. the energetic microorganism, in the reactor. There are number of ways of achieving such high cell concentration, also called high cell density. The simplest method is to add more cells at the beginning of a batch, or fed-batch fermentation. To maintain the cost of cell propagation low, cells must be reused, recycled, or retained inside the reactor (Westman & Franzén, 2015).

Through the high sugar concentrations reached from first-generation unprocessed resources, the cell concentration throughout the fermentation will conclude the volumetric production rate of ethanol (Basso et al., 2011). Concentration of cells is achieved by centrifugation, followed by removal of most of the fermentation broth and successive adding up of new medium. This is most effortlessly achieved with flocculating cells. Flocculation can be said to be a natural means of yeast immobilization, where the yeast cells are clustered in huge flocs (Bauer et al., 2010). Flocculating yeast cells generate large cell complexes of up to millimeter in size, which rapidly sediment if untouched (Silva et al., 2006). Cysewski and Wilke used a cell recycle reactor with a settler and vacuum ethanol separation, reaching an ethanol productivity of 82 g/L/h on a 33.4% glucose medium Cysewsk (1977). Tang and colleagues showed successful fermentation of a molasses medium using a flocculating yeast strain in serial bioreactors with settlers, yielding an ethanol concentration of 80 g/L and a productivity of 6.6 g/L/h for more than one month (Tang et al., 2010).

3.3.5.1 Batch Fermentation

Batch fermentation is the procedure of bioconversion of partial quantity of fermentation medium with total sterilization and management of feedstocks. The process of batch fermentation development is easier than other fermentation processes. The conversion of glucose and xylose to ethanol was carried out using yeast cells by batch fermentation process. Significance in ethanol fermentation has been focused on captivating of renewed concern in research mechanism in numerous areas such as use of enhanced mutant strains, yeast strain growth from cheaper resources, use of cheaper resources of unprocessed resources, most favorable reactor design, superior nutrients for optimum cell growth, optimization of fermentation factors, etc. Four diverse fermentation operations are presently used in industry: batch, continuous, fed batch, and semi-continuous. The batch process is the conventional technique that has stood for hundreds of years, and is at present the usually used means of ethanol production. In batch processing, cell slurry is grown separately from the fermentation substrate, and then slurry and substrate are added in a reactor, along with any essential enzymes or nutrients (Asli, 2010).

One technology that is significantly changing industrial ethanol production is VHG fermentation. In VHG fermentation, mashes with greater than 27 g dissolved solids per 100 g mash can be batch fermented with all substrates present at zero time and without the use of conditioned or genetically modified *Saccharomyces* yeasts. This technology has led to production of 23.8% v/v ethanol in the laboratory from wheat mash containing 38% w/v dissolved solids and it is gradually being applied to industries where goals of 15%–16% or more alcohol in some locations are being set in order to lower costs (Bayrock &

Ingledew, 2001). In this fermentation system, concentration of metabolites, medium composition, and inoculum changes constantly with time, in which four phases of growth of microorganisms were observed. They are lag phase, log phase, stationary phase and death phase. As a result, metabolism of cells were changing throughout the fermentation process. Batch fermentation has been widely used for bioethanol production with *S. cerevisiae* and *Kluyveromyces* species. "Rapid fermentations" are defined as fermentations in which the ethanol content increases from 0% to 9.5% (wt/vol; 12% vol/vol). The high rate of ethanol production was achieved primarily by using a yeast concentration of 3×10^9 cells per ml (D'amato et al., 2006). One yeast cell can ferment approximately its own weight of glucose per hour (Fakruddin et al., 2013). A major constraint of conventional alcohol fermentation processes is ethanol or end product inhibition. When a concentrated sugar solution is fermented and the ethanol concentration of the fermentation broth increases above 7%–10%, the specific ethanol production rate and the specific growth pace of the yeast is cruelly suppressed. In order to keep up the ethanol concentration at the most favorable level for ethanol production, concentrated sugar solutions, such as molasses, must be diluted to 10%–20% sugar (Vane, 2005). In Table 3.1, ethanol production from different fruit waste has been given.

Ethanol fermentation is influenced by diverse environmental, biological, and chemical conditions. The prominent parameters are temperature, sugar concentration, available nutrients, supplements, vitamins, pH, quantity of oxygen, and ethanol concentration (Priest & Campbell, 1996; Apiradee, 2006). Ethanolic fermentations with saccharides in excess (VHG fermentations) leads to sluggish fermentation and saccharides are only rarely completely fermented. Nutritional supplements are provided in required adequate amounts, whereby *Saccharomyces cervicea* can ferment effectively in very high sugar content media during the fermentation process to enhance ethanol production; high contents of saccharides in fermentation medium cause an increase of the osmotic pressure, which has a detrimental effect on yeast cells (Bafrncová et al., 1999). VHG ethanol production technologies that use media containing more than 25% (w/v) sugar and aim to achieve over 15% (v) ethanol concentrations have great attention. However, as a toxic metabolite, ethanol exerts a strong inhibition on yeast cell growth and ethanol production, which limits ethanol concentrations in the broth to no more than 13% (v) for most ethanol production plants at the present time. Yeast cell lysis and viability loss are serious under VHG conditions, especially when very high ethanol concentrations were developed inside those bioreactors toward the end of cascade fermentation systems. Normally, methylene blue stain technique (Smart et al., 1999), and sometimes colony forming units (CFU) methods, are used to evaluate yeast cell viability loss during ethanol production. However, as yeast cells tend to self-aggregate under high ethanol and poor nutrient environments, these techniques tend to be less effective. Based on kinetics and bioreactor engineering theories, this present investigation aimed at establishing a combined bioreactor system to produce ethanol under VHG conditions, and at the same time, developing new models to evaluate yeast cell lysis and viability loss (Bai et al., 2004). Its benefits include a decrease in process water requirements and energy costs, increased productivity and ethanol concentrations in the product. An important consideration in VHG fermentation is that the yeast is subjected to considerable osmotic stress, which decreases its growth and cell viability (Wang et al., 2007). Several authors have observed that Mg2+ (Hu et al., 2003), yeast extract (Bafrncová et al., 1999), glycine (Thomas et al., 1993), biotin (Alfenore et al., 2002) and peptone (Stewart et al., 1988) have a protective effect on yeast growth and viability, and they improve the final ethanol concentration. Others have shown that acetaldehyde is highly effective in shortening the lag phase of *Saccharomyces cerevisiae* and the fermentation time (Barber et al., 2002; Wang et al., 2007). However, Thomas and Ingledew (1990) reported that in grain mashes, which are used for fuel ethanol production, only assimilable nitrogen is limiting single-factor. Assimilable nitrogen is an important component of fermentation media, and it plays a significant role in the course of fermentation. Nitrogen is necessary for growing and multiplication; it influences ethanol tolerance of yeasts and the rate of ethanol production. The addition of free amino nitrogen (FAN) leads to higher final ethanol concentrations in the fermented media and amounts of the cell mass accumulation.

3.4 Conclusion

Renewable energy sources and technologies have potential to afford solutions to the long-standing energy evils being faced by rising countries. The renewable energy sources like wind energy, solar energy, geothermal energy, ocean energy, biomass energy, and fuel cell technology can be used to prevail over energy deficiency in India, gradually adopting more accountable renewable energy techniques and captivating positive steps toward carbon emissions, cleaning the air, and ensuring an additional sustainable future. India is a predominantly agrarian country where huge amounts of agricultural waste are generated. These agricultural waste biomasses are the potential sources of cellulosic feedstock to produce bioethanol as second-generation biofuel. Among the agricultural wastes, fruit waste constitutes a high percentage of fermentable sugars, which can be easily converted to bioethanol by yeast, as a first-generation of biofuel. Though many disadvantages arise in first-generation bioethanol production, the use of fruit waste as biomass in bioethanol production gives a solution to the media, and consequently the public, who are usually unable to discriminate between one biofuel and another—as a result, growing public and political concerns on the use of all biofuels in general related to food security, societal costs, limited returns, modest GHG, deforestation, habitat loss, and environmental issues.

Acknowledgments

We thank Professor Manjula Kumari D, for critical reading of the manuscript and also Professor K. Manjunath for his timely guidance. My sincere thanks to Professor Anu Appaiah, CFTRI, Mysore, for his valuable guidance throughout my research. This work was carried out in the Department of Microbiology and Biotechnology, BUB, JB campus, Bengaluru. The research work was supported by UGC-BSR Scheme (UGC No. No.4–1/2006(BSR)/7=322/2011(BSR) dated 23.01.2013.

NOTE

1. Few statistics parts taken from the website http://www.fao.org.

REFERENCES

Adams, J. M., Gallagher, J. A., & Donnison, I. S. 2009. Fermentation study on *Saccharina latissima* for bioethanol production considering variable pre-treatments. *Journal of Applied Phycology*, *21*(5), 569.

Adsule, R. N., & Patil, N. B. 1995. Pomegranate. In *Handbook of fruit science and technology* (pp. 471–480). Boca Raton, FL: CRC Press.

Ahmeh, J. B., Okagbue, R. N., Ahmadu, A. A., & Ikediobi, C. O. 1988. Ethanol production from corn-cob wastes and grass straw. *Nigerian Journal of Biotechnology*, 6, 110–112.

Ahorsu, R., Medina, F., & Constantí, M. 2018. Significance and challenges of biomass as a suitable feedstock for bioenergy and biochemical production: A review. *Energies*, *11*(12), 3366.

Akella, A. K., Saini, R. P., & Sharma, M. P. 2009. Social, economical and environmental impacts of renewable energy systems. *Renewable Energy*, *34*(2), 390–396.

Akin-Osanaiye, B. C., Nzelibe, H. C., & Agbaji, A. S. 2008. Ethanol production from *Carica papaya* (pawpaw) fruit waste. *Asian Journal of Biochemistry*, *3*(3), 188–193.

Akin-Osanaiye, B. C., Nzelibe, H. C., & Agbaji, A. S. 2005. Production of ethanol from *Carica papaya* (pawpaw) agro waste: Effect of saccharification and different treatments on ethanol yield. *African Journal of Biotechnology*, *4*(7), 657–659.

Alfenore, S., Molina-Jouve, C., Guillouet, S., Uribelarrea, J. L., Goma, G., & Benbadis, L. 2002. Improving ethanol production and viability of *Saccharomyces cerevisiae* by a vitamin feeding strategy during fed-batch process. *Applied Microbiology and Biotechnology*, *60*(1–2), 67–72.

Alshammari, A. M., Adnan, F. M., Mustafa, H., & Hammad, N. 2011. Bioethanol fuel production from rotten banana as an environmental waste management and sustainable energy. *African Journal of Microbiology Research*, *5*(6), 586–598.

Alvarenga, R. M., Carrara, A. G., Silva, C. M., & Oliveira, E. S. 2011. Potential application of *Saccharomyces cerevisiae* strains for the fermentation of banana pulp. *African Journal of Biotechnology*, *10*(18), 3608–3615.

Anex, R. P., Aden, A., Kazi, F. K., Fortman, J., Swanson, R. M., Wright, M. M., & Kothandaraman, G. 2010. Techno-economic comparison of biomass-to-transportation fuels via pyrolysis, gasification, and biochemical pathways. *Fuel, 89*, S29–S35.

Apiradee, S. (2006). *Isolation and characterization of thermotolerant yeast for ethanol production* (Degree of Master of Science). Suranaree University of Technology, 98pp.

Aro, E. M. (2016). From first generation biofuels to advanced solar biofuels. *Ambio, 45*(1), 24–31.

Arumugam, R., & Manikandan, M. 2011. Fermentation of pretreated hydrolyzates of banana and mango fruit wastes for ethanol production. *Asian Journal of Experimental Biological Sciences, 2*(2), 246–256.

Asli, M. S. 2010. A study on some efficient parameters in batch fermentation of ethanol using *Saccharomyces cerevesiae* SC1 extracted from fermented siahe sardasht pomace. *African Journal of Biotechnology, 9*(20).

Astedu, J. 1987. Physicochemical changes in plantain (*Musa paradidiaca*) during ripening and the effect of degree of ripeness on drying. *Tropical Science, 27*(4), 249–260.

Azhar, S. H. M., Abdulla, R., Jambo, S. A., Marbawi, H., Gansau, J. A., Faik, A. A. M., & Rodrigues, K. F. 2017. Yeasts in sustainable bioethanol production: A review. *Biochemistry and Biophysics Reports, 10*, 52–61.

Bafrncová, P., Sláviková, I., Pátková, J., & Dömény, Z. 1999. Improvement of very high gravity ethanol fermentation by media supplementation using *Saccharomyces cerevisiae*. *Biotechnology Letters, 21*(4), 337–341.

Bai, F. W., Chen, L. J., Zhang, Z., Anderson, W. A., & Moo-Young, M. 2004. Continuous ethanol production and evaluation of yeast cell lysis and viability loss under very high gravity medium conditions. *Journal of Biotechnology, 110*(3), 287–293.

Balasubramanian, K., Ambikapathy, V., & Panneerselvam, A. 2011. Studies on ethanol production from spoiled fruits by batch fermentations. *Journal of Microbiology and Biotechnology Research, 1*(4), 158–163.

Balat, M., & Balat, H. (2009). Recent trends in global production and utilization of bio-ethanol fuel. *Applied Energy, 86*(11), 2273–2282.

Barber, A. R., Henningsson, M., & Pamment, N. B. 2002. Acceleration of high gravity yeast fermentations by acetaldehyde addition. *Biotechnology Letters, 24*(11), 891–895.

Basile, A., & Dalena, F. (Eds.). 2019. *Second and third generation of feedstocks: The evolution of biofuels*. Amsterdam, Netherlands: Elsevier.

Basso, L. C., Basso, T. O., & Rocha, S. N. 2011. Ethanol production in Brazil: The industrial process and its impact on yeast fermentation. In *Biofuel production-recent developments and prospects*. Croatia: InTech.

Bauer, F. F., Govender, P., & Bester, M. C. 2010. Yeast flocculation and its biotechnological relevance. *Applied Microbiology and Biotechnology, 88*(1), 31–39.

Bayrock, D. P., & Ingledew, W. M. 2001. Application of multistage continuous fermentation for production of fuel alcohol by very-high-gravity fermentation technology. *Journal of Industrial Microbiology and Biotechnology, 27*(2), 87–93.

Berg, C. 2001. *World Ethanol Markets, Analysis and Outlook*. Ratzeburg, Germany: FO-Licht, October 2001. Available online at http://www.distill.com/world_ethanol_production.html.

Berla, B. M., Saha, R., Immethun, C. M., Maranas, C. D., Moon, T. S., & Pakrasi, H. 2013. Synthetic biology of cyanobacteria: Unique challenges and opportunities. *Frontiers in Microbiology, 4*, 246.

Blieck, L., Toye, G., Dumortier, F., Verstrepen, K. J., Delvaux, F. R., Thevelein, J. M., & Van Dijck, P. 2007. Isolation and characterization of brewer's yeast variants with improved fermentation performance under high-gravity conditions. *Applied and Environmental Microbiology, 73*(3), 815–824.

Bodansky, D. 2004. *International climate efforts beyond 2012: A survey of approaches*. Arlington, VA: Pew Center on Global Climate Change.

Boluda-Aguilar, M., García-Vidal, L., del Pilar González-Castañeda, F., & López-Gómez, A. 2010. Mandarin peel wastes pretreatment with steam explosion for bioethanol production. *Bioresource Technology, 101*(10), 3506–3513.

Boyce, N. 2014. Bioethanol production from Mango Waste (*Mangifera indica* L. cv chokanan): Biomass as renewable energy. *Australian Journal of Basic and Applied Sciences, 8*(9), 229–237.

Braddock, R. J. 1999. *Handbook of citrus by-products and processing technology*. New York: John Wiley & Sons.

Brennan, L., & Owende, P. 2010. Biofuels from microalgae—A review of technologies for production, processing, and extractions of biofuels and co-products. *Renewable and Sustainable Energy Reviews, 14*(2), 557–577.

Brooks, A. A. 2008. Ethanol production potential of local yeast strains isolated from ripe banana peels. *African Journal of Biotechnology, 7*(20).

Bugaud, C., Alter, P., Daribo, M. O., & Brillouet, J. M. 2009. Comparison of the physico-chemical characteristics of a new triploid banana hybrid, FLHORBAN 920, and the Cavendish variety. *Journal of the Science of Food and Agriculture*, *89*(3), 407–413.

Canilha, L., Chandel, A. K., Suzane Dos Santos Milessi, T., Antunes, F. A. F., Luiz Da Costa Freitas, W., Das Graças Almeida, F. M., & Da Silva, S. S. 2012. Bioconversion of sugar-cane biomass into ethanol: An overview about composition, pretreatment methods, detoxification of hydrolysates, enzymatic saccharification, and ethanol fermentation. *Journal of Biomedicine and Biotechnology*, *2012*(7): 1–16.

Cardona, C. A., & Sánchez, Ó. J. 2007. Fuel ethanol production: Process design trends and integration opportunities. *Bioresource Technology*, *98*(12), 2415–2457.

Carlisle, K. N., & Parker, A. W. 2014. Psychological distress and pain reporting in Australian coal miners. *Safety and Health at Work*, *5*(4), 203–209.

Casabar, J. T., Unpaprom, Y., & Ramaraj, R. 2019. Fermentation of pineapple fruit peel wastes for bioethanol production. *Biomass Conversion and Biorefinery*, *9*(4), 761–765.

Chahande, A. D., Gedam, V. V., Raut, P. A., & Moharkar, Y. P. 2018. Pretreatment and production of bioethanol from *Citrus reticulata* fruit waste with Baker's yeast by solid-state and submerged fermentation. In *Utilization and management of bioresources* (pp. 135–141). Singapore, Springer.

Chamnipa, N., Thanonkeo, S., Klanrit, P., & Thanonkeo, P. 2017. The potential of the newly isolated thermotolerant yeast Pichia kudriavzevii RZ8-1 for high-temperature ethanol production. *Brazilian Journal of Microbiology*, *49*(2), 378–391.

Chaudhary, N., Chand, S., & Kaur, N. 2014. Bioethanol production from fruit peels using simultaneous saccharification and fermentation. *GSTF Journal of BioSciences*, *3*(1).

Chisti, Y. 2008. Biodiesel from microalgae beats bioethanol. *Trends in Biotechnology*, *26*(3), 126–131.

Clark, M. A., & Fox, M. K. 2009. Nutritional quality of the diets of US public school children and the role of the school meal programs. *Journal of the American Dietetic Association*, *109*(2), S44–S56.

Coppola, F., Bastianoni, S., & Østergård, H. 2009. Sustainability of bioethanol production from wheat with recycled residues as evaluated by energy assessment. *Biomass and Bioenergy*, *33*(11), 1626–1642.

Cysewsk, S. J. 1977. *In mycotoxic fungi, mycotoxins and mycotoxicoses*. New York: Marcel Dekker.

D'amato, D., Corbo, M. R., Nobile, M. A. D., & Sinigaglia, M. 2006. Effects of temperature, ammonium and glucose concentrations on yeast growth in a model wine system. *International Journal of Food Science & Technology*, *41*(10), 1152–1157.

Datta, R., Maher, M. A., Jones, C., & Brinker, R. W. 2011. Ethanol—The primary renewable liquid fuel. *Journal of Chemical Technology and Biotechnology*, *86*(4), 473–480.

Desmond, Layne R. 1995. The Pawpaw (*Asimina triloba* (L.) Dunal). In *New crop: Fact sheet* (pp. 1–5). Alexandria, VA: American Society for Horticultural Science.

Dincer, I. 1999. Environmental impacts of energy. *Energy Policy*, *27*(14), 845–854.

Dincer, I., Midilli, A., & Kucuk, H. (Eds.). 2014. *Progress in exergy, energy, and the environment*. Cham: Springer.

Dincer, I., & Rosen, M. A. 2007. Energetic, exergetic, environmental and sustainability aspects of thermal energy storage systems. In *Thermal energy storage for sustainable energy consumption* (pp. 23–46). Dordrecht: Springer.

Dürre, P. 2008. Fermentative butanol production: Bulk chemical and biofuel. *Annals of the New York Academy of Sciences*, *1125*, 353–362.

Edenhofer, O., Pichs-Madruga, R., Sokona, Y., Seyboth, K., Matschoss, P., Kadner, S., & von Stechow, C. 2011. IPCC special report on renewable energy sources and climate change mitigation. *Prepared by Working Group III of the Intergovernmental Panel on Climate Change*. Cambridge: Cambridge University Press.

Emaga, T. H., Robert, C., Ronkart, S. N., Wathelet, B., & Paquot, M. 2008. Dietary fibre components and pectin chemical features of peels during ripening in banana and plantain varieties. *Bioresource Technology*, *99*(10), 4346–4354.

Fakruddin, M., Islam, M. A., Quayum, M. A., Ahmed, M. M., & Chowdhury, N. 2013. Characterization of stress tolerant high potential ethanol producing yeast from agro-industrial waste. *American Journal of BioScience*, *1*(2), 24–34.

Gaur, S., & Reed, T. B. 1998. *Thermal data for natural and synthetic fuels*. New York: Marcel Dekker.

George, B., Pancha, I., Desai, C., Chokshi, K., Paliwal, C., Ghosh, T., & Mishra, S. 2014. Effects of different media composition, light intensity and photoperiod on morphology and physiology of freshwater microalgae *Ankistrodesmus falcatus*—A potential strain for bio-fuel production. *Bioresource Technology*, *171*, 367–374.

Gnansounou, E., Dauriat, A., & Wyman, C. E. 2005. Refining sweet sorghum to ethanol and sugar: Economic trade-offs in the context of North China. *Bioresource Technology*, *96*(9), 985–1002.

Gouveia, L., & Oliveira, A. C. 2009. Microalgae as a raw material for biofuels production. *Journal of Industrial Microbiology & Biotechnology*, *36*(2), 269–274.

Graefe, S., Dufour, D., Giraldo, A., Muñoz, L. A., Mora, P., Solís, H., & Gonzalez, A. 2011. Energy and carbon footprints of ethanol production using banana and cooking banana discard: A case study from Costa Rica and Ecuador. *Biomass and Bioenergy*, *35*(7), 2640–2649.

Grahovac, J. A., Dodić, J. M., Dodić, S. N., Popov, S. D., Vučurović, D. G., & Jokić, A. I. 2012. Future trends of bioethanol co-production in Serbian sugar plants. *Renewable and Sustainable Energy Reviews*, *16*(5), 3270–3274.

Gray, K. A., Zhao, L., & Emptage, M. 2006. Bioethanol. *Current Opinion in Chemical Biology*, *10*, 141–146.

Grohmann, K., & Baldwin, E. A. 1992. Hydrolysis of orange peel with pectinase and cellulase enzymes. *Biotechnology Letters*, *14*(12), 1169–1174.

Grohmann, K., Baldwin, E. A., & Buslig, B. S. 1994. Production of ethanol from enzymatically hydrolyzed orange peel by the yeast *Saccharomyces cerevisiae*. *Applied Biochemistry and Biotechnology*, *45*(1), 315.

Hammond, J. B., Egg, R., Diggins, D., & Coble, C. G. 1996. Alcohol from bananas. *Bioresource Technology*, *56*(1), 125–130.

Hasager, C. B., Nielsen, M., Astrup, P., Barthelmie, R., Dellwik, E., Jensen, N. O., & Furevik, B. R. 2005. Offshore wind resource estimation from satellite SAR wind field maps. *Wind Energy*, *8*(4), 403–419.

Hashem, M., El-Latif Hesham, A., Alrumman, S. A., & Alamri, S. A. 2017. Production of bioethanol from spoilage date fruits by new osmotolerant yeasts. *International Journal of Agriculture and Biology*, *19*, 825–833.

Hayes, D. J., & Hayes, M. H. 2009. The role that lignocellulosic feedstocks and various biorefining technologies can play in meeting Ireland's biofuel targets. *Biofuels, Bioproducts and Biorefining*, *3*(5), 500–520.

Hayn, M., Klinger, R., & Esterbauer, H. 1993. Isolation and partial characterization of a low molecular weight endoglucanase from *Trichoderma reesei*. In P. Suominen & T. Reinikainen (Eds.), *Proceedings of the second TRICEL symposium on Trichoderma reesei cellulases and other hydrolases* (pp. 147–151). Helsinki: Foundation for Biotechnical and Industrial Fermentation Research.

Hays, S. G., & Ducat, D. C. 2015. Engineering cyanobacteria as photosynthetic feedstock factories. *Photosynthesis Research*, *123*(3), 285–295.

Hossain, A. B. M. S., & Fazliny, A. R. 2010. Creation of alternative energy by bio-ethanol production from pineapple waste and the usage of its properties for engine. *African Journal of Microbiology Research*, *4*(9), 813–819.

Hossain, A. B. M. S., Saleh, A. A., Aishah, S., Boyce, A. N., Chowdhury, P. P., & Naqiuddin, M. 2008. Bioethanol production from agricultural waste biomass as a renewable bioenergy resource in biomaterials. In *4th Kuala Lumpur international conference on biomedical engineering 2008* (pp. 300–305). Berlin, Heidelberg: Springer.

Hsueh, H. T., Chu, H., & Yu, S. T. 2007. A batch study on the bio-fixation of carbon dioxide in the absorbed solution from a chemical wet scrubber by hot spring and marine algae. *Chemosphere*, *66*(5), 878–886.

Hu, C. K., Bai, F. W., & An, L. J. 2003. Enhancing ethanol tolerance of a self-flocculating fusant of *Schizosaccharomyces pombe* and *Saccharomyces cerevisiae* by Mg 2+ via reduction in plasma membrane permeability. *Biotechnology Letters*, *25*(14), 1191–1194.

Hu, Q., Sommerfeld, M., Jarvis, E., Ghirardi, M., Posewitz, M., Seibert, M., & Darzins, A. 2008. Microalgal triacylglycerols as feedstocks for biofuel production: Perspectives and advances. *The Plant Journal*, *54*(4), 621–639.

Huron, M., Hudebine, D., Ferreira, N. L., & Lachenal, D. 2016. Impact of delignification on the morphology and the reactivity of steam exploded wheat straw. *Industrial Crops and Products*, *79*, 104–109.

Ibeto, C. N., Ofoefule, A. U., & Agbo, K. E. 2011. A global overview of biomass potentials for bioethanol production: A renewable alternative fuel. *Trends in Applied Sciences Research*, *6*(5), 410.

IEA-ETSAP, & I. R. E. N. A. (2013). *Production of bio-ethylene: Technology brief*. Abu Dhabi: International Renewable Energy Agency.

Ifeanyi, V. O. 2004. Utilization of *Saccharomyces cerevisiae* for locally processed fruit wines. *Nigerian Journal of Microbiology*, *18*(1–2), 306–315.

Jahanbakhshi, A., & Salehi, R. 2019. Processing watermelon waste using *Saccharomyces cerevisiae* yeast and the fermentation method for bioethanol production. *Journal of Food Process Engineering*, *42*(7), e13283.

Jahid, M., Gupta, A., & Sharma, D. K. 2018. Production of bioethanol from fruit wastes (banana, papaya, pineapple and mango peels) under milder conditions. *Journal of Bioprocessing & Biotechniques*, *8*(3), 327.

Janani, K., Ketzi, M., Megavathi, S., Vinothkumar, D., & Ramesh Babu, N. G. 2013. Comparative studies of ethanol production from different fruit wastes using *Saccharomyces cerevisiae*. *International Journal of Innovative Research in Science, Journal Engineering and Technology*, 2(12), 7161–7167.

Joshi, R., & Ahmed, S. 2016. Status and challenges of municipal solid waste management in India: A review. *Cogent Environmental Science*, 2(1), 1139434.

Kavanagh, E. 2006. Looking at biofuels and bioenergy—Commentary. *Science*, 312, 1743–1748.

Khambhaty, Y., Upadhyay, D., Kriplani, Y., Joshi, N., Mody, K., & Gandhi, M. R. 2013. Bioethanol from macroalgal biomass: Utilization of marine yeast for production of the same. *BioEnergy Research*, 6(1), 188–195.

Kim, K. C., Kim, S. W., Kim, M. J., & Kim, S. J. 2005. Saccharification of foodwastes using cellulolytic and amylolytic enzymes from *Trichoderma harzianum* FJ1 and its kinetics. *Biotechnology and Bioprocess Engineering*, 10(1), 52.

Klass, D. L. 1998. *Biomass for renewable energy, fuels, and chemicals*. San Diego, CA: Academic Press, Elsevier.

Kroyer, G. T. 1991. Food processing wastes. In *Bioconversion of waste materials to industrial products* (pp. 293–311). Barking: Elsevier Science Publishers Ltd.

Kumar, A., Kumar, K., Kaushik, N., Sharma, S., & Mishra, S. 2010. Renewable energy in India: Current status and future potentials. *Renewable and Sustainable Energy Reviews*, 14(8), 2434–2442.

Kumar, J. V., Mathew, R., & Shahbazi, A. 1998. Bioconversion of solid food wastes to ethanol. *Analyst*, 123(3), 497–502.

Lakhawat, S. S., Aseri, G. K., & Daur, V. S. 2011. Comparative study of ethanol production using yeast and fruit of *Vitis lanata* ROXB. *International Journal of Advanced Biotechnology and Research*, 2(2), 269–277.

Lardon, L., Hélias, A., Sialve, B., Steyer, J. P., & Bernard, O. 2009. Life-cycle assessment of biodiesel production from microalgae. *Environmental Science & Technology*, 43(17), 6475–6481.

Larsen, J., Ostergaard Petersen, M., Thirup, L., Wen Li, H., & Krogh Iversen, F. 2008. The IBUS process—Lignocellulosic bioethanol close to a commercial reality. *Chemical Engineering & Technology*, 31(5), 765–772.

Larson, D. E. 2007. Biofuel production technologies: Status and prospect. In *United Nations Conference on Trade and Development*. http://r0.unctad.org/ghg/events/biofuels/LarsonAHEM19% 20June(Vol.202007).

Lee, Y. J., Choi, Y. R., Lee, S. Y., Park, J. T., Shim, J. H., Park, K. H., & Kim, J. W. 2011. Screening wild yeast strains for alcohol fermentation from various fruits. *Mycobiology*, 39(1), 33–39.

Lundgren, A., & Hjertberg, T. 2010. Ethylene from renewable resources. *Surfactants from Renewable Resources*, 111.

Lynch, D. 2006. Brazil hopes to build on its ethanol success. *USA Today*, pp. 1–3.

Lysen, E. H., & Ouwens, C. D. 2002. Energy effects of the rapid growth of PV capacity and the urgent need to reduce the energy payback time. In *Proceedings of 17th European photovoltaic solar energy conference, München*.

Mamma, D., Kourtoglou, E., & Christakopoulos, P. 2008. Fungal multienzyme production on industrial by-products of the citrus-processing industry. *Bioresource Technology*, 99(7), 2373–2383.

Mascarenhas, G. C. C. 1999. Banana: Comercialização e mercados. *Informe Agropecuário, Belo Horizonte*, 20(196), 97–108.

Mata, T. M., Martins, A. A., & Caetano, N. S. 2010. Microalgae for biodiesel production and other applications: A review. *Renewable and Sustainable Energy Reviews*, 14(1), 217–232.

Mendez-Alvarez, S., Leisinger, U., & Eggen, R. I. 1999. Adaptive responses in *Chlamydomonas reinhardtii*. *International Microbiology*, 2(1), 15–22.

Midilli, A., Dincer, I., & Ay, M. 2006. Green energy strategies for sustainable development. *Energy Policy*, 34(18), 3623–3633.

Mojović, L., Pejin, D., Grujić, O., Markov, S., Pejin, J., Rakin, M., & Savić, D. 2009. Progress in the production of bioethanol on starch-based feedstocks. *Chemical Industry and Chemical Engineering Quarterly/CICEQ*, 15(4), 211–226.

Molina, A. B., & Kudagamage, C. 2002. The international network for the improvement of banana and plantain (INIBAP): PGR activities in South Asia. In *South Asia network on plant genetic resources (SANPGR) meeting held on December 9–11*.

Naik, S. N., Goud, V. V., Rout, P. K., & Dalai, A. K. 2010. Production of first and second generation biofuels: A comprehensive review. *Renewable and Sustainable Energy Reviews*, 14(2), 578–597.

Nigam, P. S., & Singh, A. 2011. Production of liquid biofuels from renewable resources. *Progress in Energy and Combustion Science*, 37(1), 52–68.

Noor, A. A., Hameed, A., Bhatti, K. P., & Tunio, S. A. 2003. Bio-ethanol fermentation by the bioconversion of sugar from dates by *Saccharomyces cerevisiae* strains ASN-3 and HA-4. *Biotechnology, 2*(1), 8–17.

Obire, O., Putheti, R. R., Dick, A. A., & Okigbo, R. N. 2015. Biotechnology influence for the production of ethyl alcohol (ethanol) from waste fruites. *e-Journal of Science & Technology, 3*(3), 17–32.

Okafor, U. A., Okochi, V. I., Onyegeme-Okerenta, B. M., & Nwodo-Chinedu, S. 2007. Xylanase production by *Aspergillus niger* ANL 301 using agro-wastes. *African Journal of Biotechnology, 6*(14).

Osho, A. 2005. Ethanol and sugar tolerance of wine yeasts isolated from fermenting cashew apple juice. *African Journal of Biotechnology, 4*(7), 660–662.

Oyedepo, S. O. 2012. Energy and sustainable development in Nigeria: The way forward. *Energy, Sustainability and Society, 2*(1), 15.

Palacios-Bereche, R., Ensinas, A. V., & Nebra, S. A. 2011. Energy consumption in ethanol production by enzymatic hydrolysis—The integration with the conventional process using pinch analysis. *Chemical Engineering Transactions, 24*, 1189–1194.

Parry, M., Parry, M. L., Canziani, O., Palutikof, J., Van der Linden, P., & Hanson, C. (Eds.). 2007. *Climate change 2007-impacts, adaptation and vulnerability: Working group II contribution to the fourth assessment report of the IPCC* (Vol. 4). Cambridge: Cambridge University Press.

Patil, V., Tran, K. Q., & Giselrod, H. R. 2008. Towards sustainable production of biofuels from microalgae. *International Journal of Molecular Sciences, 9*(7), 1188–1195.

Perez, V. H., Junior, E. G. S., Cubides, D. C., David, G. F., Justo, O. R., Castro, M. P., . . . & de Castro, H. F. 2014. Trends in biodiesel production: present status and future directions. In *Biofuels in Brazil* (pp. 281–302). Cham: Springer.

Priest, F. G., & Campbell, I. (Eds.). 1996. *Brewing microbiology*. London: Chapman & Hall.

Raikar, R. V. 2012. Enhanced production of ethanol from grape waste. *International Journal of Environmental Sciences, 3*(2), 776–783.

Resch, G., Held, A., Faber, T., Panzer, C., Toro, F., & Haas, R. 2008. Potentials and prospects for renewable energies at global scale. *Energy Policy, 36*(11), 4048–4056.

Saha, B. C., Iten, L. B., Cotta, M. A., & Wu, Y. V. 2005. Dilute acid pretreatment, enzymatic saccharification and fermentation of wheat straw to ethanol. *Process Biochemistry, 40*, 3693–3700.

Sakai, K., & Ezaki, Y. 2006. Open L-lactic acid fermentation of food refuse using thermophilic *Bacillus coagulans* and fluorescence in situ hybridization analysis of microflora. *Journal of Bioscience and Bioengineering, 101*(6), 457–463.

Scaife, M. A., Nguyen, G. T., Rico, J., Lambert, D., Helliwell, K. E., & Smith, A. G. 2015. Establishing *Chlamydomonas reinhardtii* as an industrial biotechnology host. *Plant Journal, 82*(3), 532–546.

Scholz, M. J., Riley, M. R., & Cuello, J. L. 2013. Acid hydrolysis and fermentation of microalgal starches to ethanol by the yeast *Saccharomyces cerevisiae*. *Biomass and Bioenergy, 48*, 59–65.

Scott, A., & Bryner, M. 2006. Alternative fuels rolling out next-generation technologies. *Chemical Week, 168*(43), 17–19.

Sevda, S. B., & Rodrigues, L. 2011. The making of pomegranate wine using yeast immobilized on sodium alginate. *African Journal of Food Science, 5*(5), 299–304.

Silva, C. L. C., Rosa, C. A., & Oliveira, E. S. 2006. Studies on the kinetic parameters for alcoholic fermentation by flocculent *Saccharomyces cerevisiae* strains and non-hydrogen sulfide-producing strains. *World Journal of Microbiology and Biotechnology, 22*(8), 857–863.

Silva, E. F. 2004. *Obtenção de Aguardente de Banana em Micro-Escala: Caracterização do Processo e do Produto* (Doctoral dissertation). Dissertação, Mestrado em Ciência de Alimentos, Federal University of Paraíba, 112pp.

Singh, N., & Cheryan, M. 1998. Membrane technology in corn refining and bioproduct processing. *StarchmStärke, 50*(1), 16–23.

Sivaramakrishnan, R., & Incharoensakdi, A. 2018. Utilization of microalgae feedstock for concomitant production of bioethanol and biodiesel. *Fuel, 217*, 458–466.

Smart, K. A., Chambers, K. M., Lambert, I., Jenkins, C., & Smart, C. A. 1999. Use of methylene violet staining procedures to determine yeast viability and vitality. *Journal of the American Society of Brewing Chemists, 57*(1), 18–23.

Somda, M. K., Savadogo, A., Ouattara, C. A. T., Ouattara, A. S., & Traore, A. S. 2011. Thermotolerant and alcohol-tolerant yeasts targeted to optimize hydrolyzation from mango peel for high bioethanol production. *Asian Journal of Biotechnology, 3*(1), 77–83.

Srinivasarao, B., Ratnam, B. V. V., Subbarao, S., Narasimharao, M., & Ayyanna, C. 2013. Ethanol production from cashew apple juice using statistical designs. *Journal of Microbial and Biochemical Technology, 1*, 8–15.

Srivastava, S., Modi, D. R., & Garg, S. K. 1997. Production of ethanol from guava pulp by yeast strains. *Bioresource Technology*, 60(3), 263–265.

Stewart, G. G., D'Amore, T., Panchal, C. J., & Russell, I. (1988). Factors that influence the ethanol tolerance of brewer's yeast strains during high gravity wort fermentations. *Technical Quarterly-Master Brewers Association of the Americas (USA)*, 25(2), 47–53.

Sudhakar, M. P., Merlyn, R., Arun kumar, K., & Perumal, K. 2016. Characterization, pretreatment and saccharification of spent seaweed biomass for bioethanol production using baker's yeast. *Biomass and Bioenergy*, 90, 148–154.

Sun, Y., & Cheng, J. 2002. Hydrolysis of lignocellulosic materials for ethanol production: A review. *Bioresource Technology*, 83(1), 1–11.

Taherzadeh, M. J., & Karimi, K. 2007. Enzymatic-based hydrolysis processes for ethanol from lignocellulosic materials: A review. *BioResources*, 2(4), 707–738.

Talebnia, F., Bafrani, M. P., Lundin, M., & Taherzadeh, M. 2008. Optimization study of citrus wastes saccharification by dilute acid hydrolysis. *BioResources*, 3(1), 108–122.

Tanaka, K., Hilary, Z. D., & Ishizaki, A. 1999. Investigation of the utility of pineapple juice and pineapple waste material as low-cost substrate for ethanol fermentation by *Zymomonas mobilis*. *Journal of Bioscience and Bioengineering*, 87(5), 642–646.

Tang, Y. Q., An, M. Z., Zhong, Y. L., Shigeru, M., Wu, X. L., & Kida, K. 2010. Continuous ethanol fermentation from non-sulfuric acid-washed molasses using traditional stirred tank reactors and the flocculating yeast strain KF-7. *Journal of Bioscience and Bioengineering*, 109(1), 41–46.

Tang, Y. Q., Koike, Y., Liu, K., An, M. Z., Morimura, S., Wu, X. L., & Kida, K. 2008. Ethanol production from kitchen waste using the flocculating yeast *Saccharomyces cerevisiae* strain KF-7. *Biomass and Bioenergy*, 32(11), 1037–1045.

Thomas, K. C., Hynes, S. H., Jones, A. M., & Ingledew, W. M. 1993. Production of fuel alcohol from wheat by VHG technology. *Applied Biochemistry and Biotechnology*, 43(3), 211–226.

Thomas, K. C., & Ingledew, W. M. 1990. Fuel alcohol production: Effects of free amino nitrogen on fermentation of very-high-gravity wheat mashes. *Applied and Environmental Microbiology*, 56(7), 2046–2050.

Tripodo, M. M., Lanuzza, F., Micali, G., Coppolino, R., & Nucita, F. 2004. Citrus waste recovery: A new environmentally friendly procedure to obtain animal feed. *Bioresource Technology*, 91(2), 111–115.

Tropea, A., Wilson, D., La Torre, L. G., Curto, R. B. L., Saugman, P., Troy-Davies, P., . . . & Waldron, K. W. 2014. Bioethanol production from pineapple wastes. *Journal of Food Research*, 3(4), 60.

Turcotte, D. L., & Schubert, G. 2002. *Geodynamics* (456pp.). Cambridge: Cambridge University Press.

Tyagi, R. D., & Ghose, T. K. 1980. Batch and multistage continuous ethanol fermentation of cellulose hydrolysate and optimum design of fermentor by graphical analysis. *Biotechnology and Bioengineering*, 22(9), 1907–1928.

Vane, L. M. 2005. A review of pervaporation for product recovery from biomass fermentation processes. *Journal of Chemical Technology and Biotechnology*, 80(6), 603–629.

Verma, L. R., & Joshi, V. K. 2000. Postharvest technology of fruits and vegetables: An overview. In *Postharvest technology of fruits and vegetables: Handling, processing, fermentation, and waste management*. New Delhi: Indus Publishing Company.

W.E.Council. 2013. *World energy resources*. London: World Energy Council.

W.E.Council. 2016. *World energy resources 2016*. London: World Energy Council.

Wang, F. Q., Gao, C. J., Yang, C. Y., & Xu, P. 2007. Optimization of an ethanol production medium in very high gravity fermentation. *Biotechnology Letters*, 29(2), 233–236.

Westman, J. O., & Franzén, C. J. 2015. Current progress in high cell density yeast bioprocesses for bioethanol production. *Biotechnology Journal*, 10(8), 1185–1195.

Yergin, D., & Gross, S. 2012. Energy for economic growth: Energy vision update 2012; Industry agenda. *World Economic Forum*.

Zhang, P., Whistler, R. L., BeMiller, J. N., & Hamaker, B. R. 2005. Banana starch: Production, physicochemical properties, and digestibility—A review. *Carbohydrate Polymers*, 59(4), 443–458.

Zhang, Q., Wei, Y., Han, H., & Weng, C. 2018. Enhancing bioethanol production from water hyacinth by new combined pretreatment methods. *Bioresource Technology*, 251, 358–363.

4
The Role of Microbes in Biofuel Production

Arpan Kumar Basak, Sayantani M. Basak, Kazimierz Strzałka, and Pradip Kumar Chatterjee

CONTENTS

4.1	Introduction	65
4.2	Characteristic Features of Microbes in Biofuel Production	66
	4.2.1 Metabolic Pathways for Direct Application	67
	4.2.2 Tools for Indirect Application of Microbes	68
4.3	Microbes in Action	79
	4.3.1 Bacteria in Action	79
	4.3.2 Role of Fungi	79
	4.3.3 Algal Revolution	81
	4.3.4 Synthetic Microbes, Extremophiles and Beyond	82
4.4	Microbial Symbionts in Biofuel Production	84
	4.4.1 Bacteria and Fungi Symbionts	84
	4.4.2 Microalgae and Fungi Symbionts	85
	4.4.3 Microalgae and Bacteria Symbionts	86
4.5	Conclusion	88
References		88

4.1 Introduction

Biofuels are generated from renewable organic matter, although some microorganisms may produce them from inorganic sources (e.g. hydrogen); these are alternative means to decrease the dependency on fossil resources and reduced emission of greenhouse gases. Microbial activity generated on organic substrates produces a variety of biofuels, namely biogas, biobutanol, biodiesel, biohydrogen, and bioethanol (Peralta-Yahya & Keasling, 2010; Srivastava et al., 2020). Among the organic sources, sugars, starch, lignocellulosic biomass, and animal waste contribute as substrates in biofuel production. Even though biofuels are an emerging and eco-friendly resource, it is expensive by cost and time to achieve a higher magnitude of yield. Therefore, the optimization of biofuel production is crucial for effective yield. This is achieved through chemical and biochemical catalytic processes. The complex cell wall of plants is comprised of cellulose, hemicellulose, and lignin, often called the lignocellulosic complex. The plant's lifecycle depends on these complexes which shape the plant's structure, stability, and function. In biofuel production, these lignocellulosic biomasses and their remnants undergo physical and chemical treatment to expose the complex sugar content. To hydrolyze the complex polysaccharides into simple monosaccharides, the complex sugars undergo enzymatic degradation by a process called saccharification (Jönsson & Martín, 2016; Lakatos et al., 2019; Xue et al., 2020). Microbes use these monosaccharides to produce volatile organic compounds by a process called fermentation. The direct role of microbes can be justified at the stage of the use of monosaccharide units or the process of fermentation. However, the functions of the microbes are not limited to fermentation. Microbes and their metabolic tools play an important role in processing biomass before fermentation (Fernandes & Murray, 2010; Wood et al., 2013). The indirect role, on the other hand, is the application of microbes and their metabolic tools before fermentation (Bhatia et al., 2020; Cantero et al., 2019; Jönsson & Martín, 2016; Passos et al., 2014). Microbes found in extreme temperature and pH conditions are often considered in these intermediate processes of biofuel production (Barnard et al., 2010). Bacteria, protozoa, and fungi throughout evolution have developed a unique set of metabolic processes and biochemical strategies to encounter plants' complex defense systems to gain access to the carbon source of the plants (Abhilash et al., 2012; Schlaeppi & Bulgarelli, 2015). In addition to plant–microbe interactions, microbes have successfully colonized the alimentary canal of animals throughout their lifecycle (Brune, 2014; Jovel et al., 2016). Herbivorous animals feed on plants that are rich in lignocellulosic fibers; animals rely on their gut microbiota to digest the complex biomass to produce energy. The gut itself gives us an understanding of the bioprocess plant in microscale (Donaldson et al., 2016). The digestive system of the animals initiates the pretreatment processes by acid hydrolysis and saccharification by amylase. The role of microbes here is to remove the lignin content from the biomass through their metabolic processes (S. Lee et al., 2019). These metabolic processes provide a toolbox that is implemented to improve biofuel production (Peralta-Yahya & Keasling, 2010). Also, the symbiosis of these organisms across the kingdom provides a necessary toolbox for improving biofuel production.

This chapter summarizes the role of microbes in production of biofuels. Microbes belonging to kingdom bacteria, protozoa, fungi, and several groups of microalgae have direct roles in biofuel production and indirect roles in the improvement of yield. A comprehensive explanation is summarized from the studies conducted over two decades.

4.2 Characteristic Features of Microbes in Biofuel Production

Taking advantage of growing demand for fuel creates an opportunity for renewable fuel (such as biofuels) with gradual diminishing fossil fuels (Pandey et al., 2012; Srivastava & Jaiswal, 2016). Agricultural industrial–derived (Sahaym & Norton, 2008) lignocellulosic biomass (Srivastava et al., 2015a, 2015b) is in high demand among researchers, owing to its ready availability. However, biofuel is not without its limitations. High cellulase enzyme costs (Bhalla et al., 2013; Srivastava et al., 2016), covering almost 40% of total cost, coupled with variable stability of cellulase at conflicting pH (Yeoman et al., 2010), calls for methods which would elevate productivity, as well as bring stability of cellulase enzyme Srivastava et al., 2016).

Among archaebacteria, bacteria, protista, microalgae, and fungi are the microbial kingdom that has the metabolic functions necessary for biofuel production from organic and inorganic (e.g. hydrogen production from water by cyanobacteria) sources. The prokaryotes from archaebacteria can withstand extreme temperature, salinity, and pH conditions, thereby contributing to the step of pretreatment. The eubacteria are often used to genetically engineer the metabolic processes for enhanced production of biofuels (Santos-Fo et al., 2011; Y. Yang et al., 2019). However, eukaryotes can degrade complex fibers from biomass. Both of these members can use carbon sources and produce components of biofuels. Importantly, cyanobacteria and other eukaryotes such as members of microalgae have developed the ability to initiate photo-fermentation that contributes to the next-generation biofuel (Brodie et al., 2017; Parmar et al., 2011).

Microbes across the kingdom have developed these tools to survive environmental stress conditions and thrive to obtain nutrients. These members have successfully colonized the animal gut and plants root compartment, forming a symbiosis. For example, termites accommodate a plethora of gut microbes that symbiotically coexist during their lifecycle (Brune, 2014). It has been shown by metagenomics and metatranscriptomics that these gut symbionts express genes encoding for enzymes like cellulase, glucosidase, nitrogenases, endoglucanases, and endoxylanases (Mathew et al., 2013). Among the members of the gut symbionts of the termites, it was observed that protozoan symbionts are involved in cellulose degradation in lower termites, but the same protozoan symbionts are involved in both cellulosic and hemicellulosic degradation in higher termites (Bastien et al., 2013).

Similar to gut symbionts, there is a diversity of microbes in the root compartment of the plants (Berendsen et al., 2012; Sugiyama, 2019). These microbes accommodate the rhizosphere and the endosphere (Abhilash et al., 2012). In addition to plant–microbe interaction, these members of the microbiota have enriched both pathogens and beneficial microbes in their lifecycle. Members of the fungi and bacteria symbiotically infiltrate plants' defense mechanisms to access their nutrients. Studies show that among the plant-associated microbes, pathogenic endophytes such as arbuscular fungi have ligninolytic enzymes and bacteria having cellulolytic enzymes. On the other hand, rhizobacteria species and few members from the fungal kingdom benefit plants by nutrient translocation and acquisition. Integration of the cellulolytic processes, ligninolytic processes, and fermentation from the existing knowledge of the gut- and root-associated microbes may be productive in producing biofuel (Parisutham et al., 2014; Xue et al., 2020).

Not only in the complex organism but also in simple eukaryotes, the symbiosis is beneficial from the context of biofuel production. Studies showed that symbiosis between microalgae and heterotrophs resulted in improved yields in biodiesel production. Bacterial and microalgal symbiosis has proven its efficiency in biofuel production. Other symbiotic factors like vitamins, secondary metabolites, and siderophores—or transporters of iron between these kingdoms—contribute to the improvement of classes of biofuels. To improve the yield efficiency, it has been suggested to implement strain-specific interactions along with optimum physiochemical factors. Further, the direct and indirect application of the metabolic pathways in microbes are discussed.

4.2.1 Metabolic Pathways for Direct Application

There is substantial demand for using C-source and produce an effective yield of biofuels. This is achieved by the anaerobic respiration process, also called fermentation. The reaction involved is the conversion of simple monosaccharide units like hexose sugar C_6 to volatile compounds such as ethanol and carbon dioxide. Additionally, microbes can convert free fatty acids to oleaginous compounds that are commonly accommodated in the production of biodiesel. The fatty acids are bioconverted by transesterification process to produce wax esters. The end products of ethanol and fatty acid biosynthetic pathways undergo transesterification to produce the wax esters or biodiesel (Parawira, 2009).

The C_3-C_4 biosynthetic pathway is crucial for the production of advanced biofuels. *Clostridium* sp. converts an equal proportion of glucose into isopropanol C_3 alcohol. Similarly, conversion of 1M of glucose and 4M of NADH yields 1M of butanol. In the biosynthetic pathway, the major step is the conversion of pyruvate. The substrate pyruvate is converted into acetyl coenzyme A (acetyl-CoA), acetaldehyde, 2-ketoisovalerate, or threonine. In ethanol production, acetyl-CoA is converted to acetaldehyde and finally the end product. In the ABE pathway, acetyl-CoA is transformed to acetoacetyl-CoA and then consequently by butanol. This acetoacetyl-CoA can also be converted into fatty acids, forming alkanes and fatty acid esters. The substrate 2-ketoisovalerate gets converted to isobutanol and 3-methyl-1-butanol. Threonine is converted to 2-ketobutyrate and forms 1-propanol. Further, 2-ketobutyrate gets converted to 2-ketovalerate, and subsequently, 2-keto-3-methylvalerate produces 1-butanol and 2-methyl-1-butanol, respectively. However, the intermediate substrate 2-keto-3-methylvalerate is converted to 2-keto-4-methyl hexanoate, forming the end product 3-methyl-1-pentanol. The end product of these metabolic processes is common ingredients or components of biofuels. Studies have shown, by pathway reconstruction experiments in *E. coli* and *S. cerevisiae* from *Clostridium* sp., that there is an improvement in the production of advanced biofuels.

In addition to C_3-C_4 biosynthesis, the amino acid biosynthetic pathway generates keto acid intermediates that are converted into "fusel" alcohols in leucine, valine, isoleucine, phenylalanine, tryptophan, and methionine pathways. In yeast, these alcohols are produced by substantial keto acid decarboxylation to the aldehyde, which is preceded by lowering aldehyde to alcohol; this pathway is known as the "Ehrlich pathway." In *E. coli*, it was shown that a substantially higher yield of alcohols is obtained by reconstituting the Ehrlich pathway.

Another important metabolic pathway for consideration is the fatty acid biosynthetic pathway, where alcohols are produced by sequential reduction of fatty acid. Additionally, the reduction of fatty acid into aldehyde can produce alkanes and consequently decarboxylation. In the context of biodiesel production, fatty acids metabolize into esters and small alcohols by a process called esterification. The enzymatic bioconversion of esters of fatty acids is conducted by immobilized-extracellular-lipase or whole-cell-catalysis as described in Parawira (2009).

Metabolic pathways for the production of third-generation or next-generation biofuel deal exclusively with light energy. Third-generation biofuel is obtained from microalgae. The bioprocessing steps involved are: (a) fermenting the pretreated microalgae biomass; (b) dark fermenting the reserved carbohydrates; and finally, (c) photo-fermenting to bioethanol from carbon dioxide by light (Lakatos et al., 2019).

Lack of light induces dark fermentation (Heyer & Krumbein, 1991; Hirano et al., 1997), whereby amylase, the maximum reserved starch is hydrolyzed into sugars which are then converted by the process of glycolysis into pyruvate (Catalanotti et al., 2013). Species such as *Chlamydomonas moewusii*, *Chlamydomonas reinhardtii*, *Chlorella fusca*, and *Chlorogonium elongatum* can use intracellular polysaccharides for fermentation. These polysaccharides, including starch, may be transformed into pyruvate (which is an intermediate compound) subsequently, followed by other metabolic end products such as H_2 and CO_2, acetate, formates, lactate, ethanol, and glycerol (Grossman et al., 2007; Mus et al., 2007). Additionally, in the fermentation process, pyruvate serves as a substratum for acetyl-CoA, which then converts to acetate or ethanol by generating ATP, as stated in *Chlamydomonas* (Hemschemeier & Happe, 2005; Mus et al., 2007; Magneschi et al., 2012).

The ethanol-producing process directly following the photosynthetic route is called photo-fermentation. Cyanobacteria has "photofermentative'" or "photanol" routes, coordinated by enzymes such as pyruvate decarboxylase and alcohol dehydrogenase II (Deng & Coleman, 1999; Dexter et al., 2015). The photo-fermentation process to produce bioethanol on genetically engineered cyanobacterial strains (*Synechocystis* sp. PCC 6803, *Synechococcus* sp. PCC 7942, *Synechococcus* sp. PCC 7002; Pasteur Culture Collection, Paris) has been done recently (Dexter & Fu, 2009; Gao et al., 2012; Dienst et al., 2014). The ethanol yield obtained was 2.5 g/L and 4.5 g/L, which repressed the growth of the strains (Kämäräinen et al., 2012). To observe the metabolism in *Synechocystis* sp. PCC 6803, proteomic analysis was conducted to have an understanding of the operation of the pathway incorporated into the strains (Song et al., 2014; Pade et al., 2017).

Therefore, microbes can use not only sugars but other macromolecules for the production of biofuel components. This suggests the list of metabolic processes in microbes that are important for the direct role of microbes in biofuel production.

4.2.2 Tools for Indirect Application of Microbes

To obtain monosaccharides from the source of complex polysaccharides, the complex needs to go through pretreatment and liquefaction processes. Physical and chemical pretreatments are often used before liquefaction and fermentation at higher temperature conditions. These processes are energy-demanding and cost-inefficient, and result in compounds that later hinder the fermentation process. An alternative approach is to combine liquefaction and fermentation by using the high-end metabolic tools of microbes. This approach is termed as consolidated bioprocessing (CBP). Additionally, products obtained from microbes include biocatalysts that are capable of removing polysaccharides from the phenolic compounds (Parisutham et al., 2014).

The process of lignin removal from biomass is termed as delignification. One challenge before fermentation is delignification (Sindhu et al., 2016). Application of enzymes during pretreatment processes has shown a substantial decrease in lignin, cellulose, and hemicellulose content (Bastien et al., 2013; Masran et al., 2016; T.-Y. Wang et al., 2011). Lignin, an aromatic heteropolymer, results from polymerization of syringyl, guiacyl, and p-hydroxyphenyl units connected by β-aryl ether, heterocyclic, and biphenyl linkages (Pollegioni et al., 2015). It has been reported that both bacteria and fungi have developed strategies to evade the plant defense system by degrading lignin during evolution (Bilal et al., 2018; Fernández-Fueyo et al., 2014; S. Lee et al., 2019; Masran et al., 2016; Wan & Li, 2012).

In fungi, lignin degradation pathway is often triggered by the enzymes such as manganese peroxidases (MnP), lignin peroxidases (LiP), and versatile peroxidases (VP) that are members of the Class II peroxidases. These enzymes belong to the superfamily of heme peroxidases that are secreted exogenously (Fernández-Fueyo et al., 2014). Laccase is a polyphenol oxidase that catalyzes benzenediol into benzo semiquinone in the delignification process (Villalba et al., 2010). Intracellular and extracellular isoenzymes produces laccases; they differ in their state and as well as in the level of glycosylation

TABLE 4.1

Direct Application in Production of Biofuel Components

Kingdom	Strains	Biodiesel	Biobutanol	Bioethanol	Biohydrogen	Biogas	Reference
Bacteria	*Azotobacter vinelandii*	+	+	+	+	+	(Knutson et al., 2018; Pagliano et al., 2017; Murugesan et al., 2017; Yoneyama et al., 2015; Segura et al., 2003; Gama-Castro et al., 2001)
	Cyanobacteria leptolyngbya	+					(Ellison et al., 2019)
	Klebsiella oxytoca	+	+	+	+		(Yu et al., 2017; Cho et al., 2013; Cao et al., 1997; Burchhardt & Ingram et al., 1992)
	Pseudomonas cepacia	+	+	+	+		(Encinar et al., 2019; Li et al., 2018; Salis et al., 2005; Tanaka et al., 1999)
	Pseudomonas fluorescens	+	+			+	(Kovacs et al., 2016; Lima et al., 2015; Miller et al., 1973)
	Saccharophagus degradans			+			(Brognaro et al., 2016)
	Spirulina maxima				+	+	(Ananyev et al., 2008, 2012; Carrieri et al., 2011; Samson & Leduy, 1982)
	Spirulina platensis			+	+	+	(Rempel et al., 2019; Onwudili et al., 2013; Morsy, 2011; Matsudo et al., 2011)
	Thermoanaerobacter pseudoethanolicus		+	+	+		(Scully et al., 2020; Scully et al., 2019; Chades et al., 2018)
	Thermus thermophilus	+				+	(Park et al., 2012; Walfridsson et al., 1996)
	Zymomonas mobilis	+	+	+	+	+	(Todhanakasem et al., 2020; Palamae et al., 2020; Geng et al., 2020; Xia et al., 2013; Duan et al., 2019; Wang et al., 2018c; Huang et al., 2018; He et al., 2014; Xiao et al., 2016)
Fungi	*Alternaria* sp.	+					(Dey et al., 2011; You et al., 2005)
	Aspergillus oryzae	+	+	+	+		(Li et al., 2016; Jin et al., 2013 Jahnke et al., 2016; Han, 2016; Han, 2015; Tamano et al., 2015; Bátori et al., 2015; Ahmad et al., 2015; Adachi et al., 2013; Adachi et al., 2011; Hama et al., 2008; Kotaka et al., 2008; Dhananjay & Mulimani, 2009; Carlsen et al., 1996)
	Candida curvata			+			(Evans & Ratledge, 1983)
	Candida intermedia	+	+	+			(Moreno et al., 2019; Wu et al., 2018; Brexó et al., 2018; Tanino et al., 2012; Olofsson et al., 2011; Runquist et al., 2010)

(*Continued*)

TABLE 4.1
Direct Application in Production of Biofuel Components (Continued)

Kingdom	Strains	Biodiesel	Biobutanol	Bioethanol	Biohydrogen	Biogas	Reference
	Candida lipolytica	+	+		+		(Souza et al., 2016; Karatay & Dönmez, 2010; Roostita & Fleet, 1996; Hussein & Jwanny, 1975)
	Candida rugosa	+	+	+		+	(Cea et al., 2019; Zeng et al., 2017; Bakkiyaraj et al., 2016; Kuo et al., 2015; Lee et al., 2010, 2011, 2008)
	Candida tenuis	+		+			(Bessadok et al., 2019; Veras et al., 2017; Eixelsberger et al., 2013; Toivola et al., 1984)
	Colletotrichum sp.	+	+	+			Dey et al., 2011)
	Cryptococcus curvatus	+		+	+	+	(Tang et al., 2020; Filippousi et al., 2019; Chatterjee & Venkata Mohan, 2018; Zhou et al., 2017; Li et al., 2017b; Carota et al., 2017; Jiru et al., 2016; Gong et al., 2016; Gong et al., 2015; Liang et al., 2014a; Gong et al., 2014; Seo et al., 2013; Ryu et al., 2013; Thiru et al., 2011; Takishita, 2010; Lian et al., 2010; Takishita et al., 2006; Greenwalt et al., 2000)
	Cryptococcus curvatus DSM 70022	+				+	(Signori et al., 2016)
	Cryptococcus curvatus MTCC 2698	+					(Chatterjee & Venkata Mohan, 2018)
	Cryptococcus curvatus MUCL 29819	+					(Huang et al., 2019; Huang et al., 2018)
	Cryptococcus sp.	+	+	+		+	Filippousi et al., 2019; Carota et al., 2017; Gong et al., 2016; Sung et al., 2014; Tanimura et al., 2014; Shobha et al., 2011; Ravella et al., 2010; Lian et al., 2010; Salama et al., 2001)
	Cryptococcus vishniaccii	+				+	(Deeba et al., 2016)
	Cunninghamella echinulata	+	+	+	+	+	(Kosa et al., 2018; Ling et al., 2015; Hassan et al., 2015; Foster et al., 1991)
	Gliocladium roseum (NRRL 50072)	+				+	(Strobel et al., 2010)
	Kluyveromyces lactis	+	+	+	+	+	(Yamaoka et al., 2014; Oda et al., 2013; Jäger et al., 2011; Kim et al., 2010)

Lipomyces starkeyi	+	+	(Spagnuolo et al., 2019; Brandenburg et al., 2018)
Microsphaeropsis sp.	+		(Holler et al., 1999)
Mortierella alliacea strain YN-15	+		(Tanaka et al., 2002)
Mortierella alpina	+		(Kosa et al., 2018; Jin et al., 2008)
Mortierella isabellina	+		(Papanikolaou & Aggelis, 2019; Papanikolaou et al., 2017)
Mucor circinelloides	+		(Reis et al., 2019; Vellanki et al., 2018; Lübbehüsen et al., 2004)
Mucor rouxii	+	+	(Abasian et al., 2020; Muniraj et al., 2015; Bernard et al., 1982)
Naganishia liquefaciens NITTS2	+		(Selvakumar & Sivashanmugam, 2018)
Pichia stipitis	+	+	(Kityo et al., 2020; Antonopoulou, 2020; Travaini et al., 2016; Rumbold et al., 2009)
Rhizomucor miehei	+		(Wang et al., 2020b; Rahman et al., 2018; Ilmi et al., 2018; Huang et al., 2015; Huang et al., 2014; Calero et al., 2014; Sánchez et al., 2000; Oliveira et al., 2000)
Rhizopus delemar	+		(Yan et al., 2016; Bora et al., 2016)
Rhizopus oryzae	+	+	(Canet et al., 2017; Duarte et al., 2015; Meussen et al., 2012; Thongchul et al., 2010; Méndez et al., 2007; Ben Salah et al., 2007; Taherzadeh et al., 2003)
Rhodosporidium kratochvilovae HIMPA1	+	+	(Patel et al., 2015; Patel et al., 2014)
Rhodosporidium toruloides Y4	+		(Thliveros et al., 2014; Zhao et al., 2011; Zhao et al., 2010)
Rhodotorula glutinis	+		(Zhang et al., 2020; Vasconcelos et al., 2018; Kot et al., 2016; Tinoi & Rakariyatham, 2016; Cescut et al., 2014)
Saccharomyces cerevisiae	+	+	(Wu et al., 2020; Azambuja et al., 2020; Favaro et al., 2019; Krivoruchko et al., 2013; Waks & Silver, 2009)
Sporidiobolus pararoseus	+		(Li et al., 2019, 2021; Chaiyaso et al., 2018)

(*Continued*)

TABLE 4.1

Direct Application in Production of Biofuel Components (Continued)

Kingdom	Strains	Biodiesel	Biobutanol	Bioethanol	Biohydrogen	Biogas	Reference
	Sporidiobolus pararoseus	+					(Chaiyaso et al., 2018)
	Talaromyces emersonii			+			(Figueroa-Espinoza, 2002)
	Trichosporon cutaneum	+	+	+			(Zhao et al., 2017; Wang et al., 2016b, 2017; Chen et al., 2016; Xiong et al., 2015; Hu et al., 2011; Qi et al., 2016)
	Trichosporon fermentans	+					(Liu et al., 2017; Huang et al., 2012; Huang et al., 2009; Zhu et al., 2008)
	Yarrowia lipolytica	+	+	+		+	(Yu et al., 2020; Ng et al., 2020; Wang et al., 2020a; Gęsicka et al., 2020; Wu et al., 2019; Yu et al., 2018; Guo et al., 2018)
Protista	*Ankistrodesmus* sp.	+					(Sassi et al., 2017; Yee, 2016; He et al., 2015; Do Nascimento et al., 2013)
	Botryococcus braunii	+		+	+		(Al-Hothaly, 2018; Sasaki et al., 2017; Ruangsomboon et al., 2017; Kavitha et al., 2016; Watanabe et al., 2015; Ruangsomboon, 2015; Hidalgo et al., 2015; Choi et al., 2013; Samorì et al., 2010; Ciudad et al. 2014)
	Chaetoceros calcitrans	+		+			(Şirin et al., 2015; Kwangdinata et al., 2014; Ríos et al., 2013)
	Chaetoceros muelleri	+					(Lin et al., 2018; Pikula et al., 2018; Wang et al., 2014; Gao et al., 2013)
	Chlamydomonas reinhardtii	+	+	+	+		(Yaisamlee C, Sirikhachornkit, 2020; Xu & Pan, 2020; Slocombe et al., 2020; Grechanik et al., 2020; Fakhimi et al., 2020; Varaprasad et al., 2019; Nagy et al. 2019; Ban et al., 2018, 2019; Yagi et al., 2016; Furuhashi et al., 2016; Eilenberg et al., 2016; Hung et al., 2016; Toepel et al., 2013; Pinto et al., 2013; Volgusheva et al., 2013; Chochois et al., 2009; Nguyen et al., 2009; Tolstygina et al., 2009; Rupprecht, 2009; Hemschemeier & Happe, 2005)
	Chlorella emersonii	+		+	+	+	(Smith-Bädorf et al., 2013)
	Chlorella protothecoides	+		+	+	+	(He et al., 2016; Gülyurt et al., 2016; Rismani-Yazdi et al., 2015; Muradov et al., 2015; Li et al., 2015; Krzemińska et al., 2015; Fei et al., 2015; Darpito et al., 2015; Ahmad et al., 2015; Santos et al., 2014; Li et al., 2013; Wen et al., 2013; Li et al., 2011a; Xiong et al., 2008; Li et al., 2007; Xu et al., 2006; Miao & Wu, 2006)

Species				References
Chlorella pyrenoidosa	+		+	+	(Ding et al., 2020; Singh et al., 2019b; Liu et al., 2019; Ratnapuram et al., 2018; Ahmad et al., 2018; Han et al., 2016; Reyna-Martínez et al., 2015; Cao et al., 2013; Wang et al., 2011; Tang et al., 2011; Li et al., 2011a; Kojima & Lin, 2004)
Chlorella sorokiniana	+		+	+	(Arora et al., 2020; Eladel et al., 2019; De Francisci et al., 2018; Córdova et al., 2018; Hena et al., 2015; Parsaeimehr et al., 2015; Ramanna et al., 2014; Ma et al., 2013)
Chlorella sp.	+	+	+	+	Du et al., 2019; Zhou et al., 2018; Li et al., 2018
Chlorella vulgaris	+	+	+	+	(Sakarika & Kornaros, 2019; Mohd-Sahib et al., 2017; Kalhor et al., 2017; Wang et al., 2013; Rashid et al., 2011; Park et al., 2014; Park et al., 2021)
Chlorella zofingiensis	+		+	+	(Zhou et al., 2018; Feng et al., 2012; Ahmad et al., 2015; Liu et al., 2012a; Feng et al., 2011; Zhu et al., 2014b; Liu et al., 2011; Zhu et al., 2014a)
Chlorococcum sp.	+		+	+	(Varaprasad et al., 2019; Rehman & Anal, 2018; Lv et al., 2018; Sassi et al., 2017; Pan et al., 2017a; Feng et al., 2016; Feng et al., 2014; Ummalyma & Sukumaran, 2014; Liu et al., 2013; Wu et al., 2012; Harwati et al., 2012; Halim et al., 2011)
Crypthecodinium cohnii	+		+	+	(de Swaaf et al., 2003; Ishida et al., 1990)
Dunaliella primolecta	+			+	(Wu et al., 2012)
Dunaliella salina	+	+			(Ahmed et al., 2017)
Dunaliella sp.	+	+	+	+	(Talebi et al., 2015; Lakaniemi et al., 2011)
Dunaliella tertiolecta	+		+	+	(Lin & Lee, 2017; Chen et al., 2012)
Euglena gracilis		+	+	+	(Takemura et al., 2020; Munir et al., 2002; Ono et al., 1995)
Haematococcus pluvialis	+		+		(Molino et al., 2018; Otero et al., 2017; Lei et al., 2012)
Isochrysis galbana	+		+		(Santos et al., 2014; Poisson & Ergan, 2001)
Isochrysis sp.	+		+		(Gnouma et al., 2018; O'Neil et al., 2021)
Nannochloris sp.	+				(Jazzar et al., 2015)
Nannochloropsis oculata	+		+		Şirin et al., 2015; Duman et al., 2014

(*Continued*)

TABLE 4.1
Direct Application in Production of Biofuel Components (Continued)

Kingdom	Strains	Biodiesel	Biobutanol	Bioethanol	Biohydrogen	Biogas	Reference
	Nannochloropsis sp.	+	+	+	+		(He et al., 2019; Qiu et al., 2019; Rodriguez-López et al., 2020; Duman et al., 2014; Navarro López et al., 2016; Pan et al., 2017b)
	Neochloris oleoabundans	+				+	(Kwak et al., 2015; Yoon et al., 2015)
	Nitzschia sp.	+		+		+	(Wang et al., 2017; Montalvão et al., 2016; Duong et al., 2015)
	Phaeodactylum tricornutum	+				+	(Butler et al., 2020; Simonazzi et al., 2019; Libralato et al., 2011)
	Porphyridium cruentum	+		+			(Hu et al., 2018; Patel et al., 2013)
	Porphyridium purpureum	+				+	(Jiao et al., 2018; Sato et al., 2017; Jiao et al., 2017)
	Pythium irregulare	+		+			(Liang et al., 2012; Athalye et al., 2009)
	Scenedesmus obliquus	+		+	+	+	(Arun et al., 2020; Choix et al., 2017; Wirth et al., 2015; Wu et al., 2014; Chu et al., 2014; Papazi et al., 2012; Miranda et al., 2012; Mandal & Mallick, 2012)
	Scenedesmus quadricauda	+				+	(Song & Pei, 2018; Kandimalla et al., 2016; Musharraf et al., 2012)
	Scenedesmus sp.	+		+	+	+	(Wang et al., 2019d; Ramos-Ibarra, 2019; Alam et al., 2019; Ban et al., 2018; Mandotra et al., 2016; Chu et al., 2014; Wu et al., 2013; Miranda et al., 2012; Mandal & Mallick, 2012)
	Schizochytrium sp.	+				+	(Wang et al., 2018b, 2019a; Ethier et al., 2011)
	Skeletonema costatum	+				+	(Gao et al., 2019)
	Skeletonema sp.	+				+	(Johansson et al., 2019; Bertozzini et al., 2013)
	Tetraselmis sp.	+		+	+	+	(Han et al., 2018; Ji et al., 2015; Guo et al., 2017; Santos-Ballardo et al., 2015; Teo et al., 2014; Ji et al., 2014; Holcomb et al., 2011; Bhosale et al., 2009)
	Tetraselmis suecica	+				+	(Kassim et al., 2017; Heo et al., 2015; Santos-Ballardo et al., 2015; Go et al., 2012)
	Thalassiosira pseudonana	+			+		(Sabia et al., 2018; Sachs & Kawka, 2015; Zendejas et al., 2012)

(Pollegioni et al., 2015). It has been reported that in the absence of a mediator, white-rot fungi *Panus tigrinus* and *P. radiata* produce laccase that catalyzes and oxidates nonphenolic β-1 lignin model compounds (Kirk & Moore, 1972; Villalba et al., 2010; H. Yu et al., 2009).

Similarly, in bacteria, CotA laccase from *Bacillus* sp. is the frequently representative enzyme. These are coated proteins in the endospore that show great temperature stability. Additionally, β-O-4 aryl -ether cleaving enzyme system involves a three-step process coordinated by LigF (β-etherase), LigD (C-α-dehydrogenase) and LigG (glutathione lyase) (Melo et al., 2016; S. Lee et al., 2019). This enzymatic system is specific to the cleavage of β-aryl ethers. Additionally, enzymes such as perhydrolase, extradiol dioxygenase, peroxidases and mycelium-associated dehydrogenases, and reductases are associated in lignin depletion. There are few bacterial laccases studied, with the current progress in omics across diverse bacterial species. The CotA laccase from *Bacillus subtilis*, shows high thermostability (Hullo et al., 2001; Martins et al., 2002), determining the crystal structure of this protein (Enguita et al., 2003). Later, *E. coli* was used to express CotA laccase from *Bacillus licheniformis* and *Bacillus pumilus*. Therefore, delignification is achieved by using microbial enzymatic processes.

Microbes have metabolic processes to degrade the polysaccharide units, primarily cellulose and hemicellulose, into simple sugars. This is achieved by saccharolytic microbes that are capable of hydrolyzing complex sugar into C_6 and C_5 components. The cellulose and hemicellulose degradation are carried forward by enzymes, cellulase, and hemicellulase, whereas endo-1,3-β-xylanases (EC 3.2.1.8), endo-1,4-β-xylanases, β-xylosidases (EC 3.2.1.37), α-glucuronidases, acetylxylanases, and α-arabinofuranosidases make up xylanase (Beg et al., 2001). Pretreatment of cellulosic, along with hemicellulosic, elements is the first step of bioconversion, together with reduced cost of biofuel production (Alvira et al., 2010). The conversion rate and carbon content are greatly affected due to the introduction of microbes during pretreatment. After pretreatment, the saccharification process is initiated by enzymatic hydrolysis of glycosidic linkages using cellulases and hemicellulases, thereby releasing monosaccharides (Bhalla et al., 2013; Rawat et al., 2014). Commercial cellulases and xylanases are used in pure form or via microbial sources directly by industries. However, it is not economical to produce these enzymes by industrial processes, due to cost constrains. It has been reported that bacteria *Clostridium* sp. produce a substantial amount of cellulases or xylanases (Thomas et al., 2014). The members of *Clostridium* have an extracellular enzymatic complex called cellulosome that digests cellulose. Multi-functional cellulases contain mannase and xylanase units. These enzymes form a complex by forming intramolecular synergism and show enhanced activity. On the other hand, microbes have discrete, multi-functional, and multi-enzyme complexes called xylanosome (McClendon et al., 2012; Srikrishnan et al., 2013). Xylanosome is absent in *Clostridium* sp., and some reports suggest that xylanosome are often found to be associated with the cellulosome (Han et al., 2003; Moraïs et al., 2010; Nölling et al., 2001). The major role of cellulases and xylanases is to conduct glucanase-like activity. The microalgae cell walls are made up of cellulose and hemicellulose, and hydrolytic enzymes are used to convert these complex structures in between the pretreatment (Córdova et al., 2018, 2019; Sato et al., 2004). There is a potential application of microbes in enhancement of cellulase production and maintain the stability.

Other enzymes—such as α-amylase, protease, and lipases—are used to hydrolyze components that are not required during the process of fermentation. These enzymes are often used when the biomass is from animal origin (Liu et al., 2013; Sanchez Rizza et al., 2017).

With regards to biofuel production from animal waste, the major source of biomass is food waste or municipal waste, and the fuel obtained is mostly biogas. The biomass undergoes physiochemical or enzymatic pretreatment, followed by solid-state fermentation, thereby yielding a substantial amount of biogas (N. Li et al., 2018; Lopes et al., 2019). Such biomass is composed of complex macromolecules including protein structures, oil and grease, lignin, and complex carbohydrate structures. Therefore, along with cellulase and hemicellulase, amylase, protease, lipase, and pectate lyase are used. The latter mentioned enzymes degrade the complex starch, protein fibers, lipid polymers, and pectins (Sanchez Rizza et al., 2017). Pretreatment of biomass using the optimum combination of these enzymes results in enhanced anaerobic bioconversion to produce a substantial amount of biogas.

TABLE 4.2
Indirect Application in Pretreatment of Biomass Prior to Bioconversion

Kingdom	Strains	Lignin Degradation	Saccharification	Animal Waste Degradation	Reference
Bacteria	*Azotobacter vinelandii*	+			(Zhang et al., 2004)
	Klebsiella oxytoca		+	+	(Edwards & Doran-Peterson, 2012; Doran et al., 1994)
	Pseudomonas cepacia	+			(Odier & Rolando, 1985)
	Pseudomonas fluorescens	+		+	(Granja-Travez & Bugg, 2018; Rahmanpour & Bugg, 2015; Xu et al., 2015)
	Saccharophagus degradans	+	+		(Lee et al., 2017b; Park et al., 2014)
	Spirulina maxima			+	(Kim et al., 2016; Wu and Pond, 1981)
	Spirulina platensis	+	+	+	(Rempel et al., 2019; Lu et al., 2019; Halmemies-Beauchet-Filleau et al., 2018; Wang et al., 2013; Chaiklahan et al., 2010)
	Thermus thermophilus		+		(Aulitto et al., 2018; Wu et al., 2014)
	Zymomonas mobilis	+	+		(Sun et al., 2020; Gu et al., 2015; Zhao et al., 2014; Yang et al., 2014; dos Santos et al., 2010)
Fungi	*Alternaria* sp.	+		+	(Jurado et al., 2014; Ahmad et al., 2013; Zhang et al., 2011; Song et al., 2010; Vibha et al., 2005; Funnell-Harris et al., 2012)
	Aspergillus oryzae	+	+		(Yasui et al., 2020; Oguro et al., 2019; Bhardwaj et al., 2019; Zhang et al., 2015; Lin et al., 2014; Guo et al., 2014; Udatha et al., 2012; Gibbons et al., 2012; Chang et al., 2012; Lin et al., 2010; Aachary & Prapulla, 2009)
	Candida intermedia	+	+		(Moreno et al., 2019; Wu et al., 2018; Ren et al., 2016; Fonseca et al., 2011)
	Candida rugosa	+		+	(Park et al., 2015; Yang et al., 2013; Gomes et al., 2005)
	Colletotrichum sp.	+	+		(Chakraborty et al., 2019; Tomas-Grau et al., 2019; Naziya et al., 2019; Muslim et al., 2019; Carvalho et al., 2019; Zimbardi et al., 2013; Stein et al., 1992)
	Cryptococcus curvatus	+	+		(Tang et al., 2020; Gong et al., 2016; Liang et al., 2014a; Gong et al., 2014; Greenwalt et al., 2000)
	Cryptococcus curvatus MTCC 2698	+			(Chatterjee & Venkata Mohan, 2018)

Kingdom	Strains	Lignin Degradation	Saccharification	Animal Waste Degradation	Reference
	Cryptococcus sp.	+			(Liang et al., 2014a; Gong et al., 2014; Singhal et al., 2013; Greenwalt et al., 2000)
	Kluyveromyces lactis		+		(Fonseca et al., 2008; Drozdíková et al., 2015)
	Lipomyces starkeyi	+	+	+	(Gou et al., 2020; Blomqvist et al., 2018; Huang et al., 2011)
	Mortierella isabellina	+		+	(Ruan et al., 2015; Zeng et al., 2013)
	Mucor circinelloides	+	+		(Qiao et al., 2018; Kato et al., 2013)
	Pichia stipitis	+			(Sunwoo et al., 2018; Kashid & Ghosalkar, 2018; Lee et al., 2009; Rudolf et al., 2008)
	Rhizomucor miehei		+	+	(Nagao et al., 2011)
	Rhizopus delemar		+		(Scholz et al., 2018)
	Rhizopus oryzae	+	+	+	(Yin et al., 2020; Scholz et al., 2018; Zhang et al., 2016; Sellami et al., 2013; Saito et al., 2012; Deng et al., 2012)
	Rhodotorula glutinis	+			(Matsuo et al., 2000)
	Saccharomyces cerevisiae		+	+	(Besada-Lombana & Da Silva, 2019; Lee et al., 2013a; Edwards & Doran-Peterson, 2012)
	Talaromyces emersonii	+	+		(Sun et al., 2018; Rahikainen et al., 2013)
	Trichosporon cutaneum	+	+	+	(Hu et al., 2018; Wang et al., 2016c; Chen et al., 2016; Liu et al., 2012b; Zheng et al., 2005)
	Trichosporon fermentans	+	+		(Shen et al., 2018; Zhan et al., 2013)
	Yarrowia lipolytica	+	+	+	(Song et al., 2017; Pokora et al., 2017; Jin & Hd Cate, 2017; Ryu et al., 2015; Tsigie et al., 2013; Wang et al., 2019c; Yano et al., 2008)
Protista	*Botryococcus braunii*	+			(Ciudad et al., 2014)
	Chlamydomonas reinhardtii		+		(Hognon et al., 2014; Frangville et al., 2012; Lee et al., 2011; Choi et al., 2010)
	Chlorella protothecoides			+	(Joe et al., 2015; Espinosa-Gonzalez et al., 2014)
	Chlorella pyrenoidosa	+	+	+	(Li et al., 2017b; Gai et al., 2015; Wang et al., 2012; Xia et al. 2013)
	Chlorella sorokiniana	+		+	(Bui et al., 2016)
	Chlorella sp.	+	+	+	(J.-Y. Park et al., 2021; Abd Ellatif et al., 2021; Li et al., 2017a; Lee et al., 2017b; Wahal et al., 2016; Bui et al., 2016; Joe et al., 2015; O. K. Lee et al., 2015a; Zhu et al., 2014a; Xia et al., 2013; Cheng et al., 2013; Lee et al., 2012)

(*Continued*)

TABLE 4.2
Indirect Application in Pretreatment of Biomass Prior to Bioconversion (Continued)

Kingdom	Strains	Lignin Degradation	Saccharification	Animal Waste Degradation	Reference
	Chlorella vulgaris	+	+	+	(Abd Ellatif et al., 2021; Lee et al., 2011; Park et al., 2009; Park et al., 2014; Park et al., 2021)
	Chlorella zofingiensis	+		+	(Zhu et al., 2012; Zhu et al., 2015)
	Crypthecodinium cohnii	+			(Karnaouri et al., 2020)
	Dunaliella salina			+	(Tafreshi & Shariati, 2009)
	Dunaliella sp.		+	+	(Lee et al., 2013; Tafreshi & Shariati, 2009)
	Dunaliella tertiolecta		+	+	(Harbi et al., 2017; Lee et al., 2013; Chen et al., 2012)
	Euglena gracilis			+	(Zhu & Wakisaka, 2020)
	Haematococcus pluvialis		+		(Amado & Vázquez, 2015)
	Nannochloropsis sp.	+			(Wang et al., 2017)
	Neochloris oleoabundans			+	(Olguín et al., 2015)
	Phaeodactylum tricornutum	+			(Libralato et al., 2011)
	Porphyridium cruentum		+		(Kim et al., 2017)
	Pythium irregulare			+	(Liang et al., 2011)
	Scenedesmus obliquus			+	(Oliveira et al., 2019; Ferreira et al., 2018)
	Scenedesmus quadricauda			+	(Kim et al., 2016)
	Scenedesmus sp.		+	+	(Oliveira et al., 2019; Pancha et al., 2016)
	Schizochytrium sp.	+			(Halmemies-Beauchet-Filleau et al., 2018; Liang et al., 2010)
	Tetraselmis sp.	+			(Antúnez-Argüelles et al., 2020; Thorsson et al., 2008; Selck et al., 2005)
	Tetraselmis suecica			+	(Slooten et al., 2015)

4.3 Microbes in Action

4.3.1 Bacteria in Action

Bacteria contribute largely in biofuel production. Therefore, they are economically the most important species in the planet. These species have metabolic pathways that specifically have a key function in biofuel production. Among the bacterial species, *Clostridium* and *E. coli* are the most important organisms in biofuel production. In nature, *Clostridium* species produce C_3-C_4 alcohols that are potential components of biofuels. It has been reported that *Bacillus macerans, Bacteroides polypragmatus* NRCC 2288, *Erwinia chrysanthemi* B374, and DMS 1574 are bacterial strains that are capable of fermenting pentose sugar such as xylose (Robak & Balcerek, 2018). However, the metabolic pathways governing the biofuel production are often overexpressed by using genetic engineering techniques.

In the context of next-generation biofuel, *Cyanobacteria* sp. can conduct photo-fermentation and are capable of performing photosynthesis with oxygen as a byproduct (Hagemann & Hess, 2018; B. Wang et al., 2018a). Initially, their application was overlooked for biofuel production in the U.S. Aquatic Species Program, as the majority do not naturally collect stored lipids as oleaginous eukaryotic microalgae in the form of triacylglycerol (TAG) (Sheehan et al., 1998). Nevertheless, since 2009, strains of cyanobacteria were developed as unique framework for the production of biofuels and biochemicals, due to photosynthesis and consistent genetic structures (Atsumi et al., 2008; Dexter & Fu, 2009; Lindberg et al., 2010; Angermayr et al., 2015). Genetically engineered strains of cyanobacteria are capable of producing compounds directly from CO_2, regardless of the presence of fermentable sugars and fertile land (Lai & Lan, 2015). The last decade has observed engineered cyanobacteria to generate oleochemicals directly from CO_2 which was motivated by successful policies on fatty acids by *E. coli*. These efforts were the first steps toward engineering cyanobacteria for oleochemical production; however, as for commercial applications, further improvements in production are required. Recently, Eungrasamee et al. (2020) reported that lipids and free fatty acids significantly enhanced in genetically engineered *Synechocystis* sp. PCC 6803 overexpress genes involved in the Calvin-Benson-Bassham cycle. However, more intensive efforts are needed to further improve cyanobacterial lipids and fatty acid production in the future.

The indirect application of bacterial species belonging to genus *Streptomyces, Rhodococcus, Pseudomonas,* and *Bacillus* strains have the set of tools necessary for lignin decomposition (Iwasaki et al., 2020; S. Lee et al., 2019; F. Meng et al., 2014; Odier & Rolando, 1985; Zhu et al., 2017). Lignin deconstruction aided by co-culture *Klebsiella* and *Citrobacter* genus has been suggested to improve the efficiency (Salvachúa et al., 2015). Considerably, delignification by other bacterial strains was observed in *Cupriavidus basilensis, Novosphingobium, Pandoraea,* and *Norcadia,* species, among others (Haider et al., 1978; Chen et al., 2012; Shi et al., 2013; Si et al., 2018). Additionally, *Pandoraea* strains were used to improve pretreatment of biomass through lignin depolymerization generated by their lignolytic ability (Kumar et al., 2015; D. Liu et al., 2018).

4.3.2 Role of Fungi

The fungal kingdom accommodates microbes belonging to *Saccharomyces* genus that are directly used for the bioconversion of sugars into biofuel components. There are classes of filamentous fungi that are used indirectly to improve the yield on implementing their metabolic tools before the fermentation process.

More than 1,500 species of *Saccharomyces* or yeasts in over 100 genera are now defined, among which some 30 species can collect more than 25% of lipids on the basis of dry weight (Beopoulos et al., 2009; Stoytcheva & Montero, 2011). The oleaginous yeasts of genera *Cryptococcus, Candida, Rhodotorula, Rhodosporidium, Lipomyces, Yarrowia, Trichosporon,* and *Pichia* are investigated thoroughly (Q. Li et al., 2008; Meng et al., 2009; Ageitos et al., 2011). Among oleaginous yeasts species, Basidiomycota are a common phylum apart from Ascomycota phylum (e.g., *Y. lipolytica* and *L. starkeyi*). It is reported that a representative of *Candida* genus yeast, *C. curvata*, can store and synthesize lipids of great amounts (Holdsworth & Ratledge, 1991; Ratledge, 1991). *Cryptococcus* yeasts are found ubiquitously in natural conditions. It is also reported that *C. curvatus* has possibilities for industrial production due to low nutrient requirement, and collects about 60% lipids that can be cultivated on diverse substrates (Meesters et al., 1996; L. Zhang et al., 2011). In addition to lipids, carotenoids—a natural pigment—can be synthesized from red yeast varieties of *R. glutinis* and *R. toruloides*, (S.-L. Wang et al., 2007). Thus, apart from pigment production, they have biotechnological application for their ability to exploit glycerol and lignocellulose for production of lipid (Easterling et al., 2009; X. Yu et al., 2011). Intracellular lipid accumulation is 70% in oleaginous yeast *L. starkeyi*. This estimation is generated from dry biomass, and they use xylose and ethanol, along with l-arabinose, to produce lipids (Zhao et al., 2008). *Y. lipolytica* has potential for bio-oil production (Beopoulos et al., 2009) and is often used as a model organism due to the ease of genetic and metabolic engineering based on the available genomic data (Beopoulos et al., 2009; Stoytcheva & Montero, 2011). Similarly, fermentation of xylose, a pentose sugar, relies on availability of appropriate xylose-fermenting fungal strains naturally or through genetic engineering, and the end product of the fermentation process generates xylitol that may lower the ethanol yield obtained from xylose-fermenting strains (Olsson & Hahn-Hägerdal, 1996). Olsson and Hahn-Hägerdal (1996) provided detailed microorganisms that can ferment xylose. In-depth research on natural xylose-fermenting yeast showed candidate species such as *Candida shehatae*, *Pichia stipitis*, and *Pachysolen tannophilus* (Toivola et al., 1984). Also, *Candida* and *Pichia* strains are reported to show better tolerance to inhibitors compared to *S. cerevisiae*.

The selection and choice of *Oleaginous fungi* and yeasts are reported by many researches. The endophytic *Oleaginous fungi Colletotrichum* sp. and *Alternaria* sp. have lipid content of 30% and 58%, respectively, under optimal and nutrient-stress environments, as reported by Dey and colleagues (2011). *Microsphaeropsis* sp., an endophytic oleaginous fungus, uses wheat straw hemicellulose hydrolysate for lipid production (X.-W. Peng & Chen, 2007). *Aspergillus* sp., often used for production of biofuel, accumulates 22% lipid, the substrate being corncob (X.-W. Peng & Chen, 2007). Kitcha and Cheirsilp (2014) were successful in obtaining many oleaginous yeast strains of *Trichosporonoides spathulata* with lipid content above 40%, from soils as well as from the wastes of palm oil mills. Tanimura and colleagues (2014) chose *Cryptococcus* sp. with known lipid content higher than 60% and reviewed their properties for further production of biofuel.

In addition to fermentation, fungi have developed the ability to infiltrate plants' defense mechanisms along the course of evolution. Members of the filamentous fungus have enzymatic strategies to degrade plants' primary defense systems. Specifically, white-rot fungi that degrades lignin hydrolyzes a significantly greater quantity of lignin than cellulose. Nevertheless, this selective behavior is species specific, and parameters such as pretreatment time, rate of degradation, and associated digestion by fungi differs with feedstocks. Keller and colleagues (2003) showed a 36% saccharification yield of was acquired from corn stover in agricultural residue when pretreated for 29-d with *C. stercoreus*. The hydrolysis of enzyme was conducted at 60 FPU/g glucan for 136 hours. On the contrary, *Irpex lacteus* resulted in a 66.4% saccharification yield when treating CD2 corn stover at a cellulase loading of 20 FPU/g solid for 25-d (Xu et al., 2010). It was understood that at the early stage (0–5 days) holocellulose degradation was dominant and degradation of lignin was absent. This indicates a rate of lignin degradation higher than holocellulose degradation during the period of 5–10 days. Contrastingly, material pretreated with *C. subvermispora* had 67% glucose yield when there was 35.61% of degradation of lignin with cellulase loading of 10 FPU/g solid (Wan & Li, 2010). These studies indicated that fungus-specific *P. ostreatus* generated effective delignification from straw materials than other fungi (Taniguchi et al., 2005). When wheat straw was pretreated with this fungus, cellulose digestibility was 27%–33% for enzymatic hydrolysis with cellulase loading of about 10 FPU/g solid over a 72-hour period (Hatakka, 1983; Taniguchi et al., 2005). *P. ostreatus*, when subjected to prolonged pretreatment, was not particularly selective in cellulose and lignin degradation. Therefore, during the later stage of cultivation, the cellulose digestibility results in saturation (Taniguchi et al., 2005; J. Yu et al., 2009). The digestibility of hardwoods—for example, birch—displayed progress when subjected to continued fungal pretreatment. The polysaccharide digestibility of aspen wood was about 50%–55% when pretreated with *Polyporus berkeleyi*, *Polyporus giganteus*, or *Polyporus resinosus* during 63–99 days. Such digestibility of was not observed in birch wood after pretreatment among similar fungi (Kirk & Moore, 1972). About 37% of the polysaccharides were converted during enzymatic hydrolysis at a cellulase loading of 20 FPU/g solid for 120 days on pretreated

Chinese willow. A similar yield was perceived under similar pretreatment and hydrolysis conditions for bamboo residues (X. Zhang et al., 2007). Softwood, although requiring longer pretreatment, did not yield an attractive rate of saccharification. China fir, when treated over 120 days, reported a saccharification yield of 17% post pretreatment with *E. taxodii* (Yu et al., 2009). Japanese red pine, when pretreated for eight weeks with *Stereum hirsutum*, showed a yield of 21% of sugar (Lee et al., 2007). This indicates that for softwood, pretreatment of thermochemical is better than fungal pretreatment. But the most effective application of fungal treatment is the strategic use of fungi with physiochemical treatment by integrating before pretreatment or after. Repeatedly, the studies dictate that filamentous fungi have lignolytic and cellulolytic enzymes to degrade the complexity of the biomass.

4.3.3 Algal Revolution

Microalgae became the first studied substrate for production of biofuel during the 1950s. During the oil crisis of the 1970s, intensive investigation was conducted on this novel energy source, with huge progresses on microalgae cultivation, an efficient mode of biofuel production. These organisms have been investigated for major components of the biofuel production. The metabolic pathways are mainly focused on the direct application of algal biomass for the production of biofuels. However, indirect applications are not extensively studied.

Microalgae have complex cell wall characteristics with cellulose and hemicellulose. Like every other cellulosic biomass, these polysaccharides need to be hydrolyzed to expose the cytoplasmic content for anaerobic fermentation (Hendriks & Zeeman, 2009; Carrère et al., 2010; Carlsson et al., 2012; De la Rubia et al., 2013; Monlau et al., 2013).

The kingdom protista, the autotrophs belonging to *Euglena*, and microalgae are fascinating organisms that are capable of undergoing bioconversion using light energy. These organisms have photosensitive organelle called stigma or eyespot that are enriched in carotenoids. Unlike lignocellulosic biomass and animal waste, algal biomass relatively takes less space and yields in biofuel components. For efficient production of biofuel, accessibility of feedstock is crucial. This makes microalgae a potential for future energy, called microalgal biorefineries (Y.-C. Lee et al., 2015b), due to its rapid rate of growth and high lipid contents (Sharma et al., 2011). Complex sugars can generate bioethanol with the help of microalgae (Nguyễn & Vu, 2012), as well as generate biodiesel and biohydrogen from lipid content (Ghirardi et al., 2000; Metzger & Largeau, 2005). Thus, microalgae has potential for next-generation bioenergy production due to its ability to grown under any environment with excess nitrogen demand (Christenson & Sims, 2011; Lam & Lee, 2012; Rashid et al., 2014). Cultivation, harvesting, lipid extraction, and conversion are the subsequential downstream processes in the microalgal biorefinery (Seo et al., 2015).

About 40% of the total cost goes for cultivation of microalgal (Kim et al., 2013), and the major limitations observed are maintaining an industrial-scale lipid production (Gordon & Seckbach, 2012) and contagion of oleaginous microalgae while outdoor cultivation (Cho et al., 2013). Cell lysis is caused due to hydrolysis of *Chlorella sorokiniana* cell walls by the enzyme endo-b-1,4-glucanase from *Cellulomonas* sp. *YJ5* after with 1–3 hours of treatment (Dexter & Fu, 2009). Amidst pretreatments of enzymes, when combined it produced enhanced methane yield (0.145 L CH_4/g TS), while cellulase alone indicated prominent results (0.133 L CH_4/g TS) (Elbehri et al., 2013). This may be attributed to a chain process, whereby the bioavailability of one compound is enriched due to the hydrolysis of the other. Therefore, co-culturing microalgae with fungi and bacteria can result in better yield.

Among diverse phytoplankton species, diatoms deserve special attention, as they are the main primary producers playing a very important role in CO_2 fixation. They accumulate carbohydrates or lipids in amounts which can account up to 25% of their dry biomass. Also of interest is the finding that diatoms may be the source of long unsaturated fatty acids (Zulu et al., 2018 and references therein).

The cell walls of microalgae are rigid, thus making lipid extraction challenging. Therefore, either intensive energy or solvents of organic nature are a prerequisite while its pretreated (Y.-C. Lee et al., 2015b). Conventional approaches for extraction of lipid from microalgae involves extraction of solvent by mechanical procedures (Ranjith Kumar et al., 2015), but the major limits are consumption of energy, process cost, quality of product, and efficiency, including stabilizing the lipid extracted (Chu, 2017; Seo et al., 2014). Another important limitation to be solved is connected with the fact that enhancement of triacylglycerols synthesis in microalgae occurs under unfavorable, stressful conditions (e.g. nitrogen deprivation), which in turn limits cell division, growth rate, and biomass accumulation (Khotimchenko & Yakovleva, 2005; Li et al., 2014; Zienkiewicz et al., 2016; Zienkiewicz et al., 2018). As demonstrated by Zienkiewicz et al. (2020), in *Nannochloropsis oceanica* (CCMP1779), nitrogen starvation activates TAG accumulation which is accompanied with transition from autotrophy to quiescence. During such a transition, the photosynthetic apparatus undergoes strong rearrangement. Addition of nitrogen reversed these processes—i.e., cell cycle and photosynthesis were activated—and simultaneously, lipid droplets were degraded.

It was reported that *Scenedesmus dimorphus*, when hydrolyzed with sulfuric acids, showed carbohydrate accumulation of 53.7 w/w. They are a good contender in production bioethanol, as they generated fermentable sugars of 80% (Chng et al., 2017). It was reported that ethanol yield was 11.7 g/L from *Chlorella vulgaris* (Ho et al., 2013), and another report suggested that *Scenedesmus* sp. fermentation generated sugars from 93% of yield of ethanol (Sivaramakrishnan & Incharoensakdi, 2018). Though co-cultivation, yield of ethanol could still be improved from two engineered strains of *Synechocystis* sp. PCC 6803 (Velmurugan & Incharoensakdi, 2020).

C. reinhardtii, cyanobacteria, *A. variabilis*, and *Anabaena cylindrica*, diverse algal strains, produce hydrogen in the presence of sunlight. These organisms are proficient in extracting protons and electrons from water to hydrogen production (Gupta & Tuohy, 2013; Kruse et al., 2005). Kruse et al. (2005, p. 20) reported that in *C. reinhardtii*, production of biohydrogen was enhanced (maximal rate of 4 mL/h). Modification in respiratory metabolism and elimination of electron competition can achieve this (Kruse et al., 2005). A 150% increase in H_2 production was reported due to the hexose uptake protein (HUP1 hexose symporter from *Chlorella kessleri*) in a hydrogen-producing mutant (Stm6) of *C. reinhardtii* (Doebbe et al., 2007). Recent studies on microalgae focus on genetic and metabolic engineering with the aim to enhance the yield of biohydrogen (Khetkorn et al., 2017).

Another group of organisms intensively studied for hydrogen production are purple non-sulfur bacteria, which exhibit several desirable traits (Tiang et al., 2020). Recently, Elkahlout et al. (2018) reported a hydrogen generation by agar-immobilized cells of *Rhodobacter capsulatus* from acetate showing much higher productivity and yield compared to suspension culture. Enhancement of biohydrogen production by *Rhodobacter capsulatus* was also observed in a pressurized photobioreactor (Magnin & Deseure, 2019). Cell immobilization in agar was also effective in increasing production of hydrogen in cyanobacterium *Aphanothece halophytica* (Pansook et al., 2019) and *Nostoc* sp. CU2561 (Sukrachan & Incharoensakdi, 2020).

In the context of emerging research in third-generation biofuels, it has been reported that algal strains *Porphyridium cruentum* (Kim et al., 2017; Hu et al., 2018), *Tetraselmis suecica* (Reyimu & Özçimen, 2017) and *Desmodesmus* sp. (Sanchez Rizza et al., 2017) are ideal candidates for studies on bioethanol production. *Synechocystis* sp., an algal strain, can phototrophically covert CO_2 into bioethanol by double homologous recombination (Dexter & Fu, 2009). *Synechocystis* sp had a higher yield of 0.696 g ethanol/g CO_2 than did *S. cerevisiae*, which had a yield of 0.51 g ethanol/g glucose. To improve the yield of bioethanol and make it cost-effective, more investigation needs to be channeled on algae-based research.

It must be noted that the viability of bioethanol production depends on conversion of sugars (pentoses or hexoses) to bioethanol. Substrates help a microorganism to ferment, which calls for careful substrate selection. As the biomass in algae are enriched in fermentable sugars, this requires an amalgamation of microorganisms to promote efficiency in fermentation (Khan et al., 2017). In a nutshell, microalgae use solar energy, CO_2, and water for cellular metabolism and produce components of biofuels more efficiently than lignocellulosic biomass or animal waste products. Therefore, algae are ideal candidates for next-generation biofuel.

4.3.4 Synthetic Microbes, Extremophiles and Beyond

Application of synthetic biology in the last two decades has been profitable to the biofuel industry. Synthetic microbes are genetically engineered to either express functions that are not native to their metabolism or suppress native functions. The typical strategy behind identifying candidate pathway is by optimization of stress tolerance and investigation of the role of the organism. In the era of next-generation sequencing, omics analysis is conducted to identify the metabolic network associated with the genes of the corresponding network. The pathway that is most applicable for the discussed direct or indirect function in biofuel production can be considered to be the key pathway. The genetic components of the respective pathway are either integrated into an organism or removed from an organism. This type of approach brings the gain of function mutants and lack of function mutants of the respective strains. Recent advances in genome editing by the CRISPR-Cas system has opened a new door in pathway modification in microbes by recombination, mutagenesis, and gene targeting. In industrial biotechnology, including biofuel production, CRISPR-Cas usage systems suggest diverse microbes across the kingdom that are genetically engineered for industrial purposes (Donohoue et al., 2018). Among these microbes, members of the bacterial kingdom *E. coli* and *Clostridium* sp., and members of the fungi *Saccharomyces* sp. and *Pichia* sp., are explicitly used for pathway reconstruction (Donohoue et al., 2018). Here, the strategy is incorporating pathways that are specific for either bioconversion or pretreatment, or integration of both.

Indirect application of synthetic microbes is explicitly used to efficiently degrade lignocellulosic biomass and animal waste. The concept of CBP microbes is incorporated into the typical bioprocess workflow. The CBP microbes reduce the steps involved in the workflow, improving the quality and quantity of the end product.

In the kingdom protista, the class *Euglena* and *Chlorophyta* includes microalgae that significantly contribute to biofuel production. These few members of protista and microalgae are used as a robust and sustainable energy feedstock; strategies to improve the production are: (a) increasing the efficiency of photosynthesis to increase yield of oil and improved rate of carbon sequestration; (b) producing energy-rich compounds by the carbon flux; and (c) growth of algae that are capable to survive large-scale cultivation, which will decrease the operational costs (R. Singh et al., 2016). The main objective is to derive novel function by reprogramming microalgae. A secure assembly is a prerequisite for creating alternatives (synthetics), which involves merging of heterologous DNA fragments to the host. There are different strategies to fabricate genes for re-creation, such as: (a) DNA is ligated and digested; (b) recombining fragments (*in vitro*); and (c) recombining by *in vivo* (Apel et al., 2017).

Extremophiles such as alkaliphiles and acidophiles are tolerable to lower and higher pH than the standard fermentation process. This unique characteristic makes them ideal candidates for CBP and integration of these extremophiles during the pretreatment step, along with alkali or acid hydrolysis. These members of extremophilic bacteria are thermophiles and halophiles that are resistant to extreme temperature and salinity (Singh et al. 2019c).

Thermophiles are capable of survival at 41–122°C with ideal growing at temperatures between 60°C and 108°C. They are useful for industries by virtue of the presence of thermostable enzymes (S. Singh et al., 2011). There are three varieties of thermophiles, namely moderate thermophiles (T_{opt}: 50–60°C), extreme thermophiles (T_{opt}: 60–80°C), and hyperthermophiles (T_{opt}: 80–110°C) (S. Singh et al., 2011). Hyperthermophiles *Methanopyrus kandleri* 116 has a greater utility in industries and thrives at 122°C (Su et al., 2013). In general, these thermostable enzymes have potential application in biomass pretreatment and advancements in biofuel production. Therefore, the indirect application of thermophiles lies in the step of pretreatment by integrating the physical and enzymatic treatment of biomass. A dearth of scientific knowledge for genetic manipulation of these thermophilic bacteria is responsible for the lack of their utility in producing biofuels.

Halophiles are a kind of extremophilic microorganism capable of thriving in saline and conditions like in deep sea sediments, saline lakes, saline soils, and seawater (Kumar & Khare, 2012), whereas halotolerant microbes can survive both in saline and natural conditions. This makes them promising candidates for producing biofuel and extra associated processes (P. Singh et al., 2019). Many hydrolytic enzymes of halophilic and halotolerant bacteria—like α-amylase, β-amylase, ligninase, lipase, esterase, and glucoamylase—play a vital role in the biofuel industry (Amoozegar et al., 2019; Elmansy et al., 2018; Schreck & Grunden, 2014).

These extremophiles and their enzymes have prospective lignin use. An example is the halotolerant and alkalophilic bacterium *Bacillus ligniniphilus* L1, which reduces lignin to provide aromatic compounds (vanillic acid) at pH 9 (Zhu et al., 2017). *Bacillus* sp. B1, a thermophilic strain, can degrade cinnamic acid into catechol, ferulic acid into protocatechuic, and coumaric acid into gentisic acid (X. Peng et al., 2003). Additionally, some thermotolerant and halotolerant laccases acquired from *Bacillus* sp. SS4, *Thermobifida fusca*, and *Trametes trogii*, showed promising utility for valorization of lignin (C.-Y. Chen et al., 2013; Christopher et al., 2014; Singh et al., 2019a; X. Yang et al., 2020). This makes the use of those extremozymes one of the breakthroughs in lignin valorization, by converting lignin into a series of chemicals.

From the hyperthermophilic archaeon *Sulfolobus solfataricus*, endoglucanase SSO1354 is isolated. It possesses distinctive features and aids in conversion of biomass. Klose and colleagues (2012) have successfully demonstrated expression of the enzyme *in planta* and activation in post-harvest, thereby stating that the enzyme is seemly correct for both pretreatment and applications of hydrolysis.

Production of second-generation biofuels is expensive, with many physical, chemical, and biological pretreatment operations associated. Slow enzymatic hydrolysis of lignocellulosic biomass is responsible for making the process of obtaining biofuels more costly than the more accessible fossil fuels. Therefore, this conversion of lignocellulose biomass conducted at temperature ≤50°C is not without limitations. To overcome the disadvantage, Bhalla et al. (2013) stated the role of thermophilic bacteria and thermostable enzymes. Extremophilic CBP is a solution to overcome the challenges in the mentioned processes (Bhalla et al., 2013).

4.4 Microbial Symbionts in Biofuel Production

Recent studies on understanding the role of microbes in association with higher organisms revealed that there is a substantial contribution of these microbes and their genes in developing the function of the host. Together with the microbes, these complex organisms have co-evolved with specific members of the microbiota as a holobiont. This pre-knowledge provides an insight that host–microbe interaction is partly governed by inter- and intrakingdom microbe–microbe interaction. The members of the microbe–microbe interaction involve symbionts that support the growth and biochemical processes (Mathew et al., 2013). The inter-relationship between these members suggests that a co-culture system of microbes can substantially improve the biomass and quality of the biofuels. The components underlying the molecular mechanism of the symbionts can provide us with an understanding of stress management (Brune, 2014). The mode of the symbiotic interactions is either mutualism or commensalism. Further interkingdom symbiosis is described from the reported studies.

4.4.1 Bacteria and Fungi Symbionts

The knowledge of members of the plant–microbe interactions and gut symbionts dictates that the most important role of bacteria–fungi symbiosis is from the perspective of pretreatment and fermentation of lignocellulosic biomass. Wood-rotting fungi can degrade lignocellulose (Kirk & Moore, 1972; J.-W. Lee et al., 2007; Warnecke et al., 2007; Ohm et al., 2014). Many aerobic bacteria, as well as anaerobic bacteria like *Bacillus* or *Cellulomonas*, have enzymes that degrade lignocellulose. Given that the metabolic process is well understood, it is possible to express the components of these metabolic processes within these organisms through techniques such as genetic engineering (Donohoue et al., 2018; Javed et al., 2019).

The plethora of microbes inside termite gut gives an insight about the symbiosis between microbes to cope up with the challenges in digesting lignocellulosic biomass. Mathew and colleagues (2012) report that hydrogen production using gut symbionts biomimics the termite gut of *Odontotermes formosanus*. Solid-state NMR can detect the fungus comb aging, as well as *in vitro* lignocellulosic degradation of the mango tree substrates, through the process of interaction of *Bacillus*, *Termitomyces*, and *Clostridium*. Substantial degradation of lignocellulosic biomass in trees were observed in the presence of *Clostridium*, *Bacillus*, and *Termitomyces*. Therefore, these microbes act as a potential mutualist (Perisin & Sund, 2018; Ryan et al., 2008; F. Wang et al., 2019b).

As discussed earlier, extremophiles have the appropriate toolbox for degrading plant-based biomass that is coherent with the members of the kingdom fungi. Fungi secretomes attained through the phyla Ascomycota, Basidiomycota, Zygomycota, and Neocallimastigomycota represent those fungi possessing

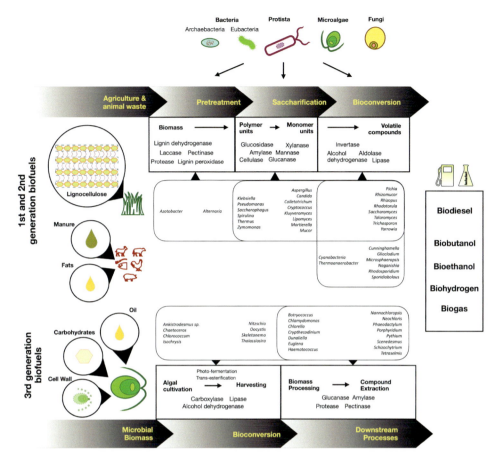

FIGURE 4.1 An overview of the role of microbes in the biofuel production pipeline.

delignification strategies and various metabolic cues (nutrient sensing, nitrogen metabolism, and coordination of carbon) that change their structure (Guerriero et al., 2015).

4.4.2 Microalgae and Fungi Symbionts

Mortierella elongate, a fungus, breeds in soils and produces a generous quantity of oils useful for industrial purposes (Bonfante, 2019; Du et al., 2019). *Mortierella*'s life system is still unknown, and it is believed that the fungus survives by obtaining nutrition through decayed matter. Any symbiotic relation with algae is yet to be identified.

However, recently in a laboratory, *M. elongata*'s growth with a marine algae *Nannochloropsis oceanica* has been reported by a team of researchers (Du et al., 2018) that indicated symbiosis. These algae and fungi successfully produced a greater quantity of oil through symbiosis than individually. Yet, whether facilitation of nutrient exchange or stress control is achieved through this symbiosis is unclear.

Lichen are the most common mutualisms that exist between algae and fungi, dating from 480 million years ago (Lutzoni et al., 2018). Lichen have adaptive symbiosis, permitting mycobiont and photobiont symbionts to colonize habitats and environments facultatively favorable to both species, such as on a rock protrusion or in a desert, including harsh polar areas (Haranczyk et al., 2017). Lichenized fungi possess multiple neutral origins in phyla Ascomycota and Basidiomycota, and are metaorganisms in addition to yeasts (Spribille et al., 2016).

The mutualisms among photobionts and mycobionts involve nutrient exchange. A latent capacity of ascomycetous yeasts and filamentous fungi's interaction with algae is demonstrated by the reciprocal transfer of nitrogen and carbon for *Chlamydomonas reinhardtii* and a diverse panel of ascomycete fungi (Hom & Murray, 2014). Other research showed that *Alternaria infectoria*, a filamentous ascomycetous fungus, provide nitrogen to *C. reinhardtii* in a long-lived split system (Simon et al., 2017). An example of a non-lichen algal–fungal mutualism involves *Rhizidium phycophilum*, a fungus, and *Bracteacoccus*, a green algae. This gives evidence of early diverging fungi having evolved mutualisms (Picard et al., 2013). Algal cells are external to fungal hyphae and are not known to enter cells of living fungus. However, in the soil, the interaction between *Mortierella elongata Nannochloropsis oceanica*, a marine algae, alters the metabolism of both associates (Du et al., 2018). This is an indication of greater specificity as a result contact among the associates. Another study describes the apparent symbiosis in detail by experiments (isotope tracer) exhibiting bidirectional nutrient exchange in algal–fungal interactions. The ecology of *Mortierella* sp. has eluded mycologists for years (Summerbell, 2005). In Antarctica, they were isolated directly under green macroalgae from strata (Furbino et al., 2014). *Mortierella* was abundant in cryptobiotic desert (together with bacteria, algae, and fungi) (Bates et al., 2010), and it was found together with red algae in alpine region (Brown & Jumpponen, 2014). *N. oceanica* is still the nearest algae to terrestrial plants yet identified. *Chromalveolate alga* has formed biotrophic mutualism with an adapted fungus that explains the diversity between algae and fungi, and indicating that their mutualism is more ancient than previously recognized.

4.4.3 Microalgae and Bacteria Symbionts

Perception has transformed in identifying microorganisms' interactions which enhance biochemical processes. The bacteria and algae interactions are front runners for biotechnological applications. There has been a positive outcome of algae–bacteria symbiosis on the former's growth and flocculation methods, which are landmarks in algal biotechnology (Fuentes et al., 2016). This symbiotic algae–bacteria interactions are called parasitism, mutualism and commensalism (Ramanan et al., 2016). Yao and colleagues (2019) discussed comprehensively the mutualism among microalgae and bacteria. Symbiotic bacteria are associated with stimulating microalgal growth, as their symbiosis is crucial for biofuel production. The underlying mechanism of the symbiosis is mainly focused on nutrient acquisition and bioavailability of vitamin B. Vitamin B complexes are crucial for eukaryotic phytoplankton. The result of the symbiosis is the growth promotion of microalgae. The factors governing the symbiosis are called symbiotic factors.

The hydrogen production in the microalga *Chlamydomonas* was enhanced by *Brevundimonas* sp., *Leifsonia* sp., and *Rhodococcus* sp., all of which are bacterial symbionts (Lakatos et al., 2019). It resulted from oxygen elimination by efficient bacterial respiration, which is mandatory for the activation of a iron-dependent hydrogenase in *Chlamydomonas* (Lakatos et al., 2019). This phenomenon can be also determined by bacterial–algal communities. Interestingly, when *Escherichia coli*, a hydrogenase-deficient, was applied to *Chlamydomonas* as an artificial symbiotic bacterium, highest hydrogen yield was attained (Lakatos et al., 2019). Thus, due to bacterial aid which uses the oxygen, along with the algae that exploit light energy concurrently, can result in significant hydrogen yield without further manipulation of the system.

Organic matter of various kinds can be used by electroactive bacteria to produce electricity or hydrogen (Cheng & Logan, 2007; Logan, 2009). Such bacteria may also use surplus of electricity produced e.g. by photovoltaics, wind farms, etc. to fix CO_2 and to produce various carbon-based compounds (Costa et al., 2018; Su & Ajo-Franklin, 2019). Microbial fuel cells can use organic wastes and biomass to produce electricity without input of external energy (Logan, 2009; Santoro et al., 2017). Without any external help of exogenous organics, solar energy is successfully transformed into electricity through the process of synergistic cooperation. This transformation takes place between heterotrophic bacteria and photosynthetic microalgae through microalgal fuel cells (He et al., 2009). Such algae–microbial fuel cells, apart from electricity production, may also be used for wastewater treatment, CO_2 fixation resulting in lowering its level in atmosphere with concomitant production of many useful carbon-containing compounds (Reddy et al., 2019). Extracellular electron transfer processes are a promising and fast-developing area which may be useful for many applications.

In algal biodiesel production, *Chlorella vulgaris*, a species with high lipid content (Chisti, 2013; Griffiths et al., 2012) has its importance. Various classes of *Pseudomonas* were related with *C. vulgaris*

(Sapp et al., 2007; Guo & Tong, 2014). A symbiotic relationship may be identified with growing of *Pseudomonas* populations along with *C. vulgaris* in open ponds (Bell et al., 2015). Additionally, growth of green microalga, *Auxenochlorella protothecoides*, along with *E. coli* is another example of symbiosis when it leads to a significant rise in algal growth and twice the neutral lipid content (Higgins & VanderGheynst, 2014). This can indicate a positive implication related to *E. coli* providing thiamine derivatives, together with degradation of production of *A. protothecoides* (Higgins et al., 2016). The marine microalga *Tetraselmis striata*, a high lipid–content marine microalga, makes a fine contender in biodiesel production, owing to its rapid growth process (Chisti, 2007). *Pelagibaca bermudensis* and *Stappia* sp., separated from *T. striata*, had growth-promoting outcomes on mentioned microalgae (Park et al., 2017). These symbionts and their underlying molecular mechanisms provide a base for large-scale biodiesel production.

Biogas can be produced from microalgae after collecting or extracting useful products. The composition of microalgae is about 20%–40% of carbohydrates, 30%–50% of proteins and 8%–15% of lipids grown under ambient conditions (Henry, 2004). Additionally, microalgae are capable of producing 40% starch granules (dry weight), which can be exploited by several bacteria for ethanol production (Ramanan et al., 2016). Marine microalgae can produce ethanol from bacteria, *Pseudoalteromonas undina*, through saccharification (Matsumoto et al., 2003). Microalga *C. reinhardtii* can aid in enzymatic hydrolysis and fermentation from *Bacillus licheniformis* bacteria. This is succeeded by fermenting *Saccharomyces cerevisiae*, a brewer's yeast (de Farias Silva & Bertucco, 2016). Thus, production of bioethanol as a result of symbiosis of algal-bacterial co-cultures needs further exploration, owing to its novelty.

Biogas is the end product of microalgal biomass residue of anaerobic digestion. Microalgae—by virtue of having cobalt, iron, and zinc—aids in the requirements of anaerobic microbiota's nutrients (Richmond, 2008). Methanogenesis is created by microalgae's incubation of residual biomass and with anaerobic microbes (Sialve et al., 2009). This biogas amount is reliant on the type of microalgal species (Illman et al., 2000). Additionally, the protease resistance can influence the methanogenic ability of microalgae, which controls the efficiency of the primary microbial degradation (Angelidaki & Sanders, 2004). The cell walls of *Botryococcus braunii* and *Nannochloropsis gaditana*, a microalgae, can be degraded by some bacteria with endoglucanase potential (Muñoz et al., 2014). The degradation of the biomass of *Chlorella vulgaris* was enhanced by hydrogenogenic bacterium *Clostridium thermocellum*. This concluded in greater methane yield and production of hydrogen, resulting in total increase in biogas yield (Lü et al., 2013).

A two-step method with anaerobic digestion and dark fermentation was conducted by Abreu and colleagues, providing evidence of the potential production of energy from garden waste. *Caldicellulosiruptor saccharolyticus* and *Thermotoga maritima*'s co-culture presented greater yields of hydrogen from xylose (2.7 ± 0.1 molmol^{-1} total sugar) and cellobiose (4.8 ± 0.3 molmol^{-1} total sugar) than individual cultures. The hydrogen yield from garden waste was (98.3 ± 6.9 Lkg^{-1}) greater as a result of co-culture of *C. saccharolyticus* and *Caldicellulosiruptor bescii* than the individual cultures and co-culture of *T. maritima* and *C. saccharolyticus*. Potential energy of 22.2 MJkg^{-1} was generated from biohythane, a biogas with 15% hydrogen (Abreu et al., 2016).

Cyanobacteria are also capable of symbiosis microalgae. This symbiosis between cyanobacteria and microalgae provide long-term benefits though increased microalgal growth, metabolites, and enhanced nutrients (V.K. Gupta et al., 2017; Subashchandrabose et al., 2011). These growth facilities have an excellent potential in different biotechnological fields, more so in biofuel industries. Another instance of mutualism is between microalgae (*Chlorella vulgaris*), a native of Louisiana with cyanobacterium (*Leptolyngbya* sp.) (Silaban et al., 2014). Source of carbon and the C:N ratio had an effect of growth and productivity of biomass. Mixotrophic cultures having sodium acetate (C:N = 15:1) caused greater biomass production (134 g m^{-3} d^{-1}) with neutral lipid efficiency (24.07 g m^{-3} d^{-1}) in comparison to its autotrophic growth (66 g m^{-3} d^{-1} and 8.2 g m^{-3} d^{-1}, respectively). A greater mixotrophic growth with sodium acetate addition (18.2%) was observed in the lipid content of Louisiana co-culture than its autotrophic growth (8.7.%). These experiments indicate mixotrophic growth with sodium acetate (C:N = 15:1) as the ideal circumstance for cultivation for increase production of biomass and lipid. This symbiosis between cyanobacteria and microalgae paves way for efficient cultivation (Lutzu & Turgut Dunford, 2018).

4.5 Conclusion

Microorganisms play a crucial role in alternative sources of fuel production and provide a wide range of application in improving yield conditions. Members of the microbial kingdom belonging to fungi, protista, microalgae and bacteria have a direct application by fermentation or bioconversion of substrates from plants, animals, or another microbial origin. Simultaneously, these microbes improve the production indirectly by improving the pretreatment of the biomass. These applications are mostly due to the diverse metabolic processes of the microbial system. Further, we have observed that by engineering the metabolic processes of these microbes, we can obtain a double-edged sword with better pretreatment of biomass and efficient bioconversion of the substrate from a single genetically modified strain. Also, there is evidence that suggests that symbiotic cultures of microbes can significantly increase in yield and better bioconversion. The evidence altogether suggests that the microbial kingdom and its metabolic processes can be implemented extensively for the development of green energy solutions.

REFERENCES

Aachary, A. A., & Prapulla, S. G. (2009). Value addition to corncob: Production and characterization of xylooligosaccharides from alkali pretreated lignin-saccharide complex using *Aspergillus oryzae* MTCC 5154. *Bioresource Technology, 100*(2), 991–995. https://doi.org/10.1016/j.biortech.2008.06.050

Abasian, L., Shafiei Alavijeh, R., Satari, B., & Karimi, K. (2020). Sustainable and effective chitosan production by dimorphic fungus *Mucor rouxii* via replacing yeast extract with fungal extract. *Applied Biochemistry and Biotechnology, 191*(2), 666–678. https://doi.org/10.1007/s12010-019-03220-w

Abd Ellatif, S., El-Sheekh, M. M., & Senousy, H. H. (2021). Role of microalgal ligninolytic enzymes in industrial dye decolorization. *International Journal of Phytoremediation, 23*(1), 41–52. https://doi.org/10.1080/15226514.2020.1789842

Abhilash, P. C., Powell, J. R., Singh, H. B., & Singh, B. K. (2012). Plant-microbe interactions: Novel applications for exploitation in multipurpose remediation technologies. *Trends in Biotechnology, 30*(8), 416–420. https://doi.org/10.1016/j.tibtech.2012.04.004

Abreu, A. A., Tavares, F., Alves, M. M., & Pereira, M. A. (2016). Boosting dark fermentation with co-cultures of extreme thermophiles for biohythane production from garden waste. *Bioresource Technology, 219*, 132–138. https://doi.org/10.1016/j.biortech.2016.07.096

Adachi, D., Hama, S., Nakashima, K., Bogaki, T., Ogino, C., & Kondo, A. (2013). Production of biodiesel from plant oil hydrolysates using an *Aspergillus oryzae* whole-cell biocatalyst highly expressing *Candida antarctica* Lipase B. *Bioresource Technology, 135*(May), 410–416. https://doi.org/10.1016/j.biortech.2012.06.092

Adachi, D., Hama, S., Numata, T., Nakashima, K., Ogino, C., Fukuda, H., & Kondo, A. (2011). Development of an *Aspergillus oryzae* whole-cell biocatalyst coexpressing triglyceride and partial glyceride lipases for biodiesel production. *Bioresource Technology, 102*(12), 6723–6729. https://doi.org/10.1016/j.biortech.2011.03.066

Ageitos, J. M., Vallejo, J. A., Veiga-Crespo, P., & Villa, T. G. (2011). Oily yeasts as oleaginous cell factories. *Applied Microbiology and Biotechnology, 90*(4), 1219–1227. https://doi.org/10.1007/s00253-011-3200-z

Ahmad, A., Shafique, S., & Shafique, S. (2013). Cytological and physiological basis for tomato varietal resistance against *Alternaria alternata*. *Journal of the Science of Food and Agriculture, 93*(9), 2315–2322. https://doi.org/10.1002/jsfa.6045

Ahmad, F. B., Zhang, Z., Doherty, W. O. S., & O'Hara, I. M. (2015). A multi-criteria analysis approach for ranking and selection of microorganisms for the production of oils for biodiesel production. *Bioresource Technology, 190*(August), 264–273. https://doi.org/10.1016/j.biortech.2015.04.083

Ahmad, S., Pathak, V. V., Kothari, R., Kumar, A., & Naidu Krishna, S. B. (2018). Optimization of nutrient stress using *C. Pyrenoidosa* for lipid and biodiesel production in integration with remediation in dairy industry wastewater using response surface methodology. *3 Biotech, 8*(8), 326. https://doi.org/10.1007/s13205-018-1342-8

Ahmed, R. A., He, M., Aftab, R. A., Zheng, S., Nagi, M., Bakri, R., & Wang, C. (2017). Bioenergy application of *Dunaliella salina* SA 134 grown at various salinity levels for lipid production. *Scientific Reports, 7*(1), 8118. https://doi.org/10.1038/s41598-017-07540-x

Alam, Md A., Yuan, T., Xiong, W., Zhang, B., Lv, Y., & Xu, J. (2019). Process optimization for the production of high-concentration ethanol with *Scenedesmus raciborskii* biomass. *Bioresource Technology*, *294*(December), 122219. https://doi.org/10.1016/j.biortech.2019.122219

Al-Hothaly, K. A. (2018). An optimized method for the bio-harvesting of microalgae, *Botryococcus braunii*, using *Aspergillus* sp. in large-scale studies. *MethodsX*, *5*, 788–794. https://doi.org/10.1016/j.mex.2018.07.010

Alvira, P., Tomás-Pejó, E., Ballesteros, M., & Negro, M. J. (2010). Pretreatment technologies for an efficient bioethanol production process based on enzymatic hydrolysis: A review. *Bioresource Technology*, *101*(13), 4851–4861. https://doi.org/10.1016/j.biortech.2009.11.093

Amado, I. R., & Vázquez, J. A. (2015). Mussel processing wastewater: A low-cost substrate for the production of astaxanthin by *Xanthophyllomyces dendrorhous*. *Microbial Cell Factories*, *14*(November), 177. https://doi.org/10.1186/s12934-015-0375-5

Amoozegar, M. A., Safarpour, A., Noghabi, K. A., Bakhtiary, T., & Ventosa, A. (2019). Halophiles and their vast potential in biofuel production. *Frontiers in Microbiology*, *10*. https://doi.org/10.3389/fmicb.2019.01895

Angelidaki, I., & Sanders, W. (2004). Assessment of the anaerobic biodegradability of macropollutants. *Re/Views in Environmental Science & Bio/Technology*, *3*(2), 117–129. https://doi.org/10.1007/s11157-004-2502-3

Ananyev, G., Carrieri, D., & Charles Dismukes, G. (2008). Optimization of metabolic capacity and flux through environmental cues to maximize hydrogen production by the cyanobacterium "*Arthrospira (Spirulina) maxima*." *Applied and Environmental Microbiology*, *74*(19), 6102–6113. https://doi.org/10.1128/AEM.01078-08

Ananyev, G. M., Skizim, N. J., & Charles Dismukes, G. (2012). Enhancing biological hydrogen production from cyanobacteria by removal of excreted products. *Journal of Biotechnology*, *162*(1), 97–104. https://doi.org/10.1016/j.jbiotec.2012.03.026

Angermayr, S. A., Gorchs Rovira, A., & Hellingwerf, K. J. (2015). Metabolic engineering of cyanobacteria for the synthesis of commodity products. *Trends in Biotechnology*, *33*(6), 352–361. https://doi.org/10.1016/j.tibtech.2015.03.009

Antonopoulou, G. (2020). Designing efficient processes for sustainable bioethanol and bio-hydrogen production from grass lawn waste. *Molecules (Basel, Switzerland)*, *25*(12). https://doi.org/10.3390/molecules25122889

Antúnez-Argüelles, E., Herrera-Bulnes, M., Torres-Ariño, A., Mirón-Enríquez, C., Soriano-García, M., & Robles-Gómez, E. (2020). Enzymatic-assisted polymerization of the lignin obtained from a macroalgae consortium, using an extracellular laccase-like enzyme (Tg-Laccase) from *Tetraselmis gracilis*. *Journal of Environmental Science and Health. Part A, Toxic/Hazardous Substances & Environmental Engineering*, *55*(6), 739–747. https://doi.org/10.1080/10934529.2020.1738171

Apel, A. C., Pfaffinger, C. E., Basedahl, N., Mittwollen, N., Göbel, J., Sauter, J., Brück, T., & Weuster-Botz, D. (2017). Open thin-layer cascade reactors for saline microalgae production evaluated in a physically simulated Mediterranean summer climate. *Algal Research*, *25*, 381–390. https://doi.org/10.1016/j.algal.2017.06.004

Arora, N., Jaiswal, K. K., Kumar, V., Vlaskin, M. S., Nanda, M., Pruthi, V., & Chauhan, P. K. (2020). Small-scale phyco-mitigation of raw urban wastewater integrated with biodiesel production and its utilization for aquaculture. *Bioresource Technology*, *297*(February), 122489. https://doi.org/10.1016/j.biortech.2019.122489

Arun, J., Gopinath, K. P., SundarRajan, P. S., Malolan, R., Adithya, S., Jayaraman, R. S., & Ajay, P. S. (2020). Hydrothermal liquefaction of *Scenedesmus obliquus* using a novel catalyst derived from clam shells: Solid residue as catalyst for hydrogen production. *Bioresource Technology*, *310*(August), 123443. https://doi.org/10.1016/j.biortech.2020.123443

Athalye, S. K., Garcia, R. A., & Wen, Z. (2009). Use of biodiesel-derived crude glycerol for producing Eicosapentaenoic Acid (EPA) by the fungus *Pythium irregulare*. *Journal of Agricultural and Food Chemistry*, *57*(7), 2739–2744. https://doi.org/10.1021/jf803922w

Atsumi, S., Hanai, T., & Liao, J. C. (2008). Non-fermentative pathways for synthesis of branched-chain higher alcohols as biofuels. *Nature*, *451*(7174), 86–89. https://doi.org/10.1038/nature06450

Aulitto, M., Fusco, F. A., Fiorentino, G., Bartolucci, S., Contursi, P., & Limauro, D. (2018). A thermophilic enzymatic cocktail for galactomannans degradation. *Enzyme and Microbial Technology*, *111*(April), 7–11. https://doi.org/10.1016/j.enzmictec.2017.12.008

Azambuja, S. P. H., & Goldbeck, R. (2020). Butanol production by *Saccharomyces cerevisiae*: Perspectives, strategies and challenges. *World Journal of Microbiology & Biotechnology*, *36*(3), 48. https://doi.org/10.1007/s11274-020-02828-z

Bakkiyaraj, S., Syed, M. B., Devanesan, M. G., & Thangavelu, V. (2016). Production and optimization of biodiesel using mixed immobilized biocatalysts in packed bed reactor. *Environmental Science and Pollution Research International*, *23*(10), 9276–9283. https://doi.org/10.1007/s11356-015-4583-7

Ban, S., Lin, W., Luo, Z., & Luo, J. (2019). Improving hydrogen production of *Chlamydomonas reinhardtii* by reducing chlorophyll content via atmospheric and room temperature plasma. *Bioresource Technology*, *275*(March), 425–429. https://doi.org/10.1016/j.biortech.2018.12.062

Ban, S., Lin, W., Wu, F., & Luo, J. (2018). Algal-bacterial cooperation improves algal photolysis-mediated hydrogen production. *Bioresource Technology*, *251*(March), 350–357. https://doi.org/10.1016/j.biortech.2017.12.072

Barnard, D., Casanueva, A., Tuffin, M., & Cowan, D. (2010). Extremophiles in biofuel synthesis. *Environmental Technology*, *31*(8–9), 871–888. https://doi.org/10.1080/09593331003710236

Bastien, G., Arnal, G., Bozonnet, S., Laguerre, S., Ferreira, F., Fauré, R., Henrissat, B., Lefèvre, F., Robe, P., Bouchez, O., Noirot, C., Dumon, C., & O'Donohue, M. (2013). Mining for hemicellulases in the fungus-growing termite *Pseudacanthotermes militaris* using functional metagenomics. *Biotechnology for Biofuels*, *6*(1), 78. https://doi.org/10.1186/1754-6834-6-78

Bates, S. T., Nash, T. H., Sweat, K. G., & Garcia-Pichel, F. (2010). Fungal communities of lichen-dominated biological soil crusts: Diversity, relative microbial biomass, and their relationship to disturbance and crust cover. *Journal of Arid Environments*, *74*(10), 1192–1199. https://doi.org/10.1016/j.jaridenv.2010.05.033

Bátori, V., Ferreira, J. A., Taherzadeh, M. J., & Lennartsson, P. R. (2015). Ethanol and protein from ethanol plant by-products using edible fungi *Neurospora intermedia* and *Aspergillus oryzae*. *BioMed Research International*, *2015*, 176371. https://doi.org/10.1155/2015/176371

Beg, Q. K., Kapoor, M., Mahajan, L., & Hoondal, G. S. (2001). Microbial xylanases and their industrial applications: A review. *Applied Microbiology and Biotechnology*, *56*(3–4), 326–338. https://doi.org/10.1007/s002530100704

Bell, T. A. S., Prithiviraj, B., Wahlen, B. D., Fields, M. W., & Peyton, B. M. (2015). A lipid-accumulating alga maintains growth in outdoor, alkaliphilic raceway pond with mixed microbial communities. *Frontiers in Microbiology*, *6*, 1480. https://doi.org/10.3389/fmicb.2015.01480

Ben Salah, R., Ghamghui, H., Miled, N., Mejdoub, H., & Gargouri, Y. (2007). Production of butyl acetate ester by lipase from novel strain of *Rhizopus oryzae*. *Journal of Bioscience and Bioengineering*, *103*(4), 368–372. https://doi.org/10.1263/jbb.103.368

Beopoulos, A., Chardot, T., & Nicaud, J.-M. (2009). *Yarrowia lipolytica*: A model and a tool to understand the mechanisms implicated in lipid accumulation. *Biochimie*, *91*(6), 692–696. https://doi.org/10.1016/j.biochi.2009.02.004

Berendsen, R. L., Pieterse, C. M. J., & Bakker, P. A. H. M. (2012). The rhizosphere microbiome and plant health. *Trends in Plant Science*, *17*(8), 478–486. https://doi.org/10.1016/j.tplants.2012.04.001

Bernard, E. A., Guaragna, R., Amaral, B. B., Perry, M. L., Pereira, I. R., Ielpi, L., & Couso, R. O. (1982). Formation of lipid-linked sugars in mycelial and yeast-like forms of *Mucor rouxii*. *Molecular and Cellular Biochemistry*, *45*(1), 41–48. https://doi.org/10.1007/BF01283162

Bertozzini, E., Galluzzi, L., Ricci, F., Penna, A., & Magnani, M. (2013). Neutral lipid content and biomass production in *Skeletonema marinoi* (Bacillariophyceae) culture in response to nitrate limitation. *Applied Biochemistry and Biotechnology*, *170*(7), 1624–1636. https://doi.org/10.1007/s12010-013-0290-3

Besada-Lombana, P. B., & Da Silva, N. A. (2019). Engineering the early secretory pathway for increased protein secretion in *Saccharomyces cerevisiae*. *Metabolic Engineering*, *55*(September), 142–151. https://doi.org/10.1016/j.ymben.2019.06.010

Bessadok, B., Santulli, A., Brück, T., Breuck, T., & Sadok, S. (2019). Species disparity response to mutagenesis of marine yeasts for the potential production of biodiesel. *Biotechnology for Biofuels*, *12*, 129. https://doi.org/10.1186/s13068-019-1459-y

Bhalla, A., Bansal, N., Kumar, S., Bischoff, K. M., & Sani, R. K. (2013). Improved lignocellulose conversion to biofuels with thermophilic bacteria and thermostable enzymes. *Bioresource Technology*, *128*, 751–759. https://doi.org/10.1016/j.biortech.2012.10.145

Bhardwaj, N., Kumar, B., Agarwal, K., Chaturvedi, V., & Verma, P. (2019). Purification and characterization of a thermo-acid/alkali stable xylanases from *Aspergillus oryzae* LC1 and its application in Xylo-Oligosaccharides production from lignocellulosic agricultural wastes. *International Journal of Biological Macromolecules*, *122*(February), 1191–1202. https://doi.org/10.1016/j.ijbiomac.2018.09.070

Bhatia, S. K., Jagtap, S. S., Bedekar, A. A., Bhatia, R. K., Patel, A. K., Pant, D., Rajesh Banu, J., Rao, C. V., Kim, Y.-G., & Yang, Y.-H. (2020). Recent developments in pretreatment technologies on lignocellulosic biomass: Effect of key parameters, technological improvements, and challenges. *Bioresource Technology*, *300*, 122724. https://doi.org/10.1016/j.biortech.2019.122724

Bhosale, S. H., Pant, A., & Khan, M. I. (2009). Purification and characterization of putative alkaline [Ni-Fe] hydrogenase from unicellular marine green alga, *Tetraselmis kochinensis* NCIM 1605. *Microbiological Research*, *164*(2), 131–137. https://doi.org/10.1016/j.micres.2006.11.006

Bilal, M., Nawaz, M. Z., Iqbal, H. M. N., Hou, J., Mahboob, S., Al-Ghanim, K. A., & Cheng, H. (2018). Engineering ligninolytic consortium for bioconversion of lignocelluloses to ethanol and chemicals. *Protein and Peptide Letters*, *25*(2), 108–119. https://doi.org/10.2174/0929866525666180122105835

Blomqvist, J., Pickova, J., Tilami, S. K., Sampels, S., Mikkelsen, N., Brandenburg, J., Sandgren, M., & Passoth, V. (2018). Oleaginous yeast as a component in fish feed. *Scientific Reports*, *8*(1), 15945. https://doi.org/10.1038/s41598-018-34232-x

Bonfante, P. (2019). Algae and fungi move from the past to the future. *ELife*, *8*. https://doi.org/10.7554/eLife.49448

Bora, S. S., Keot, J., Das, S., Sarma, K., & Barooah, M. (2016). Metagenomics analysis of microbial communities associated with a traditional rice wine starter culture (*Xaj-pitha*) of Assam, India. *3 Biotech*, *6*(2), 153. https://doi.org/10.1007/s13205-016-0471-1

Brandenburg, J., Poppele, I., Blomqvist, J., Puke, M., Pickova, J., Sandgren, M., Rapoport, A., Vedernikovs, N., & Passoth, V. (2018). Bioethanol and lipid production from the enzymatic hydrolysate of wheat straw after furfural extraction. *Applied Microbiology and Biotechnology*, *102*(14), 6269–6277. https://doi.org/10.1007/s00253-018-9081-7

Brexó, R. P., Andrietta, M. G. S., & Sant'Ana, A. S. (2018). *Artisanal cachaça* and brewer's spent grain as sources of yeasts with promising biotechnological properties. *Journal of Applied Microbiology*, *125*(2), 409–421. https://doi.org/10.1111/jam.13778

Brodie, J., Chan, C. X., De Clerck, O., Cock, J. M., Coelho, S. M., Gachon, C., Grossman, A. R., Mock, T., Raven, J. A., Smith, A. G., Yoon, H. S., & Bhattacharya, D. (2017). The algal revolution. *Trends in Plant Science*, *22*(8), 726–738. https://doi.org/10.1016/j.tplants.2017.05.005

Brognaro, H., Almeida, V. M., de Araujo, E. A., Piyadov, V., Santos, M. A. M., Marana, S. R., & Polikarpov, I. 2016. Biochemical Characterization and Low-Resolution SAXS Molecular Envelope of GH1 β-Glycosidase from *Saccharophagus degradans*. *Molecular Biotechnology* 58 (12): 777–788. https://doi.org/10.1007/s12033-016-9977-3

Brown, S. P., & Jumpponen, A. (2014). Contrasting primary successional trajectories of fungi and bacteria in retreating glacier soils. *Molecular Ecology*, *23*(2), 481–497. https://doi.org/10.1111/mec.12487

Brune, A. (2014). Symbiotic digestion of lignocellulose in termite guts. *Nature Reviews. Microbiology*, *12*(3), 168–180. https://doi.org/10.1038/nrmicro3182

Bui, H.-H., Tran, K.-Q., & Chen, W.-H. (2016). Pyrolysis of microalgae residues: a kinetic study. *Bioresource Technology*, *199*(January), 362–366. https://doi.org/10.1016/j.biortech.2015.08.069

Burchhardt, G., & Ingram, L. O. (1992). Conversion of Xylan to ethanol by ethanologenic strains of *Escherichia coli* and *Klebsiella oxytoca*. *Applied and Environmental Microbiology*, *58*(4), 1128–1133. https://doi.org/10.1128/aem.58.4.1128-1133.1992

Butler, T., Kapoore, R. V., & Vaidyanathan, S. (2020). *Phaeodactylum tricornutum*: A diatom cell factory. *Trends in Biotechnology*, *38*(6), 606–622. https://doi.org/10.1016/j.tibtech.2019.12.023

Calero, J., Verdugo, C., Luna, D., Sancho, E. D., Luna, C., Posadillo, A., Bautista, F. M., & Romero, A. A. (2014). Selective ethanolysis of sunflower oil with lipozyme RM IM, an immobilized *Rhizomucor miehei* lipase, to obtain a biodiesel-like biofuel, which avoids glycerol production through the monoglyceride formation. *New Biotechnology*, *31*(6), 596–601. https://doi.org/10.1016/j.nbt.2014.02.008

Canet, A., Dolors Benaiges, M., Valero, F., & Adlercreutz, P. (2017). Exploring substrate specificities of a recombinant *Rhizopus oryzae* lipase in biodiesel synthesis. *New Biotechnology*, *39*(Pt A), 59–67. https://doi.org/10.1016/j.nbt.2017.07.003

Cantero, D., Jara, R., Navarrete, A., Pelaz, L., Queiroz, J., Rodríguez-Rojo, S., & Cocero, M. J. (2019). Pretreatment processes of biomass for biorefineries: Current status and prospects. *Annual Review of Chemical and Biomolecular Engineering*, *10*, 289–310. https://doi.org/10.1146/annurev-chembioeng-060718-030354

Cao, H., Zhang, Z., Wu, X., & Miao, X. (2013). Direct biodiesel production from wet microalgae biomass of *Chlorella pyrenoidosa* through in situ transesterification. *BioMed Research International*, *2013*, 930686. https://doi.org/10.1155/2013/930686

Cao, N., Xia, Y., Gong, C. S., & Tsao, G. T. (1997). Production of 2,3-butanediol from pretreated corn cob by *Klebsiella oxytoca* in the presence of fungal cellulase. *Applied Biochemistry and Biotechnology, 63–65*, 129–139. https://doi.org/10.1007/BF02920419

Carlsen, M., Spohr, A. B., Nielsen, J., & Villadsen, J. (1996). Morphology and physiology of an alpha-amylase producing strain of *Aspergillus oryzae* during batch cultivations. *Biotechnology and Bioengineering, 49*(3), 266–276. https://doi.org/10.1002/(SICI)1097-0290(19960205)49:3<266::AID-BIT4>3.0.CO;2-I

Carlsson, M., Lagerkvist, A., & Morgan-Sagastume, F. (2012). The effects of substrate pre-treatment on anaerobic digestion systems: A review. *Waste Management, 32*(9), 1634–1650. https://doi.org/10.1016/j.wasman.2012.04.016

Carota, E., Crognale, S., D'Annibale, A., Gallo, A. M., Stazi, S. R., & Petruccioli, M. (2017). A Sustainable use of ricotta cheese whey for microbial biodiesel production. *The Science of the Total Environment, 584–585*(April), 554–560. https://doi.org/10.1016/j.scitotenv.2017.01.068

Carrère, H., Dumas, C., Battimelli, A., Batstone, D. J., Delgenès, J. P., Steyer, J. P., & Ferrer, I. (2010). Pretreatment methods to improve sludge anaerobic degradability: A review. *Journal of Hazardous Materials, 183*(1), 1–15. https://doi.org/10.1016/j.jhazmat.2010.06.129

Carrieri, D., Ananyev, G., Lenz, O., Bryant, D. A., & Charles Dismukes, G. (2011). Contribution of a sodium ion gradient to energy conservation during fermentation in the cyanobacterium *Arthrospira (Spirulina) maxima* CS-328. *Applied and Environmental Microbiology, 77*(20), 7185–7194. https://doi.org/10.1128/AEM.00612-11

Carvalho, A. S., Melo, M. P., Silva, J. P., Matos, K. S., & Beserra, J. E. A. (2019). Identification of *Colletotrichum* species associated with anthracnose of *Spondias spp.* in Brazil. *Forest Pathology, 49*(6), e12554. https://doi.org/10.1111/efp.12554

Catalanotti, C. P., Yang, W., Posewitz, M. C., & Grossman, A. R. (2013). Fermentation metabolism and its evolution in algae. *Frontiers in Plant Science, 4*. https://doi.org/10.3389/fpls.2013.00150

Cea, M., González, M. E., Abarzúa, M., & Navia, R. (2019). Enzymatic esterification of oleic acid by *Candida rugosa* lipase immobilized onto biochar. *Journal of Environmental Management, 242*(July), 171–177. https://doi.org/10.1016/j.jenvman.2019.04.013

Cescut, J., Fillaudeau, L., Molina-Jouve, C., & Uribelarrea, J.-L. (2014). Carbon accumulation in *Rhodotorula glutinis* induced by nitrogen limitation. *Biotechnology for Biofuels, 7*(1), 164. https://doi.org/10.1186/s13068-014-0164-0

Chades, T., Scully, S. M., Ingvadottir, E. M., & Orlygsson, J. (2018). Fermentation of mannitol extracts from brown macro algae by *Thermophilic clostridia*. *Frontiers in Microbiology, 9*, 1931. https://doi.org/10.3389/fmicb.2018.01931

Chaiklahan, R., Chirasuwan, N., Siangdung, W., Paithoonrangsarid, K., & Bunnag, B. (2010). Cultivation of *Spirulina platensis* using pig wastewater in a semi-continuous process. *Journal of Microbiology and Biotechnology, 20*(3), 609–614. https://doi.org/10.4014/jmb.0907.07026

Chaiyaso, T., Srisuwan, W., Techapun, C., Watanabe, M., & Takenaka, S. (2018). Direct bioconversion of rice residue from canteen waste into lipids by new amylolytic oleaginous yeast *Sporidiobolus pararoseus* KX709872. *Preparative Biochemistry & Biotechnology, 48*(4), 361–371. https://doi.org/10.1080/10826068.2018.1446155

Chakraborty, N., Mukherjee, K., Sarkar, A., & Acharya, K. (2019). Interaction between bean and colletotrichum gloeosporioides: Understanding through a biochemical approach. *Plants (Basel, Switzerland), 8*(9). https://doi.org/10.3390/plants8090345

Chang, J., Cheng, W., Yin, Q., Zuo, R., Song, A., Zheng, Q., Wang, P., Wang, X., & Liu, J. (2012). Effect of steam explosion and microbial fermentation on cellulose and lignin degradation of corn stover. *Bioresource Technology, 104*(January), 587–592. https://doi.org/10.1016/j.biortech.2011.10.070

Chatterjee, S., & Venkata Mohan, S. (2018). Microbial lipid production by *Cryptococcus curvatus* from vegetable waste hydrolysate. *Bioresource Technology, 254*(April), 284–289. https://doi.org/10.1016/j.biortech.2018.01.079

Chen, C.-Y., Huang, Y.-C., Wei, C.-M., Meng, M., Liu, W.-H., & Yang, C.-H. (2013). Properties of the newly isolated extracellular thermo-alkali-stable laccase from thermophilic actinomycetes, *Thermobifida fusca* and its application in dye intermediates oxidation. *AMB Express, 3*(1), 49. https://doi.org/10.1186/2191-0855-3-49

Chen, W., Xie, T., Shao, Y., & Chen, F. (2012). Genomic characteristics comparisons of 12 food-related filamentous fungi in TRNA gene set, codon usage and amino acid composition. *Gene, 497*(1), 116–124. https://doi.org/10.1016/j.gene.2012.01.016

Chen, X.-F., Huang, C., Xiong, L., Wang, B., Qi, G.-X., Lin, X.-Q., Wang, C., & Chen, X.-D. (2016). Use of Elephant Grass (*Pennisetum purpureum*) acid hydrolysate for microbial oil production by *Trichosporon cutaneum*. *Preparative Biochemistry & Biotechnology*, *46*(7), 704–708. https://doi.org/10.1080/10826068.2015.1135453

Chen, Y., Chai, L., Tang, C., Yang, Z., Zheng, Y., Shi, Y., & Zhang, H. (2012). Kraft lignin biodegradation by *Novosphingobium* sp. B-7 and analysis of the degradation process. *Bioresource Technology*, *123*, 682–685. https://doi.org/10.1016/j.biortech.2012.07.028

Chen, Y., Wu, Y., Zhang, P., Hua, D., Yang, M., Li, C., Chen, Z., & Liu, J. (2012). Direct liquefaction of *Dunaliella tertiolecta* for bio-oil in sub/supercritical ethanol-water. *Bioresource Technology*, *124*(November), 190–198. https://doi.org/10.1016/j.biortech.2012.08.013

Cheng, J., Huang, Y., Feng, J., Sun, J., Zhou, J., & Cen, K. (2013). Improving CO_2 fixation efficiency by optimizing *Chlorella PY-ZU1* culture conditions in sequential bioreactors. *Bioresource Technology*, *144*(September), 321–327. https://doi.org/10.1016/j.biortech.2013.06.122

Cheng, S., Logan, B. E. (2007): Sustainable and efficient biohydrogen production via electrohydrogenesis. *Proceedings of the National Academy of Sciences USA*, *104*, 18871–18873.

Chisti, Y. (2007). Biodiesel from microalgae. *Biotechnology Advances*, *25*(3), 294–306. https://doi.org/10.1016/j.biotechadv.2007.02.001

Chisti, Y. (2013). Constraints to commercialization of algal fuels. *Journal of Biotechnology*, *167*(3), 201–214. https://doi.org/10.1016/j.jbiotec.2013.07.020

Chng, L. M., Lee, K. T., & Chan, D. C. J. (2017). Evaluation on Microalgae Biomass for Bioethanol Production. *IOP Conference Series: Materials Science and Engineering*, *206*, 012018. https://doi.org/10.1088/1757-899X/206/1/012018

Cho, S., Lee, N., Park, S., Yu, J., Luong, T. T., Oh, Y.-K., & Lee, T. (2013). Microalgae cultivation for bioenergy production using wastewaters from a municipal WWTP as nutritional sources. *Bioresource Technology*, *131*, 515–520. https://doi.org/10.1016/j.biortech.2012.12.176

Chochois, V., Dauvillée, D., Beyly, A., Tolleter, D., Cuiné, S., Timpano, H., Ball, S., Cournac, L., & Peltier, G. (2009). Hydrogen production in *Chlamydomonas*: Photosystem II-dependent and -independent pathways differ in their requirement for starch metabolism. *Plant Physiology*, *151*(2), 631–640. https://doi.org/10.1104/pp.109.144576

Choi, S. P., Bahn, S.-H., & Sim, S. J. (2013). Improvement of hydrocarbon recovery by spouting solvent into culture of *Botryococcus braunii*. *Bioprocess and Biosystems Engineering*, *36*(12), 1977–1985. https://doi.org/10.1007/s00449-013-0974-7

Choi, S. P., Nguyen, M. T., & Sim, S. J. (2010). Enzymatic pretreatment of *Chlamydomonas reinhardtii* biomass for ethanol production. *Bioresource Technology*, *101*(14), 5330–5336. https://doi.org/10.1016/j.biortech.2010.02.026

Choix, F. J., Polster, E., Corona-González, R. S.-C., & Méndez-Acosta, H. O. (2017). Nutrient composition of culture media induces different patterns of CO_2 fixation from biogas and biomass production by the Microalga *Scenedesmus obliquus* U169. *Bioprocess and Biosystems Engineering*, *40*(12), 1733–1742. https://doi.org/10.1007/s00449-017-1828-5

Christenson, L., & Sims, R. (2011). Production and harvesting of microalgae for wastewater treatment, biofuels, and bioproducts. *Biotechnology Advances*, *29*(6), 686–702. https://doi.org/10.1016/j.biotechadv.2011.05.015

Christopher, L. P., Yao, B., & Ji, Y. (2014). Lignin biodegradation with laccase-mediator systems. *Frontiers in Energy Research*, *2*. https://doi.org/10.3389/fenrg.2014.00012

Chu, F.-F., Chu, P.-N., Shen, X.-F. Lam, P. K. S., & Zeng, R. J. (2014). Effect of phosphorus on biodiesel production from *Scenedesmus obliquus* under nitrogen-deficiency stress. *Bioresource Technology*, *152*, 241–246. https://doi.org/10.1016/j.biortech.2013.11.013

Chu, W.-L. (2017). Strategies to enhance production of microalgal biomass and lipids for biofuel feedstock. *European Journal of Phycology*, *52*(4), 419–437. https://doi.org/10.1080/09670262.2017.1379100

Ciudad, G., Rubilar, O., Azócar, L., Toro, C., Cea, M., Torres, Á., Ribera, A., & Navia, R. (2014). Performance of an enzymatic extract in *Botrycoccus braunii* cell wall disruption. *Journal of Bioscience and Bioengineering*, *117*(1), 75–80. https://doi.org/10.1016/j.jbiosc.2013.06.012

Córdova, O., Passos, F., & Chamy, R. (2018). Physical pretreatment methods for improving microalgae anaerobic biodegradability. *Applied Biochemistry and Biotechnology*, *185*(1), 114–126. https://doi.org/10.1007/s12010-017-2646-6

Córdova, O., Passos, F., & Chamy, R. (2019). Enzymatic pretreatment of microalgae: Cell wall disruption, biomass solubilisation and methane yield increase. *Applied Biochemistry and Biotechnology*, *189*(3), 787–797. https://doi.org/10.1007/s12010-019-03044-8

Costa, N. L., Clarke, T. A., Philipp, L.-A., Gescher, J., Louro, R. O., Paquete, C. M. (2018). Electron transfer process in microbial electrochemical technologies: The role of cell-surface exposed conductive proteins. *Bioresource Technology, 255*, 308–317.

Darpito, C., Shin, W.-S., Jeon, S., Lee, H., Nam, K., Kwon, J.-H., & Yang, J.-W. (2015). Cultivation of *Chlorella protothecoides* in anaerobically treated brewery wastewater for cost-effective biodiesel production. *Bioprocess and Biosystems Engineering, 38*(3), 523–530. https://doi.org/10.1007/s00449-014-1292-4

Deeba, F., Pruthi, V., & Negi, Y. S. (2016). Converting paper mill sludge into neutral lipids by oleaginous yeast *Cryptococcus vishniaccii* for biodiesel production. *Bioresource Technology, 213*(August), 96–102. https://doi.org/10.1016/j.biortech.2016.02.105

de Farias Silva, C. E., & Bertucco, A. (2016). Bioethanol from microalgae and cyanobacteria: A review and technological outlook. *Process Biochemistry, 51*(11), 1833–1842. https://doi.org/10.1016/j.procbio.2016.02.016

De la Rubia, M. A., Fernández-Cegrí, V., Raposo, F., & Borja, R. (2013). Anaerobic digestion of sunflower oil cake: A current overview. *Water Science and Technology, 67*(2), 410–417. https://doi.org/10.2166/wst.2012.586

Deng, M.-D., & Coleman, J. R. (1999). Ethanol synthesis by genetic engineering in cyanobacteria. *Applied and Environmental Microbiology, 65*(2), 523–528. https://doi.org/10.1128/AEM.65.2.523-528.1999

Deng, Y., Li, S., Xu, Q., Gao, M., & Huang, H. (2012). Production of fumaric acid by simultaneous saccharification and fermentation of starchy materials with 2-deoxyglucose-resistant mutant strains of *Rhizopus oryzae*. *Bioresource Technology, 107*(March), 363–367. https://doi.org/10.1016/j.biortech.2011.11.117

De Francisci, D., Su, Y., Iital, A., & Angelidaki, I. (2018). Evaluation of microalgae production coupled with wastewater treatment. *Environmental Technology, 39*(5), 581–592. https://doi.org/10.1080/09593330.2017.1308441

Dexter, J., Armshaw, P., Sheahan, C., & Pembroke, J. T. (2015). The state of autotrophic ethanol production in Cyanobacteria. *Journal of Applied Microbiology, 119*(1), 11–24. https://doi.org/10.1111/jam.12821

Dexter, J., & Fu, P. (2009). Metabolic engineering of cyanobacteria for ethanol production. *Energy & Environmental Science, 2*(8), 857–864. https://doi.org/10.1039/B811937F

Dey, P., Banerjee, J., & Maiti, M. K. (2011). Comparative lipid profiling of two endophytic fungal isolates—*Colletotrichum* sp. and *Alternaria* sp. having potential utilities as biodiesel feedstock. *Bioresource Technology, 102*(10), 5815–5823. https://doi.org/10.1016/j.biortech.2011.02.064

Dhananjay, S. K., & Mulimani, V. H. (2009). Three-phase partitioning of alpha-galactosidase from fermented media of *Aspergillus oryzae* and comparison with conventional purification techniques. *Journal of Industrial Microbiology & Biotechnology, 36*(1), 123–128. https://doi.org/10.1007/s10295-008-0479-6

Dienst, D., Georg, J., Abts, T., Jakorew, L., Kuchmina, E., Börner, T., Wilde, A., Dühring, U., Enke, H., & Hess, W. R. (2014). Transcriptomic response to prolonged ethanol production in the cyanobacterium *Synechocystis* sp. PCC6803. *Biotechnology for Biofuels, 7*, 21. https://doi.org/10.1186/1754-6834-7-21

Ding, W., Jin, W., Zhou, X., Li, S.-F., Tu, R., Han, S.-F., Chen, C., Feng, X., & Huang, Y. (2020). Enhanced lipid extraction from the biodiesel-producing microalga *Chlorella pyrenoidosa* cultivated in municipal wastewater via daphnia ingestion and digestion. *Bioresource Technology, 306*(March), 123162. https://doi.org/10.1016/j.biortech.2020.123162

Do Nascimento, M., de Los Angeles Dublan, M., Ortiz-Marquez, J. C. F., & Curatti, L. (2013). High lipid productivity of an ankistrodesmus-rhizobium artificial consortium. *Bioresource Technology, 146*(October), 400–407. https://doi.org/10.1016/j.biortech.2013.07.085

Doebbe, A., Rupprecht, J., Beckmann, J., Mussgnug, J. H., Hallmann, A., Hankamer, B., & Kruse, O. (2007). Functional integration of the HUP1 hexose symporter gene into the genome of *C. reinhardtii*: Impacts on biological H2 production. *Journal of Biotechnology, 131*(1), 27–33. https://doi.org/10.1016/j.jbiotec.2007.05.017

Donaldson, G. P., Lee, S. M., & Mazmanian, S. K. (2016). Gut biogeography of the bacterial microbiota. *Nature Reviews. Microbiology, 14*(1), 20. https://doi.org/10.1038/nrmicro3552

Donohoue, P. D., Barrangou, R., & May, A. P. (2018). Advances in Industrial Biotechnology Using CRISPR-Cas Systems. *Trends in Biotechnology, 36*(2), 134–146. https://doi.org/10.1016/j.tibtech.2017.07.007

Doran, J. B., Aldrich, H. C., & Ingram, L. O. (1994). Saccharification and fermentation of sugar cane bagasse by *Klebsiella oxytoca* P2 containing chromosomally integrated genes encoding the *Zymomonas mobilis* ethanol pathway. *Biotechnology and Bioengineering, 44*(2), 240–247. https://doi.org/10.1002/bit.260440213

Drozdíková, E., Garaiová, M., Csáky, Z., Obernauerová, M., & Hapala, I. (2015). Production of squalene by lactose-fermenting yeast *Kluyveromyces lactis* with reduced squalene epoxidase activity. *Letters in Applied Microbiology, 61*(1), 77–84. https://doi.org/10.1111/lam.12425

Du, Z.-Y., Alvaro, J., Hyden, B., Zienkiewicz, K., Benning, N., Zienkiewicz, A., Bonito, G., & Benning, C. (2018). Enhancing oil production and harvest by combining the marine alga *Nannochloropsis oceanica* and the oleaginous fungus *Mortierella elongata*. *Biotechnology for Biofuels*, *11*. https://doi.org/10.1186/s13068-018-1172-2

Du, Z.-Y., Zienkiewicz, K., Vande Pol, N., Ostrom, N. E., Benning, C., & Bonito, G. M. (2019). Algal-fungal symbiosis leads to photosynthetic mycelium. *ELife*, *8*. https://doi.org/10.7554/eLife.47815

Duan, G., Wu, B., Qin, H., Wang, W., Tan, Q., Dai, Y., Qin, Y., Tan, F., Hu, G., & He, M. (2019). Replacing water and nutrients for ethanol production by ARTP derived biogas slurry tolerant *Zymomonas mobilis* strain. *Biotechnology for Biofuels*, *12*, 124. https://doi.org/10.1186/s13068-019-1463-2

Duarte, S. H., del Peso Hernández, G. L., Canet, A., Benaiges, M. D., Maugeri, F., & Valero, F. (2015). Enzymatic biodiesel synthesis from yeast oil using immobilized recombinant *Rhizopus oryzae* lipase. *Bioresource Technology*, *183*(May), 175–180. https://doi.org/10.1016/j.biortech.2015.01.133

Duman, G., Azhar Uddin, Md., & Yanik, J. (2014). Hydrogen production from algal biomass via steam gasification. *Bioresource Technology*, *166*(August), 24–30. https://doi.org/10.1016/j.biortech.2014.04.096

Duong, V. T., Thomas-Hall, S. R., & Schenk, P. M. (2015). Growth and lipid accumulation of microalgae from fluctuating brackish and sea water locations in South East Queensland-Australia. *Frontiers in Plant Science*, *6*, 359. https://doi.org/10.3389/fpls.2015.00359

Easterling, E. R., French, W. T., Hernandez, R., & Licha, M. (2009). The effect of glycerol as a sole and secondary substrate on the growth and fatty acid composition of *Rhodotorula glutinis*. *Bioresource Technology*, *100*(1), 356–361. https://doi.org/10.1016/j.biortech.2008.05.030

Edwards, M. C., & Doran-Peterson, J. (2012). Pectin-rich biomass as feedstock for fuel ethanol production. *Applied Microbiology and Biotechnology*, *95*(3), 565–575. https://doi.org/10.1007/s00253-012-4173-2

Eilenberg, H., Weiner, I., Ben-Zvi, O., Pundak, C., Marmari, A., Liran, O., Wecker, M. S., Milrad, Y., & Yacoby, I. (2016). The dual effect of a ferredoxin-hydrogenase fusion protein in vivo: Successful divergence of the photosynthetic electron flux towards hydrogen production and elevated oxygen tolerance. *Biotechnology for Biofuels*, *9*(1), 182. https://doi.org/10.1186/s13068-016-0601-3

Eixelsberger, T., Woodley, J. M., Nidetzky, B., & Kratzer, R. (2013). Scale-up and intensification of (S)-1-(2-chlorophenyl)ethanol bioproduction: Economic evaluation of whole cell-catalyzed reduction of o-Chloroacetophenone. *Biotechnology and Bioengineering*, *110*(8), 2311–2315. https://doi.org/10.1002/bit.24896

Eladel, H., El-Fatah Abomohra, A., Battah, M., Mohmmed, S., Radwan, A., & Abdelrahim, H. (2019). Evaluation of *Chlorella sorokiniana* isolated from local municipal wastewater for dual application in nutrient removal and biodiesel production. *Bioprocess and Biosystems Engineering*, *42*(3), 425–433. https://doi.org/10.1007/s00449-018-2046-5

Elbehri, A., Segerstedt, A., & Liu, P. (2013). Biofuels and the sustainability challenge: A global assessment of sustainability issues, trends and policies for biofuels and related feedstocks. *Biofuels and the Sustainability Challenge: A Global Assessment of Sustainability Issues, Trends and Policies for Biofuels and Related Feedstocks*. www.cabdirect.org/cabdirect/abstract/20133054982

Elkahlout, K., Sagir, E., Alipour, S., Koku, H., Gunduz, U., Eroglu, I., Yucel, M. (2018). Long-term stable hydrogen production from acetate using immobilized *Rhodobacter capsulatus* in a panel photobioreactor. *International Journal of Hydrogen Energy*, *44*(34), 18801–18810. https://doi.org/10.1016/j.ijhydene.2018.10.133

Ellison, C. R., Overa, S., & Boldor, D. (2019). Central composite design parameterization of microalgae/cyanobacteria co-culture pretreatment for enhanced lipid extraction using an external clamp-on ultrasonic transducer. *Ultrasonics Sonochemistry*, *51*(March), 496–503. https://doi.org/10.1016/j.ultsonch.2018.05.006

Elmansy, E. A., Asker, M. S., El-Kady, E. M., Hassanein, S. M., & El-Beih, F. M. (2018). Production and optimization of α-amylase from thermo-halophilic bacteria isolated from different local marine environments. *Bulletin of the National Research Centre*, *42*(1), 31. https://doi.org/10.1186/s42269-018-0033-2

Encinar, J. M., González, J. F., Sánchez, N., Nogales-Delgado, S. (2019). Sunflower oil transesterification with methanol using immobilized lipase enzymes. *Bioprocess and Biosystems Engineering*, *42*(1), 157–166. https://doi.org/10.1007/s00449-018-2023-z

Enguita, F. J., Martins, L. O., Henriques, A. O., & Carrondo, M. A. (2003). Crystal structure of a bacterial endospore coat component. A laccase with enhanced thermostability properties. *Journal of Biological Chemistry*, *278*(21), 19416–19425. https://doi.org/10.1074/jbc.M301251200

Espinosa-Gonzalez, I., Parashar, A., & Bressler, D. C. (2014). Heterotrophic growth and lipid accumulation of *Chlorella protothecoides* in whey permeate, a dairy by-product stream, for biofuel production. *Bioresource Technology*, *155*(March), 170–176. https://doi.org/10.1016/j.biortech.2013.12.028

Ethier, S., Woisard, K., Vaughan, D., & Wen, Z. (2011). Continuous culture of the microalgae *Schizochytrium limacinum* on biodiesel-derived crude glycerol for producing docosahexaenoic acid. *Bioresource Technology*, *102*(1), 88–93. https://doi.org/10.1016/j.biortech.2010.05.021

Eungrasamee, K., Incharoensakdi, A., Lindblad, P., Jantaro, S. (2020). *Synechocystis* sp. PCC 6803 over-expressing genes involved in CBB cycle and free fatty acid cycling enhances the significant levels of intracellular lipids and secreted free fatty acids. *Scientific Reports*, *10*, 4515. https://doi.org/10.1038/s41598-020-61100-4 1

Evans, C. T., & Ratledge, C. (1983). A comparison of the oleaginous yeast, *Candida curvata*, grown on different carbon sources in continuous and batch culture. *Lipids*, *18*(9), 623–629. https://doi.org/10.1007/BF02534673

Fakhimi, N., Gonzalez-Ballester, D., Fernández, E., Galván, A., & Dubini, A. (2020). Algae-bacteria consortia as a strategy to enhance H2 production. *Cells*, *9*(6). https://doi.org/10.3390/cells9061353

Favaro, L., Jansen, T., & van Zyl, W. H. (2019). Exploring industrial and natural *Saccharomyces cerevisiae* strains for the bio-based economy from biomass: The case of bioethanol. *Critical Reviews in Biotechnology*, *39*(6), 800–816. https://doi.org/10.1080/07388551.2019.1619157

Fei, Q., Fu, R., Shang, L., Brigham, C. J., & Nam Chang, H. (2015). Lipid production by microalgae *Chlorella protothecoides* with Volatile Fatty Acids (VFAs) as carbon sources in heterotrophic cultivation and its economic assessment. *Bioprocess and Biosystems Engineering*, *38*(4), 691–700. https://doi.org/10.1007/s00449-014-1308-0

Feng, J., Guo, Y., Zhang, X., Wang, G., Lv, J., Liu, Q., & Xie, S. (2016). Identification and characterization of a symbiotic alga from soil bryophyte for lipid profiles. *Biology Open*, *5*(9), 1317–1323. https://doi.org/10.1242/bio.019992

Feng, P., Deng, Z., Fan, L., & Hu, Z. (2012). Lipid accumulation and growth characteristics of *Chlorella zofingiensis* under different nitrate and phosphate concentrations. *Journal of Bioscience and Bioengineering*, *114*(4), 405–410. https://doi.org/10.1016/j.jbiosc.2012.05.007

Feng, P., Deng, Z., Hu, Z., & Fan, L. (2011). Lipid accumulation and growth of *Chlorella zofingiensis* in flat plate photobioreactors outdoors. *Bioresource Technology*, *102*(22), 10577–10584. https://doi.org/10.1016/j.biortech.2011.08.109

Feng, P., Deng, Z., Hu, Z., Wang, Z., & Fan, L. (2014). Characterization of *Chlorococcum pamirum* as a potential biodiesel feedstock. *Bioresource Technology*, *162*(June), 115–122. https://doi.org/10.1016/j.biortech.2014.03.076

Fernandes, S., & Murray, P. (2010). Metabolic engineering for improved microbial pentose fermentation. *Bioengineered Bugs*, *1*(6), 424–428. https://doi.org/10.4161/bbug.1.6.12724

Fernández-Fueyo, E., Ruiz-Dueñas, F. J., & Martínez, A. T. (2014). Engineering a fungal peroxidase that degrades lignin at very acidic pH. *Biotechnology for Biofuels*, *7*, 114. https://doi.org/10.1186/1754-6834-7-114

Ferreira, A., Marques, P., Ribeiro, B., Assemany, P., de Mendonça, H. V., Barata, A., Oliveira, A. C., Reis, A., Pinheiro, H. M., & Gouveia, L. (2018). Combining biotechnology with circular bioeconomy: From poultry, swine, cattle, brewery, dairy and urban wastewaters to biohydrogen. *Environmental Research*, *164*(July), 32–38. https://doi.org/10.1016/j.envres.2018.02.007

Figueroa-Espinoza, M. C., Poulsen, C., Borch Søe, J., Zargahi, M. R., & Rouau, X. (2002). Enzymatic solubilization of arabinoxylans from isolated rye pentosans and rye flour by different endo-xylanases and other hydrolyzing enzymes. Effect of a fungal caccase on the flour extracts oxidative gelation. *Journal of Agricultural and Food Chemistry*, *50*(22), 6473–6484. https://doi.org/10.1021/jf0255026

Filippousi, R., Antoniou, D., Tryfinopoulou, P., Nisiotou, A. A., Nychas, G.-J., Koutinas, A. A., & Papanikolaou, S. (2019). Isolation, identification and screening of yeasts towards their ability to assimilate biodiesel-derived crude glycerol: Microbial production of polyols, endopolysaccharides and lipid. *Journal of Applied Microbiology*, *127*(4), 1080–1100. https://doi.org/10.1111/jam.14373

Fonseca, C., Olofsson, K., Ferreira, C., Runquist, D., Fonseca, L. L., Hahn-Hägerdal, B., & Lidén, G. (2011). The glucose/xylose facilitator Gxf1 from *Candida intermedia* expressed in a xylose-fermenting industrial strain of *Saccharomyces cerevisiae* increases xylose uptake in SSCF of wheat straw. *Enzyme and Microbial Technology*, *48*(6–7), 518–525. https://doi.org/10.1016/j.enzmictec.2011.02.010

Fonseca, G. G., Heinzle, E., Wittmann, C., & Gombert, A. K. (2008). The yeast *Kluyveromyces marxianus* and its biotechnological potential. *Applied Microbiology and Biotechnology*, *79*(3), 339–354. https://doi.org/10.1007/s00253-008-1458-6

Foster, B. C., Litster, D. L., & Lodge, B. A. (1991). Biotransformation of 2-, 3-, and 4-methoxy-amphetamines by *Cunninghamella echinulata*. *Xenobiotica; the Fate of Foreign Compounds in Biological Systems*, *21*(10), 1337–1346. https://doi.org/10.3109/00498259109043208

Frangville, C., Rutkevičius, M., Richter, A. P., Velev, O. D., Stoyanov, S. D., & Paunov, V. N. (2012). Fabrication of environmentally biodegradable lignin nanoparticles. *Chemphyschem: A European Journal of Chemical Physics and Physical Chemistry*, *13*(18), 4235–4243. https://doi.org/10.1002/cphc.201200537

Fuentes, J. L., Garbayo, I., Cuaresma, M., Montero, Z., González-del-Valle, M., & Vílchez, C. (2016). Impact of microalgae-bacteria interactions on the production of algal biomass and associated compounds. *Marine Drugs, 14*(5), 100. https://doi.org/10.3390/md14050100

Funnell-Harris, D. L., Prom, L. K., & Pedersen, J. F. (2012). Isolation and characterization of the grain mold fungi *Cochliobolus* and *Alternaria* spp. from sorghum using semiselective media and DNA sequence analyses. *Canadian Journal of Microbiology, 59*(2), 87–96. https://doi.org/10.1139/cjm-2012-0649

Furbino, L. E., Godinho, V. M., Santiago, I. F., Pellizari, F. M., Alves, T. M. A., Zani, C. L., Junior, P. A. S., Romanha, A. J., Carvalho, A. G. O., Gil, L. H. V. G., Rosa, C. A., Minnis, A. M., & Rosa, L. H. (2014). Diversity patterns, ecology and biological activities of fungal communities associated with the endemic macroalgae across the Antarctic peninsula. *Microbial Ecology, 67*(4), 775–787. https://doi.org/10.1007/s00248-014-0374-9

Furuhashi, T., Nakamura, T., Fragner, L., Roustan, V., Schön, V., & Weckwerth, W. (2016). Biodiesel and polyunsaturated fatty acids production from algae and crop plants: A rapid and comprehensive workflow for lipid analysis. *Biotechnology Journal, 11*(10), 1262–1267. https://doi.org/10.1002/biot.201400197

Gai, C., Li, Y., Peng, N., Fan, A., & Liu, Z. (2015). Co-liquefaction of microalgae and lignocellulosic biomass in subcritical water. *Bioresource Technology, 185*(June), 240–245. https://doi.org/10.1016/j.biortech.2015.03.015

Gama-Castro, S., Núñez, C., Segura, D., Moreno, S., Guzmán, J., & Espín, G. (2001). *Azotobacter vinelandii* aldehyde dehydrogenase regulated by sigma(54): Role in alcohol catabolism and encystment. *Journal of Bacteriology, 183*(21), 6169–6174. https://doi.org/10.1128/JB.183.21.6169-6174.2001

Gao, G., Wu, M., Fu, Q., Li, X., & Xu, J. (2019). A two-stage model with nitrogen and silicon limitation enhances lipid productivity and biodiesel features of the marine bloom-forming diatom *Skeletonema costatum*. *Bioresource Technology, 289*(October), 121717. https://doi.org/10.1016/j.biortech.2019.121717

Gao, Y., Yang, M., & Wang, C. (2013). Nutrient deprivation enhances lipid content in marine microalgae. *Bioresource Technology, 147*(November), 484–491. https://doi.org/10.1016/j.biortech.2013.08.066

Gao, Z., Zhao, H., Li, Z., Tan, X., & Lu, X. (2012). Photosynthetic production of ethanol from carbon dioxide in genetically engineered cyanobacteria. *Energy & Environmental Science, 5*(12), 9857–9865. https://doi.org/10.1039/C2EE22675H

Gęsicka, A., Borkowska, M., Białas, W., Kaczmarek, P., & Celińska, E. (2020). Production of raw starch-digesting amylolytic preparation in *Yarrowia lipolytica* and its application in biotechnological synthesis of lactic acid and ethanol. *Microorganisms, 8*(5). https://doi.org/10.3390/microorganisms8050717

Ghirardi, M. L., Zhang, L., Lee, J. W., Flynn, T., Seibert, M., Greenbaum, E., & Melis, A. (2000). Microalgae: A green source of renewable H(2). *Trends in Biotechnology, 18*(12), 506–511. https://doi.org/10.1016/s0167-7799(00)01511-0

Gibbons, J. G., Salichos, L., Slot, J. C., Rinker, D. C., McGary, K. L., King, J. G., Klich, M. A., Tabb, D. L., McDonald, W. H., & Rokas, A. (2012). The evolutionary imprint of domestication on genome variation and function of the filamentous fungus *Aspergillus oryzae*. *Current Biology: CB, 22*(15), 1403–1409. https://doi.org/10.1016/j.cub.2012.05.033

Gnouma, A., Sehli, E., Medhioub, W., Dhieb, R. B., Masri, M., Mehlmer, N., Slimani, W., et al. (2018). Strain selection of microalgae isolated from Tunisian coast: Characterization of the lipid profile for potential biodiesel production. *Bioprocess and Biosystems Engineering, 41*(10), 1449–1459. https://doi.org/10.1007/s00449-018-1973-5

Go, S., Lee, S.-J., Jeong, G.-T., & Kim, S.-K. (2012). Factors affecting the growth and the oil accumulation of marine microalgae, *Tetraselmis suecica*. *Bioprocess and Biosystems Engineering, 35*(1–2), 145–150. https://doi.org/10.1007/s00449-011-0635-7

Gomes, F. M., Silva, G. S., Pinatti, D. G., Conte, R. A., & de Castro, H. F. (2005). Wood cellulignin as an alternative matrix for enzyme immobilization—PubMed. *Applied Biochemistry & Biotechnology, 121–124*, 255–268. https://pubmed.ncbi.nlm.nih.gov/15917604/

Gong, Z., Shen, H., Yang, X., Wang, Q., Xie, H., & Zhao, Z. K. (2014). Lipid production from corn stover by the oleaginous yeast *Cryptococcus curvatus*. *Biotechnology for Biofuels, 7*(1), 158. https://doi.org/10.1186/s13068-014-0158-y

Gong, Z., Shen, H., Zhou, W., Wang, Y., Yang, X., & Zhao, Z. K. (2015). Efficient conversion of acetate into lipids by the oleaginous yeast *Cryptococcus curvatus*. *Biotechnology for Biofuels, 8*, 189. https://doi.org/10.1186/s13068-015-0371-3

Gong, Z., Wang, Q., Shen, H., Wang, L., Xie, H., & Zhao, Z. K. (2014). Conversion of biomass-derived oligosaccharides into lipids. *Biotechnology for Biofuels, 7*(1), 13. https://doi.org/10.1186/1754-6834-7-13

Gong, Z., Zhou, W., Shen, H., Yang, Z., Wang, G., Zuo, Z., Hou, Y., & Zhao, Z. K. (2016). Co-Fermentation of acetate and sugars facilitating microbial lipid production on acetate-rich biomass hydrolysates. *Bioresource Technology, 207*(May), 102–108. https://doi.org/10.1016/j.biortech.2016.01.122

Gong, Z., Zhou, W., Shen, H., Zhao, Z. K., Yang, Z., Yan, J., & Zhao, M. (2016). Co-utilization of corn stover hydrolysates and biodiesel-derived glycerol by *Cryptococcus curvatus* for lipid production. *Bioresource Technology, 219*(November), 552–558. https://doi.org/10.1016/j.biortech.2016.08.021

Gordon, R., & Seckbach, J. (Eds.). (2012). *The Science of Algal Fuels: Phycology, Geology, Biophotonics, Genomics and Nanotechnology*. Dordrecht, Netherlands: Springer. https://doi.org/10.1007/978-94-007-5110-1

Gou, Q., Tang, M., Wang, Y., Zhou, W., Liu, Y., & Gong, Z. (2020). Deficiency of β-Glucosidase beneficial for the simultaneous saccharification and lipid production by the oleaginous yeast *Lipomyces starkeyi*. *Applied Biochemistry and Biotechnology, 190*(2), 745–757. https://doi.org/10.1007/s12010-019-03129-4

Granja-Travez, R. S., & Bugg, T. D. H. (2018). Characterization of multicopper oxidase CopA from *Pseudomonas putida* KT2440 and *Pseudomonas fluorescens* Pf-5: Involvement in bacterial lignin oxidation. *Archives of Biochemistry and Biophysics, 660*(December), 97–107. https://doi.org/10.1016/j.abb.2018.10.012

Grechanik, V., Romanova, A., Naydov, I., & Tsygankov, A. (2020). Photoautotrophic cultures of *Chlamydomonas reinhardtii*: Sulfur deficiency, anoxia, and hydrogen production. *Photosynthesis Research, 143*(3), 275–286. https://doi.org/10.1007/s11120-019-00701-1

Greenwalt, C. J., Hunter, J. B., Lin, S., McKenzie, S., & Denvir, A. (2000). Ozonation and alkaline-peroxide pretreatment of wheat straw for *Cryptococcus curvatus* fermentation. *Life Support & Biosphere Science: International Journal of Earth Space, 7*(3), 243–249.

Griffiths, M. J., van Hille, R. P., & Harrison, S. T. L. (2012). Lipid productivity, settling potential and fatty acid profile of 11 microalgal species grown under nitrogen replete and limited conditions. *Journal of Applied Phycology, 24*(5), 989–1001. https://doi.org/10.1007/s10811-011-9723-y

Grossman, A. R., Croft, M., Gladyshev, V. N., Merchant, S. S., Posewitz, M. C., Prochnik, S., & Spalding, M. H. (2007). Novel metabolism in *Chlamydomonas* through the lens of genomics. *Current Opinion in Plant Biology, 10*(2), 190–198. https://doi.org/10.1016/j.pbi.2007.01.012

Gu, H., Zhang, J., & Bao, J. (2015). High tolerance and physiological mechanism of *Zymomonas mobilis* to phenolic inhibitors in ethanol fermentation of corncob residue. *Biotechnology and Bioengineering, 112*(9), 1770–1782. https://doi.org/10.1002/bit.25603

Guerriero, G., Hausman, J.-F., Strauss, J., Ertan, H., & Siddiqui, K. S. (2015). Destructuring plant biomass: Focus on fungal and extremophilic cell wall hydrolases. *Plant Science: An International Journal of Experimental Plant Biology, 234*, 180–193. https://doi.org/10.1016/j.plantsci.2015.02.010

Gülyurt, M. Ö., Özçimen, D., & İnan, B. (2016). Biodiesel production from *Chlorella protothecoides* oil by microwave-assisted transesterification. *International Journal of Molecular Sciences, 17*(4). https://doi.org/10.3390/ijms17040579

Guo, D., Zhang, Z., Liu, D., Zheng, H., Chen, H., & Chen, K. (2014). A comparative study on the degradation of gallic acid by *Aspergillus oryzae* and *Phanerochaete chrysosporium*. *Water Science and Technology: A Journal of the International Association on Water Pollution Research, 70*(1), 175–181. https://doi.org/10.2166/wst.2014.213

Guo, Z., & Tong, Y. W. (2014). The interactions between *Chlorella vulgaris* and algal symbiotic bacteria under photoautotrophic and photoheterotrophic conditions. *Journal of Applied Phycology, 26*(3), 1483–1492. https://doi.org/10.1007/s10811-013-0186-1

Guo, Z., Li, Y., & Guo, H. (2017). Effect of light/dark regimens on hydrogen production by *Tetraselmis subcordiformis* coupled with an alkaline fuel cell system. *Applied Biochemistry and Biotechnology, 183*(4), 1295–1303. https://doi.org/10.1007/s12010-017-2498-0

Guo, Z.-P., Robin, J., Duquesne, S., O'Donohue, M. J., Marty, A., & Bordes, F. (2018). Developing cellulolytic *Yarrowia lipolytica* as a platform for the production of valuable products in consolidated bioprocessing of cellulose. *Biotechnology for Biofuels, 11*, 141. https://doi.org/10.1186/s13068-018-1144-6

Gupta, V. K., & Tuohy, M. G. (Eds.). (2013). *Biofuel Technologies*. Berlin & Heidelberg: Springer. https://doi.org/10.1007/978-3-642-34519-7

Gupta, V. K., Zellinger, S., Ferreira Filho, E. X., & Durián-Dominguez-de-Bazua, M. C. (2017). *Microbial Applications: Recent Advancement and Future Developments*. Berlin, Germany: Watler de Gruyter GmbH & Co KG.

Hagemann, M., & Hess, W. R. (2018). Systems and synthetic biology for the biotechnological application of cyanobacteria. *Current Opinion in Biotechnology, 49*, 94–99. https://doi.org/10.1016/j.copbio.2017.07.008

Haider, K., Trojanowski, J., & Sundman, V. (1978). Screening for lignin degrading bacteria by means of 14C-labelled lignins. *Archives of Microbiology, 119*(1), 103–106. https://doi.org/10.1007/BF00407936

Halim, R., Gladman, B., Danquah, M. K., & Webley, P. A. (2011). Oil extraction from microalgae for biodiesel production. *Bioresource Technology, 102*(1), 178–185. https://doi.org/10.1016/j.biortech.2010.06.136

Halmemies-Beauchet-Filleau, A., Rinne, M., Lamminen, M., Mapato, C., Ampapon, T., Wanapat, M., & Vanhatalo, A. (2018). Review: Alternative and novel feeds for ruminants: Nutritive value, product quality and environmental aspects. *Animal: An International Journal of Animal Bioscience, 12*(s2), s295–309. https://doi.org/10.1017/S1751731118002252

Hama, S., Tamalampudi, S., Suzuki, Y., Yoshida, A., Fukuda, H., & Kondo, A. (2008). Preparation and comparative characterization of immobilized *Aspergillus oryzae* expressing *Fusarium heterosporum* lipase for enzymatic biodiesel production. *Applied Microbiology and Biotechnology, 81*(4), 637–645. https://doi.org/10.1007/s00253-008-1689-6

Han, M.-A., Hong, S.-J., Kim, Z.-H., Cho, B.-K., Lee, H., Choi, H.-K., & Lee, C.-G. (2018). Enhanced production of fatty acids via redirection of carbon flux in marine microalga *Tetraselmis* sp. *Journal of Microbiology and Biotechnology, 28*(2), 267–274. https://doi.org/10.4014/jmb.1702.02064

Han, S. O., Yukawa, H., Inui, M., & Doi, R. H. (2003). Transcription of *Clostridium cellulovorans* cellulosomal cellulase and hemicellulase genes. *Journal of Bacteriology, 185*(8), 2520–2527. https://doi.org/10.1128/JB.185.8.2520-2527.2003

Han, S., Jin, W., Chen, Y., Tu, R., & El-Fatah Abomohra, A. (2016). Enhancement of lipid production of *Chlorella pyrenoidosa* cultivated in municipal wastewater by magnetic treatment. *Applied Biochemistry and Biotechnology, 180*(6), 1043–1055. https://doi.org/10.1007/s12010-016-2151-3

Han, W., Huang, J., Zhao, H., & Li, Y. (2016). Continuous biohydrogen production from waste bread by anaerobic sludge. *Bioresource Technology, 212*(July), 1–5. https://doi.org/10.1016/j.biortech.2016.04.007

Han, W., Ye, M., Zhu, A. J., Zhao, H. T., & Li, Y. F. (2015). Batch dark fermentation from enzymatic hydrolyzed food waste for hydrogen production. *Bioresource Technology, 191*(September), 24–29. https://doi.org/10.1016/j.biortech.2015.04.120

Haranczyk, H., Casanova-Katny, A., Olech, M., & Strzalka, K. (2017). Dehydration and freezing resistance of Lichenized Fungi. In V. Shukla et al. (Eds.), *Plant Adaptation Strategies in Changing Environment*. Singapore: Springer Nature. https://doi.org/10.1007/978-981-10-6744-0_3

Harbi, K., Makridis, P., Koukoumis, C., Papadionysiou, M., Vgenis, T., Kornaros, M., Ntaikou, I., Giokas, S., & Dailianis, S. (2017). Evaluation of a battery of marine species-based bioassays against raw and treated municipal wastewaters. *Journal of Hazardous Materials, 321*(January), 537–546. https://doi.org/10.1016/j.jhazmat.2016.09.036

Harwati, T. U., Willke, T., & Vorlop, K. D. (2012). Characterization of the lipid accumulation in a tropical freshwater microalgae *Chlorococcum* sp. *Bioresource Technology, 121*(October), 54–60. https://doi.org/10.1016/j.biortech.2012.06.098

Hassan, E. A., Abd-Alla, M. H., Khalil Bagy, M. M., & Mohamed Morsy, F. (2015). In situ hydrogen, acetone, butanol, ethanol and microdiesel production by *Clostridium acetobutylicum* ATCC 824 from oleaginous fungal biomass. *Anaerobe, 34*(August), 125–131. https://doi.org/10.1016/j.anaerobe.2015.05.007

Hatakka, A. I. (1983). Pretreatment of wheat straw by white-rot fungi for enzymic saccharification of cellulose. *European Journal of Applied Microbiology and Biotechnology, 18*(6), 350–357. https://doi.org/10.1007/BF00504744

He, M. X., Wu, B., Qin, H., Ruan, Z. Y., Tan, F. R., Wang, J. L., Shui, Z. X., et al. (2014). *Zymomonas mobilis*: A novel platform for future biorefineries. *Biotechnology for Biofuels, 7*, 101. https://doi.org/10.1186/1754-6834-7-101

He, Q., Yang, H., & Hu, C. (2015). Optimizing light regimes on growth and lipid accumulation in *Ankistrodesmus fusiformis* H1 for biodiesel production. *Bioresource Technology, 198*(December), 876–883. https://doi.org/10.1016/j.biortech.2015.09.085

He, X., Dai, J., & Wu, Q. (2016). Identification of sporopollenin as the outer layer of cell wall in microalga *Chlorella protothecoides*. *Frontiers in Microbiology, 7*, 1047. https://doi.org/10.3389/fmicb.2016.01047

He, Y., Wang, X., Wei, H., Zhang, J., Chen, B., & Chen, F. (2019). Direct enzymatic ethanolysis of potential *Nannochloropsis* biomass for co-production of sustainable biodiesel and nutraceutical eicosapentaenoic acid. *Biotechnology for Biofuels, 12*, 78. https://doi.org/10.1186/s13068-019-1418-7

He, Z., Kan, J., Mansfeld, F., Angenent, L. T., & Nealson, K. H. (2009). Self-sustained phototrophic microbial fuel cells based on the synergistic cooperation between photosynthetic microorganisms and heterotrophic bacteria. *Environmental Science & Technology, 43*(5), 1648–1654. https://doi.org/10.1021/es803084a

Hemschemeier, A., & Happe, T. (2005). The exceptional photofermentative hydrogen metabolism of the green alga *Chlamydomonas reinhardtii*. *Biochemical Society Transactions*, *33*(Pt 1), 39–41. https://doi.org/10.1042/BST0330039

Hena, S., Fatihah, N., Tabassum, S., & Ismail, N. (2015). Three stage cultivation process of facultative strain of *Chlorella sorokiniana* for treating dairy farm effluent and lipid enhancement. *Water Research*, *80*(September), 346–356. https://doi.org/10.1016/j.watres.2015.05.001

Hendriks, A. T. W. M., & Zeeman, G. (2009). Pretreatments to enhance the digestibility of lignocellulosic biomass. *Bioresource Technology*, *100*(1), 10–18. https://doi.org/10.1016/j.biortech.2008.05.027

Henry, E. C. (2004). Handbook of microalgal culture: Biotechnology and applied phycology. *Journal of Phycology*, *40*(5), 1001–1002. https://doi.org/10.1111/j.1529-8817.2004.40502.x

Heo, S.-W., Ryu, B.-G., Nam, K., Kim, W., & Yang, J.-W. (2015). Simultaneous treatment of food-waste recycling wastewater and cultivation of *Tetraselmis suecica* for biodiesel production. *Bioprocess and Biosystems Engineering*, *38*(7), 1393–1398. https://doi.org/10.1007/s00449-015-1380-0

Heyer, H., & Krumbein, W. E. (1991). Excretion of fermentation products in dark and anaerobically incubated cyanobacteria. *Archives of Microbiology*, *155*(3), 284–287. https://doi.org/10.1007/BF00252213

Hidalgo, P., Ciudad, G., Schober, S., Mittelbach, M., & Navia, R. (2015). Biodiesel synthesis by direct transesterification of microalga *Botryococcus braunii* with continuous methanol reflux. *Bioresource Technology*, *181*(April), 32–39. https://doi.org/10.1016/j.biortech.2015.01.047

Higgins, B. T., Gennity, I., Samra, S., Kind, T., Fiehn, O., & VanderGheynst, J. S. (2016). Cofactor symbiosis for enhanced algal growth, biofuel production, and wastewater treatment. *Algal Research*, *17*, 308–315. https://doi.org/10.1016/j.algal.2016.05.024

Higgins, B. T., & VanderGheynst, J. S. (2014). Effects of *Escherichia coli* on mixotrophic growth of *Chlorella minutissima* and production of biofuel precursors. *PLoS One*, *9*(5), e96807. https://doi.org/10.1371/journal.pone.0096807

Hirano, A., Ueda, R., Hirayama, S., & Ogushi, Y. (1997). CO_2 fixation and ethanol production with microalgal photosynthesis and intracellular anaerobic fermentation. *Energy*, *22*(2), 137–142. https://doi.org/10.1016/S0360-5442(96)00123-5

Ho, S.-H., Huang, S.-W., Chen, C.-Y., Hasunuma, T., Kondo, A., & Chang, J.-S. (2013). Bioethanol production using carbohydrate-rich microalgae biomass as feedstock. *Bioresource Technology*, *135*, 191–198. https://doi.org/10.1016/j.biortech.2012.10.015

Hognon, C., Dupont, C., Grateau, M., & Delrue, F. (2014). Comparison of steam gasification reactivity of algal and lignocellulosic biomass: Influence of inorganic elements. *Bioresource Technology*, *164*(July), 347–353. https://doi.org/10.1016/j.biortech.2014.04.111

Holcomb, R. E., Mason, L. J., Reardon, K. F., Cropek, D. M., & Henry, C. S. (2011). Culturing and investigation of stress-induced lipid accumulation in microalgae using a microfluidic device. *Analytical and Bioanalytical Chemistry*, *400*(1), 245–253. https://doi.org/10.1007/s00216-011-4710-3

Holdsworth, J. E., & Ratledge, C. (1991). Triacylglycerol synthesis in the oleaginous yeast *Candida curvata D*. *Lipids*, *26*(2), 111–118. https://doi.org/10.1007/BF02544004

Holler, U., Konig, G. M., & Wright, A. D. (1999). Three new metabolites from marine-derived fungi of the genera coniothyrium and microsphaeropsis. *Journal of Natural Products*, *62*(1), 114–118. https://doi.org/10.1021/np980341e

Hom, E. F. Y., & Murray, A. W. (2014). Niche engineering demonstrates a latent capacity for fungal-algal mutualism. *Science (New York, NY)*, *345*(6192), 94–98. https://doi.org/10.1126/science.1253320

Hu, C., Wu, S., Wang, Q., Jin, G., Shen, H., & Zhao, Z. K. (2011). Simultaneous utilization of glucose and xylose for lipid production by *Trichosporon cutaneum*. *Biotechnology for Biofuels*, *4*(1), 25. https://doi.org/10.1186/1754-6834-4-25

Hu, H., Wang, H.-F., Ma, L.-L., Shen, X.-F., & Zeng, R. J. (2018). Effects of nitrogen and phosphorous stress on the formation of high value LC-PUFAs in *Porphyridium cruentum*. *Applied Microbiology and Biotechnology*, *102*(13), 5763–5773. https://doi.org/10.1007/s00253-018-8943-3

Hu, M., Wang, J., Gao, Q., & Bao, J. (2018). Converting lignin derived phenolic aldehydes into microbial lipid by *Trichosporon cutaneum*. *Journal of Biotechnology*, *281*(September), 81–86. https://doi.org/10.1016/j.jbiotec.2018.06.341

Huang, C., Wu, H., Li, R.-F., & Zong, M.-H. (2012). Improving lipid production from bagasse hydrolysate with *Trichosporon fermentans* by response surface methodology. *New Biotechnology*, *29*(3), 372–378. https://doi.org/10.1016/j.nbt.2011.03.008

Huang, C., Zong, M.-H., Wu, H., & Liu, Q.-P. (2009). Microbial oil production from rice straw hydrolysate by *Trichosporon fermentans*. *Bioresource Technology*, *100*(19), 4535–4538. https://doi.org/10.1016/j.biortech.2009.04.022

Huang, J., Xia, J., Jiang, W., Li, Y., & Li, J. (2015). Biodiesel production from microalgae oil catalyzed by a Recombinant Lipase. *Bioresource Technology*, *180*(March), 47–53. https://doi.org/10.1016/j.biortech.2014.12.072

Huang, J., Xia, J., Yang, Z., Guan, F., Cui, D., Guan, G., Jiang, W., & Li, Y. (2014). Improved production of a recombinant *Rhizomucor miehei* lipase expressed in *Pichia pastoris* and its application for conversion of microalgae oil to biodiesel. *Biotechnology for Biofuels*, *7*, 111. https://doi.org/10.1186/1754-6834-7-111

Huang, L., Zhang, B., Gao, B., & Sun, G. (2011). Application of fishmeal wastewater as a potential low-cost medium for lipid production by *Lipomyces starkeyi* HL. *Environmental Technology*, *33*(15–16), 1975–1981. https://doi.org/10.1080/09593330.2011.562551

Huang, X.-F., Wang, Y.-H., Shen, Y., Peng, K.-M., Lu, L.-J., & Liu, J. (2019). Using non-ionic surfactant as an accelerator to increase extracellular lipid production by Oleaginous Yeast *Cryptococcus curvatus* MUCL 29819. *Bioresource Technology*, *274*(February), 272–280. https://doi.org/10.1016/j.biortech.2018.11.100

Hullo, M. F., Moszer, I., Danchin, A., & Martin-Verstraete, I. (2001). CotA of *Bacillus subtilis* is a copper-dependent laccase. *Journal of Bacteriology*, *183*(18), 5426–5430. https://doi.org/10.1128/jb.183.18.5426-5430.2001

Hung, C.-H., Kanehara, K., & Nakamura, Y. (2016). Isolation and characterization of a mutant defective in triacylglycerol accumulation in nitrogen-starved *Chlamydomonas reinhardtii*. *Biochimica Et Biophysica Acta*, *1861*(9 Pt B), 1282–1293. https://doi.org/10.1016/j.bbalip.2016.04.001

Hussein, M. M., & Jwanny, E. W. (1975). The production of extracellular polysaccharides by a hydrocarbon assimilating yeast. *Acta Microbiologica Polonica. Series B: Microbiologia Applicata*, *7*(4), 253–258

Illman, A. M., Scragg, A. H., & Shales, S. W. (2000). Increase in *Chlorella* strains calorific values when grown in low nitrogen medium. *Enzyme and Microbial Technology*, *27*(8), 631–635. https://doi.org/10.1016/s0141-0229(00)00266-0

Ilmi, M., Abduh, M. Y., Hommes, A., Winkelman, J. G. M., Hidayat, C., & Heeres, H. J. (2018). Process intensification of enzymatic fatty acid butyl ester synthesis using a continuous centrifugal contactor separator. *Industrial & Engineering Chemistry Research*, *57*(2), 470–482. https://doi.org/10.1021/acs.iecr.7b03297

Ishida, Y., Ogihara, Y., & Okabe, S. (1990). Effects of a crude extract of a marine dinoflagellate, containing dimethyl-beta-propiothetin, on HCl.ethanol-induced gastric lesions and gastric secretion in rats. *Japanese Journal of Pharmacology*, *54*(3), 333–338. https://doi.org/10.1254/jjp.54.333

Iwasaki, Y., Ichino, T., & Saito, A. (2020). Transition of the bacterial community and culturable chitinolytic bacteria in chitin-treated upland soil: From *Streptomyces* to methionine-auxotrophic *Lysobacter* and other genera. *Microbes and Environments*, *35*(1). https://doi.org/10.1264/jsme2.ME19070

Jäger, G., Girfoglio, M., Dollo, F., Rinaldi, R., Bongard, H., Commandeur, U., Fischer, R., Spiess, A. C., & Büchs, J. (2011). How recombinant swollenin from *Kluyveromyces lactis* affects cellulosic substrates and accelerates their hydrolysis. *Biotechnology for Biofuels*, *4*(1), 33. https://doi.org/10.1186/1754-6834-4-33

Jahnke, J. P., Hoyt, T., LeFors, H. M., Sumner, J. J., & Mackie, D. M. (2016). *Aspergillus oryzae-Saccharomyces cerevisiae* consortium allows bio-hybrid fuel cell to run on complex carbohydrates. *Microorganisms*, *4*(1). https://doi.org/10.3390/microorganisms4010010

Javed, M. R., Noman, M., Shahid, M., Ahmed, T., Khurshid, M., Rashid, M. H., Ismail, M., Sadaf, M., & Khan, F. (2019). Current situation of biofuel production and its enhancement by CRISPR/Cas9-mediated genome engineering of microbial cells. *Microbiological Research*, *219*, 1–11. https://doi.org/10.1016/j.micres.2018.10.010

Jazzar, S., Quesada-Medina, J., Olivares-Carrillo, P., Néjib Marzouki, M., Acién-Fernández, F. G., Fernández-Sevilla, J. M., Molina-Grima, E., & Smaali, I. (2015). A whole biodiesel conversion process combining isolation, cultivation and in situ supercritical methanol transesterification of native microalgae. *Bioresource Technology*, *190*(August), 281–288. https://doi.org/10.1016/j.biortech.2015.04.097

Ji, C., Cao, X., Liu, H., Qu, J., Yao, C., Zou, H., & Xue, S. (2015). Investigating cellular responses during photohydrogen production by the marine microalga *Tetraselmis subcordiformis* by quantitative proteome analysis. *Applied Biochemistry and Biotechnology*, *177*(3), 649–661. https://doi.org/10.1007/s12010-015-1769-x

Ji, C., Cao, X., Yao, C., Xue, S., & Xiu, Z. (2014). Protein-protein interaction network of the marine microalga *Tetraselmis subcordiformis*: Prediction and application for starch metabolism analysis. *Journal of Industrial Microbiology & Biotechnology*, *41*(8), 1287–1296. https://doi.org/10.1007/s10295-014-1462-z

Jiao, K., Chang, J., Zeng, X., Ng, I.-S., Xiao, Z., Sun, Y., Tang, X., & Lin, L. (2017). 5-aminolevulinic acid promotes arachidonic acid biosynthesis in the red microalga *Porphyridium purpureum*. *Biotechnology for Biofuels*, *10*, 168. https://doi.org/10.1186/s13068-017-0855-4

Jiao, K., Xiao, W., Xu, Y., Zeng, X., Ho, S.-H., Laws, E. A., Lu, Y., et al. (2018). Using a trait-based approach to optimize mixotrophic growth of the red microalga *Porphyridium purpureum* towards fatty acid production. *Biotechnology for Biofuels*, *11*, 273. https://doi.org/10.1186/s13068-018-1277-7

Jin, B., Duan, P., Xu, Y., Wang, F., & Fan, Y. (2013). Co-liquefaction of micro- and macroalgae in subcritical water. *Bioresource Technology*, *149*(December), 103–110. https://doi.org/10.1016/j.biortech.2013.09.045

Jin, M.-J., Huang, H., Xiao, A.-H., Zhang, K., Liu, X., Li, S., & Peng, C. (2008). A novel two-step fermentation process for improved arachidonic acid production by *Mortierella alpina*. *Biotechnology Letters*, *30*(6), 1087–1091. https://doi.org/10.1007/s10529-008-9661-1

Jin, Y.-S., & Hd Cate, J. (2017). Metabolic engineering of yeast for lignocellulosic biofuel production. *Current Opinion in Chemical Biology*, *41*(December), 99–106. https://doi.org/10.1016/j.cbpa.2017.10.025

Jiru, T. M., Abate, D., Kiggundu, N., Pohl, C., & Groenewald, M. (2016). Oleaginous yeasts from Ethiopia. *AMB Express*, *6*(1), 78. https://doi.org/10.1186/s13568-016-0242-8

Joe, M.-H., Kim, J.-Y., Lim, S., Kim, D.-H., Bai, S., Park, H., Lee, S. G., Han, S. J., & Choi, J.-I. (2015). Microalgal lipid production using the hydrolysates of rice straw pretreated with gamma irradiation and alkali solution. *Biotechnology for Biofuels*, *8*, 125. https://doi.org/10.1186/s13068-015-0308-x

Johansson, O. N., Pinder, M. I. M., Ohlsson, F., Egardt, J., Töpel, M., & Clarke, A. K. (2019). Friends with benefits: Exploring the phycosphere of the marine diatom *Skeletonema marinoi*. *Frontiers in Microbiology*, *10*, 1828. https://doi.org/10.3389/fmicb.2019.01828

Jönsson, L. J., & Martín, C. (2016). Pretreatment of lignocellulose: Formation of inhibitory by-products and strategies for minimizing their effects. *Bioresource Technology*, *199*, 103–112. https://doi.org/10.1016/j.biortech.2015.10.009

Jovel, J., Patterson, J., Wang, W., Hotte, N., O'Keefe, S., Mitchel, T., Perry, T., Kao, D., Mason, A. L., Madsen, K. L., & Wong, G. K.-S. (2016). Characterization of the gut microbiome using 16S or shotgun metagenomics. *Frontiers in Microbiology*, *7*. https://doi.org/10.3389/fmicb.2016.00459

Jurado, M., López, M. J., Suárez-Estrella, F., Vargas-García, M. C., López-González, J. A., & Moreno, J. (2014). Exploiting composting biodiversity: Study of the persistent and biotechnologically relevant microorganisms from lignocellulose-based composting. *Bioresource Technology*, *162*(June), 283–293. https://doi.org/10.1016/j.biortech.2014.03.145

Kalhor, A. X., Movafeghi, A., Mohammadi-Nassab, Ad., Abedi, E., & Bahrami, A. (2017). Potential of the green alga *Chlorella vulgaris* for biodegradation of crude oil hydrocarbons. *Marine Pollution Bulletin*, *123*(1–2), 286–290. https://doi.org/10.1016/j.marpolbul.2017.08.045

Kämäräinen, J., Knoop, H., Stanford, N. J., Guerrero, F., Akhtar, M. K., Aro, E.-M., Steuer, R., & Jones, P. R. (2012). Physiological tolerance and stoichiometric potential of cyanobacteria for hydrocarbon fuel production. *Journal of Biotechnology*, *162*(1), 67–74. https://doi.org/10.1016/j.jbiotec.2012.07.193

Kandimalla, P., Desi, S., & Vurimindi, H. (2016). Mixotrophic cultivation of microalgae using industrial flue gases for biodiesel production. *Environmental Science and Pollution Research International*, *23*(10), 9345–9354. https://doi.org/10.1007/s11356-015-5264-2

Karatay, S. E., & Dönmez, G. (2010). Improving the lipid accumulation properties of the yeast cells for biodiesel production using molasses. *Bioresource Technology*, *101*(20), 7988–7990. https://doi.org/10.1016/j.biortech.2010.05.054

Karnaouri, A., Chalima, A., Kalogiannis, K. G., Varamogianni-Mamatsi, D., Lappas, A., & Topakas, E. (2020). Utilization of lignocellulosic biomass towards the production of Omega-3 fatty acids by the heterotrophic marine microalga *Crypthecodinium cohnii*. *Bioresource Technology*, *303*(May), 122899. https://doi.org/10.1016/j.biortech.2020.122899

Kashid, M., & Ghosalkar, A. (2018). Critical factors affecting ethanol production by immobilized *Pichia stipitis* using corn cob hemicellulosic hydrolysate. *Preparative Biochemistry & Biotechnology*, *48*(3), 288–295. https://doi.org/10.1080/10826068.2018.1425715

Kassim, M. A., & Meng, T. K. (2017). Carbon dioxide (CO_2) biofixation by microalgae and its potential for biorefinery and biofuel production. *The Science of the Total Environment*, *584–585*(April), 1121–1129. https://doi.org/10.1016/j.scitotenv.2017.01.172

Kato, Y., Nomura, T., Ogita, S., Takano, M., & Hoshino, K. (2013). Two new β-glucosidases from ethanol-fermenting fungus *Mucor circinelloides* NBRC 4572: Enzyme purification, functional characterization, and molecular cloning of the gene. *Applied Microbiology and Biotechnology*, *97*(23), 10045–10056. https://doi.org/10.1007/s00253-013-5210-5

Kavitha, G., Kurinjimalar, C., Sivakumar, K., Palani, P., & Rengasamy, R. (2016). Biosynthesis, purification and characterization of polyhydroxybutyrate from *Botryococcus braunii kütz*. *International Journal of Biological Macromolecules*, *89*(August), 700–706. https://doi.org/10.1016/j.ijbiomac.2016.04.086

Keller, F. A., Hamilton, J. E., & Nguyen, Q. A. (2003). Microbial pretreatment of biomass: Potential for reducing severity of thermochemical biomass pretreatment. *Applied Biochemistry and Biotechnology*, *105–108*, 27–41. https://doi.org/10.1385/abab:105:1-3:27

Khan, M. I., Lee, M. G., Shin, J. H., & Kim, J. D. (2017). Pretreatment optimization of the biomass of *Microcystis aeruginosa* for efficient bioethanol production. *AMB Express*, *7*(1), 19. https://doi.org/10.1186/s13568-016-0320-y

Khetkorn, W., Rastogi, R. P., Incharoensakdi, A., Lindblad, P., Madamwar, D., Pandey, A., & Larroche, C. (2017). Microalgal hydrogen production—A review. *Bioresource Technology*, *243*, 1194–1206.

Khotimchenko, S. V., & Yakovleva, I. M. (2005). Lipid composition of the red alga *Tichocarpus crinitus* exposed to different levels of photon irradiance. *Phytochemistry*, *66*, 73–79.

Kim, G.-E., Lee, J.-H., Jung, S.-H., Seo, E.-S., Jin, S.-D., Kim, G. J., Cha, J., Kim, E.-J., Park, K.-D., & Kim, D. (2010). Enzymatic synthesis and characterization of hydroquinone galactoside using *Kluyveromyces lactis* lactase. *Journal of Agricultural and Food Chemistry*, *58*(17), 9492–9497. https://doi.org/10.1021/jf101748j

Kim, H. M., Oh, C. H., & Bae, H.-J. (2017). Comparison of red microalgae (*Porphyridium cruentum*) culture conditions for bioethanol production. *Bioresource Technology*, *233*(June), 44–50. https://doi.org/10.1016/j.biortech.2017.02.040

Kim, H.-C., Choi, W. J., Chae, A. N., Park, J., Kim, H. J., & Song, K. G. (2016). Evaluating integrated strategies for robust treatment of high saline piggery wastewater. *Water Research*, *89*(February), 222–231. https://doi.org/10.1016/j.watres.2015.11.054

Kim, J., Yoo, G., Lee, H., Lim, J., Kim, K., Kim, C. W., Park, M. S., & Yang, J.-W. (2013). Methods of downstream processing for the production of biodiesel from microalgae. *Biotechnology Advances*, *31*(6), 862–876. https://doi.org/10.1016/j.biotechadv.2013.04.006

Kirk, T. K., & Moore, W. E. (1972). Removing lignin from wood with white-rot fungi and digestibility of resulting wood. *Wood and Fiber Science*, *4*(2), 72–79.

Kitcha, S., & Cheirsilp, B. (2014). Bioconversion of lignocellulosic palm byproducts into enzymes and lipid by newly isolated oleaginous fungi. *Biochemical Engineering Journal*, *88*, 95–100. https://doi.org/10.1016/j.bej.2014.04.006

Kityo, M. K., Sunwoo, I., Kim, S. H., Park, Y. R., Jeong, G.-T., & Kim, S.-K. (2020). Enhanced bioethanol fermentation by sonication using three yeasts species and Kariba Weed (*Salvinia molesta*) as biomass collected from Lake Victoria, Uganda. *Applied Biochemistry and Biotechnology*, *192*(1), 180–195. https://doi.org/10.1007/s12010-020-03305-x

Klose, H., Röder, J., Girfoglio, M., Fischer, R., & Commandeur, U. (2012). Hyperthermophilic endoglucanase for in planta lignocellulose conversion. *Biotechnology for Biofuels*, *5*(1), 63. https://doi.org/10.1186/1754-6834-5-63

Knutson, C. M., Plunkett, M. H., Liming, R. A., & Barney, B. M. (2018). Efforts toward optimization of aerobic biohydrogen reveal details of secondary regulation of biological nitrogen fixation by nitrogenous compounds in *Azotobacter vinelandii*. *Applied Microbiology and Biotechnology*, *102*(23), 10315–10325. https://doi.org/10.1007/s00253-018-9363-0.

Kojima, E., & Lin, B. (2004). Effect of partial shading on photoproduction of hydrogen by chlorella. *Journal of Bioscience and Bioengineering*, *97*(5), 317–321. https://doi.org/10.1016/S1389-1723(04)70212-1

Kosa, G., Zimmermann, B., Kohler, A., Ekeberg, D., Afseth, N. K., Mounier, J., & Shapaval, V. (2018). High-throughput screening of *Mucoromycota fungi* for production of low- and high-value lipids. *Biotechnology for Biofuels*, *11*, 66. https://doi.org/10.1186/s13068-018-1070-7

Kot, A. M., Błażejak, S., Kurcz, A., Gientka, I., & Kieliszek, M. (2016). *Rhodotorula glutinis*-: Potential source of lipids, carotenoids, and enzymes for use in industries. *Applied Microbiology and Biotechnology*, *100*(14), 6103–6117. https://doi.org/10.1007/s00253-016-7611-8

Kotaka, A., Bando, H., Kaya, M., Kato-Murai, M., Kuroda, K., Sahara, H., Hata, Y., Kondo, A., & Ueda, M. (2008). Direct ethanol production from barley beta-glucan by sake yeast displaying *Aspergillus oryzae* beta-glucosidase and endoglucanase. *Journal of Bioscience and Bioengineering*, *105*(6), 622–627. https://doi.org/10.1263/jbb.105.622

Kovacs, K. F., Xu, Y., West, G. H., & Popp, M. (2016). The tradeoffs between market returns from agricultural crops and non-market ecosystem service benefits on an irrigated agricultural landscape in the presence of groundwater overdraft. *Water, 8*(11), 501. https://doi.org/10.3390/w8110501

Krivoruchko, A., Serrano-Amatriain, C., Chen, Y., Siewers, V., & Nielsen, J. (2013). Improving biobutanol production in engineered *Saccharomyces cerevisiae* by manipulation of Acetyl-CoA metabolism. *Journal of Industrial Microbiology & Biotechnology, 40*(9), 1051–1056. https://doi.org/10.1007/s10295-013-1296-0

Kruse, O., Rupprecht, J., Bader, K.-P., Thomas-Hall, S., Schenk, P. M., Finazzi, G., & Hankamer, B. (2005). Improved photobiological H2 production in engineered green algal cells. *Journal of Biological Chemistry, 280*(40), 34170–34177. https://doi.org/10.1074/jbc.M503840200

Krzemińska, I., Piasecka, A., Nosalewicz, A., Simionato, D., & Wawrzykowski, J. (2015). Alterations of the lipid content and fatty acid profile of *Chlorella protothecoides* under different light intensities. *Bioresource Technology, 196*(November), 72–77. https://doi.org/10.1016/j.biortech.2015.07.043

Kumar, M., Singhall, A., & Thakur, I. S. (2015). Comparison of submerged and solid state pretreatment of sugarcane bagasse by *Pandoraea* sp. ISTKB: Enzymatic and structural analysis. *Bioresource Technology, 203*, 18–25. https://doi.org/10.1016/j.biortech.2015.12.034

Kumar, S., & Khare, S. K. (2012). Purification and characterization of maltooligosaccharide-forming α-amylase from moderately halophilic *Marinobacter* sp. EMB8. *Bioresource Technology, 116*, 247–251. https://doi.org/10.1016/j.biortech.2011.11.109

Kuo, C.-H., Peng, L.-T., Kan, S.-C., Liu, Y.-C., & Shieh, C.-J. (2013). Lipase-immobilized biocatalytic membranes for biodiesel production. *Bioresource Technology, 145*(October), 229–232. https://doi.org/10.1016/j.biortech.2012.12.054

Kuo, T.-C., Shaw, J.-F., & Lee, G.-C. (2015). Conversion of crude *Jatropha curcas* seed oil into biodiesel using liquid recombinant *Candida rugosa* lipase isozymes. *Bioresource Technology, 192*(September), 54–59. https://doi.org/10.1016/j.biortech.2015.05.008

Kwak, H. S., Kim, J. Y. H., & Sim, S. J. (2015). A microscale approach for simple and rapid monitoring of cell growth and lipid accumulation in *Neochloris oleoabundans*. *Bioprocess and Biosystems Engineering, 38*(10), 2035–2043. https://doi.org/10.1007/s00449-015-1444-1

Kwangdinata, R., Raya, I., & Zakir, M. (2014). Production of biodiesel from lipid of phytoplankton *Chaetoceros calcitrans* through ultrasonic method. *TheScientificWorldJournal, 2014*, 231361. https://doi.org/10.1155/2014/231361

Lai, M. C., & Lan, E. I. (2015). Advances in metabolic engineering of cyanobacteria for photosynthetic biochemical production. *Metabolites, 5*(4), 636–658. https://doi.org/10.3390/metabo5040636

Lakaniemi, A.-M., Hulatt, C. J., Thomas, D. N., Tuovinen, O. H., & Puhakka, J. A. (2011). Biogenic hydrogen and methane production from *Chlorella vulgaris* and *Dunaliella tertiolecta* biomass. *Biotechnology for Biofuels, 4*(1), 34. https://doi.org/10.1186/1754-6834-4-34

Lakatos, G. E., Ranglová, K., Manoel, J. C., Grivalský, T., Kopecký, J., & Masojídek, J. (2019). Bioethanol production from microalgae polysaccharides. *Folia Microbiologica, 64*(5), 627–644. https://doi.org/10.1007/s12223-019-00732-0

Lam, M. K., & Lee, K. T. (2012). Microalgae biofuels: A critical review of issues, problems and the way forward. *Biotechnology Advances, 30*(3), 673–690. https://doi.org/10.1016/j.biotechadv.2011.11.008

Lee, D.-J., Liao, G.-Y., Chang, Y.-R., & Chang, J.-S. (2012). Coagulation-membrane filtration of *Chlorella vulgaris*. *Bioresource Technology, 108*(March), 184–189. https://doi.org/10.1016/j.biortech.2011.12.098

Lee, H. J., Kim, I. J., Youn, H. J., Yun, E. J., Choi, I.-G., & Kim, K. H. (2017a). Cellotriose-Hydrolyzing activity conferred by truncating the carbohydrate-binding modules of Cel5 from *Hahella chejuensis*. *Bioprocess and Biosystems Engineering, 40*(2), 241–249. https://doi.org/10.1007/s00449-016-1692-8

Lee, J. H., Kim, S. B., Park, C., Tae, B., Han, S. O., & Kim, S. W. (2010). Development of batch and continuous processes on biodiesel production in a packed-bed reactor by a mixture of immobilized *Candida rugosa* and *Rhizopus oryzae* lipases. *Applied Biochemistry and Biotechnology, 161*(1–8), 365–371. https://doi.org/10.1007/s12010-009-8829-z

Lee, J. H., Lee, D. H., Lim, J. S., Um, B.-H., Park, C., Kang, S. W., & Kim, S. W. (2008). Optimization of the process for biodiesel production using a mixture of immobilized *Rhizopus oryzae* and *Candida rugosa* lipases. *Journal of Microbiology and Biotechnology, 18*(12), 1927–1931

Lee, J.-S., Park, E.-H., Kim, J.-W., Yeo, S.-H., & Kim, M.-D. (2013a). Growth and fermentation characteristics of *Saccharomyces cerevisiae* NK28 isolated from kiwi fruit. *Journal of Microbiology and Biotechnology, 23*(9), 1253–1259. https://doi.org/10.4014/jmb.1307.07050

Lee, J.-W., Gwak, K.-S., Park, J.-Y., Park, M.-J., Choi, D.-H., Kwon, M., & Choi, I.-G. (2007). Biological pretreatment of softwood pinus densiflora by three white rot fungi. *Journal of Microbiology, 8*.

Lee, J.-W., Rodrigues, R. C. L. B., & Jeffries, T. W. (2009). Simultaneous saccharification and ethanol fermentation of oxalic acid pretreated corncob assessed with response surface methodology. *Bioresource Technology, 100*(24), 6307–6311. https://doi.org/10.1016/j.biortech.2009.06.088

Lee, L.-C., Yen, C.-C., Malmis, C. C., Chen, L.-F., Chen, J.-C., Lee, G.-C., & Shaw, J.-F. (2011). Characterization of codon-optimized recombinant *Candida rugosa* lipase 5 (LIP5). *Journal of Agricultural and Food Chemistry, 59*(19), 10693–10698. https://doi.org/10.1021/jf202161a

Lee, O. K., Kim, A. L., Seong, D. H., Lee, C. G., Jung, Y. T., Lee, J. W., & Lee, E. Y. (2013b). Chemo-enzymatic saccharification and bioethanol fermentation of lipid-extracted residual biomass of the microalga, *Dunaliella tertiolecta*. *Bioresource Technology, 132*(March), 197–201. https://doi.org/10.1016/j.biortech.2013.01.007

Lee, O. K., Oh, Y.-K., & Lee, E. Y. (2015a). Bioethanol production from carbohydrate-enriched residual biomass obtained after lipid extraction of *Chlorella* sp. KR-1. *Bioresource Technology, 196*(November), 22–27. https://doi.org/10.1016/j.biortech.2015.07.040

Lee, S. Y., Cho, J. M., Chang, Y. K., & Oh, Y.-K. (2017b). Cell disruption and lipid extraction for microalgal biorefineries: A review. *Bioresource Technology, 244*(Pt 2), 1317–1328. https://doi.org/10.1016/j.biortech.2017.06.038

Lee, S., Kang, M., Bae, J.-H., Sohn, J.-H., & Sung, B. H. (2019). Bacterial valorization of lignin: Strains, enzymes, conversion pathways, biosensors, and perspectives. *Frontiers in Bioengineering and Biotechnology, 7*, 209. https://doi.org/10.3389/fbioe.2019.00209

Lee, S., Oh, Y., Kim, D., Kwon, D., Lee, C., & Lee, J. (2011). Converting carbohydrates extracted from marine algae into ethanol using various ethanolic *Escherichia xoli* strains. *Applied Biochemistry and Biotechnology, 164*(6), 878–888. https://doi.org/10.1007/s12010-011-9181-7

Lee, Y.-C., Lee, K., & Oh, Y.-K. (2015b). Recent nanoparticle engineering advances in microalgal cultivation and harvesting processes of biodiesel production: A review. *Bioresource Technology, 184*, 63–72. https://doi.org/10.1016/j.biortech.2014.10.145

Lei, A., Chen, H., Shen, G., Hu, Z., Chen, L., & Wang, J. (2012). Expression of fatty acid synthesis genes and fatty acid accumulation in *Haematococcus pluvialis* under different stressors. *Biotechnology for Biofuels, 5*(1), 18. https://doi.org/10.1186/1754-6834-5-18

Li, C., Li, B., Zhang, N., Wang, Q., Wang, W., & Zou, H. (2019). Comparative transcriptome analysis revealed the improved β-carotene production in *Sporidiobolus pararoseus* Yellow Mutant MuY9. *Journal of General and Applied Microbiology, 65*(3), 121–128. https://doi.org/10.2323/jgam.2018.07.002

Li, C., Zhao, D., Yan, J., Zhang, N., & Li, B. (2021). Metabolomics integrated with transcriptomics: Assessing the central metabolism of marine red yeast *Sporobolomyces pararoseus* under salinity stress. *Archives of Microbiology, 203*(3), 889–899. https://doi.org/10.1007/s00203-020-02082-9

Li, D., Bi, R., Chen, H., Mu, L., Zhang, L., Chen, Q., Xie, H., Luo, Y., & Xie, L. (2017). The acute toxicity of Bisphenol A and lignin-derived bisphenol in algae, daphnids, and Japanese medaka. *Environmental Science and Pollution Research International, 24*(30), 23872–23879. https://doi.org/10.1007/s11356-017-0018-y

Li, J., Han, D., Wang, D., Ning, K., Jia, J., Wei, L., Jing, X., Huang, S., Chen, J., Li, Y., Hu, Q., Xu, J. (2014). Choreography of transcriptomes and lipidomes of *Nannochloropsis* reveals the mechanisms of oil synthesis in microalgae. *Plant Cell, 26*, 1645–1665.

Li, N., Yang, F., Xiao, H., Zhang, J., & Ping, Q. (2018). Effect of feedstock concentration on biogas production by inoculating rumen microorganisms in biomass solid waste. *Applied Biochemistry and Biotechnology, 184*(4), 1219–1231. https://doi.org/10.1007/s12010-017-2615-0

Li, P., Miao, X., Li, R., & Zhong, J. (2011a). In situ biodiesel production from fast-growing and high oil content *Chlorella pyrenoidosa* in rice straw hydrolysate. *Journal of Biomedicine & Biotechnology, 2011*, 141207. https://doi.org/10.1155/2011/141207

Li, Q., Du, W., & Liu, D. (2008). Perspectives of microbial oils for biodiesel production. *Applied Microbiology and Biotechnology, 80*(5), 749–756. https://doi.org/10.1007/s00253-008-1625-9

Li, R., Duan, N., Zhang, Y., Liu, Z., Li, B., Zhang, D., & Dong, T. (2017a). Anaerobic co-digestion of chicken manure and microalgae *Chlorella* sp.: Methane potential, microbial diversity and synergistic impact evaluation. *Waste Management (New York, N.Y.), 68*(October), 120–127. https://doi.org/10.1016/j.wasman.2017.06.028

Li, T., Li, C.-T., Butler, K., Hays, S. G., Guarnieri, M. T., Oyler, G. A., & Betenbaugh, M. J. (2017b). Mimicking Lichens: Incorporation of yeast strains together with sucrose-secreting cyanobacteria improves survival, growth, ROS removal, and lipid production in a stable mutualistic co-culture production platform. *Biotechnology for Biofuels*, *10*, 55. https://doi.org/10.1186/s13068-017-0736-x

Li, X., Xu, H., & Wu, Q. (2007). Large-scale biodiesel production from microalga *Chlorella protothecoides* through heterotrophic cultivation in bioreactors. *Biotechnology and Bioengineering*, *98*(4), 764–771. https://doi.org/10.1002/bit.21489

Li, Y., Mu, J., Chen, D., Han, F., Xu, H., Kong, F., Xie, F., & Feng, B. (2013). Production of biomass and lipid by the microalgae *Chlorella protothecoides* with Heterotrophic-Cu(II) Stressed (HCuS) coupling cultivation. *Bioresource Technology*, *148*(November), 283–292. https://doi.org/10.1016/j.biortech.2013.08.153

Li, Y., Mu, J., Chen, D., Xu, H., & Han, F. (2015). Enhanced lipid accumulation and biodiesel production by oleaginous *Chlorella protothecoides* under a structured heterotrophic-iron (II) induction strategy. *World Journal of Microbiology & Biotechnology*, *31*(5), 773–783. https://doi.org/10.1007/s11274-015-1831-4

Li, Y., Zhou, W., Hu, B., Min, M., Chen, P., & Ruan, R. R. (2011b). Integration of algae cultivation as biodiesel production feedstock with municipal wastewater treatment: Strains screening and significance evaluation of environmental factors. *Bioresource Technology*, *102*(23), 10861–10867. https://doi.org/10.1016/j.biortech.2011.09.064

Li, Z., Pradeep, K.G., Deng, Y., Raabe, D., & Tasan, C. C. (2016). Metastable high-entropy dual-phase alloys overcome the strength-ductility trade-off. *Nature*, *534*(7606), 227–230. https://doi.org/10.1038/nature17981

Lian, J., Chen, S., Zhou, S., Wang, Z., O'Fallon, J., Li, C.-Z., & Garcia-Perez, M. (2010). Separation, hydrolysis and fermentation of pyrolytic sugars to produce ethanol and lipids. *Bioresource Technology*, *101*(24), 9688–9699. https://doi.org/10.1016/j.biortech.2010.07.071

Liang, Y., Garcia, R. A., Piazza, G. J., & Wen, Z. (2011). Nonfeed application of rendered animal proteins for microbial production of eicosapentaenoic acid by the fungus *Pythium irregulare*. *Journal of Agricultural and Food Chemistry*, *59*(22), 11990–11996. https://doi.org/10.1021/jf2031633

Liang, Y., Jarosz, K., Wardlow, A. T., Zhang, J., & Cui, Y. (2014a). Lipid production by *Cryptococcus curvatus* on hydrolysates derived from corn fiber and sweet sorghum bagasse following dilute acid pretreatment. *Applied Biochemistry and Biotechnology*, *173*(8), 2086–2098. https://doi.org/10.1007/s12010-014-1007-y

Liang, Y., Perez, I., Goetzelmann, K., & Trupia, S. (2014b). Microbial lipid production from pretreated and hydrolyzed corn fiber. *Biotechnology Progress*, *30*(4), 945–951. https://doi.org/10.1002/btpr.1927

Liang, Y., Sarkany, N., Cui, Y., Yesuf, J., Trushenski, J., & Blackburn, J. W. (2010). Use of sweet sorghum juice for lipid production by *Schizochytrium limacinum* SR21. *Bioresource Technology*, *101*(10), 3623–3627. https://doi.org/10.1016/j.biortech.2009.12.087

Liang, Y., Zhao, X., Strait, M., & Wen, Z. (2012). Use of dry-milling derived thin stillage for producing Eicosapentaenoic Acid (EPA) by the fungus *Pythium irregulare*. *Bioresource Technology*, *111*(May), 404–409. https://doi.org/10.1016/j.biortech.2012.02.035

Libralato, G., Avezzù, F., & Ghirardini, A. V. (2011). Lignin and tannin toxicity to *Phaeodactylum tricornutum* (Bohlin). *Journal of Hazardous Materials*, *194*(October), 435–439. https://doi.org/10.1016/j.jhazmat.2011.07.103

Lima, L. N., Oliveira, G. C., Rojas, M. J., Castro, H. F., Da Rós, P. C. M., Mendes, A. A., Giordano, R. L. C., & Tardioli, P. W. (2015). Immobilization of *Pseudomonas fluorescens* lipase on hydrophobic supports and application in biodiesel synthesis by transesterification of vegetable oils in solvent-free systems. *Journal of Industrial Microbiology & Biotechnology*, *42*(4), 523–535. https://doi.org/10.1007/s10295-015-1586-9

Lin, H., Cheng, W., Ding, H. T., Chen, X. J., Zhou, Q. F., & Zhao, Y. H. (2010). Direct microbial conversion of wheat straw into lipid by a cellulolytic fungus of *Aspergillus oryzae* A-4 in solid-state fermentation. *Bioresource Technology*, *101*(19), 7556–7562. https://doi.org/10.1016/j.biortech.2010.04.027

Lin, H., & Lee, Y. K. (2017). Genetic engineering of medium-chain-length fatty acid synthesis in *Dunaliella tertiolecta* for improved biodiesel production. *Journal of Applied Phycology*, *29*(6), 2811–2819. https://doi.org/10.1007/s10811-017-1210-7

Lin, H., Wang, Q., Shen, Q., Ma, J., Fu, J., & Zhao, Y. (2014). Engineering *Aspergillus oryzae* A-4 through the chromosomal insertion of foreign cellulase expression cassette to improve conversion of cellulosic biomass into lipids. *PLoS One*, *9*(9), e108442. https://doi.org/10.1371/journal.pone.0108442

Lin, Q., Zhuo, W.-H., Wang, X.-W., Chen, C.-P., Gao, Y.-H., & Liang, J.-R. (2018). Effects of fundamental nutrient stresses on the lipid accumulation profiles in two diatom species *Thalassiosira weissflogii* and *Chaetoceros muelleri*. *Bioprocess and Biosystems Engineering*, *41*(8), 1213–1224. https://doi.org/10.1007/s00449-018-1950-z

Lindberg, P., Park, S., & Melis, A. (2010). Engineering a platform for photosynthetic isoprene production in cyanobacteria, using Synechocystis as the model organism. *Metabolic Engineering*, *12*(1), 70–79. https://doi.org/10.1016/j.ymben.2009.10.001

Ling, X., Guo, J., Zheng, C., Ye, C., Lu, Y., Pan, X., Chen, Z., & Ng, I.-S. (2015). Simple, effective protein extraction method and proteomics analysis from polyunsaturated fatty acids-producing micro-organisms. *Bioprocess and Biosystems Engineering*, *38*(12), 2331–2341. https://doi.org/10.1007/s00449-015-1467-7

Liu, D., Yan, X., Zhuo, S., Si, M., Liu, M., Wang, S., Ren, L., Chai, L., & Shi, Y. (2018). *Pandoraea* sp. B-6 assists the deep eutectic solvent pretreatment of rice straw via promoting lignin depolymerization. *Bioresource Technology*, *257*, 62–68. https://doi.org/10.1016/j.biortech.2018.02.029

Liu, J., Ge, Y., Cheng, H., Wu, L., & Tian, G. (2013). Aerated swine lagoon wastewater: A promising alternative medium for *Botryococcus braunii* cultivation in open system. *Bioresource Technology*, *139*, 190–194. https://doi.org/10.1016/j.biortech.2013.04.036

Liu, J., Huang, J., Sun, Z., Zhong, Y., Jiang, Y., & Chen, F. (2011). Differential lipid and fatty acid profiles of photoautotrophic and heterotrophic *Chlorella zofingiensis*: Assessment of algal oils for biodiesel production. *Bioresource Technology*, *102*(1), 106–110. https://doi.org/10.1016/j.biortech.2010.06.017

Liu, J., Sun, Z., Zhong, Y., Huang, J., Hu, Q., & Chen, F. (2012a). Stearoyl-acyl carrier protein desaturase gene from the oleaginous microalga *Chlorella zofingiensis*: Cloning, characterization and transcriptional analysis. *Planta*, *236*(6), 1665–1676. https://doi.org/10.1007/s00425-012-1718-7

Liu, L., Zhang, S., Dai, W., Bi, X., & Zhang, D. (2019). Comparing effects of berberine on the growth and photosynthetic activities of *Microcystis aeruginosa* and *Chlorella pyrenoidosa*. *Water Science and Technology: A Journal of the International Association on Water Pollution Research*, *80*(6), 1155–1162. https://doi.org/10.2166/wst.2019.357

Liu, L.-P., Hu, Y., Lou, W.-Y., Li, N., Wu, H., & Zong, M.-H. (2017). Use of crude glycerol as sole carbon source for microbial lipid production by oleaginous yeasts. *Applied Biochemistry and Biotechnology*, *182*(2), 495–510. https://doi.org/10.1007/s12010-016-2340-0

Liu, W., Wang, Y., Yu, Z., & Bao, J. (2012b). Simultaneous saccharification and microbial lipid fermentation of corn stover by oleaginous yeast *Trichosporon cutaneum*. *Bioresource Technology*, *118*(August), 13–18. https://doi.org/10.1016/j.biortech.2012.05.038

Liu, Z., Zhang, F., & Chen, F. (2013). High throughput screening of CO_2-tolerating microalgae using GasPak Bags. *Aquatic Biosystems*, *9*(1), 23. https://doi.org/10.1186/2046-9063-9-23

Logan, B. E. (2009). Exoelectrogenic bacteria that power microbial fuel cells. *Nature Review of Microbiology*, *7*, 375–381.

Lopes, M., Baptista, P., Duarte, E., & Moreira, A. L. N. (2019). Enhanced biogas production from anaerobic co-digestion of pig slurry and horse manure with mechanical pre-treatment. *Environmental Technology*, *40*(10), 1289–1297. https://doi.org/10.1080/09593330.2017.1420698

Lü, F., Ji, J., Shao, L., & He, P. (2013). Bacterial bioaugmentation for improving methane and hydrogen production from microalgae. *Biotechnology for Biofuels*, *6*(1), 92. https://doi.org/10.1186/1754-6834-6-92

Lu, Q., Han, P., Chen, F., Liu, T., Li, J., Leng, L., Li, J., & Zhou, W. (2019). A novel approach of using zeolite for ammonium toxicity mitigation and value-added spirulina cultivation in wastewater. *Bioresource Technology*, *280*(May), 127–135. https://doi.org/10.1016/j.biortech.2019.02.042

Lübbehüsen, T. L., Nielsen, J., & McIntyre, M. (2004). Aerobic and anaerobic ethanol production by *Mucor circinelloides* during submerged growth. *Applied Microbiology and Biotechnology*, *63*(5), 543–548. https://doi.org/10.1007/s00253-003-1394-4

Lutzoni, F., Nowak, M. D., Alfaro, M. E., Reeb, V., Miadlikowska, J., Krug, M., Arnold, A. E., Lewis, L. A., Swofford, D. L., Hibbett, D., Hilu, K., James, T. Y., Quandt, D., & Magallón, S. (2018). Contemporaneous radiations of fungi and plants linked to symbiosis. *Nature Communications*, *9*. https://doi.org/10.1038/s41467-018-07849-9

Lutzu, G. A., & Turgut Dunford, N. (2018). Interactions of microalgae and other microorganisms for enhanced production of high-value compounds. *Frontiers in Bioscience (Landmark Edition)*, *23*, 1487–1504. https://doi.org/10.2741/4656

Lv, J., Wang, X., Liu, W., Feng, J., Liu, Q., Nan, F., Jiao, X., & Xie, S. (2018). The performance of a self-flocculating microalga *Chlorococcum* sp. GD in wastewater with different ammonia concentrations. *International Journal of Environmental Research and Public Health*, *15*(3). https://doi.org/10.3390/ijerph15030434

Ma, Q., Wang, J., Lu, S., Lv, Y., & Yuan, Y. (2013). Quantitative proteomic profiling reveals photosynthesis responsible for inoculum size dependent variation in *Chlorella sorokiniana*. *Biotechnology and Bioengineering*, *110*(3), 773–784. https://doi.org/10.1002/bit.24762

Magneschi, L., Catalanotti, C., Subramanian, V., Dubini, A., Yang, W., Mus, F., Posewitz, M. C., Seibert, M., Perata, P., & Grossman, A. R. (2012). A Mutant in the ADH1 gene of *Chlamydomonas reinhardtii* elicits metabolic restructuring during anaerobiosis. *Plant Physiology*, *158*(3), 1293–1305. https://doi.org/10.1104/pp.111.191569

Magnin, J.-P., & Deseure, J. (2019). Hydrogen generation in a pressurized photobioreactor: Unexpected enhancement of biohydrogen production by the phototrophic bacterium *Rhodobacter capsulatus*. *Applied Energy*, *239*, 635–643.

Mandal, S., & Mallick, N. (2012). Biodiesel production by the green microalga *Scenedesmus obliquus* in a recirculatory aquaculture system. *Applied and Environmental Microbiology*, *78*(16), 5929–5934. Retrieved from https://pubmed.ncbi.nlm.nih.gov/22660702/

Mandotra, S. K., Kumar, P., Suseela, M. R., Nayaka, S., & Ramteke, P. W. (2016). Evaluation of fatty acid profile and biodiesel properties of microalga *Scenedesmus abundans* under the influence of phosphorus, PH and light intensities. *Bioresource Technology*, *201*(February), 222–229. https://doi.org/10.1016/j.biortech.2015.11.042

Martins, L. O., Soares, C. M., Pereira, M. M., Teixeira, M., Costa, T., Jones, G. H., & Henriques, A. O. (2002). Molecular and biochemical characterization of a highly stable bacterial laccase that occurs as a structural component of the *Bacillus subtilis* endospore coat. *Journal of Biological Chemistry*, *277*(21), 18849–18859. https://doi.org/10.1074/jbc.M200827200

Masran, R., Zanirun, Z., Bahrin, E. K., Ibrahim, M. F., Lai Yee, P., & Abd-Aziz, S. (2016). Harnessing the potential of ligninolytic enzymes for lignocellulosic biomass pretreatment. *Applied Microbiology and Biotechnology*, *100*(12), 5231–5246. https://doi.org/10.1007/s00253-016-7545-1

Mathew, G. M., Ju, Y.-M., Lai, C.-Y., Mathew, D. C., & Huang, C. C. (2012). Microbial community analysis in the termite gut and fungus comb of *Odontotermes formosanus*: The implication of *Bacillus* as mutualists. *FEMS Microbiology Ecology*, *79*(2), 504–517. https://doi.org/10.1111/j.1574-6941.2011.01232.x

Mathew, G. M., Mathew, D. C., Lo, S.-C., Alexios, G. M., Yang, J.-C., Sashikumar, J. M., Shaikh, T. M., & Huang, C.-C. (2013). Synergistic collaboration of gut symbionts in *Odontotermes formosanus* for lignocellulosic degradation and bio-hydrogen production. *Bioresource Technology*, *145*, 337–344. https://doi.org/10.1016/j.biortech.2012.12.055

Matsumoto, M., Yokouchi, H., Suzuki, N., Ohata, H., & Matsunaga, T. (2003). Saccharification of marine microalgae using marine bacteria for ethanol production. *Applied Biochemistry and Biotechnology*, *105–108*, 247–254. https://doi.org/10.1385/abab:105:1-3:247

Matsuo, K., Isogai, E., & Araki, Y. (2000). Utilization of exocellular mannan from *Rhodotorula glutinis* as an immunoreactive antigen in diagnosis of leptospirosis. *Journal of Clinical Microbiology*, *38*(10), 3750–3754. https://doi.org/10.1128/JCM.38.10.3750-3754.2000

Matsudo, M. C., Bezerra, R. P., Converti, A., Sato, S., Carvalho, J. C. M. (2011). CO_2 from alcoholic fermentation for continuous cultivation of *Arthrospira (Spirulina) platensis* in tubular photobioreactor using urea as nitrogen source. *Biotechnology Progress*, *27*(3), 650–656. https://doi.org/10.1002/btpr.581

McClendon, S. D., Mao, Z., Shin, H.-D., Wagschal, K., & Chen, R. R. (2012). Designer Xylanosomes: Protein nanostructures for enhanced xylan hydrolysis. *Applied Biochemistry and Biotechnology*, *167*(3), 395–411. https://doi.org/10.1007/s12010-012-9680-1

Meesters, P. A. E. P., Huijberts, G. N. M., & Eggink, G. (1996). High-cell-density cultivation of the lipid accumulating yeast *Cryptococcus curvatus* using glycerol as a carbon source. *Applied Microbiology and Biotechnology*, *45*(5), 575–579. https://doi.org/10.1007/s002530050731

Melo, A. L. de A., Soccol, V. T., & Soccol, C. R. (2016). *Bacillus thuringiensis*: Mechanism of action, resistance, and new applications: A review. *Critical Reviews in Biotechnology*, *36*(2), 317–326. https://doi.org/10.3109/07388551.2014.960793

Méndez, J. J., Oromi, M., Cervero, M., Balcells, M., Torres, M., & Canela, R. (2007). Combining regio- and enantioselectivity of lipases for the preparation of (R)-4-chloro-2-butanol. *Chirality*, *19*(1), 44–50. https://doi.org/10.1002/chir.20339

Meng, F., Ma, L., Ji, S., Yang, W., & Cao, B. (2014). Isolation and characterization of *Bacillus subtilis* strain BY-3, a thermophilic and efficient cellulase-producing bacterium on untreated plant biomass. *Letters in Applied Microbiology*, *59*(3), 306–312. https://doi.org/10.1111/lam.12276

Meng, X., Yang, J., Xu, X., Zhang, L., Nie, Q., & Xian, M. (2009). Biodiesel production from oleaginous microorganisms. *Renewable Energy*, *34*(1), 1–5. https://doi.org/10.1016/j.renene.2008.04.014

Metzger, P., & Largeau, C. (2005). *Botryococcus braunii*: A rich source for hydrocarbons and related ether lipids. *Applied Microbiology and Biotechnology*, *66*(5), 486–496. https://doi.org/10.1007/s00253-004-1779-z

Meussen, B. J., de Graaff, L. H., Sanders, J. P. M., & Weusthuis, R. A. (2012). Metabolic engineering of *Rhizopus oryzae* for the production of platform chemicals. *Applied Microbiology and Biotechnology, 94*(4), 875–886. https://doi.org/10.1007/s00253-012-4033-0

Miao, X., & Wu, Q. (2006). Biodiesel production from heterotrophic microalgal oil. *Bioresource Technology, 97*(6), 841–846. https://doi.org/10.1016/j.biortech.2005.04.008

Miller, A., Scanlan, R. A., Lee, J. S., & Libbey, L. M. (1973). Volatile compounds produced in sterile fish muscle (*Sebastes melanops*) by *Pseudomonas putrefaciens*, *Pseudomonas fluorescens*, and an *Achromobacter* species. *Applied Microbiology, 26*(1), 18–21. https://doi.org/10.1128/am.26.1.18-21.1973

Miranda, J. R., Passarinho, P. C., & Gouveia, L. (2012). Pre-treatment optimization of *Scenedesmus obliquus* microalga for bioethanol production. *Bioresource Technology, 104*(January), 342–348. https://doi.org/10.1016/j.biortech.2011.10.059

Mohd-Sahib, A.-A., Lim, J.-W., Lam, M.-K., Uemura, Y., Isa, M. H., Ho, C.-D., Rahman Mohamed Kutty, S., Wong, C.-Y., & Rosli, S.-S. (2017). Lipid for biodiesel production from attached growth *Chlorella vulgaris* biomass cultivating in fluidized bed bioreactor packed with polyurethane foam material. *Bioresource Technology, 239*(September), 127–136. https://doi.org/10.1016/j.biortech.2017.04.118

Molino, A., Mehariya, S., Iovine, A., Larocca, V., Di Sanzo, G., Martino, M., Casella, P., Chianese, S., & Musmarra, D. (2018). Extraction of astaxanthin and lutein from microalga *Haematococcus pluvialis* in the red phase using CO_2 supercritical fluid extraction technology with ethanol as co-solvent. *Marine Drugs, 16*(11). https://doi.org/10.3390/md16110432

Monlau, F., Barakat, A., Trably, E., Dumas, C., Steyer, J.-P., & Carrère, H. (2013). Lignocellulosic materials into biohydrogen and biomethane: Impact of structural features and pretreatment. *Critical Reviews in Environmental Science and Technology, 43*(3), 260–322. https://doi.org/10.1080/10643389.2011.604258

Montalvão, S., Demirel, Z., Devi, P., Lombardi, V., Hongisto, V., Perälä, M., Hattara, J., et al. (2016). Large-scale bioprospecting of cyanobacteria, micro- and macroalgae from the Aegean Sea. *New Biotechnology, 33*(3), 399–406. https://doi.org/10.1016/j.nbt.2016.02.002

Moraïs, S., Barak, Y., Caspi, J., Hadar, Y., Lamed, R., Shoham, Y., Wilson, D. B., & Bayer, E. A. (2010). Cellulase-xylanase synergy in designer cellulosomes for enhanced degradation of a complex cellulosic substrate. *MBio, 1*(5). https://doi.org/10.1128/mBio.00285-10

Moreno, A. D., Carbone, A., Pavone, R., Olsson, L., & Geijer, C. (2019). Evolutionary engineered *Candida intermedia* exhibits improved xylose utilization and robustness to lignocellulose-derived inhibitors and ethanol. *Applied Microbiology and Biotechnology, 103*(3), 1405–1416. https://doi.org/10.1007/s00253-018-9528-x

Morsy, F. M. (2011). Acetate versus sulfur deprivation role in creating anaerobiosis in light for hydrogen production by *Chlamydomonas reinhardtii* and *Spirulina platensis*: Two different organisms and two different mechanisms. *Photochemistry and Photobiology, 87*(1), 137–142. https://doi.org/10.1111/j.1751-1097.2010.00823.x

Munir, I., Nakazawa, M., Harano, K., Yamaji, R., Inui, H., Miyatake, K., & Nakano, Y. (2002). Occurrence of a novel NADP(+)-linked alcohol dehydrogenase in *Euglena gracilis*. *Comparative Biochemistry and Physiology. Part B, Biochemistry & Molecular Biology, 132*(3), 535–540. https://doi.org/10.1016/s1096-4959(02)00068-4

Muniraj, I. K., Xiao, L., Liu, H., & Zhan, X. (2015). Utilisation of potato processing wastewater for microbial lipids and γ-Linolenic acid production by *Oleaginous fungi*. *Journal of the Science of Food and Agriculture, 95*(15), 3084–3090. https://doi.org/10.1002/jsfa.7044

Muñoz, C., Hidalgo, C., Zapata, M., Jeison, D., Riquelme, C., & Rivas, M. (2014). Use of cellulolytic marine bacteria for enzymatic pretreatment in microalgal biogas production. *Applied and Environmental Microbiology, 80*(14), 4199–4206. https://doi.org/10.1128/AEM.00827-14

Muradov, N., Taha, M., Miranda, A. F., Wrede, D., Kadali, K., Gujar, A., Stevenson, T., Ball, A. S., & Mouradov, A. (2015). Fungal-assisted algal flocculation: Application in wastewater treatment and biofuel production. *Biotechnology for Biofuels, 8*, 24. https://doi.org/10.1186/s13068-015-0210-6

Murugesan, S., Iyyaswami, R., Vijay Kumar, S., & Surendran, A. (2017). Anionic surfactant based reverse micellar extraction of L-Asparaginase synthesized by *Azotobacter vinelandii*. *Bioprocess and Biosystems Engineering, 40*(8), 1163–1171. https://doi.org/10.1007/s00449-017-1777-z

Mus, F., Dubini, A., Seibert, M., Posewitz, M. C., & Grossman, A. R. (2007). Anaerobic acclimation in *Chlamydomonas reinhardtii*: Anoxic gene expression, hydrogenase induction, and metabolic pathways. *Journal of Biological Chemistry, 282*(35), 25475–25486. https://doi.org/10.1074/jbc.M701415200

Musharraf, S. G., Ahmed, M. A., Zehra, N., Kabir, N., Iqbal Choudhary, M. & Rahman, A.-U. (2012). Biodiesel production from microalgal isolates of Southern Pakistan and quantification of FAMEs by GC-MS/MS analysis. *Chemistry Central Journal*, *6*(1), 149. https://doi.org/10.1186/1752-153X-6-149

Muslim, A., Hyakumachi, M., Kageyama, K., & Suwandi, S. (2019). Induction of systemic resistance in cucumber by hypovirulent *Binucleate rhizoctonia* against anthracnose caused by *Colletotrichum orbiculare*. *Tropical Life Sciences Research*, *30*(1), 109–122. https://doi.org/10.21315/tlsr2019.30.1.7

Nagao, T., Watanabe, Y., Maruyama, K., Momokawa, Y., Kishimoto, N., & Shimada, Y. (2011). One-pot enzymatic synthesis of docosahexaenoic acid-rich triacylglycerols at the SN-1(3) position using by-product from selective hydrolysis of tuna oil. *New Biotechnology*, *28*(1), 7–13. https://doi.org/10.1016/j.nbt.2010.07.021

Nagy, V., Vidal-Meireles, A., Podmaniczki, A., Szentmihályi, K., G. Rákhely, Zsigmond, L., Kovács, L., & Tóth, S. Z. (2018). The mechanism of photosystem-II inactivation during sulphur deprivation-induced H2 production in *Chlamydomonas reinhardtii*. *The Plant Journal: For Cell and Molecular Biology*, *94*(3), 548–561. https://doi.org/10.1111/tpj.13878

Navarro López, E., Medina, A. R., González Moreno, P. A., Cerdán, L. E., Martín Valverde, L., & Molina Grima, E. (2016). Biodiesel production from *Nannochloropsis gaditana* lipids through transesterification catalyzed by *Rhizopus oryzae* lipase. *Bioresource Technology*, *203*(March), 236–244. https://doi.org/10.1016/j.biortech.2015.12.036

Naziya, B., Murali, M., & Amruthesh, K. N. (2019). Plant growth-promoting fungi (PGPF) instigate plant growth and induce disease resistance in *Capsicum annuum* L. upon infection with *Colletotrichum capsici* (Syd.) Butler & Bisby. *Biomolecules*, *10*(1), E41. https://doi.org/10.3390/biom10010041

Ng, T.-K., Yu, A.-Q., Ling, H., Pratomo Juwono, N. K., Choi, W. J., Jan Leong, S. S., & Chang, M. W. (2020). Engineering *Yarrowia lipolytica* towards food waste bioremediation: Production of fatty acid ethyl esters from vegetable cooking oil. *Journal of Bioscience and Bioengineering*, *129*(1), 31–40. https://doi.org/10.1016/j.jbiosc.2019.06.009

Nguyen, M. T., Choi, S. P., Lee, J., Lee, J. H., & Sim, S. J. (2009). Hydrothermal acid pretreatment of *Chlamydomonas reinhardtii* biomass for ethanol production. *Journal of Microbiology and Biotechnology*, *19*(2), 161–166. https://doi.org/10.4014/jmb.0810.578

Nguyễn, T. H. M., & Vu, V. H. (2012). Bioethanol production from marine algae biomass: Prospect and troubles. *Journal of Vietnamese Environment*, *3*(2). https://doi.org/10.13141/JVE.VOL3.NO1.PP25-29

Nölling, J., Breton, G., Omelchenko, M. V., Makarova, K. S., Zeng, Q., Gibson, G., Hong, M. L., Dubois, J., Qiu, D., Hitti, J., Aldredge, T., Ayers, M., Bashirzadeh, R., Bochner, H., Boivin, M., Bross, S., Bush, D., Butler, C., Caron, A., ... & Smith, D. R. (2001). Genome sequence and comparative analysis of the solvent-producing bacterium *Clostridium acetobutylicum*. *Journal of Bacteriology*, *183*(16), 4823–4838. https://doi.org/10.1128/JB.183.16.4823-4838.2001

O'Neil, G. W., Gale, A. C., Nelson, R. K., Dhaliwal, H. K., & Reddy, C. M. (2021). Unusual shorter-chain C35 and C36 alkenones from commercially grown *Isochrysis* sp. microalgae. *Journal of the American Oil Chemists' Society*, n/a(n/a). Accessed June 4, 2021. https://doi.org/10.1002/aocs.12481

Oda, T., Oda, K., Yamamoto, H., Matsuyama, A., Ishii, M., Igarashi, Y., & Nishihara, H. (2013). Hydrogen-driven asymmetric reduction of hydroxyacetone to (R)-1,2-propanediol by *Ralstonia* Eutropha transformant expressing alcohol dehydrogenase from *Kluyveromyces lactis*. *Microbial Cell Factories*, *12*(1), 2. https://doi.org/10.1186/1475-2859-12-2

Odier, E., & Rolando, C. (1985). Catabolism of arylglycerol-beta-aryl ethers lignin model compounds by *Pseudomonas cepacia* 122. *Biochimie*, *67*(2), 191–197. https://doi.org/10.1016/s0300-9084(85)80047-x

Oguro, Y., Nakamura, A., & Kurahashi, A. (2019). Effect of temperature on saccharification and oligosaccharide production efficiency in *Koji amazake*. *Journal of Bioscience and Bioengineering*, *127*(5), 570–574. https://doi.org/10.1016/j.jbiosc.2018.10.007

Ohm, R. A., Riley, R., Salamov, A., Min, B., Choi, I.-G., & Grigoriev, I. V. (2014). Genomics of wood-degrading fungi. *Fungal Genetics and Biology: FG & B*, *72*, 82–90. https://doi.org/10.1016/j.fgb.2014.05.001

Olguín, E. J., Castillo, O. S., Mendoza, A., Tapia, K., González-Portela, R. E., & Hernández-Landa, V. J. (2015). Dual purpose system that treats anaerobic effluents from pig waste and produce *Neochloris oleoabundans* as lipid rich biomass. *New Biotechnology*, *32*(3), 387–395. https://doi.org/10.1016/j.nbt.2014.12.004

Oliveira, A. C., Barata, A., Batista, A. P., & Gouveia, L. (2019). *Scenedesmus obliquus* in poultry wastewater bioremediation. *Environmental Technology*, *40*(28), 3735–3744. https://doi.org/10.1080/09593330.2018.1488003

Oliveira, A. C., Rosa, M. F., Aires-Barros, M. R., & Cabral, J. M. S. (2000). Enzymatic esterification of ethanol by an immobilised *Rhizomucor miehei* lipase in a perforated rotating disc bioreactor. *Enzyme and Microbial Technology*, *26*(5–6), 446–450. https://doi.org/10.1016/s0141-0229(99)00185-4

Olofsson, K., Runquist, D., Hahn-Hägerdal, B., & Lidén, G. (2011). A mutated xylose reductase increases bioethanol production more than a glucose/xylose facilitator in simultaneous fermentation and co-fermentation of wheat straw. *AMB Express*, *1*(1), 4. https://doi.org/10.1186/2191-0855-1-4

Olsson, L., & Hahn-Hägerdal, B. (1996). Fermentation of lignocellulosic hydrolysates for ethanol production. *Enzyme and Microbial Technology*, *18*(5), 312–331. https://doi.org/10.1016/0141-0229(95)00157-3

Ono, K., Kawanaka, Y., Izumi, Y., Inui, H., Miyatake, K., Kitaoka, S., & Nakano, Y. (1995). Mitochondrial alcohol dehydrogenase from ethanol-grown *Euglena gracilis*. *Journal of Biochemistry*, *117*(6), 1178–1182. https://doi.org/10.1093/oxfordjournals.jbchem.a124841

Onwudili, J. A., Lea-Langton, A. R., Ross, A. B., & Williams, P. T. (2013). Catalytic hydrothermal gasification of algae for hydrogen production: Composition of reaction products and potential for nutrient recycling. *Bioresource Technology*, *127*(January), 72–80. https://doi.org/10.1016/j.biortech.2012.10.020

Otero, P., Saha, S. K., Mc Gushin, J., Moane, S., Barron, J., & Murray, P. (2017). Identification of optimum fatty acid extraction methods for two different microalgae *Phaeodactylum tricornutum* and *Haematococcus pluvialis* for food and biodiesel applications. *Analytical and Bioanalytical Chemistry*, *409*(19), 4659–4667. https://doi.org/10.1007/s00216-017-0412-9

Pade, N., Mikkat, S., & Hagemann, M. (2017). Ethanol, glycogen and glucosylglycerol represent competing carbon pools in ethanol-producing cells of *Synechocystis* sp. PCC 6803 under high-salt conditions. *Microbiology (Reading, England)*, *163*(3), 300–307. https://doi.org/10.1099/mic.0.000433

Pagliano, G., Ventorino, V., Panico, A., & Pepe, O. (2017). Integrated systems for biopolymers and bioenergy production from organic waste and by-products: A review of microbial processes. *Biotechnology for Biofuels*, *10*, 113. https://doi.org/10.1186/s13068-017-0802-4

Palamae, S., Choorit, W., Chatsungnoen, T., & Chisti, Y. (2020). Simultaneous nitrogen fixation and ethanol production by *Zymomonas mobilis*. *Journal of Biotechnology*, *314–315*(May), 41–52. https://doi.org/10.1016/j.jbiotec.2020.03.016

Pan, Y., Alam, M. A., Wang, Z., Huang, D., Hu, K., Chen, H., & Yuan, Z. (2017a). One-step production of biodiesel from wet and unbroken microalgae biomass using deep eutectic solvent. *Bioresource Technology*, *238*(August), 157–163. https://doi.org/10.1016/j.biortech.2017.04.038

Pan, Z., Huang, Y., Wang, Y., & Wu, Z. (2017b). Disintegration of *Nannochloropsis* sp. cells in an improved turbine bead mill. *Bioresource Technology*, *245*(Pt A), 641–648. https://doi.org/10.1016/j.biortech.2017.08.146

Pancha, I., Chokshi, K., Maurya, R., Bhattacharya, S., Bachani, P., & Mishra, S. (2016). Comparative evaluation of chemical and enzymatic saccharification of mixotrophically grown de-oiled microalgal biomass for reducing sugar production. *Bioresource Technology*, *204*(March), 9–16. https://doi.org/10.1016/j.biortech.2015.12.078

Pandey, A., Srivastava, N., & Sinha, P. (2012). Optimization of Hydrogen production by *Rhodobacter sphaeroides* NMBL-01. *Biomass and Bioenergy*, *37*(February), 251–256. https://doi.org/10.1016/j.biombioe.2011.12.005

Pansook, S., Incharoensakdi, A., & Phunpruch, S. (2019). Enhanced dark fermentative H2 production by agar-immobilized cyanobacterium *Aphanothece halophytica*. *Journal of Applied Phycology*, *31*, 2869–2879. https://doi.org/10.1007/s10811-019-01822-9

Papanikolaou, S., & Aggelis, G. (2019). Sources of microbial oils with emphasis to *Mortierella (umbelopsis) Isabellina* fungus. *World Journal of Microbiology & Biotechnology*, *35*(4), 63. https://doi.org/10.1007/s11274-019-2631-z

Papanikolaou, S., Rontou, M., Belka, A., Athenaki, M., Gardeli, C., Mallouchos, A., Kalantzi, O., et al. (2017). Conversion of biodiesel-derived glycerol into biotechnological products of industrial significance by yeast and fungal strains. *Engineering in Life Sciences*, *17*(3), 262–281. https://doi.org/10.1002/elsc.201500191

Papazi, A., Andronis, E., Ioannidis, N. E., Chaniotakis, N., & Kotzabasis, K. (2012). High yields of hydrogen production induced by meta-substituted dichlorophenols biodegradation from the green alga *Scenedesmus obliquus*. *PloS One*, *7*(11), e49037. https://doi.org/10.1371/journal.pone.0049037

Parawira, W. (2009). Biotechnological production of biodiesel fuel using biocatalyzed transesterification: A review. *Critical Reviews in Biotechnology*, *29*(2), 82–93. https://doi.org/10.1080/07388550902823674

Parisutham, V., Kim, T. H., & Lee, S. K. (2014). Feasibilities of consolidated bioprocessing microbes: From pretreatment to biofuel production. *Bioresource Technology*, *161*, 431–440. https://doi.org/10.1016/j.biortech.2014.03.114

Park, D., Jagtap, S., & Nair, S. K. (2014). Structure of a PL17 family alginate lyase demonstrates functional similarities among exotype depolymerases. *Journal of Biological Chemistry*, *289*(12), 8645–8655. https://doi.org/10.1074/jbc.M113.531111

Park, J., Park, B. S., Wang, P., Patidar, S. K., Kim, J. H., Kim, S.-H., & Han, M.-S. (2017). Phycospheric native bacteria *Pelagibaca bermudensis* and *Stappia* sp. Ameliorate biomass productivity of *Tetraselmis striata* (KCTC1432BP) in co-cultivation system through mutualistic interaction. *Frontiers in Plant Science*, *8*, 289. https://doi.org/10.3389/fpls.2017.00289

Park, J. I., Steen, E. J., Burd, H., Evans, S. S., Redding-Johnson, A. M., Batth, T., Benke, P. I. et al. (2012). A thermophilic ionic liquid-tolerant cellulase cocktail for the production of cellulosic biofuels. *PloS One*, *7*(5), e37010. https://doi.org/10.1371/journal.pone.0037010

Park, J.-Y., Kim, M.-C., Nam, B., Chang, H., & Kim, D.-K. (2021). Behavior of surfactants in oil extraction by surfactant-assisted acidic hydrothermal process from *Chlorella vulgaris*. *Applied Biochemistry and Biotechnology*, *193*(2), 319–334. https://doi.org/10.1007/s12010-020-03426-3

Park, K. Y., Lim, B.-R., & Lee, K. (2009). Growth of microalgae in diluted process water of the animal wastewater treatment plant. *Water Science and Technology: A Journal of the International Association on Water Pollution Research*, *59*(11), 2111–2116. https://doi.org/10.2166/wst.2009.233

Park, S., Kim, S. H., Won, K., Choi, J. W., Kim, Y. H., Kim, H. J., Yang, Y.-H., & Hyun Lee, S. (2015). Wood mimetic hydrogel beads for enzyme immobilization. *Carbohydrate Polymers*, *115*(January), 223–229. https://doi.org/10.1016/j.carbpol.2014.08.096

Parmar, A., Singh, N. K., Pandey, A., Gnansounou, E., & Madamwar, D. (2011). Cyanobacteria and microalgae: A positive prospect for biofuels. *Bioresource Technology*, *102*(22), 10163–10172. https://doi.org/10.1016/j.biortech.2011.08.030

Parsaeimehr, A., Sun, Z., Dou, X., & Chen, Y.-F. (2015). Simultaneous improvement in production of microalgal biodiesel and high-value alpha-linolenic acid by a single regulator acetylcholine. *Biotechnology for Biofuels*, *8*, 11. https://doi.org/10.1186/s13068-015-0196-0

Passos, F., Uggetti, E., Carrère, H., & Ferrer, I. (2014). Pretreatment of microalgae to improve biogas production: A review. *Bioresource Technology*, *172*, 403–412. https://doi.org/10.1016/j.biortech.2014.08.114

Patel, A. K., Laroche, C., Marcati, A., Violeta Ursu, A., Jubeau, S., Marchal, L., Petit, E., Djelveh, G., & Michaud, P. (2013). Separation and fractionation of exopolysaccharides from *Porphyridium cruentum*. *Bioresource Technology*, *145*(October), 345–350. https://doi.org/10.1016/j.biortech.2012.12.038

Patel, A., Pravez, M., Deeba, F., Pruthi, V., Singh, R. P., & Pruthi, P. A. (2014). Boosting accumulation of neutral lipids in *Rhodosporidium kratochvilovae* HIMPA1 grown on hemp (*Cannabis sativa* Linn.) Seed aqueous extract as feedstock for biodiesel production. *Bioresource Technology*, *165*(August), 214–222. https://doi.org/10.1016/j.biortech.2014.03.142

Patel, A., Pruthi, V., Singh, R. P., & Pruthi, P. A. (2015). Synergistic effect of fermentable and non-fermentable carbon sources enhances TAG Accumulation in oleaginous yeast *Rhodosporidium kratochvilovae* HIMPA1. *Bioresource Technology*, *188*, 136–144. https://doi.org/10.1016/j.biortech.2015.02.062

Peng, X., Misawa, N., & Harayama, S. (2003). Isolation and characterization of thermophilic bacilli degrading cinnamic, 4-coumaric, and ferulic acids. *Applied and Environmental Microbiology*, *69*(3), 1417–1427. https://doi.org/10.1128/AEM.69.3.1417-1427.2003

Peng, X.-W., & Chen, H.-Z. (2007). Microbial oil accumulation and cellulase secretion of the endophytic fungi from oleaginous plants. *Annals of Microbiology*, *57*(2), 239. https://doi.org/10.1007/BF03175213

Peralta-Yahya, P. P., & Keasling, J. D. (2010). Advanced biofuel production in microbes. *Biotechnology Journal*, *5*(2), 147–162. https://doi.org/10.1002/biot.200900220

Perisin, M. A., & Sund, C. J. (2018). Human gut microbe co-cultures have greater potential than monocultures for food waste remediation to commodity chemicals. *Scientific Reports*, *8*(1), 15594. https://doi.org/10.1038/s41598-018-33733-z

Picard, K. T., Letcher, P. M., & Powell, M. J. (2013). Evidence for a facultative mutualist nutritional relationship between the green coccoid alga *Bracteacoccus* sp. (*Chlorophyceae*) and the zoosporic fungus *Rhizidium phycophilum* (*Chytridiomycota*). *Fungal Biology*, *117*(5), 319–328. https://doi.org/10.1016/j.funbio.2013.03.003

Pikula, K. S., Zakharenko, A. M., Chaika, V. V., Stratidakis, A. K., Kokkinakis, M., Waissi, G., Rakitskii, V. N., et al. (2019). Toxicity bioassay of waste cooking oil-based biodiesel on marine microalgae. *Toxicology Reports*, *6*, 111–117. https://doi.org/10.1016/j.toxrep.2018.12.007

Pinto, T. S., Malcata, F. X., Arrabaça, J. D., Silva, J. M., Spreitzer, R. J., & Esquível, M. G. (2013). Rubisco mutants of *Chlamydomonas reinhardtii* enhance photosynthetic hydrogen production. *Applied Microbiology and Biotechnology, 97*(12), 5635–5643. https://doi.org/10.1007/s00253-013-4920-z

Poisson, L., & Ergan, F. (2001). Docosahexaenoic acid ethyl esters from *Isochrysis galbana*. *Journal of Biotechnology, 91*(1), 75–81. https://doi.org/10.1016/s0168-1656(01)00295-4

Pokora, M., Zambrowicz, A., Zabłocka, A., Dąbrowska, A., Szołtysik, M., Babij, K., Eckert, E., Trziszka, T., & Chrzanowska, J. (2017). The use of serine protease from *Yarrowia lipolytica* yeast in the production of biopeptides from denatured egg white proteins. *Acta Biochimica Polonica, 64*(2), 245–253. https://doi.org/10.18388/abp.2016_1316

Pollegioni, L., Tonin, F., & Rosini, E. (2015). Lignin-degrading enzymes. *The FEBS Journal, 282*(7), 1190–1213. https://doi.org/10.1111/febs.13224

Qi, G.-X., Huang, C., Chen, X.-F., Xiong, L., Wang, C., Lin, X.-Q., Shi, S.-L., Yang, D., & Chen, X.-D. (2016). Semi-pilot scale microbial oil production by *Trichosporon cutaneum* using medium containing corncob acid hydrolysate. *Applied Biochemistry and Biotechnology, 179*(4), 625–632. https://doi.org/10.1007/s12010-016-2019-6

Qiao, W., Tao, J., Luo, Y., Tang, T., Miao, J., & Yang, Q. (2018). Microbial oil production from solid-state fermentation by a newly isolated oleaginous fungus, *Mucor circinelloides* Q531 from mulberry branches. *Royal Society Open Science, 5*(11), 180551. https://doi.org/10.1098/rsos.180551

Qiu, C., He, Y., Huang, Z., Li, S., Huang, J., Wang, M., & Chen, B. (2019). Lipid extraction from wet *Nannochloropsis* biomass via enzyme-assisted three phase partitioning. *Bioresource Technology, 284*(July), 381–390. https://doi.org/10.1016/j.biortech.2019.03.148

Rahikainen, J. L., Moilanen, U., Nurmi-Rantala, S., Lappas, A., Koivula, A., Viikari, L., & Kruus, K. (2013). Effect of temperature on lignin-derived inhibition studied with three structurally different cellobiohydrolases. *Bioresource Technology, 146*(October), 118–125. https://doi.org/10.1016/j.biortech.2013.07.069

Rahman, I. N. A., Attan, N., Mahat, N. A., Jamalis, J., Abdul Keyon, A. S., Kurniawan, C., & Wahab, R. A. (2018). Statistical optimization and operational stability of *Rhizomucor miehei* lipase supported on magnetic chitosan/chitin nanoparticles for synthesis of *Pentyl valerate*. *International Journal of Biological Macromolecules, 115*(August), 680–695. https://doi.org/10.1016/j.ijbiomac.2018.04.111

Rahmanpour, R., & Bugg, T. D. H. (2015). Characterisation of Dyp-type peroxidases from *Pseudomonas fluorescens* Pf-5: Oxidation of Mn(II) and polymeric lignin by Dyp1B. *Archives of Biochemistry and Biophysics, 574*(May), 93–98. https://doi.org/10.1016/j.abb.2014.12.022

Raj, J. V. A., Bharathiraja, B., Vijayakumar, B., Arokiyaraj, S., Iyyappan, J., & Praveen Kumar, R. (2019). Biodiesel production from microalgae *Nannochloropsis oculata* using heterogeneous Poly Ethylene Glycol (PEG) encapsulated ZnOMn2+ nanocatalyst. *Bioresource Technology, 282*(June), 348–352. https://doi.org/10.1016/j.biortech.2019.03.030

Ramanan, R., Kim, B.-H., Cho, D.-H., Oh, H.-M., & Kim, H.-S. (2016). Algae-bacteria interactions: Evolution, ecology and emerging applications. *Biotechnology Advances, 34*(1), 14–29. https://doi.org/10.1016/j.biotechadv.2015.12.003

Ramanna, L., Guldhe, A., Rawat, I., & Bux, F. (2014). The optimization of biomass and lipid yields of *Chlorella sorokiniana* when using wastewater supplemented with different nitrogen sources. *Bioresource Technology, 168*(September), 127–135. https://doi.org/10.1016/j.biortech.2014.03.064

Ramos-Ibarra, J. R., Snell-Castro, R., Neria-Casillas, J. A., & Choix, F. J. (2019). Biotechnological potential of *Chlorella* sp. and *Scenedesmus* sp. microalgae to endure high CO_2 and methane concentrations from biogas. *Bioprocess and Biosystems Engineering, 42*(10), 1603–1610. https://doi.org/10.1007/s00449-019-02157-y

Ranjith Kumar, R., Hanumantha Rao, P., & Arumugam, M. (2015). Lipid extraction methods from microalgae: A comprehensive review. *Frontiers in Energy Research, 2*. https://doi.org/10.3389/fenrg.2014.00061

Rashid, N., Lee, K., & Mahmood, Q. (2011). Bio-hydrogen production by *Chlorella vulgaris* under diverse photoperiods. *Bioresource Technology, 102*(2), 2101–2104. https://doi.org/10.1016/j.biortech.2010.08.032

Rashid, N., Ur Rehman, M. S., Sadiq, M., Mahmood, T., & Han, J.-I. (2014). Current status, issues and developments in microalgae derived biodiesel production. *Renewable and Sustainable Energy Reviews, 40*(C), 760–778.

Ratledge, C. (1991). Microorganisms for lipids. *Acta Biotechnologica, 11*(5), 429–438. https://doi.org/10.1002/abio.370110506

Ratnapuram, H. P., Vutukuru, S. S., & Yadavalli, R. (2018). Mixotrophic transition induced lipid productivity in *Chlorella pyrenoidosa* under stress conditions for biodiesel production. *Heliyon, 4*(1), e00496. https://doi.org/10.1016/j.heliyon.2017.e00496

Ravella, S. R., James, S. A., Bond, C. J., Roberts, I. N., Cross, K., Retter, A., & Hobbs, P. J. (2010). *Cryptococcus shivajii* sp. Nov.: A novel basidiomycetous yeast isolated from biogas reactor. *Current Microbiology*, *60*(1), 12–16. https://doi.org/10.1007/s00284-009-9493-9

Rawat, R., Srivastava, N., Chadha, B. S., & Oberoi, H. S. (2014). Generating fermentable sugars from rice straw using functionally active cellulolytic enzymes from *Aspergillus niger* HO. *Energy & Fuels*, *28*(8), 5067–5075. https://doi.org/10.1021/ef500891g

Reddy, C. N., Nguyen, H. T. H., Noori, Md. T., & Booki Min, B. (2019). Potential applications of algae in the cathode of microbial fuel cells for enhanced electricity generation with simultaneous nutrient removal and algae biorefinery: Current status and future perspectives. *Bioresource Technology*, *292*. https://doi.org/10.1016/j.biortech.2019.122010

Rehman, Z. U., & Anal, A. K. (2019). Enhanced lipid and starch productivity of Microalga (*Chlorococcum* sp. TISTR 8583) with nitrogen limitation following effective pretreatments for biofuel production. *Biotechnology Reports (Amsterdam, Netherlands)*, *21*(March), e00298. https://doi.org/10.1016/j.btre.2018.e00298

Reis, C. E. R., Bento, H. B. S., Carvalho, A. K. F., Rajendran, A., Hu, B., & De Castro, H. F. (2019). Critical applications of *Mucor circinelloides* within a biorefinery context. *Critical Reviews in Biotechnology*, *39*(4), 555–570. https://doi.org/10.1080/07388551.2019.1592104

Rempel, A., de Souza Sossella, F., Margarites, A. C., Astolfi, A. L., Steinmetz, R. L. R., Kunz, A., Treichel, H., & Colla, L. M. (2019). Bioethanol from *Spirulina platensis* biomass and the use of residuals to produce biomethane: An energy efficient approach. *Bioresource Technology*, *288*(September), 121588. https://doi.org/10.1016/j.biortech.2019.121588

Ren, X., Wang, J., Yu, H., Peng, C., Hu, J., Ruan, Z., Zhao, S., Liang, Y., & Peng, N. (2016). Anaerobic and sequential aerobic production of high-titer ethanol and single cell protein from NaOH-pretreated corn stover by a genome shuffling-modified *Saccharomyces cerevisiae* strain. *Bioresource Technology*, *218*(October), 623–630. https://doi.org/10.1016/j.biortech.2016.06.118

Reyimu, Z., & Özçimen, D. (2017). Batch cultivation of marine microalgae *Nannochloropsis oculata* and *Tetraselmis suecica* in treated municipal wastewater toward bioethanol production. *Journal of Cleaner Production*, *150*, 40–46. https://doi.org/10.1016/j.jclepro.2017.02.189

Reyna-Martínez, R., Gomez-Flores, R., López-Chuken, U. J., González-González, R., Fernández-Delgadillo, S. & Balderas-Rentería, I. (2015). Lipid production by pure and mixed cultures of *Chlorella pyrenoidosa* and *Rhodotorula mucilaginosa* isolated in Nuevo Leon, Mexico. *Applied Biochemistry and Biotechnology*, *175*(1), 354–359. https://doi.org/10.1007/s12010-014-1275-6

Richmond, A. (2008). *Handbook of Microalgal Culture: Biotechnology and Applied Phycology*. Hoboken, NJ: John Wiley & Sons.

Ríos, S. D., Castañeda, J., Torras, C. Farriol, X., & Salvadó, J. (2013). Lipid extraction methods from microalgal biomass harvested by two different paths: Screening studies toward biodiesel production. *Bioresource Technology*, *133*(April), 378–388. https://doi.org/10.1016/j.biortech.2013.01.093

Rismani-Yazdi, H., Hampel, K. H., Lane, C. D., Kessler, B. A., White, N. M., Moats, K. M., & Thomas Allnutt, F. C. (2015). High-productivity lipid production using mixed trophic state cultivation of *Auxenochlorella (Chlorella) protothecoides*. *Bioprocess and Biosystems Engineering*, *38*(4), 639–650. https://doi.org/10.1007/s00449-014-1303-5

Robak, K., & Balcerek, M. (2018). Review of second generation bioethanol production from residual biomass. *Food Technology and Biotechnology*, *56*(2), 174–187. https://doi.org/10.17113/ftb.56.02.18.5428

Rodríguez-López, A., Fernández-Acero, F. J., Andrés-Vallejo, R., Guarnizo-García, P., Macías-Sánchez, M. D., Gutiérrez-Díaz, M. & Burgos-Rodríguez, S. (2020). Optimization of outdoor cultivation of the marine microalga *Nannochloropsis gaditana* in flat-panel reactors using industrial exhaust flue gases. *Journal of Applied Phycology*, *32*(2), 809–819. https://doi.org/10.1007/s10811-019-01990-8

Roostita, R., & Fleet, G. H. (1996). The occurrence and growth of yeasts in Camembert and Blue-Veined cheeses. *International Journal of Food Microbiology*, *28*(3), 393–404. https://doi.org/10.1016/0168-1605(95)00018-6

Ruan, Z., Hollinshead, W., Isaguirre, C., Tang, Y. J., Liao, W., & Liu, Y. (2015). Effects of inhibitory compounds in lignocellulosic hydrolysates on *Mortierella isabellina* growth and carbon utilization. *Bioresource Technology*, *183*(May), 18–24. https://doi.org/10.1016/j.biortech.2015.02.026

Ruangsomboon, S. (2015). Effects of different media and nitrogen sources and levels on growth and lipid of green microalga *Botryococcus braunii* KMITL and its biodiesel properties based on fatty acid composition. *Bioresource Technology, 191*(September), 377–384. https://doi.org/10.1016/j.biortech.2015.01.091

Ruangsomboon, S., Prachom, N., & Sornchai, P. (2017). Enhanced growth and hydrocarbon production of *Botryococcus braunii* KMITL 2 by optimum carbon dioxide concentration and concentration-dependent effects on its biochemical composition and biodiesel properties. *Bioresource Technology, 244*(Pt 2), 1358–1366. https://doi.org/10.1016/j.biortech.2017.06.042

Rudolf, A., Baudel, H., Zacchi, G., Hahn-Hägerdal, B. & Lidén, G. (2008). Simultaneous saccharification and fermentation of steam-pretreated bagasse using *Saccharomyces cerevisiae* TMB3400 and *Pichia stipitis* CBS6054. *Biotechnology and Bioengineering, 99*(4), 783–790. https://doi.org/10.1002/bit.21636

Rumbold, K., van Buijsen, H. J. J., Overkamp, K. M., van Groenestijn, J. W., Punt, P. J., & van der Werf, M. J. (2009). Microbial production host selection for converting second-generation feedstocks into bioproducts. *Microbial Cell Factories, 8*(December), 64. https://doi.org/10.1186/1475-2859-8-64

Runquist, D., Hahn-Hägerdal, B., & Rådström, P. (2010). Comparison of heterologous xylose transporters in recombinant *Saccharomyces cerevisiae*. *Biotechnology for Biofuels, 3*(March), 5. https://doi.org/10.1186/1754-6834-3-5

Rupprecht, J. (2009). From systems biology to fuel: *Chlamydomonas reinhardtii* as a model for a systems biology approach to improve biohydrogen production. *Journal of Biotechnology, 142*(1), 10–20. https://doi.org/10.1016/j.jbiotec.2009.02.008

Ryan, R. P., Germaine, K., Franks, A., Ryan, D. J., & Dowling, D. N. (2008). Bacterial endophytes: Recent developments and applications. *FEMS Microbiology Letters, 278*(1), 1–9. https://doi.org/10.1111/j.1574-6968.2007.00918.x

Ryu, B.-G., Kim, J., Kim, K., Choi, Y.-E., Han, J.-I., & Yang, J.-W. (2013). High-cell-density cultivation of oleaginous yeast *Cryptococcus curvatus* for biodiesel production using organic waste from the brewery industry. *Bioresource Technology, 135*(May), 357–364. https://doi.org/10.1016/j.biortech.2012.09.054

Ryu, S., Labbé, N., & Trinh, C. T. (2015). Simultaneous saccharification and fermentation of cellulose in ionic liquid for efficient production of α-Ketoglutaric acid by *Yarrowia lipolytica*. *Applied Microbiology and Biotechnology, 99*(10), 4237–4244. https://doi.org/10.1007/s00253-015-6521-5

Sabia, A., Clavero, E., Pancaldi, S., & Rovira, J. S. (2018). Effect of different CO_2 concentrations on biomass, pigment content, and lipid production of the marine diatom *Thalassiosira pseudonana*. *Applied Microbiology and Biotechnology, 102*(4), 1945–1954. https://doi.org/10.1007/s00253-017-8728-0

Sachs, J. P., & Kawka, O. E. (2015). The influence of growth rate on 2H/1H fractionation in continuous cultures of the *Coccolithophorid emiliania huxleyi* and the diatom *Thalassiosira pseudonana*. *PloS One, 10*(11), e0141643. https://doi.org/10.1371/journal.pone.0141643

Sahaym, U., & Grant Norton, M. (2008). Advances in the application of nanotechnology in enabling a 'hydrogen economy.' *Journal of Materials Science, 43*(16), 5395–5429. https://doi.org/10.1007/s10853-008-2749-0

Saito, K., Hasa, Y., & Abe, H. (2012). Production of lactic acid from xylose and wheat straw by *Rhizopus oryzae*. *Journal of Bioscience and Bioengineering, 114*(2), 166–169. https://doi.org/10.1016/j.jbiosc.2012.03.007

Sakarika, M., & Kornaros, M. (2019). *Chlorella vulgaris* as a green biofuel factory: Comparison between biodiesel, biogas and combustible biomass production. *Bioresource Technology, 273*(February), 237–243. https://doi.org/10.1016/j.biortech.2018.11.017

Salama, S. M., Atwal, H., Gandhi, A., Simon, J., Poglod, M., Montaseri, H., Khan, J. K., et al. (2001). In vitro and in vivo activities of Syn2836, Syn2869, Syn2903, and Syn2921: New series of triazole antifungal agents. *Antimicrobial Agents and Chemotherapy, 45*(9), 2420–2426. https://doi.org/10.1128/AAC.45.9.2420-2426.2001

Salis, A., Pinna, M., Monduzzi, M., & Solinas, V. (2005). Biodiesel production from triolein and short chain alcohols through biocatalysis. *Journal of Biotechnology, 119*(3), 291–299. https://doi.org/10.1016/j.jbiotec.2005.04.009.

Salvachúa, D., Karp, E. M., Nimlos, C. T., Vardon, D. R., & Beckham, G. T. (2015). Towards lignin consolidated bioprocessing: Simultaneous lignin depolymerization and product generation by bacteria. *Green Chemistry, 17*(11), 4951–4967. https://doi.org/10.1039/C5GC01165E

Samorì, C., Torri, C., Samorì, G., Fabbri, D., Galletti, P., Guerrini, F., Pistocchi, R., & Tagliavini, E. (2010). Extraction of hydrocarbons from microalga *Botryococcus braunii* with switchable solvents. *Bioresource Technology, 101*(9), 3274–3279. https://doi.org/10.1016/j.biortech.2009.12.068

Samson, R., & Leduy, A. (1982). Biogas production from anaerobic digestion of *Spirulina maxima* algal biomass. *Biotechnology and Bioengineering*, *24*(8), 1919–1924. https://doi.org/10.1002/bit.260240822

Sanchez Rizza, L., Sanz Smachetti, M. E., Do Nascimento, M., Salerno, G. L., & Curatti, L. (2017). Bioprospecting for native microalgae as an alternative source of sugars for the production of bioethanol. *Algal Research*, *22*, 140–147. https://doi.org/10.1016/j.algal.2016.12.021

Sánchez, A., Valero, F., Lafuente, J., & Solà, C. (2000). Highly enantioselective esterification of Racemic Ibuprofen in a packed bed reactor using immobilised *Rhizomucor miehei* lipase. *Enzyme and Microbial Technology*, *27*(1–2), 157–166. https://doi.org/10.1016/s0141-0229(00)00207-6

Santoro, C., Arbizzani, C., Benjamin Erable, B., & Ieropoulos, I. (2017). Microbial fuel cells: From fundamentals to applications. A review. *Journal of Power Sources*, *356*, 225–244.

Santos, C. A., Nobre, B., Lopes da Silva, T., Pinheiro, H. M., & Reis, A. (2014). Dual-mode cultivation of *Chlorella protothecoides* applying inter-reactors gas transfer improves microalgae biodiesel production. *Journal of Biotechnology*, *184*(August), 74–83. https://doi.org/10.1016/j.jbiotec.2014.05.012

Santos, D. da Silveira dos, Camelo, A. C., Rodrigues, K. C. P., Carlos, L. C., & Pereira, N. (2010). Ethanol production from sugarcane bagasse by *Zymomonas mobilis* using Simultaneous Saccharification and Fermentation (SSF) process. *Applied Biochemistry and Biotechnology*, *161*(1–8), 93–105. https://doi.org/10.1007/s12010-009-8810-x

Santos, N. O., Oliveira, S. M., Alves, L. C., & Cammarota, M. C. (2014). Methane production from marine microalgae *Isochrysis galbana*. *Bioresource Technology*, *157*(April), 60–67. https://doi.org/10.1016/j.biortech.2014.01.091

Santos-Ballardo, D. U., Font-Segura, X., Ferrer, A. S., Barrena, R., Rossi, S., & Valdez-Ortiz, A. (2015). Valorisation of biodiesel production wastes: Anaerobic digestion of residual *Tetraselmis suecica* biomass and co-digestion with glycerol. *Waste Management & Research: The Journal of the International Solid Wastes and Public Cleansing Association, ISWA*, *33*(3), 250–257. https://doi.org/10.1177/0734242X15572182

Santos-Fo, F., Fill, T. P., Nakamura, J., Monteiro, M. R., & Rodrigues-Fo, E. (2011). Endophytic fungi as a source of biofuel precursors. *Journal of Microbiology and Biotechnology*, *21*(7), 728–733. https://doi.org/10.4014/jmb.1010.10052

Sapp, M., Schwaderer, A. S., Wiltshire, K. H., Hoppe, H.-G., Gerdts, G., & Wichels, A. (2007). Species-specific bacterial communities in the phycosphere of microalgae? *Microbial Ecology*, *53*(4), 683–699. https://doi.org/10.1007/s00248-006-9162-5

Sasaki, K., Othman, M. B., Demura, M., Watanabe, M., & Isoda, H. (2017). Modulation of neurogenesis through the promotion of energy production activity is behind the antidepressant-like effect of colonial green alga, *Botryococcus braunii*. *Frontiers in Physiology*, *8*, 900. https://doi.org/10.3389/fphys.2017.00900

Sassi, P. G. P., Calixto, C. D., da Silva Santana, J. K., Sassi, R., Sassi, C. F. C., & Abrahão, R. (2017). Cultivation of freshwater microalgae in biodiesel wash water. *Environmental Science and Pollution Research International*, *24*(22), 18332–18340. https://doi.org/10.1007/s11356-017-9351-4

Sato, M., Murata, Y., Mizusawa, M., Iwahashi, H., & Oka, S. (2004). A simple and rapid dual-fluorescence viability assay for microalgae. *Microbiology and Culture Collections*, *20*(2), 7.

Sato, N., Moriyama, T., Mori, N., & Toyoshima, M. (2017). Lipid metabolism and potentials of biofuel and high added-value oil production in red algae. *World Journal of Microbiology & Biotechnology*, *33*(4), 74. https://doi.org/10.1007/s11274-017-2236-3

Schlaeppi, K., & Bulgarelli, D. (2015). The plant microbiome at work. *Molecular Plant-Microbe Interactions*, *28*(3), 212–217. https://doi.org/10.1094/MPMI-10-14-0334-FI

Scholz, S. A., Graves, I., Minty, J. J., & Lin, X. N. (2018). Production of cellulosic organic acids via synthetic fungal consortia. *Biotechnology and Bioengineering*, *115*(4), 1096–1100. https://doi.org/10.1002/bit.26509

Schreck, S. D., & Grunden, A. M. (2014). Biotechnological applications of halophilic lipases and thioesterases. *Applied Microbiology and Biotechnology*, *98*(3), 1011–1021. https://doi.org/10.1007/s00253-013-5417-5

Scully, Sean M., Aaron Brown, Andrew B. Ross, & Johann Orlygsson. 2019. Biotransformation of Organic Acids to Their Corresponding Alcohols by *Thermoanaerobacter pseudoethanolicus*. *Anaerobe,* 57: 28–31. https://doi.org/10.1016/j.anaerobe.2019.03.004

Scully, S. M., & Orlygsson, J. (2020). Branched-chain amino acid catabolism of *Thermoanaerobacter pseudoethanolicus* reveals potential route to branched-chain alcohol formation. *Extremophiles: Life Under Extreme Conditions*, *24*(1), 121–133. https://doi.org/10.1007/s00792-019-01140-5

Segura, D., Guzmán, J., & Espín, G. (2003). *Azotobacter vinelandii* mutants that overproduce poly-beta-hydroxybutyrate or alginate. *Applied Microbiology and Biotechnology*, *63*(2), 159–163. https://doi.org/10.1007/s00253-003-1397-1

Selck, H., Granberg, M. E., & Forbes, V. E. (2005). Impact of sediment organic matter quality on the fate and effects of fluoranthene in the infaunal brittle star *Amphiura filiformis*. *Marine Environmental Research*, *59*(1), 19–45. https://doi.org/10.1016/j.marenvres.2003.01.001

Sellami, M., Kedachi, S., Frikha, F., Miled, N., & Rebah, F. B. (2013). Optimization of marine waste based-growth media for microbial lipase production using mixture design methodology. *Environmental Technology*, *34*(13–16), 2259–2266. https://doi.org/10.1080/09593330.2013.765920

Selvakumar, P., & Sivashanmugam, P. (2018). Study on lipid accumulation in novel oleaginous yeast *Naganishia liquefaciens* NITTS2 utilizing pre-digested municipal waste activated sludge: A low-cost feedstock for biodiesel production. *Applied Biochemistry and Biotechnology*, *186*(3), 731–749. https://doi.org/10.1007/s12010-018-2777-4

Seo, Y. H., Cho, C., Lee, J.-Y., & Han, J.-I. (2014). Enhancement of growth and lipid production from microalgae using fluorescent paint under the solar radiation. *Bioresource Technology*, *173*, 193–197. https://doi.org/10.1016/j.biortech.2014.09.012

Seo, Y. H., Lee, I. G., & Han, J. I. (2013). Cultivation and lipid production of yeast *Cryptococcus curvatus* using pretreated waste active sludge supernatant. *Bioresource Technology*, *135*(May), 304–308. https://doi.org/10.1016/j.biortech.2012.10.024

Seo, Y. H., Lee, Y., Jeon, D. Y., & Han, J.-I. (2015). Enhancing the light utilization efficiency of microalgae using organic dyes. *Bioresource Technology*, *181*(April), 355–359. https://doi.org/10.1016/j.biortech.2015.01.031

Sharma, Y. C., Singh, B., & Korstad, J. (2011). A critical review on recent methods used for economically viable and eco-friendly development of microalgae as a potential feedstock for synthesis of biodiesel. *Green Chemistry*, *13*(11), 2993–3006. https://doi.org/10.1039/C1GC15535K

Sheehan, J., Dunahay, T., Benemann, J., & Roessler, P. (1998). *Look Back at the U.S. Department of Energy's Aquatic Species Program: Biodiesel from Algae; Close-Out Report* (NREL/TP-580-24190, 15003040; p. NREL/TP-580-24190, 15003040). https://doi.org/10.2172/15003040

Shen, Q., Chen, Y., Lin, H., Wang, Q., & Zhao, Y. (2018). Agro-industrial waste recycling by *Trichosporon fermentans*: Conversion of waste sweetpotato vines alone into lipid. *Environmental Science and Pollution Research International*, *25*(9), 8793–8799. https://doi.org/10.1007/s11356-018-1231-z

Shi, Y., Chai, L., Tang, C., Yang, Z., Zheng, Y., Chen, Y., & Jing, Q. (2013). Biochemical investigation of kraft lignin degradation by *Pandoraea* sp. B-6 isolated from bamboo slips. *Bioprocess and Biosystems Engineering*, *36*(12), 1957–1965. https://doi.org/10.1007/s00449-013-0972-9

Shobha, K. S., & Onkarappa, R. (2011). In vitro susceptibility of C. Albicans and C. neoformens to potential metabolites from streptomycetes. *Indian Journal of Microbiology*, *51*(4), 445–449. https://doi.org/10.1007/s12088-011-0097-2

Si, M., Yan, X., Liu, M., Shi, M., Wang, Z., Wang, S., Zhang, J., Gao, C., Chai, L., & Shi, Y. (2018). In situ lignin bioconversion promotes complete carbohydrate conversion of rice straw by *Cupriavidus basilensis* B-8. *ACS Sustainable Chemistry & Engineering*, *6*(6), 7969–7978. https://doi.org/10.1021/acssuschemeng.8b01336

Sialve, B., Bernet, N., & Bernard, O. (2009). Anaerobic digestion of microalgae as a necessary step to make microalgal biodiesel sustainable. *Biotechnology Advances*, *27*(4), 409–416. https://doi.org/10.1016/j.biotechadv.2009.03.001

Signori, L., Ami, D., Posteri, R., Giuzzi, A., Mereghetti, P., Porro, D., & Branduardi, P. (2016). Assessing an effective feeding strategy to optimize crude glycerol utilization as sustainable carbon source for lipid accumulation in oleaginous yeasts. *Microbial Cell Factories*, *15*(1), 75. https://doi.org/10.1186/s12934-016-0467-x

Silaban, A., Bai, R., Gutierrez-Wing, M. T., Negulescu, I. I., & Rusch, K. A. (2014). Effect of organic carbon, C:N ratio and light on the growth and lipid productivity of microalgae/cyanobacteria coculture. *Engineering in Life Sciences*, *14*(1), 47–56. https://doi.org/10.1002/elsc.201200219

Simon, J., Kósa, A., Bóka, K., Vági, P., Simon-Sarkadi, L., Mednyánszky, Z., Horváth, Á. N., Nyitrai, P., Böddi, B., & Preininger, É. (2017). Self-supporting artificial system of the green alga *Chlamydomonas reinhardtii* and the ascomycetous fungus *Alternaria infectoria*. *Symbiosis*, *71*(3), 199–209. https://doi.org/10.1007/s13199-016-0430-y

Simonazzi, M., Pezzolesi, L., Guerrini, F., Vanucci, S., Samorì, C., & Pistocchi, R. (2019). Use of waste carbon dioxide and pre-treated liquid digestate from biogas process for *Phaeodactylum tricornutum* cultivation in photobioreactors and open ponds. *Bioresource Technology*, *292*(November), 121921. https://doi.org/10.1016/j.biortech.2019.121921

Sindhu, R., Binod, P., & Pandey, A. (2016). Biological pretreatment of lignocellulosic biomass—An overview. *Bioresource Technology*, *199*, 76–82. https://doi.org/10.1016/j.biortech.2015.08.030

Singh, G., Singh, S., Kaur, K., Arya, S. K., & Sharma, P. (2019a). Thermo and halo tolerant laccase from *Bacillus* sp. SS4: Evaluation for its industrial usefulness. *Journal of General and Applied Microbiology*, *65*(1), 26–33. https://doi.org/10.2323/jgam.2018.04.002

Singh, N. K., Naira, V. R., & Maiti, S. K. (2019b). Production of biodiesel by autotrophic *Chlorella pyrenoidosa* in a sintered disc lab scale bubble column photobioreactor under natural sunlight. *Preparative Biochemistry & Biotechnology*, *49*(3), 255–269. https://doi.org/10.1080/10826068.2018.1536991

Singh, P., Jain, K., Desai, C., Tiwari, O., & Madamwar, D. (2019c). Chapter 18—Microbial community dynamics of extremophiles/extreme environment. In S. Das & H. R. Dash (Eds.), *Microbial Diversity in the Genomic Era* (pp. 323–332). Cambridge, MA: Academic Press. https://doi.org/10.1016/B978-0-12-814849-5.00018-6

Singh, R., Mattam, A. J., Jutur, P., & Yazdani, S. S. (2016). Synthetic biology in biofuels production. In *Reviews in Cell Biology and Molecular Medicine* (pp. 144–176). Atlanta, GA: American Cancer Society. https://doi.org/10.1002/3527600906.mcb.201600003

Singh, S., Shukla, L., Nain, L., & Khare, S. (2011). Detection and characterization of new thermostable endoglucanase from *Aspergillus awamori* strain F 18. *Journal of Mycology and Plant Pathology*, *41*, 97–103.

Singhal, A., Jaiswal, P. K., Jha, P. K., Thapliyal, A., & Thakur, I. S. (2013). Assessment of *Cryptococcus albidus* for biopulping of eucalyptus. *Preparative Biochemistry & Biotechnology*, *43*(8), 735–749. https://doi.org/10.1080/10826068.2013.771784

Şirin, S., Clavero, E., & Salvadó, J. (2015). Efficient harvesting of *Chaetoceros calcitrans* for biodiesel production. *Environmental Technology*, *36*(13–16), 1902–1912. https://doi.org/10.1080/09593330.2015.1015456

Şirin, S., & Sillanpää, M. (2015). Cultivating and harvesting of marine alga *Nannochloropsis oculata* in local municipal wastewater for biodiesel. *Bioresource Technology*, *191*(September), 79–87. https://doi.org/10.1016/j.biortech.2015.04.094

Sivaramakrishnan, R., & Incharoensakdi, A. (2018). Utilization of microalgae feedstock for concomitant production of bioethanol and biodiesel. *Fuel*, *217*, 458–466. https://doi.org/10.1016/j.fuel.2017.12.119

Slocombe, S. P., Zúñiga-Burgos, T., Chu, L., Wood, N. J., Alonso Camargo-Valero, M., & Baker, A. (2020). Fixing the broken phosphorus cycle: Wastewater remediation by microalgal polyphosphates. *Frontiers in Plant Science*, *11*, 982. https://doi.org/10.3389/fpls.2020.00982

Slooten, van, C., Peperzak, L., & Buma, A. G. J. (2015). Assessment of didecyldimethylammonium chloride as a ballast water treatment method. *Environmental Technology*, *36*(1–4), 435–449. https://doi.org/10.1080/09593330.2014.951401

Smith-Bädorf, H. D., Chuck, C. J., Mokebo, K. R., Macdonald, H., Davidson, M. G., & Scott, R. J. (2013). Bioprospecting the thermal waters of the Roman Baths: Isolation of oleaginous species and analysis of the FAME profile for biodiesel production. *AMB Express*, *3*(1), 9. https://doi.org/10.1186/2191-0855-3-9

Song, F., Tian, X., Fan, X., & He, X. (2010). Decomposing ability of filamentous fungi on litter is involved in a subtropical mixed forest. *Mycologia*, *102*(1), 20–26. https://doi.org/10.3852/09-047

Song, H.-T., Yang, Y.-M., Liu, D.-K., Xu, X.-Q., Xiao, W.-J., Liu, Z.-L., Xia, W.-C., Wang, C.-Y., Yu, X., & Jiang, Z.-B. (2017). Construction of recombinant *Yarrowia lipolytica* and its application in bio-transformation of lignocellulose. *Bioengineered*, *8*(5), 624–629. https://doi.org/10.1080/21655979.2017.1293219

Song, M., & Pei, H. (2018). The growth and lipid accumulation of *Scenedesmus quadricauda* during batch mixotrophic/heterotrophic cultivation using xylose as a carbon source. *Bioresource Technology*, *263*(September), 525–531. https://doi.org/10.1016/j.biortech.2018.05.020

Song, Z., Chen, L., Wang, J., Lu, Y., Jiang, W., & Zhang, W. (2014). A transcriptional Regulator Sll0794 regulates tolerance to biofuel ethanol in photosynthetic synechocystis sp. PCC 6803. *Molecular & Cellular Proteomics: MCP*, *13*(12), 3519–3532. https://doi.org/10.1074/mcp.M113.035675

Souza, A. F., Rodriguez, D. M., Ribeaux, D. R., Luna, M. A. C., Lima E Silva, T. A., Silva Andrade, R. F., Gusmão, N. B., & Campos-Takaki, G. M. (2016). Waste soybean oil and corn steep liquor as economic substrates for bioemulsifier and biodiesel production by *Candida lipolytica* UCP 0998. *International Journal of Molecular Sciences*, *17*(10). https://doi.org/10.3390/ijms17101608

Spagnuolo, M., Yaguchi, A., & Blenner, M. (2019). Oleaginous yeast for biofuel and oleochemical production. *Current Opinion in Biotechnology*, *57*(June), 73–81. https://doi.org/10.1016/j.copbio.2019.02.011

Spribille, T., Tuovinen, V., Resl, P., Vanderpool, D., Wolinski, H., Aime, M. C., Schneider, K., Stabentheiner, E., Toome-Heller, M., Thor, G., Mayrhofer, H., Johannesson, H., & McCutcheon, J. P. (2016). Basidiomycete yeasts in the cortex of ascomycete macrolichens. *Science (New York, N.Y.)*, *353*(6298), 488–492. https://doi.org/10.1126/science.aaf8287

Srikrishnan, S., Chen, W., & Silva, N. A. D. (2013). Functional assembly and characterization of a modular xylanosome for hemicellulose hydrolysis in yeast. *Biotechnology and Bioengineering*, *110*(1), 275–285. https://doi.org/10.1002/bit.24609

Srivastava, N., & Jaiswal, P. (2016). Production of cellulases using agriculture waste: Application in biofuels production. In R. K. Gupta & S. S. Singh (Eds.), *Environmental Biotechnology: A New Approach* (chapter 15, pp. 233–244). New Delhi: Daya Publishing House

Srivastava, N., Rawat, R., Oberoi, H. S., & Ramteke, P. W. (2015a). Review on fuel ethanol production from lignocellulosic biomass. *International Journal of Green Energy*. https://agris.fao.org/agris-search/search.do?recordID=US201600035363

Srivastava, N., Singh, J., Ramteke, P. W., Mishra, P. K., & Srivastava, M. (2015b). Improved production of reducing sugars from rice straw using crude cellulase activated with Fe3O4/alginate nanocomposite. *Bioresource Technology*, *183*(May), 262–266. https://doi.org/10.1016/j.biortech.2015.02.059

Srivastava, R. K., Shetti, N. P., Reddy, K. R., & Aminabhavi, T. M. (2020). Biofuels, biodiesel and biohydrogen production using bioprocesses. A review. *Environmental Chemistry Letters*, *18*, 1049–1072

Stein, B. D., Klomparens, K. L., & Hammerschmidt, R. (1992). Comparison of bromine and permanganate as ultrastructural stains for lignin in plants infected by the fungus *Colletotrichum lagenarium*. *Microscopy Research and Technique*, *23*(3), 201–206. https://doi.org/10.1002/jemt.1070230302

Stoytcheva, M., & Montero, G. (2011). *Biodiesel: Feedstocks and Processing Technologies*. London, United Kingdom: IntechOpen

Strobel, G. A., Knighton, B., Kluck, K., Ren, Y., Livinghouse, T., Griffin, M., Spakowicz, D., & Sears, J. (2010). The production of myco-diesel hydrocarbons and their derivatives by the endophytic fungus *Gliocladium roseum* (NRRL 50072). *Microbiology*, *156*(12), 3830–3833. Accessed June 3, 2021. https://doi.org/10.1099/mic.0.30824-0

Su, A. A. H., Tripp, V., & Randau, L. (2013). RNA-Seq analyses reveal the order of tRNA processing events and the maturation of C/D box and CRISPR RNAs in the hyperthermophile *Methanopyrus kandleri*. *Nucleic Acids Research*, *41*(12), 6250–6258. https://doi.org/10.1093/nar/gkt317

Su, L., & Ajo-Franklin, C. M. (2019). Reaching full potential: Bioelectrochemical systems for storing renewable energy in chemical bonds. *Current Opinion in Biotechnology*, *57*, 66–72.

Subashchandrabose, S. R., Ramakrishnan, B., Megharaj, M., Venkateswarlu, K., & Naidu, R. (2011). Consortia of cyanobacteria/microalgae and bacteria: Biotechnological potential. *Biotechnology Advances*, *29*(6), 896–907. https://doi.org/10.1016/j.biotechadv.2011.07.009

Sugiyama, A. (2019). The soybean rhizosphere: Metabolites, microbes, and beyond—A review. *Journal of Advanced Research*, *19*, 67–73. https://doi.org/10.1016/j.jare.2019.03.005

Sukrachan, T., & Aran Incharoensakdi, A. (2020). Enhanced hydrogen production by *Nostoc* sp. CU2561 immobilized in a novel agar bead. *Journal of Applied Phycology*, *32*, 1103–1115. https://doi.org/10.1007/s10811-019-02032-z

Summerbell, R. C. (2005). From Lamarckian fertilizers to fungal castles: Recapturing the pre-1985 literature on endophytic and saprotrophic fungi associated with ectomycorrhizal root systems. *Studies in Mycology*, *53*, 191–256. https://doi.org/10.3114/sim.53.1.191

Sun, L., Wang, L., & Chen, H. (2020). High productivity ethanol from solid-state fermentation of steam-exploded corn stover using *Zymomonas mobilis* by N2 periodic pulsation process intensification. *Applied Biochemistry and Biotechnology*, *192*(2), 466–481. https://doi.org/10.1007/s12010-020-03318-6

Sun, N., Qian, Y., Wang, W., Zhong, Y., & Dai, M. (2018). Heterologous expression of *Talaromyces emersonii* cellobiohydrolase Cel7A in *Trichoderma reesei* increases the efficiency of corncob residues saccharification. *Biotechnology Letters*, *40*(7), 1119–1126. https://doi.org/10.1007/s10529-018-2564-x

Sung, M., Seo, Y. H., Han, S., & Han, J.-I. (2014). Biodiesel production from yeast *Cryptococcus* sp. using Jerusalem Artichoke. *Bioresource Technology*, *155*(March), 77–83. https://doi.org/10.1016/j.biortech.2013.12.024

Sunwoo, I. Y., Nguyen, T. H., Sukwong, P., Jeong, G.-T., & Kim, S.-K. (2018). Enhancement of ethanol production via hyper thermal acid hydrolysis and co-fermentation using waste seaweed from Gwangalli Beach, Busan, Korea. *Journal of Microbiology and Biotechnology*, *28*(3), 401–408. https://doi.org/10.4014/jmb.1708.08041

Swaaf, de, M. E., de Rijk, T. C., van der Meer, P., Eggink, G., & Sijtsma, L. (2003). Analysis of docosahexaenoic acid biosynthesis in *Crypthecodinium cohnii* by 13C labelling and desaturase inhibitor experiments. *Journal of Biotechnology*, *103*(1), 21–29. https://doi.org/10.1016/s0168-1656(03)00070-1

Tafreshi, A. H., & Shariati, M. (2009). Dunaliella biotechnology: Methods and applications. *Journal of Applied Microbiology*, *107*(1), 14–35. https://doi.org/10.1111/j.1365-2672.2009.04153.x

Taherzadeh, M. J., Fox, M., Hjorth, H., & Edebo, L. (2003). Production of mycelium biomass and ethanol from paper pulp sulfite liquor by *Rhizopus oryzae*. *Bioresource Technology, 88*(3), 167–177. https://doi.org/10.1016/s0960-8524(03)00010-5

Takemura, K., Endo, R., & Kitaya, Y. (2020). Possibility of co-culturing *Euglena gracilis* and *Lactuca sativa* L. with biogas digestate. *Environmental Technology, 41*(8), 1007–1014. https://doi.org/10.1080/09593330.2018.1516803

Takishita, K., Kakizoe, N., Yoshida, T., & Maruyama, T. (2010). Molecular evidence that phylogenetically diverged ciliates are active in microbial mats of deep-sea cold-seep sediment. *Journal of Eukaryotic Microbiology, 57*(1), 76–86. https://doi.org/10.1111/j.1550-7408.2009.00457.x

Takishita, K., Tsuchiya, M., Reimer, J. D., & Maruyama, T. (2006). Molecular evidence demonstrating the basidiomycetous fungus *Cryptococcus curvatus* is the dominant microbial eukaryote in sediment at the Kuroshima Knoll methane seep. *Extremophiles: Life Under Extreme Conditions, 10*(2), 165–169. https://doi.org/10.1007/s00792-005-0495-7

Talebi, A. F., Tohidfar, M., Mousavi Derazmahalleh, S. M., Sulaiman, A., Baharuddin, A. S., & Tabatabaei, M. (2015). Biochemical modulation of lipid pathway in microalgae *Dunaliella* sp. for biodiesel production. *BioMed Research International, 2015*, 597198. https://doi.org/10.1155/2015/597198

Tamano, K., Bruno, K. S., Koike, H., Ishii, T., Miura, A., Umemura, M., Culley, D. E., Baker, S. E., & Machida, M. (2015). Increased production of free fatty acids in *Aspergillus oryzae* by disruption of a predicted Acyl-CoA synthetase gene. *Applied Microbiology and Biotechnology, 99*(7), 3103–3113. https://doi.org/10.1007/s00253-014-6336-9

Tanaka, A., Okuda, K., Senoo, K., Obata, H., & Inouye, K. (1999). Guanidine hydrochloride-induced denaturation of *Pseudomonas cepacia* lipase. *Journal of Biochemistry, 126*(2), 382–386. https://doi.org/10.1093/oxfordjournals.jbchem.a022461.

Tanaka, Y., Aki, T., Hidaka, Y., Furuya, Y., Kawamoto, S., Shigeta, S., Ono, K., & Suzuki, O. (2002). Purification and characterization of a novel fungal alpha-glucosidase from *Mortierella alliacea* with high starch-hydrolytic activity. *Bioscience, Biotechnology, and Biochemistry, 66*(11), 2415–2423. https://doi.org/10.1271/bbb.66.2415

Tang, D., Han, W., Li, P., Miao, X., & Zhong, J. (2011). CO_2 biofixation and fatty acid composition of *Scenedesmus obliquus* and *Chlorella pyrenoidosa* in response to different CO_2 levels. *Bioresource Technology, 102*(3), 3071–3076. https://doi.org/10.1016/j.biortech.2010.10.047

Tang, S., Dong, Q., Fang, Z., Cong, W.-J., & Zhang, H. (2020). Microbial lipid production from rice straw hydrolysates and recycled pretreated glycerol. *Bioresource Technology, 312*(September), 123580. https://doi.org/10.1016/j.biortech.2020.123580

Taniguchi, M., Suzuki, H., Watanabe, D., Sakai, K., Hoshino, K., & Tanaka, T. (2005). Evaluation of pretreatment with *Pleurotus ostreatus* for enzymatic hydrolysis of rice straw. *Journal of Bioscience and Bioengineering, 100*(6), 637–643. https://doi.org/10.1263/jbb.100.637

Tanimura, A., Takashima, M., Sugita, T., Endoh, R., Kikukawa, M., Yamaguchi, S., Sakuradani, E., Ogawa, J., & Shima, J. (2014). Selection of oleaginous yeasts with high lipid productivity for practical biodiesel production. *Bioresource Technology, 153*, 230–235. https://doi.org/10.1016/j.biortech.2013.11.086

Tanimura, A., Takashima, M., Sugita, T., Endoh, R., Kikukawa, M., Yamaguchi, S., Sakuradani, E., Ogawa, J., Ohkuma, M., & Shima, J. (2014). *Cryptococcus terricola* is a promising oleaginous yeast for biodiesel production from starch through consolidated bioprocessing. *Scientific Reports, 4*(April), 4776. https://doi.org/10.1038/srep04776

Tanino, T., Ito, T., Ogino, C., Ohmura, N., Ohshima, T., & Kondo, A. (2012). Sugar consumption and ethanol fermentation by transporter-overexpressed xylose-metabolizing *Saccharomyces cerevisiae* harboring a xyloseisomerase pathway. *Journal of Bioscience and Bioengineering, 114*(2), 209–211. https://doi.org/10.1016/j.jbiosc.2012.03.004

Teo, C. L., Atta, M., Bukhari, A., Taisir, M., Yusuf, A. M., & Idris, A. (2014). Enhancing growth and lipid production of marine microalgae for biodiesel production via the use of different LED wavelengths. *Bioresource Technology, 162*(June), 38–44. https://doi.org/10.1016/j.biortech.2014.03.113

Thiru, M., Sankh, S., & Rangaswamy, V. (2011). Process for biodiesel production from *Cryptococcus curvatus*. *Bioresource Technology, 102*(22), 10436–10440. https://doi.org/10.1016/j.biortech.2011.08.102

Thliveros, P., Kiran, E. U., & Webb, C. (2014). Microbial biodiesel production by direct methanolysis of oleaginous biomass. *Bioresource Technology, 157*(April), 181–187. https://doi.org/10.1016/j.biortech.2014.01.111

Thomas, L., Joseph, A., & Gottumukkala, L. D. (2014). Xylanase and cellulase systems of *Clostridium* sp.: An insight on molecular approaches for strain improvement. *Bioresource Technology*, *158*, 343–350. https://doi.org/10.1016/j.biortech.2014.01.140

Thongchul, N., Navankasattusas, S., & Yang, S.-T. (2010). Production of lactic acid and ethanol by *Rhizopus oryzae* integrated with cassava pulp hydrolysis. *Bioprocess and Biosystems Engineering*, *33*(3), 407–416. https://doi.org/10.1007/s00449-009-0341-x

Thorsson, M. H., Hedman, J. E., Bradshaw, C., Gunnarsson, J. S., & Gilek, M. (2008). Effects of settling organic matter on the Bioaccumulation of Cadmium and BDE-99 by Baltic Sea benthic invertebrates. *Marine Environmental Research*, *65*(3), 264–281. https://doi.org/10.1016/j.marenvres.2007.11.004

Tiang, M. F., Hanipa, M. A. F., Abdul, P. M., Jahim, J. M. d., Mahmod, S. S., Takriff, M. S., Lay, C.-H., Reungsang, A., & Wu, S.-Y. (2020). Recent advanced biotechnological strategies to enhance photo-fermentative biohydrogen production by purple non-sulphur bacteria: An overview. *International Journal of Hydrogen Energy*, *45*, 13211–13230

Tinoi, J., & Rakariyatham, N. (2016). Optimization of pineapple pulp residue hydrolysis for lipid production by *Rhodotorula glutinis* TISTR5159 using as biodiesel feedstock. *Bioscience, Biotechnology, and Biochemistry*, *80*(8), 1641–1649. https://doi.org/10.1080/09168451.2016.1177444

Todhanakasem, T., Wu, B., & Simeon, S. (2020). Perspectives and new directions for bioprocess optimization using *Zymomonas mobilis* in the ethanol production. *World Journal of Microbiology & Biotechnology*, *36*(8), 112. https://doi.org/10.1007/s11274-020-02885-4

Toepel, J., Illmer-Kephalides, M., Jaenicke, S., Straube, J., May, P., Goesmann, A., & Kruse, O. (2013). New insights into *Chlamydomonas reinhardtii* hydrogen production processes by combined microarray/RNA-Seq transcriptomics. *Plant Biotechnology Journal*, *11*(6), 717–733. https://doi.org/10.1111/pbi.12062

Toivola, A., Yarrow, D., Bosch, E. van den, Dijken, J. P. van, & Scheffers, W. A. (1984). Alcoholic fermentation of d-Xylose by yeasts. *Applied and Environmental Microbiology*, *47*(6), 1221–1223.

Tolstygina, I. V., Antal, T. K., Kosourov, S. N., Krendeleva, T. E., Rubin, A. B., & Tsygankov, A. A. (2009). Hydrogen production by photoautotrophic sulfur-deprived *Chlamydomonas reinhardtii* pre-grown and incubated under high light. *Biotechnology and Bioengineering*, *102*(4), 1055–1061. https://doi.org/10.1002/bit.22148

Tomas-Grau, R. H., Peto, P. D., Chalfoun, N. R., Grellet-Bournonville, C. F., Martos, G. G., Debes, M., Arias, M. E., & Díaz-Ricci, J. C. (2019). *Colletotrichum acutatum M11* can suppress the defence response in strawberry plants. *Planta*, *250*(4), 1131–1145. https://doi.org/10.1007/s00425-019-03203-5

Travaini, R., Barrado, E., & Bolado-Rodríguez, S. (2016). Effect of ozonolysis parameters on the inhibitory compound generation and on the production of ethanol by *Pichia stipitis* and acetone-butanol-ethanol by clostridium from ozonated and water washed sugarcane bagasse. *Bioresource Technology*, *218*(October), 850–858. https://doi.org/10.1016/j.biortech.2016.07.028

Tsigie, Y. A., Wu, C.-H., Huynh, L. H., Ismadji, S., & Ju, Y.-H. (2013). Bioethanol production from *Yarrowia lipolytica* Po1g biomass. *Bioresource Technology*, *145*(October), 210–216. https://doi.org/10.1016/j.biortech.2012.11.091

Udatha, D. B. R. K. G., Mapelli, V., Panagiotou, G., & Olsson, L. (2012). Common and distant structural characteristics of feruloyl esterase families from *Aspergillus oryzae*. *PloS One*, *7*(6), e39473. https://doi.org/10.1371/journal.pone.0039473

Ummalyma, S. B., & Sukumaran, R. K. (2014). Cultivation of microalgae in dairy effluent for oil production and removal of organic pollution load. *Bioresource Technology*, *165*(August), 295–301. https://doi.org/10.1016/j.biortech.2014.03.028

Varaprasad, D., Narasimham, D., Paramesh, K., Sudha, N. R., Himabindu, Y., Kumari, M. K., Parveen, S. N., & Chandrasekhar, T. (2021a). Improvement of ethanol production using green alga *Chlorococcum minutum*. *Environmental Technology*, *42*(9), 1383–1391. https://doi.org/10.1080/09593330.2019.1669719

Vasconcelos, B., Teixeira, J. C., Dragone, G., & Teixeira, J. A. (2018). Optimization of lipid extraction from the oleaginous yeasts *Rhodotorula glutinis* and *Lipomyces kononenkoae*. *AMB Express*, *8*(1), 126. https://doi.org/10.1186/s13568-018-0658-4

Vazana, Y., Moraïs, S., Barak, Y., Lamed, R., & Bayer, E. A. (2012). Chapter twenty-three—Designer cellulosomes for enhanced hydrolysis of cellulosic substrates. In H. J. Gilbert (Ed.), *Methods in Enzymology* (Vol. 510, pp. 429–452). Cambridge, MA: Academic Press. https://doi.org/10.1016/B978-0-12-415931-0.00023-9

Vellanki, S., Navarro-Mendoza, M. I., Garcia, A., Murcia, L., Perez-Arques, C., Garre, V., Nicolas, F. E., & Lee, S. C. (2018). *Mucor circinelloides*: Growth, maintenance, and genetic manipulation. *Current Protocols in Microbiology*, *49*(1), e53. https://doi.org/10.1002/cpmc.53

Velmurugan, R., & Incharoensakdi, A. (2020). Co-cultivation of two engineered strains of *Synechocystis* sp. PCC 6803 results in improved bioethanol production. *Renewable Energy, 146*, 1124–1133. https://doi.org/10.1016/j.renene.2019.07.025

Veras, H. C. T., Parachin, N. S., & Almeida, J. R. M. (2017). Comparative assessment of fermentative capacity of different xylose-consuming yeasts. *Microbial Cell Factories, 16*(1), 153. https://doi.org/10.1186/s12934-017-0766-x

Vibha, & Sinha, A. (2005). Production of soluble crude protein using cellulolytic fungi on rice stubble as substrate under waste program management. *Mycobiology, 33*(3), 147–149. https://doi.org/10.4489/MYCO.2005.33.3.147

Villalba, L. L., Fonseca, M. I., Giorgio, M., & Zapata, P. D. (2010). White rot fungi laccases for biotechnological applications. *Recent Patents on DNA & Gene Sequences, 4*(2), 106–112. https://doi.org/10.2174/187221510793205728

Volgusheva, A., Styring, S., & Mamedov, F. (2013). Increased photosystem II stability Promotes H2 production in sulfur-deprived *Chlamydomonas reinhardtii*. *Proceedings of the National Academy of Sciences of the United States of America, 110*(18), 7223–7228. https://doi.org/10.1073/pnas.1220645110

Wahal, S., & Viamajala, S. (2016). Uptake of inorganic and organic nutrient species during cultivation of a chlorella isolate in anaerobically digested dairy waste. *Biotechnology Progress, 32*(5), 1336–1342. https://doi.org/10.1002/btpr.2313

Waks, Z., & Silver, P. A. (2009). Engineering a synthetic dual-organism system for hydrogen production. *Applied and Environmental Microbiology, 75*(7), 1867–1875. https://doi.org/10.1128/AEM.02009-08

Walfridsson, M., Bao, X., Anderlund, M., Lilius, G., Bülow, L., & Hahn-Hägerdal, B. (1996). Ethanolic fermentation of xylose with *Saccharomyces cerevisiae* harboring the *Thermus thermophilus XylA* gene, which expresses an active xylose (glucose) isomerase. *Applied and Environmental Microbiology, 62*(12), 4648–4651. https://doi.org/10.1128/aem.62.12.4648-4651.1996

Wan, C., & Li, Y. (2010). Microbial delignification of corn stover by *Ceriporiopsis subvermispora* for improving cellulose digestibility. *Enzyme and Microbial Technology, 47*(1), 31–36. https://doi.org/10.1016/j.enzmictec.2010.04.001

Wan, C., & Li, Y. (2012). Fungal pretreatment of lignocellulosic biomass. *Biotechnology Advances, 30*(6), 1447–1457. https://doi.org/10.1016/j.biotechadv.2012.03.003

Wang, B., Eckert, C., Maness, P.-C., & Yu, J. (2018a). A genetic toolbox for modulating the expression of heterologous genes in the *Cyanobacterium synechocystis* sp. PCC 6803. *ACS Synthetic Biology, 7*(1), 276–286. https://doi.org/10.1021/acssynbio.7b00297

Wang, F., Bi, Y., Diao, J., Lv, M., Cui, J., Chen, L., & Zhang, W. (2019a). Metabolic engineering to enhance biosynthesis of both docosahexaenoic acid and odd-chain fatty acids in *Schizochytrium* sp. S31. *Biotechnology for Biofuels, 12*, 141. https://doi.org/10.1186/s13068-019-1484-x

Wang, F., Wang, M., Zhao, Q., Niu, K., Liu, S., He, D., Liu, Y., Xu, S., & Fang, X. (2019b). Exploring the relationship between *Clostridium thermocellum* JN4 and *Thermoanaerobacterium thermosaccharolyticum* GD17. *Frontiers in Microbiology, 10*, 2035. https://doi.org/10.3389/fmicb.2019.02035

Wang, H., Xiong, H., Hui, Z., & Zeng, X. (2012). Mixotrophic cultivation of *Chlorella pyrenoidosa* with diluted primary piggery wastewater to produce lipids. *Bioresource Technology, 104*(January), 215–220. https://doi.org/10.1016/j.biortech.2011.11.020

Wang, J., Gao, Q., & Bao, J. (2016b). Genome sequence of *Trichosporon cutaneum* ACCC 20271: An oleaginous yeast with excellent lignocellulose derived inhibitor tolerance. *Journal of Biotechnology, 228*(June), 50–51. https://doi.org/10.1016/j.jbiotec.2016.04.043

Wang, J., Gao, Q., Zhang, H., & Bao, J. (2016c). Inhibitor degradation and lipid accumulation potentials of oleaginous yeast *Trichosporon cutaneum* using lignocellulose feedstock. *Bioresource Technology, 218*(October), 892–901. https://doi.org/10.1016/j.biortech.2016.06.130

Wang, J., Hu, M., Zhang, H., & Bao, J. (2017). Converting Chemical Oxygen Demand (COD) of cellulosic ethanol fermentation wastewater into microbial lipid by oleaginous yeast *Trichosporon cutaneum*. *Applied Biochemistry and Biotechnology, 182*(3), 1121–1130. https://doi.org/10.1007/s12010-016-2386-z

Wang, J., Ledesma-Amaro, R., Wei, Y., Ji, B., & Ji, X.-J. (2020a). Metabolic engineering for increased lipid accumulation in *Yarrowia lipolytica*: A review. *Bioresource Technology, 313*(October), 123707. https://doi.org/10.1016/j.biortech.2020.123707

Wang, K., Sun, T., Cui, J., Liu, L., Bi, Y., Pei, G., Chen, L., & Zhang, W. (2018b). Screening of chemical modulators for lipid accumulation in *Schizochytrium* sp. S31. *Bioresource Technology, 260*(July), 124–129. https://doi.org/10.1016/j.biortech.2018.03.104

Wang, L., Zong, Z., Liu, Y., Zheng, M., Li, D., Wang, C., Zheng, F., Madzak, C., & Liu, Z. (2019c). Metabolic engineering of *Yarrowia lipolytica* for the biosynthesis of crotonic acid. *Bioresource Technology*, *287*(September), 121484. https://doi.org/10.1016/j.biortech.2019.121484

Wang, S., Yerkebulan, M., El-Fatah Abomohra, A., El-Khodary, S., & Wang, Q. (2019d). Microalgae harvest influences the energy recovery: A case study on chemical flocculation of *Scenedesmus obliquus* for biodiesel and crude bio-oil production. *Bioresource Technology*, *286*(August), 121371. https://doi.org/10.1016/j.biortech.2019.121371

Wang, S.-L., Sun, J.-S., Han, B.-Z., & Wu, X.-Z. (2007). Optimization of beta-carotene production by *Rhodotorula glutinis* using high hydrostatic pressure and response surface methodology. *Journal of Food Science*, *72*(8), M325–329. https://doi.org/10.1111/j.1750-3841.2007.00495.x

Wang, T.-Y., Chen, H.-L., Lu, M.-Y. J., Chen, Y.-C., Sung, H.-M., Mao, C.-T., Cho, H.-Y., Ke, H.-M., Hwa, T.-Y., Ruan, S.-K., Hung, K.-Y., Chen, C.-K., Li, J.-Y., Wu, Y.-C., Chen, Y.-H., Chou, S.-P., Tsai, Y.-W., Chu, T.-C., Shih, C.-C. A., . . . & Shih, M.-C. (2011). Functional characterization of cellulases identified from the cow rumen fungus *Neocallimastix patriciarum* W5 by transcriptomic and secretomic analyses. *Biotechnology for Biofuels*, *4*, 24. https://doi.org/10.1186/1754-6834-4-24

Wang, W., Zhou, W., Liu, J., Li, Y., & Zhang, Y. (2013). Biodiesel production from hydrolysate of *Cyperus esculentus* waste by *Chlorella vulgaris*. *Bioresource Technology*, *136*(May), 24–29. https://doi.org/10.1016/j.biortech.2013.03.075

Wang, W.-L., Moore, J. K., Martiny, A. C., & Primeau, F. W. (2019e). Convergent estimates of marine nitrogen fixation. *Nature*, *566*(7743), 205–211. https://doi.org/10.1038/s41586-019-0911-2

Wang, X., Bao, K., Cao, W., Zhao, Y., & Wei Hu, C. (2017). Screening of microalgae for integral biogas slurry nutrient removal and biogas upgrading by different microalgae cultivation technology. *Scientific Reports*, *7*(1), 5426. https://doi.org/10.1038/s41598-017-05841-9

Wang, X., He, Q., Yang, Y., Wang, J., Haning, K., Hu, Y., Wu, B., et al. (2018c). Advances and prospects in metabolic engineering of *Zymomonas mobilis*. *Metabolic Engineering*, *50*(November), 57–73. https://doi.org/10.1016/j.ymben.2018.04.001

Wang, X.-W., Liang, J.-R., Luo, C.-S., Chen, C.-P., & Gao, Y.-H. (2014). Biomass, total lipid production, and fatty acid composition of the marine diatom *Chaetoceros muelleri* in response to different CO_2 levels. *Bioresource Technology*, *161*(June), 124–130. https://doi.org/10.1016/j.biortech.2014.03.012

Wang, Y.-C., Hu, H.-F., Ma, J.-W., Yan, Q.-J., Liu, H.-J., & Jiang, Z.-Q. (2020b). A novel high maltose-forming α-Amylase from *Rhizomucor miehei* and its application in the food industry. *Food Chemistry*, *305*(February), 125447. https://doi.org/10.1016/j.foodchem.2019.125447

Warnecke, F., Luginbühl, P., Ivanova, N., Ghassemian, M., Richardson, T. H., Stege, J. T., Cayouette, M., McHardy, A. C., Djordjevic, G., Aboushadi, N., Sorek, R., Tringe, S. G., Podar, M., Martin, H. G., Kunin, V., Dalevi, D., Madejska, J., Kirton, E., Platt, D., . . . & Leadbetter, J. R. (2007). Metagenomic and functional analysis of hindgut microbiota of a wood-feeding higher termite. *Nature*, *450*(7169), 560–565. https://doi.org/10.1038/nature06269

Watanabe, H., Li, D., Nakagawa, Y., Tomishige, K., & Watanabe, M. M. (2015). Catalytic gasification of oil-extracted residue biomass of *Botryococcus braunii*. *Bioresource Technology*, *191*(September), 452–459. https://doi.org/10.1016/j.biortech.2015.03.034

Wen, Q., Chen, Z., Li, P., Duan, R., & Ren, N. (2013). Lipid production for biofuels from hydrolyzate of waste activated sludge by *Heterotrophic chlorella* protothecoides. *Bioresource Technology*, *143*(September), 695–698. https://doi.org/10.1016/j.biortech.2013.06.085

Wirth, R., Lakatos, G., Böjti, T., G. Maróti, Bagi, Z., Kis, M., Kovács, A., Ács, N., Rákhely, G., & Kovács, K. L. (2015). Metagenome changes in the mesophilic biogas-producing community during fermentation of the green alga *Scenedesmus obliquus*. *Journal of Biotechnology*, *215*(December), 52–61. https://doi.org/10.1016/j.jbiotec.2015.06.396

Wood, B. M., Jader, L. R., Schendel, F. J., Hahn, N. J., Valentas, K. J., McNamara, P. J., Novak, P. M., & Heilmann, S. M. (2013). Industrial symbiosis: Corn ethanol fermentation, hydrothermal carbonization, and anaerobic digestion. *Biotechnology and Bioengineering*, *110*(10), 2624–2632. https://doi.org/10.1002/bit.24924

Wu, H., & Miao, X. (2014). Biodiesel quality and biochemical changes of microalgae *Chlorella pyrenoidosa* and *Scenedesmus obliquus* in response to nitrate levels. *Bioresource Technology*, *170*(October), 421–427. https://doi.org/10.1016/j.biortech.2014.08.017

Wu, H., Yu, X., Chen, L., & Wu, G. (2014). Cloning, overexpression and characterization of a thermostable pullulanase from *Thermus thermophilus* HB27. *Protein Expression and Purification*, *95*(March), 22–27. https://doi.org/10.1016/j.pep.2013.11.010

Wu, J., Hu, J., Zhao, S., He, M., Hu, G., Ge, X., & Peng, N. (2018). Single-cell protein and xylitol production by a novel yeast strain *Candida intermedia* FL023 from lignocellulosic hydrolysates and xylose. *Applied Biochemistry and Biotechnology*, *185*(1), 163–178. https://doi.org/10.1007/s12010-017-2644-8

Wu, J. F., & Pond, W. G. (1981). Amino acid composition and microbial contamination of *Spirulina maxima*, a blue-green alga, grown on the effluent of different fermented animal wastes. *Bulletin of Environmental Contamination and Toxicology, 27*(2), 151–159. https://doi.org/10.1007/BF01611001

Wu, R., Chen, D., Cao, S., Lu, Z., Huang, J., Lu, Q., Chen, Y., et al. (2020). Enhanced ethanol production from sugarcane molasses by industrially engineered *Saccharomyces cerevisiae* via replacement of the PHO4 gene. *RSC Advances*, *10*(4), 2267–2276. https://doi.org/10.1039/C9RA08673K

Wu, Y., Xu, S., Gao, X., Li, M., Li, D., & Lu, W. (2019). Enhanced protopanaxadiol production from xylose by engineered *Yarrowia lipolytica*. *Microbial Cell Factories*, *18*(1), 83. https://doi.org/10.1186/s12934-019-1136-7

Wu, Y.-H., Yu, Y., & Hu, H.-Y. (2013). Potential biomass yield per phosphorus and lipid accumulation property of seven microalgal species. *Bioresource Technology*, *130*(February), 599–602. https://doi.org/10.1016/j.biortech.2012.12.116

Wu, Z., Zhu, Y., Huang, W., Zhang, C., Li, T., Zhang, Y., & Li, A. (2012). Evaluation of flocculation induced by PH Increase for harvesting microalgae and reuse of flocculated medium. *Bioresource Technology*, *110*(April), 496–502. https://doi.org/10.1016/j.biortech.2012.01.101

Xia, A., Cheng, J., Ding, L., Lin, R., Huang, R., Zhou, J., & Cen, K. (2013). Improvement of the energy conversion efficiency of *Chlorella pyrenoidosa* biomass by a three-stage process comprising dark fermentation, photofermentation, and methanogenesis. *Bioresource Technology*, *146*(October), 436–443. https://doi.org/10.1016/j.biortech.2013.07.077

Xiao, S., Xu, J., Chen, X., Li, X., Zhang, Y., & Yuan, Z. (2016). 3-Methyl-1-Butanol biosynthesis in an engineered corynebacterium glutamicum. *Molecular Biotechnology, 58*(5), 311–318. https://doi.org/10.1007/s12033-016-9929-y

Xiong, L., Huang, C., Li, X.-M., Chen, X.-F., Wang, B., Wang, C., Zeng, X.-A., & Chen, X.-D. (2015). Acetone-Butanol-Ethanol (ABE) fermentation wastewater treatment by oleaginous yeast *Trichosporon cutaneum*. *Applied Biochemistry and Biotechnology*, *176*(2), 563–571. https://doi.org/10.1007/s12010-015-1595-1

Xiong, W., Li, X., Xiang, J., & Wu, Q. (2008). High-density fermentation of microalga *Chlorella protothecoides* in bioreactor for microbio-diesel production. *Applied Microbiology and Biotechnology*, *78*(1), 29–36. https://doi.org/10.1007/s00253-007-1285-1

Xu, C., Ma, F., Zhang, X., & Chen, S. (2010). Biological pretreatment of Corn Stover by Irpex lacteus for enzymatic hydrolysis. *Journal of Agricultural and Food Chemistry*, *58*(20), 10893–10898. https://doi.org/10.1021/jf1021187

Xu, F., & Pan, J. (2020). Potassium channel KCN11 is required for maintaining cellular osmolarity during nitrogen starvation to control proper cell physiology and TAG accumulation in *Chlamydomonas reinhardtii*. *Biotechnology for Biofuels*, *13*, 129. https://doi.org/10.1186/s13068-020-01769-x

Xu, H., Miao, X., & Wu, Q. (2006). High quality biodiesel production from a microalga *Chlorella protothecoides* by heterotrophic growth in fermenters. *Journal of Biotechnology*, *126*(4), 499–507. https://doi.org/10.1016/j.jbiotec.2006.05.002

Xu, Y.-B., Zhou, Y., Ruan, J.-J., Xu, S.-H., Gu, J.-D., Huang, S.-S., Zheng, L., Yuan, B.-H., & Wen, L.-H. (2015). Endogenous nitric oxide in *Pseudomonas fluorescens* ZY2 as mediator against the combined exposure to zinc and cefradine. *Ecotoxicology (London, England)*, *24*(4), 835–843. https://doi.org/10.1007/s10646-015-1428-6

Xue, D., Yao, D., Sukumaran, R. K., You, X., Wei, Z., & Gong, C. (2020). Tandem integration of aerobic fungal cellulase production, lignocellulose substrate saccharification and anaerobic ethanol fermentation by a modified gas lift bioreactor. *Bioresource Technology*, *302*, 122902. https://doi.org/10.1016/j.biortech.2020.122902

Yagi, T., Yamashita, K., Okada, N., Isono, T., Momose, D., Mineki, S., & Tokunaga, E. (2016). Hydrogen photoproduction in green algae *Chlamydomonas reinhardtii* sustainable over 2 weeks with the original cell culture without supply of fresh cells nor exchange of the whole culture medium. *Journal of Plant Research*, *129*(4), 771–779. https://doi.org/10.1007/s10265-016-0825-0

Yaisamlee, C., & Sirikhachornkit, A. (2020). Characterization of chlamydomonas very high light-tolerant mutants for enhanced lipid production. *Journal of Oleo Science, 69*(4), 359–368. https://doi.org/10.5650/jos.ess19270

Yamaoka, C., Kurita, O., & Kubo, T. (2014). Improved ethanol tolerance of *Saccharomyces cerevisiae* in mixed cultures with *Kluyveromyces lactis* on high-sugar fermentation. *Microbiological Research, 169*(12), 907–914. https://doi.org/10.1016/j.micres.2014.04.007

Yan, Q., Duan, X., Liu, Y., Jiang, Z., & Yang, S. (2016). Expression and characterization of a novel 1,3-regioselective cold-adapted lipase from *Rhizomucor endophyticus* suitable for biodiesel synthesis. *Biotechnology for Biofuels, 9*, 86. https://doi.org/10.1186/s13068-016-0501-6

Yang, Q., Zhang, H., Li, X., Wang, Z., Xu, Y., Ren, S., Chen, X., Xu, Y., Hao, H., & Wang, H. (2013). Extracellular enzyme production and phylogenetic distribution of yeasts in wastewater treatment systems. *Bioresource Technology, 129*(February), 264–273. https://doi.org/10.1016/j.biortech.2012.11.101

Yang, S., Pan, C., Hurst, G. B., Dice, L., Davison, B. H., & Brown, S. D. (2014). Elucidation of *Zymomonas mobilis* physiology and stress responses by quantitative proteomics and transcriptomics. *Frontiers in Microbiology, 5*, 246. https://doi.org/10.3389/fmicb.2014.00246

Yang, X., Wu, Y., Zhang, Y., Yang, E., Qu, Y., Xu, H., Chen, Y., Irbis, C., & Yan, J. (2020). A Thermo-active laccase isoenzyme from *Trametes trogii* and its potential for dye decolorization at high temperature. *Frontiers in Microbiology, 11*. https://doi.org/10.3389/fmicb.2020.00241

Yang, Y., Heidari, F., & Hu, B. (2019). Fungi (Mold)-based lipid production. *Methods in Molecular Biology (Clifton, N.J.), 1995*, 51–89. https://doi.org/10.1007/978-1-4939-9484-7_3

Yano, Y., Oikawa, H., & Satomi, M. (2008). Reduction of lipids in fish meal prepared from fish waste by a yeast *Yarrowia lipolytica*. *International Journal of Food Microbiology, 121*(3), 302–307. https://doi.org/10.1016/j.ijfoodmicro.2007.11.012

Yao, S., Lyu, S., An, Y., Lu, J., Gjermansen, C., & Schramm, A. (2019). Microalgae-bacteria symbiosis in microalgal growth and biofuel production: A review. *Journal of Applied Microbiology, 126*(2), 359–368. https://doi.org/10.1111/jam.14095

Yasui, M., Oda, K., Masuo, S., Hosoda, S., Katayama, T., Maruyama, J.-I., Takaya, N., & Takeshita, N. (2020). Invasive growth of *Aspergillus oryzae* in rice koji and increase of nuclear number. *Fungal Biology and Biotechnology, 7*, 8. https://doi.org/10.1186/s40694-020-00099-9

Yee, W. (2016). Microalgae from the selenastraceae as emerging candidates for biodiesel production: A mini review. *World Journal of Microbiology & Biotechnology, 32*(4), 64. https://doi.org/10.1007/s11274-016-2023-6

Yeoman, C. J., Han, Y., Dodd, D., Schroeder, C. M., Mackie, R. I., & Cann, I. K. O. (2010). Chapter 1 – Thermostable enzymes as biocatalysts in the biofuel industry. *Advances in Applied Microbiology, 70*, 1–55. https://doi.org/10.1016/S0065-2164(10)70001-0

Yin, X., Yoshizaki, Y., Ikenaga, M., Han, X.-L., Okutsu, K., Futagami, T., Tamaki, H., & Takamine, K. (2020). Manufactural impact of the solid-state saccharification process in rice-flavor baijiu production. *Journal of Bioscience and Bioengineering, 129*(3), 315–321. https://doi.org/10.1016/j.jbiosc.2019.09.017

Yoneyama, F., Yamamoto, M., Hashimoto, W., & Murata, K. (2015). Production of polyhydroxybutyrate and alginate from glycerol by *Azotobacter vinelandii* under nitrogen-free conditions. *Bioengineered, 6*(4), 209–217. https://doi.org/10.1080/21655979.2015.1040209

Yoon, S. Y., Hong, M. E., Chang, W. S., & Sim, S. J. (2015). Enhanced biodiesel production in *Neochloris oleoabundans* by a semi-continuous process in two stage photobioreactors. *Bioprocess and Biosystems Engineering, 38*(7), 1415–1421. https://doi.org/10.1007/s00449-015-1383-x

You, M. P., Simoneau, P., Dongo, A., Barbetti, M. J., Li, H., & Sivasithamparam, K. (2005). First report of an alternaria leaf spot caused by *Alternaria brassicae* on *Crambe abyssinicia* in Australia. *Plant Disease, 89*(4), 430. https://doi.org/10.1094/PD-89-0430A

Yu, A., Zhao, Y., Li, J., Li, S., Pang, Y., Zhao, Y., Zhang, C., & Xiao, D. (2020). Sustainable production of FAEE biodiesel using the oleaginous yeast *Yarrowia lipolytica*. *MicrobiologyOpen, 9*(7), e1051. https://doi.org/10.1002/mbo3.1051

Yu, H., Guo, G., Zhang, X., Yan, K., & Xu, C. (2009). The effect of biological pretreatment with the selective white-rot fungus *Echinodontium taxodii* on enzymatic hydrolysis of softwoods and hardwoods. *Bioresource Technology, 100*(21), 5170–5175. https://doi.org/10.1016/j.biortech.2009.05.049

Yu, J., Landberg, J., Shavarebi, F., Bilanchone, V., Okerlund, A., Wanninayake, U., Zhao, L., Kraus, G., & Sandmeyer, S. (2018). Bioengineering triacetic acid lactone production in *Yarrowia lipolytica* for pogostone synthesis. *Biotechnology and Bioengineering, 115*(9), 2383–2388. https://doi.org/10.1002/bit.26733

Yu, J., Zhang, J., He, J., Liu, Z., & Yu, Z. (2009). Combinations of mild physical or chemical pretreatment with biological pretreatment for enzymatic hydrolysis of rice hull. *Bioresource Technology*, *100*(2), 903–908. https://doi.org/10.1016/j.biortech.2008.07.025

Yu, L., Cao, M.-Y., Wang, P.-T., Wang, S., Yue, Y.-R., Yuan, W.-D., Qiao, W.-C., Wang, F., & Song, X. (2017). Simultaneous decolorization and biohydrogen production from xylose by *Klebsiella oxytoca* GS-4-08 in the presence of azo dyes with sulfonate and carboxyl groups. *Applied and Environmental Microbiology*, *83*(10). https://doi.org/10.1128/AEM.00508-17

Yu, X., Zheng, Y., Dorgan, K. M., & Chen, S. (2011). Oil production by oleaginous yeasts using the hydrolysate from pretreatment of wheat straw with dilute sulfuric acid. *Bioresource Technology*, *102*(10), 6134–6140. https://doi.org/10.1016/j.biortech.2011.02.081

Zendejas, F. J., Benke, P. I., Lane, P. D., Simmons, B. A., & Lane, T. W. (2012). Characterization of the acylglycerols and resulting biodiesel derived from vegetable oil and microalgae (*Thalassiosira pseudonana* and *Phaeodactylum tricornutum*). *Biotechnology and Bioengineering*, *109*(5), 1146–1154. https://doi.org/10.1002/bit.24395

Zeng, J., Zheng, Y., Yu, X., Yu, L., Gao, D., & Chen, S. (2013). Lignocellulosic biomass as a carbohydrate source for lipid production by *Mortierella isabellina*. *Bioresource Technology*, *128*(January), 385–391. https://doi.org/10.1016/j.biortech.2012.10.079

Zeng, L., He, Y., Jiao, L., Li, K., & Yan, Y. (2017). Preparation of biodiesel with liquid synergetic lipases from rapeseed oil deodorizer distillate. *Applied Biochemistry and Biotechnology*, *183*(3), 778–791. https://doi.org/10.1007/s12010-017-2463-y

Zhan, J., Lin, H., Shen, Q., Zhou, Q., & Zhao, Y. (2013). Potential utilization of waste sweetpotato vines hydrolysate as a new source for single cell oils production by *Trichosporon fermentans*. *Bioresource Technology*, *135*(May), 622–629. https://doi.org/10.1016/j.biortech.2012.08.068

Zhang, L., Ch. Jia, Liu, L., Zhang, Z., Li, C., & Wang, Q. (2011). The involvement of jasmonates and ethylene in *Alternaria alternata f.* sp. *lycopersici* toxin-induced tomato cell death. *Journal of Experimental Botany*, *62*(15), 5405–5418. https://doi.org/10.1093/jxb/err217

Zhang, L., Chao, B., & Zhang, X. (2020). Modeling and optimization of microbial lipid fermentation from cellulosic ethanol wastewater by *Rhodotorula glutinis* based on the support vector machine. *Bioresource Technology*, *301*(April), 122781. https://doi.org/10.1016/j.biortech.2020.122781

Zhang, L., Li, X., Yong, Q., Yang, S.-T., Ouyang, J., & Yu, S. (2016). Impacts of lignocellulose-derived inhibitors on L-lactic acid fermentation by *Rhizopus oryzae*. *Bioresource Technology*, *203*(March), 173–180. https://doi.org/10.1016/j.biortech.2015.12.014

Zhang, L., Tang, Y., Guo, Z., Ding, Z., & Shi, G. (2011). Improving the ethanol yield by reducing glycerol formation using cofactor regulation in *Saccharomyces cerevisiae*. *Biotechnology Letters*, *33*(7), 1375–1380. https://doi.org/10.1007/s10529-011-0588-6

Zhang, X., Yu, H., Huang, H., & Liu, Y. (2007). Evaluation of biological pretreatment with white rot fungi for the enzymatic hydrolysis of bamboo culms. *International Biodeterioration & Biodegradation*, *60*(3), 159–164. https://doi.org/10.1016/j.ibiod.2007.02.003

Zhang, X., Zhao, H., Zhang, J., & Li, Z. (2004). Growth of *Azotobacter vinelandii* in a solid-state fermentation of technical lignin. *Bioresource Technology*, *95*(1), 31–33. https://doi.org/10.1016/j.biortech.2003.10.011

Zhang, Z., Xia, L., Wang, F., Lv, P., Zhu, M., Li, J., & Chen, K. (2015). Lignin degradation in corn stalk by combined method of H2O2 hydrolysis and *Aspergillus oryzae* CGMCC5992 liquid-state fermentation. *Biotechnology for Biofuels*, *8*, 183. https://doi.org/10.1186/s13068-015-0362-4

Zhao, C., Fang, H., & Chen, S. (2017). Single cell oil production by *Trichosporon cutaneum* from steam-exploded corn stover and its upgradation for production of long-chain α,ω-dicarboxylic acids. *Biotechnology for Biofuels*, *10*, 202. https://doi.org/10.1186/s13068-017-0889-7

Zhao, N., Bai, Y., Liu, C.-G., Zhao, X.-Q., Xu, J.-F., & Bai, F.-W. (2014). Flocculating *Zymomonas mobilis* is a promising host to be engineered for fuel ethanol production from lignocellulosic biomass. *Biotechnology Journal*, *9*(3), 362–371. https://doi.org/10.1002/biot.201300367

Zhao, X., Hu, C., Wu, S., Shen, H., & Zhao, Z. K. (2011). Lipid production by *Rhodosporidium toruloides* Y4 using different substrate feeding strategies. *Journal of Industrial Microbiology & Biotechnology*, *38*(5), 627–632. https://doi.org/10.1007/s10295-010-0808-4

Zhao, X., Kong, X., Hua, Y., Feng, B., & Zhao, Z. (Kent). (2008). Medium optimization for lipid production through co-fermentation of glucose and xylose by the oleaginous yeast *Lipomyces starkeyi*. *European Journal of Lipid Science and Technology*, *110*(5), 405–412. https://doi.org/10.1002/ejlt.200700224

Zhao, X., Wu, S., Hu, C., Wang, Q., Hua, Y., & Zhao, Z. K. (2010). Lipid production from Jerusalem Artichoke by *Rhodosporidium toruloides* Y4. *Journal of Industrial Microbiology & Biotechnology*, *37*(6), 581–585. https://doi.org/10.1007/s10295-010-0704-y

Zheng, S., Yang, M., & Yang, Z. (2005). Biomass production of yeast isolate from salad oil manufacturing wastewater. *Bioresource Technology*, *96*(10), 1183–1187. https://doi.org/10.1016/j.biortech.2004.09.022

Zhou, L., Santomauro, F., Fan, J., Macquarrie, D., Clark, J., Chuck, C. J., & Budarin, V. (2017). Fast microwave-assisted acidolysis: A new biorefinery approach for the zero-waste utilisation of lignocellulosic biomass to produce high quality lignin and fermentable saccharides. *Faraday Discussions*, *202*(September), 351–370. https://doi.org/10.1039/c7fd00102a

Zhou, W., Wang, Z., Xu, J., & Ma, L. (2018). Cultivation of microalgae *Chlorella zofingiensis* on municipal wastewater and biogas slurry towards bioenergy. *Journal of Bioscience and Bioengineering*, *126*(5), 644–648. https://doi.org/10.1016/j.jbiosc.2018.05.006

Zhu, D., Zhang, P., Xie, C., Zhang, W., Sun, J., Qian, W.-J., & Yang, B. (2017). Biodegradation of alkaline lignin by *Bacillus ligniniphilus* L1. *Biotechnology for Biofuels*, *10*(1), 44. https://doi.org/10.1186/s13068-017-0735-y

Zhu, J., & Wakisaka, M. (2020). Finding of phytase: Understanding growth promotion mechanism of phytic acid to freshwater microalga *Euglena gracilis*. *Bioresource Technology*, *296*(January), 122343. https://doi.org/10.1016/j.biortech.2019.122343

Zhu, L. Y., Zong, M. H., & Wu, H. (2008). Efficient lipid production with *Trichosporon fermentans* and its use for biodiesel preparation. *Bioresource Technology*, *99*(16), 7881–7885. https://doi.org/10.1016/j.biortech.2008.02.033

Zhu, S., Huang, W., Xu, J., Wang, Z., Xu, J., & Yuan, Z. (2014a). Metabolic changes of starch and lipid triggered by nitrogen starvation in the microalga *Chlorella zofingiensis*. *Bioresource Technology*, *152*, 292–298. https://doi.org/10.1016/j.biortech.2013.10.092

Zhu, S., Wang, Y., Huang, W., Xu, J., Wang, Z., Xu, J., & Yuan, Z. (2014b). Enhanced accumulation of carbohydrate and starch in *Chlorella zofingiensis* induced by nitrogen starvation. *Applied Biochemistry and Biotechnology*, *174*(7), 2435–2445. https://doi.org/10.1007/s12010-014-1183-9

Zhu, S., Wang, Y., Shang, C., Wang, Z., Xu, J., & Yuan, Z. (2015). Characterization of lipid and fatty acids composition of *Chlorella zofingiensis* in response to nitrogen starvation. *Journal of Bioscience and Bioengineering*, *120*(2), 205–209. https://doi.org/10.1016/j.jbiosc.2014.12.018

Zienkiewicz, A., Zienkiewicz, K., Poliner, E., Pulman, J. A., Du, Z.-Y., Stefano, G., Tsai, C.-H., et al. (2020). The microalga *Nannochloropsis* during transition from quiescence to autotrophy in response to nitrogen availability1 [OPEN]. *Plant Physiology*, *182*(2), 819–839. https://doi.org/10.1104/pp.19.00854

Zienkiewicz, K., Benning, U., Siegler, H., & Feussner, I. (2018). The type 2 acyl-CoA:diacylglycerol acyltransferase family of the oleaginous microalga *Lobosphaera incisa*. *BMC Plant Biology*, *18*. https://doi.org/10.1186/s12870-018-1510-3

Zienkiewicz, K., Du, Z. Y., Ma, W., Vollheyde, K., & Benning, C. (2016). Stress-induced neutral lipid biosynthesis in microalgae: Molecular, cellular and physiological insights. *Biochimica et Biophysica Acta*, *1861*, 1269–1281.

Zimbardi, A. L. R. L., Sehn, C., Meleiro, L. P., Souza, F. H. M., Masui, D. C., Nozawa, M. S. F., Guimarães, L. H. S., Jorge, J. A., & Furriel, R. P. M. (2013). Optimization of β-glucosidase, β-xylosidase and xylanase production by *Colletotrichum graminicola* under solid-state fermentation and application in raw sugarcane trash saccharification. *International Journal of Molecular Sciences*, *14*(2), 2875–2902. https://doi.org/10.3390/ijms14022875

Zulu, N. N., Zienkiewicz, K., Vollheyde, K., & Feussner, I. (2018). Current trends to comprehend lipid metabolism in diatoms. *Progress in Lipid Research*, *70*, 1–16.

5 A Comparative Account on Biodiesel Production from Forest Seeds

Jigna G. Tank and Rohan V. Pandya

CONTENTS

5.1	Introduction	130
5.2	Biodiesels from Forest Seeds Which Are Evaluated in Different Diesel Engines	132
	5.2.1 *Jatropha curcus*	132
	5.2.2 *Pongamia pinnata*	133
	5.2.3 *Simmondsia chinensis* (Jojoba)	133
	5.2.4 *Linum usitatissimum*	134
	5.2.5 *Moringa oleifera*	134
	5.2.6 *Sapindus mukorossi*	135
	5.2.7 *Zanthoxylum bungeanum*	135
	5.2.8 *Sterculia foetida*	136
	5.2.9 *Balanites aegyptiaca*	136
	5.2.10 *Cannabis sativa*	137
	5.2.11 *Camelina sativa*	137
	5.2.12 *Terminalia belerica*	138
	5.2.13 *Schleichera oleosa*	138
	5.2.14 *Madhuca longifolia*	138
	5.2.15 *Azadirachta indica*	138
	5.2.16 *Hibiscus cannabinus*	139
	5.2.17 *Luffa cylindrica* and *Blighia unijugata*	139
	5.2.18 *Simarouba glauca*	140
	5.2.19 *Ceiba pentandra*	140
	5.1.20 *Calophyllum inophyllum*	141
	5.2.21 *Raphanus sativus*	141
	5.2.22 *Cleome viscosa*	142
	5.2.23 *Manilkara zapota*	142
	5.2.24 *Aegle marmelos*	143
	5.2.25 *Thevetia peruviana*	143
	5.2.26 *Aleurites moluccanus*	144
	5.2.27 *Syzygium cumini*	144
	5.2.28 *Hevea brasiliensis*	144
	5.2.29 *Guizotia abyssinica*	144
	5.2.30 *Annona reticulata*	145
	5.2.31 *Citrullus colocynthis*	145
	5.2.32 *Pistacia khinjuk*	145
5.3	Biodiesels That Remain to Be Evaluated in Diesel Engines	146
	5.3.1 *Terminalia catappa*	146
	5.3.2 *Limnanthes alba*	146
	5.3.3 *Vernicia montana*	146

DOI: 10.1201/9780429262975-5

    5.3.4   *Annona cherimola* 146
    5.3.5   *Maclura pomifera* 147
    5.3.6   *Euphorbia lathyris* and *Sapium sebiferum* 147
    5.3.7   *Licania rigida* 147
    5.3.8   *Treculia africana* 147
    5.3.9   *Dipteryx alata* 147
    5.3.10 *Xanthoceras sorbifolia* 148
    5.3.11 *Lepidium sativum* 148
    5.3.12 *Pentaclethra macrophylla* 148
    5.3.13 *Xanthium sibiricum* 148
    5.3.14 *Amygdalus pedunculata* 148
    5.3.15 *Forsythia suspense* 149
    5.3.16 *Reutealis trisperma* 149
    5.3.17 *Aleurites trisperma* 149
    5.3.18 *Silybum marianum* 149
    5.3.19 *Crotalaria juncea* 150
    5.3.20 *Gliricidia sepium* 150
    5.3.21 *Pangium edule* 150
    5.3.22 *Senna obtusifolia* and *Panicum virgatum* 150
    5.3.23 *Cascabela thevetia* 150
    5.3.24 *Annona muricata* 151
    5.3.25 *Leucaena leucocephala* 151
    5.3.26 *Ocimum basilicum* 151
    5.3.27 *Cucumis melo* var. *agrestis* 151
    5.3.28 *Salvadora persica* 152
    5.3.29 *Koelreuteria paniculata* 152
5.4 Conclusion 152
References 157

5.1 Introduction

There is a wide range of forest plant species which produce seeds rich in oil content. These oil-containing seeds can be used for biodiesel production to patch up the demand of petroleum diesel at the global level. The world is endowed with enormous forest wealth, and only part of this forest wealth is used by tribal peoples living in and around forest areas. The rest of the forest products undergo decomposition due to unorganized forest products collection. Therefore, forest seeds rich in oil content can be used as a feedstock for biodiesel production. Various research scientists have optimized biodiesel production from oil-containing forest seeds. There are various methods used for biodiesel production from forest seed oils. These methods are transesterification, blending, microemulsions, pyrolysis, enzymatic transesterification, extraction, and transesterification using supercritical instrument, etc. (Ma and Hanna, 1999; Srivastava and Prasad, 2000). The main problem faced during biodiesel production was the high content of free fatty acids and low oxidative stability of oils. The high amount of free fatty acids present in oil forms soap with alkaline catalyst and prevents separation of biodiesel from the glycerin fractions (Demirbas, 2003). In order to solve this problem, pre-transesterification of oil was done using methanol and acid catalysts (H_2SO_4 or HCl). This converts free fatty acids in to esters and reduces the acid value of oil (Canakci and Van Gerpen, 2001; Dorado et al., 2002). To improve oxidative stability of oil, antioxidants and low polarity solvents were mixed with it (Baranitharan et al., 2019; Kolli et al., 2020). Finally, the transesterification process was carried out for biodiesel production.

Another problem faced during biodiesel production was how to improve the physico-chemical parameters of biodiesel and its blends? These parameters were the density of biodiesel, kinematic viscosity of biodiesel, level of free fatty acids formed, high or low acid number, iodine number, flash point, sulfated ash content, cloud point, pour point, and carbon residues deposition. The seed oils rich in unsaturated fatty acids and having less chain length and a greater number of double bonds has high density (Alptekin et al., 2008). Density regulates atomization efficiency of biodiesel. High density decreases atomization efficiency of biodiesel and low density increases atomization efficiency of biodiesel. The seed oils having higher amounts of trans fatty acids and saturated fatty acids have high viscosity. The seed oils which have high amounts of unsaturated fatty acids have low viscosity (Knothe, 2005; Knothe, 2006; Demirbas, 2008; Demirbas, 1997). Kinematic viscosity affects the flow of biodiesel through the pipelines, injector nozzles, and orifices. It influences during atomization of fuel in the cylinder. Highly viscous biodiesel causes pumping pressure, decreases pump clearance, and seizes the pump (Tesfa et al., 2010). Carbon residue deposition should be low in the combustion chamber, as high deposition reduces the working efficiency of an engine. It gets deposited on the piston top and cylinder head surfaces that increase compression ratio and surface temperature, which in turn affects the engine performance and drivability (Diaby et al., 2009). Cloud and pour point determines the freezing condition of the biodiesel. A biodiesel with high cloud and pour point forms wax crystals at low temperatures and cannot be used in cold climate regions. Flash point determines the presence of highly flammable volatile compounds in biodiesel. It helps to predict explosion hazards during storage and handling. A high cetane number ensures excellent cold start properties and minimizes white smoke formation (Wang et al., 2012). The cetane number is the relative measure of delay in ignition of any fuel. Oil containing more saturated fatty acids has a higher cetane number, and more monounsaturated fatty acids have a medium cetane number. Biodiesel with a high cetane number has a short ignition delay and has excellent cold start quality. Biodiesel with a low cetane number has a long ignition delay and has poor cold start quality (Knothe et al., 2003; Harrington, 1986; Van Gerpen, 1996). Oil containing higher unsaturated fatty acids develops biodiesel with better lubricity than oil containing higher saturated fatty acids (Demirbas, 2008).

Biodiesel was also tested in different models of diesel engines to observe the effect of biodiesel on engine speed, power injector coking, release of energy, and engine compatibility. Consequences of using pure biodiesel (100%) and biodiesel blends on working efficiency of diesel engines was analyzed in various biodiesels prepared from forest seed oils. Even emission rates of hazardous exhaust gases like carbon dioxide, carbon monoxide, smoke, and nitrogen oxide from diesel engine were analyzed after use of biodiesel and its blends. It was found that the direct use of biodiesel in diesel engines causes numerous problems in engines such as poor fuel atomization, incomplete combustion of fuel, deposition of carbon residues on fuel injectors, and engine fouling (Sridharan and Mathai, 1974; Encinar et al., 2002; Williamson and Badr, 1998; Karaosmanoğlu et al., 2000). Biodiesel produced from seed oils having high free fatty acids, high viscosity, high pour and cloud point results in choking of power injectors, decreases in engine speed, decreases in release of energy, and engine compatibility is reduced (Ali et al., 2016). The peak torque remains less when viscous biodiesel is used in diesel engines. This decreases power of the diesel engine up to 5% at rate load as compared to petroleum diesel, which in turn leads to high engine maintenance charges (Upadhyay and Sharma, 2013). These parameters were standardized by pyrolysis, blending of biodiesels or oils, microemulsification, and repetition of transesterification processes using different catalysts (Ma and Hanna, 1999). However, after standardization, it was observed that the use of pure biodiesel and its blends in diesel engines increases nitrogen oxide emissions compared to that of petroleum based diesel due to more accumulation of fuel and air mixture during beginning of combustion. This in turn increases nitrogen oxide formation and the temperature of the combustion chamber because of rapid release of heat (Papagiannakis et al., 2007). The biodiesel which increases working efficiency of engines with decreased emission of hazardous gases is considered for use at regular basis in diesel engines. However, the steps of biodiesel production, reduction of oil viscosity, reduction of carbon residues deposition in the combustion chamber, reduction in release of hazardous gases, and improvement in the working efficiency of diesel engine requires input of resources and time, which in turn increases the cost of biodiesel and engine maintenance charges in the market.

Therefore, the present book chapter aims to give a comparative account on biodiesel production optimized from forest seeds. It will compare the physico-chemical, performance, combustion, and emission parameters of different biodiesels produced from forest seed oils. It will help researchers to understand the problems associated with biodiesel production from forest seed oils and develop an understanding among researchers to modify the process of biodiesel production to develop an efficient biodiesel which can replace petroleum diesel in the near future. It will help researchers to judge the scenario of biodiesel production from forest seeds and its use for large-scale production of biodiesel.

5.2 Biodiesels from Forest Seeds Which Are Evaluated in Different Diesel Engines

5.2.1 *Jatropha curcus*

Jatropha curcus was considered for its potential for biodiesel production, as this plant is drought tolerant, can grow well in rocky regions and low nutrient soils, and can be easily grown in arid and semi-arid regions (Openshaw, 2000; Divakara et al., 2010). It prevents soil erosion and can be used for reclamation of waste lands (Achten et al., 2007). The seeds of *Jatropha curcus* contain 40%–60% of oil which can be used for biodiesel production (Foidl et al., 1996; Makkar et al., 1997). The oil of *Jatropha curcus* contains 14% palmic acid, 6.7% stearic acid, 47% oleic acid, and 31% linoleic acid, which are rich in monounsaturated oleic acids, and polyunsaturated linoleic acids, which are efficient sources for biodiesel production (Augustus et al., 2002). Biodiesel production from *Jatropha curcus* oil is done either through chemical or enzymatic transesterification. Biodiesel production through chemical transesterification has several challenges such as purification of biodiesel, revival of glycerol, requirement of waste treatment before disposal, and energy consuming phenomenon (Rathore and Madras, 2007). However, biodiesel production through enzymatic transesterification was beneficial. It helped in synthesis of specific alkyl esters, recovery of glycerol from biodiesel was easy, purification of biodiesel was not required, waste treatment before disposal was not required, and transesterification of oil having high free fatty acids was achieved (Nelson et al., 1996). Hence, transesterification of *Jatropha curcus* seed oil using lipases enzyme was a more successful method than chemical transesterification. For large-scale biodiesel production from *Jatropha curcus*, the major challenges were the cost of enzymes and standardization of enzyme immobilization techniques to improve biodiesel quality. Biodiesel produced from *Jatropha curcus* seed oil is a potential alternative for diesel because its performance facilitates continuous operation of a diesel engine without any change in its design. Its physico-chemical parameters were similar to those of petroleum diesel (Sayyar et al., 2009; Abdulla et al., 2011). Lim and colleagues (2010) optimized production of biodiesel from *Jatropha curcus* seed oil with the help of supercritical extraction and reaction method. It was suggested that supercritical reactive extraction technology can successfully extract oil from seeds and subsequently did transesterification to produce fatty acid methyl esters in short period of time (45–80 minutes). Since no catalyst is used in this process, it reduces the steps of catalyst separation and washing of biodiesel. Particle size of seeds (0.5–2 mm) and reaction temperature (300°C) were critical parameters for maximum yield of fatty acid methyl esters. Wang and colleagues (2011) optimized production of biodiesel from *Jatropha curcus* seed oil and suggested that the yield obtained after transesterification process was 98.27%. The physico-chemical parameters such as cetane number (55.4), oxidative stability (8 h at 110°C), cold filter plug point (−8°C), density (880.2 kg/m^3 at 20°C), kinematic viscosity (4.312 mm^2/s at 40°C), flash point (147°C), sulfated ash content (0.01%), acid value (0.04 mg KOH/g), sulfur content (<3.0 mg/kg), copper strip corrosion (grade 1A for 3 h at 50°C), linolenic acid content (0.22%), free glycerol (0.02%), and total glycerol (0.14%) were in range as per ASTM and EN (European Committee for Standardization) standard specifications. Hailegiorgis and colleagues (2012) optimized production of biodiesel from *Jatropha curcus* using cetyltrimethylammoniumbromide as phase transfer catalyst. It was suggested after investigation that the maximum yield of 99.5% of fatty acids methyl esters can be obtained after 150 minutes of reaction time. The physico-chemical parameters such as kinematic viscosity (4.8 mm^2/s), density (867 kg/m^3), acid value (0.21 mg KOH/g), and flash point (173°C) of biodiesel were almost equivalent to that of the ASTM D6751 and EN 14214 standards. Palash and colleagues (2013) reported that the *Jatropha curcus* biodiesel blended with petroleum diesel emitted less hydrocarbons up to 48%, carbon monoxide up to 52.6% and high nitrogen oxide up to 11.82%. There was remarkable increase in consumption of brake-specific fuel up to 8.33% in diesel engine with use of biodiesel. It was suggested that 20% *Jatropha curcus* biodiesel blend has potential to work as fuel in compression ignition diesel engines without any modifications in engines. Kavitha and colleagues (2019) investigated the combustion and emission characteristics of *Jatropha curcus* biodiesel in a Kirloskar single-cylinder VCR research engine by blending it with ethanol and petroleum diesel in various combinations. They reported that the emission of exhaust gases such as carbon monoxide, nitrogen oxide, hydrocarbons, and smoke were less when the engine was running with biodiesel blends as compared to petroleum diesel. When the proportion of biodiesel increases in the blend, emission of carbon monoxide reduces due to changes in supply of oxygen and proper combustion of fuel in the diesel engine. They suggested that the mixture of 98% diesel, 1.5% biodiesel and 0.5% ethanol had significant improvement in the performance of diesel engine as compared to petroleum diesel. it can be a better alternate for diesel engines in the near future, but some parameters like timing of injection, changing compression ratio, and pressure of injection should be optimized further.

5.2.2 *Pongamia pinnata*

Karmee and Chadha (2005) optimized production of biodiesel from *Pongamia pinnata* seed oil and compared yields of biodiesel when different catalysts (potassium hydroxide, zinc oxide, Hb-zeolite, and montmorillonite) were used. They suggested that maximum biodiesel yield can be obtained up to 92% when potassium hydroxide was used as catalyst at optimum ratio of oil and methanol (1:10) at 60°C in the transesterification process. However, yield was low when zinc oxide (47–59%), Hb-zeolite, and montmorillonite (83%) were used as catalysts in the same conditions. Even the reaction time taken by potassium hydroxide was less as compared to other catalysts. The viscosity (4.8 mm²/s at 40°C) and flash point (150°C) of biodiesel was in range as per ASTM and German specifications. Agarwal and Bajaj (2009) optimized production of biodiesel from *Pongamia pinnata* and observed its physico-chemical parameters. They observed effect of temperature, time, molar ratio of alcohol to oil, and concentration of catalyst on the yield of biodiesel. They concluded that the optimum molar ratio of methanol and oil (9:1) can yield maximum biodiesel only in the presence of catalyst potassium hydroxide (0.75%). The transesterification reaction should be carried out at 60°C for one hour. The biodiesel can be purified by gently washing it with warm distilled water three to four times at 60°C. The physico-chemical parameters (viscosity, flash point, percentage of sulfated ash, neutralization value, appearance, percentage of water, and sediments content) of biodiesel were almost in range as per ASTM, EN, and BIS standards. Kumar and colleagues (2011) optimized production of biodiesel from *Pongamia pinnata* seed oil using a microwave assisted transesterification process. The transesterification process was optimized in the presence of sodium hydroxide and potassium hydroxide as catalysts. The molar ratio of methanol and oil used during transesterification was 6:1 at 60°C in a microwave oven. After analysis, they suggested that the optimum catalyst concentration required for production of biodiesel from *Pongamia pinnata* seed oil using microwave oven is 0.5% sodium hydroxide and 1% potassium hydroxide. The *Pongamia pinnata* biodiesel can be produced successfully using microwave assisted transesterification process in 5–10 minutes as compared to conventional heating process which takes three hours for reaction time. The use of the microwave assisted process helps in time and cost savings. Gogoi and Baruah (2011) evaluated performance of *Pongamia pinnata* biodiesel blends (10%–40%) in a single-cylinder, four-stroke diesel engine. They observed various parameters and suggested that *Pongamia pinnata* biodiesel blends increase exhaust gas temperature and consumption of brake specific fuel in diesel engines. It decreases brake thermal efficiency and brake power of diesel engines, which indicates more energy loss and incomplete fuel combustion. Hence, further optimization was required for complete combustion of biodiesel and to decrease the energy loss in the diesel engine. Gogoi (2013) further did exergy analysis in a single-cylinder, four-stroke diesel engine and suggested that the exergetic efficiency of diesel engine remains low with use of *Pongamia pinnata* biodiesel blends (10%–40%) as compared to petroleum diesel. Hence, further improvement in engine operation should be optimized using biodiesel blends with higher percentage of biodiesel (50% or more).

5.2.3 *Simmondsia chinensis* (Jojoba)

Canoira and colleagues (2006) optimized biodiesel production from *Simmondsia chinensis* seed oil and suggested that the biodiesel yield of 79% can be obtained by using sodium methoxide (1%) as catalyst at 60°C with 7.5:1 ratio of methanol and oil in the transesterification process. The physico-chemical parameters such as density (863.5 kg/m³), calorific value (41.52 kJ/g) and cold filter plugging point (−14°C) were in range as per EN standards, except kinematic viscosity (9 mm²/s), which was above maximum value as prescribed by European standards. Bouaid and colleagues (2007) optimized biodiesel production from *Simmondsia chinensis* and suggested that 83% yield of biodiesel can be obtained when 1.35% catalyst is used in transesterification process under 25°C. The physico-chemical parameters (acid value, viscosity, and iodine value) of biodiesel were suitable for use in diesel engine. Similarly, Shah and colleagues (2010) optimized biodiesel production from *Simmondsia chinensis* seed oil and evaluated the physico-chemical properties of biodiesel and its blends (5% and 20% biodiesel mixed with ultra low sulfur diesel fuel). Based on their analysis, it was suggested that the quality of *Simmondsia chinensis* biodiesel was better than soyabean biodiesel as per ASTM D6751 and EN 14214 standards. The blends of *Simmondsia chinensis* biodiesel had improved low temperature performance properties compared to ultra low sulfur diesel fuel and soyabean biodiesel blends. Al-Widyan and Al-Muhtaseb (2010) optimized biodiesel production from *Simmondsia chinensis* seed oil and evaluated the performance of *Simmondsia chinensis* 100% biodiesel and its 50% blend in a single-cylinder diesel engine. From comparative analysis, they suggested that the biodiesel and its blend (50%) reduced the brake thermal efficiency, brake power, and brake effective pressure of the diesel engine. The 50% blend of biodiesel increases the consumption of brake-specific fuel and decreases emissions of carbon monoxide, carbon dioxide, and nitrogen oxide from the diesel engine. Shehata and Razek (2011) also observed similar results in a direct injection diesel engine, when they evaluated the performance of 100% and 20% *Simmondsia chinensis* biodiesel blends. Azad and colleagues (2019) evaluated the combustion and emission parameters of *Simmondsia chinensis* biodiesel and its blends (5%, 10%, and 20%) in a four-cylinder Kubota V3300 tractor engine. They suggested that the combustion parameter of 10% *Simmondsia chinensis* biodiesel blend was similar to that of petroleum diesel. However, the emission of nitrogen oxide and carbon dioxide was more, which can be improved by applying alternate combustion strategies.

5.2.4 *Linum usitatissimum*

Agrawal and Agrawal (2004) suggested that a 50% blend of *Linum usitatissimum* biodiesel is more thermally efficient than other biodiesel blends. The improved thermal efficiency of *Linum usitatissimum* biodiesel blends was due to complete combustion of biodiesel and its lubricity.

Agarwal (2005) investigated the performance of *Linum usitatissimum* biodiesel in compression ignition diesel engine and suggested that it is compatible with low emission of exhaust gases. Ghadge and Raheman (2005) suggested that the thermal efficiency of all blends of *Linum usitatissimum* biodiesel was similar to diesel. Nabi and colleagues (2008) suggested that *Linum usitatissimum* biodiesel blends (10% and 20%) have similar consumption of brake-specific fuel efficiency as that of conventional diesel. However, it shows higher exhaust gas temperature and nitrogen oxide emission than diesel. Jaichandar and Annamalai (2011) also tested *Linum usitatissimum* biodiesel blends with high sulfur in a single-cylinder portable diesel engine and observed increase in brake thermal efficiency.

5.2.5 *Moringa oleifera*

Rashid and colleagues (2008) optimized production of biodiesel from *Moringa oleifera* seed oil and suggested that the kinematic viscosity (4.83 mm²/s), oxidative stability (3.61 h), and lubricity (138.5 μm) of *Moringa oleifera* biodiesel is in range as per ASTM and EN standard specifications. However, cetane number (67.07), cloud point (18°C), and pour point (17°C) were higher than the standard specifications. Mofijur and colleagues (2014) evaluated the performance of *Moringa oleifera* biodiesel blends (10% and 20%) in a multi-cylinder diesel engine and suggested that at the entire range of speeds, there was decrease in brake power and increase in consumption of brake-specific fuel as compared to petroleum diesel. There was reduction in carbon dioxide and hydrocarbons emission with use of 10% and 20% biodiesel blends. However, there was slight increase in nitric oxide emission as compared to petroleum diesel. Rashed and colleagues (2016) compared performance of 20% blends of *Jatropha curcus*, *Moringa oleifera*, and palm biodiesel in a multi-cylinder, four-stroke, direct injection diesel engine at different speeds and full load conditions. They suggested that all the biodiesel samples reduced the brake power of the diesel engine and increased the consumption of brake-specific fuel of diesel engine as compared to petroleum diesel. The emission of carbon dioxide and hydrocarbons was low in 20% blended biodiesel, but nitrogen oxide emission was higher than the petroleum diesel. Among the three biodiesel samples, palm biodiesel had better performance and reduced emission of exhaust gases than *Jatropha curcus* and *Moringa oleifera* biodiesel. Teoh and colleagues (2019) evaluated the combustion, emission, and performance characteristics of *Moringa oleifera* biodiesel blends (10%, 20%, 30%, and 40%) in a multi-cylinder high pressure common rail diesel engine. They suggested that the engine torque and brake power was lower in this engine than the petroleum diesel. All biodiesel blends and petroleum diesel had similar brake thermal efficiencies. However, there was a remarkable increase in peak cylinder pressure, heat release rate, and consumption of brake-specific fuel with increase in the proportion of biodiesel in the blends. The emission of carbon monoxide, hydrocarbons, and smoke was low in all blends except nitrogen oxide, which was slightly higher than the petroleum diesel. In order to improve fuel consumption and emission characteristics of *Moringa oleifera* biodiesel and its blends, Karthickeyan (2019) modified a Kirloskar TV1 model direct injection water cooled diesel engine by thermally coating its combustion chamber components such as piston head, cylinder head, intake valve, and exhaust valve with yttria-stabilized zirconia to convert it in to low heat rejection engine. Even 1% pyrogallol (antioxidant) was added to *Moringa oleifera* methyl esters to improve its physico-chemical parameters. After analysis, it was suggested that there was remarkable improvement in brake thermal efficiency, reduced consumption of brake specific fuel, low smoke emission, low carbon monoxide and hydrocarbons emission, and even low nitrogen oxide emission with use of coated engine and modified biodiesel blends than when using petroleum diesel.

5.2.6 *Sapindus mukorossi*

Chhetri and colleagues (2008) optimized biodiesel production from *Sapindus mukorossi* and characterized its fatty acids methyl esters using gas chromatography and mass spectrometry. From analysis, they observed that *Sapindus* mukorossi seed oil contained free fatty acids (9.1%), triglycerides (84.43%), and sterols (4.88%) which are lower as compared to *Jatropha curcus* seed oil. *Sapindus mukorossi* biodiesel contains 85% unsaturated fatty acids, which was higher than *Jatropha curcus* biodiesel (80% unsaturated fatty acids). Oleic acid content was high, and yield of biodiesel was 97% from *Sapindus mukorossi* seed oil. Misra and Murthy (2011) evaluated the combustion, emission, and performance parameters of *Sapindus mukorossi* biodiesel blends (10%, 20%, 30%, and 40%) in a single-cylinder (DI) constant speed diesel engine by conducting engine tests at 25% load, full load, and no load conditions. They suggested from evaluation that among all biodiesel blends, 10% biodiesel blend had better performance with respect to consumption of specific fuel, thermal efficiency, and emission of exhaust gases. All blends had lower nitrogen oxide emission but higher hydrocarbons emission after 75% load. Similarly, Padmanabhan and colleagues (2017) evaluated the combustion, emission, and performance characteristics of *Sapindus mukorossi* biodiesel blends (10%, 20%, and 30%) in a single-cylinder (DI) constant speed diesel engine. They suggested that at maximum load, the consumption of brake-specific fuel was high in all blends of *Sapindus mukorossi* biodiesel as compared to petroleum diesel. The consumption of brake specific fuel value of 10% blend of biodiesel was slightly high, but nearer to the petroleum diesel. There was only 2% variation in the brake thermal efficiency of biodiesel blends and petroleum diesel. In all the blends, the nitrogen oxide emission was low and carbon monoxide emission was high. To reduce the emission of carbon monoxide, surplus oxygen was supplied in the combustion chamber, which converted carbon monoxide into carbon dioxide after complete combustion.

5.2.7 *Zanthoxylum bungeanum*

Zhang and Jiang (2008) optimized biodiesel production from *Zanthoxylum bungeanum* seed oil using pretreatment of acid catalyzed transesterification process to reduce free fatty acids from the oil. Further, they carried out biodiesel production from *Zanthoxylum bungeanum* seed oil using alkali catalyzed transesterification process. They suggested that pretreatment of acid catalyzed transesterification reduces the acid value of *Zanthoxylum bungeanum* seed oil up to 2 mg KOH/g in one step. The yield of biodiesel obtained after alkali catalyzed transesterification process was 98%.

However, it is necessary to analyze the fuel properties of *Zanthoxylum bungeanum* biodiesel further. Yang and colleagues (2008) optimized biodiesel production from *Zanthoxylum bungeanum* seed oil and suggested that the high level of free fatty acids present in *Zanthoxylum bungeanum* seed oil can be reduced through initial acid catalyzed (2% H_2SO_4) transesterification process of one hour at 25:1 ratio of methanol and oil. The yield obtained after alkali catalyzed transesterification was 96% and physico-chemical parameters (density 882.6 kg/m³ at 15°C, kinematic viscosity 4 mm²/s at 40°C, acid value 0.3 mg KOH/g, cloud point 2°C, flash point >174°C, cetane number 47, copper corrosion 1A, 90% distillation temperature 337.5°C) were almost in range as per China's standards. Jian and colleagues (2015) evaluated the combustion, emission, and performance parameters of *Zanthoxylum bungeanum* biodiesel in the compression ignition diesel engine. They suggested that *Zanthoxylum bungeanum* biodiesel and its blends had slight influence on engine operation. It reduces pressure lifting ratio and heat release rate up to 38% at partial load of 1,500 rpm and 2,000 rpm. It reduces granule and nitrogen oxide emission at 1,500 rpm, but nitrogen oxide emission increases at 2,000 rpm.

5.2.8 *Sterculia foetida*

Devan and Mahalakshmi (2009) optimized biodiesel production from *Sterculia foetida* seed oil and evaluated its physico-chemical parameters such as density (875 kg/m³), kinematic viscosity (6 mm²/s at 40°C), pour point (1°C), calorific value (40.211 MJ/kg), and flash point (162°C). They suggested that all physico-chemical parameters were as per ASTM standards. Similar results were observed by Manurung and colleagues (2012) when they observed physico-chemical parameters such as density (873 kg/m³), kinematic viscosity (4.92 mm²/s at 40°C), pour point (−3°C), calorific value (40.167 MJ/kg), and flash point (160.5°C) of *Sterculia foetida* methyl esters. Bindhu and colleagues (2012) also optimized production of biodiesel from *Sterculia foetida* seed oil and evaluated its physico-chemical parameters. They suggested that the *Sterculia foetida* biodiesel has physico-chemical parameters such as iodine value of 72.6 g I_2/100 g, free fatty acids 0.17%, phosphorus content 0 ppm, flash point 179°C, cloud point 3°C, pour point 3°C, kinematic viscosity 4.72 mm²/s at 40°C, oxidative stability 3.42 h at 110°C, and density 0.850 g/cm³ at 15°C that were similar to the biodiesel of sunflower, soyabean, and rapeseed, except the cold temperature operation ability. Similarly, Silitonga and colleagues (2013) optimized biodiesel production from *Sterculia foetida* seed oil and suggested that the physico-chemical parameters of biodiesel were almost in range as per ASTM standards; however, it cannot be used in cold temperature countries. Sambasivam and Murugavelh (2019) also optimized biodiesel production from *Sterculia foetida* seed oil and suggested that physico-chemical parameters were almost in range as per ASTM D6751 standards. They observed that the catalyst required 37.91 KJ/mol of activation energy to carry out the transesterification process. Since the enthalpy value is positive, the reaction is endothermic and production of biodiesel requires energy. Kavitha and Murugavelh (2019) optimized production of biodiesel from *Sterculia foetida* seed oil and evaluated combustion, performance, and emission parameters of biodiesel and its blends in a Kirloskar TV1 model single-cylinder, water-cooled, four-stroke diesel engine. They suggested that the biodiesel yield of 90.2% can be obtained after transesterification of *Sterculia foetida* oil at 55°C at 1:12 oil and methanol ratio. The brake thermal efficiency of biodiesel blends was lower than the pure biodiesel (100%) due to reduced calorific value, viscosity, and spraying ability (Raman and colleagues, 2019). The specific fuel consumption was higher than petroleum diesel with use of biodiesel and its blends. In biodiesel and its blends, emission of carbon dioxide and hydrocarbons was low but nitrogen oxide emission was high as compared to petroleum diesel.

5.2.9 *Balanites aegyptiaca*

Chapagain and colleagues (2009) optimized biodiesel production from *Balanites aegyptiaca* seed oil and observed its performance in a 2.5-liter Ford diesel engine. They observed that the oil content in seed kernel of *Balanites aegyptiaca* seeds was 46.7%. The fatty acids composition of *Balanites aegyptiaca* seed oil was similar to that of soyabean seed oil. The four major fatty acids present in *Balanites aegyptiaca* seed oil were palmitic acid, stearic acid, oleic acid. and linoleic acid. Biodiesel yield after *in situ* transesterification process was 90%, and physico-chemical parameters of biodiesel were in range as per EN 14214 standard specifications. The diesel engine test analysis of pure biodiesel and its 5% blend suggested that the performance of the engine was better when 5% biodiesel blend was used as compared to pure biodiesel. The combustion, emission, and performance parameters of 5% blend of biodiesel were similar to that of petroleum diesel.

5.2.10 Cannabis sativa

Li and colleagues (2010) studied optimized biodiesel production from *Cannabis sativa* seed oil and suggested that the yield of biodiesel after transesterification and purification is 97%. Among the physico-chemical parameters, acid number (0.25 mg KOH/g), sulfur content (0.4 ppm), density (884 kg/m^3), and total glycerin (0.10%) were in range as per ASTM D6751 standards. It had high flash point (162°C) and low kinematic viscosity (3.48 mm^2/s) and cloud point (−5°C). They suggested that the improved quality of physico-chemical parameters is due to high content of polyunsaturated fatty acids and 3:1 ratio of linoleic to linolenic acid in *Cannabis sativa* seed oil. Mohammed and colleagues (2020) evaluated the physico-chemical parameters of *Cannabis sativa* biodiesel by blending it with diethyl ether, euro diesel, or butanol. From the analysis, they observed that the iodine value and oxidative stability of *Cannabis sativa* biodiesel was unsatisfactory due to elevated content of unsaturated fatty acids. Mixing of biodiesel with butanol, euro diesel, or diethyl ether improves the density, cloud point, and kinematic viscosity of biodiesel, but decreases the flash point (especially when diethyl ether was blended). They suggested that 20% blend of biodiesel with diethyl ether, butanol, or euro diesel can be considered for use in compression ignition diesel engines with further investigation of engine combustion, emission, and performance parameters. Afif and Biradar (2019) evaluated the combustion, emission, and performance characteristics of *Cannabis sativa* biodiesel and its blends in a Kirloskar TV1 single-cylinder, four-stroke, direct injection diesel engine under different load conditions. They suggested that the brake thermal and fuel consumption efficiency was high when pure biodiesel and its blends (30% and 50%) were used, but it was similar to petroleum diesel when 10% or 20% biodiesel blends were used. There was reduction in emission of carbon monoxide and carbon dioxide as compared to petroleum diesel, but emission of nitrogen oxide was high. The opacity of smoke improved up to 10% at full load conditions.

5.2.11 Camelina sativa

Moser and Vaughn (2010) optimized biodiesel production from *Camelina sativa* seed oil using homogenous base catalyzed transesterification process. They reported that the ethyl ester and methyl ester of *Camelina sativa* seed oil had poor oxidative stability and high iodine value as compared to canola, palm, and soyabean methyl esters. Other physico-chemical parameters (acid value, kinematic viscosity, low temperature operation, cetane number, lubricity, sulfur and phosphorus content) were similar as that of canola, palm and soyabean, methyl esters.

Soriano and Narani (2012) observed similar results when they optimized biodiesel production from *Camelina sativa* seed oil. They suggested that the physico-chemical parameters such as flash point, kinematic viscosity (40°C), cloud point, oil stability index, and cold filter plugging point of *Camelina sativa* biodiesel is in accordance with ASTM D6751 standards and is similar with that of sunflower methyl esters. It exhibited poor oxidative stability and high distillation temperature, and has high potential to form coke during combustion due to the presence of n-3 fatty acids in *Camelina sativa* oil. To improve these properties, they used a 20% blend of Camelina sativa biodiesel and suggested that it has all physico-chemical parameters similar to that of soyabean methyl esters. Oni and Oluwatosin (2020) evaluated the performance of *Azadirachta indica* and *Camelina sativa* biodiesel and its blends (5% and 10%) in a 1.9 multiple-injection diesel engine. They suggested that a 10% blend of *Camelina sativa* had better brake power and consumption of brake-specific fuel as compared to biodiesel of *Azadiracta indica* through all ranges of engine speeds. There was reduced emission of carbon dioxide and hydrocarbons in all blends tested. However, nitrogen oxide emission was high in all blends as compared to petroleum diesel.

5.2.12 Terminalia belerica

Sarin and colleagues (2010) optimized biodiesel production from *Terminalia belerica* seed oil and observed its physico-chemical parameters. The physico-chemical parameters of *Terminalia belerica* biodiesel was compared with *Jatropha*, sunflower, soyabean, and rapeseed methyl esters as per ASTM D 6751 specifications of biodiesel. However, the density of biodiesel was slightly higher than the standard values. The density of biodiesel was improved by blending the biodiesel with petroleum diesel (5%, 10%, and 20%). They suggested that the blended biodiesel meets the physico-chemical parameters as per Indian standards 1460 and ASTM D6751 specifications. Rudreshaiah and colleagues (2020) evaluated the performance of *Terminalia belerica* biodiesel and its blends (10%, 20%, 30%, and 50%) in the compression ignition diesel engine. They suggested that the combustion and performance of biodiesel blends was similar to that of petroleum diesel, with minor decrements. There was reduction in carbon dioxide and hydrocarbons emissions, but nitrogen oxide emission was high as compared to petroleum diesel.

5.2.13 Schleichera oleosa

Acharya and colleagues (2011) suggested the use of *Schleichera oleosa* seed oil in stationary diesel engine by reducing viscosity of oil to almost the same as that of petroleum diesel. They reduced the viscosity of *Schleichera oleosa* seed oil by preheating at 90°C–130°C using a heat exchanger and tested its performance in a stationary diesel engine at different loads and constant speed of 1,500 rpm. They observed that there were no operational difficulties in the diesel engine when 20% blended *Schleichera oleosa* oil or preheated *Schleichera oleosa* oil having less viscosity was used. There was remarkable increase in consumption of brake-specific fuel, decrease in brake thermal efficiency, and decrease in emission of carbon monoxide, carbon dioxide, hydrocarbons, and nitrogen oxides when 20% blended or preheated *Schleichera oleosa* oil was used in a diesel engine. Finally, they concluded that the *Schleichera oleosa* oil either as 20% blended with diesel or preheated *Schleichera oleosa* oil using a heat exchanger can be used as fuel in stationary diesel powered machines such as electricity generators, irrigation machines, and machines processing agriculture products. Yadav and colleagues (2017b) reported the combustion, performance, and emission of *Schleichera oleosa* biodiesel and its blends (5%, 10%, 15%, and 20%) in a multi-cylinder transportation diesel engine. They suggested that there was reduction in emission of carbon monoxide, hydrocarbons, and smoke, with an increase in proportion of biodiesel in blends. However, there was slight increase in fuel consumption and nitrogen oxide emission as compared to petroleum diesel. It was observed that when operated with 10% biodiesel blended with petroleum diesel, the engine had slightly higher thermal efficiency and lower brake-specific energy consumption.

5.2.14 Madhuca longifolia

Kapilan and colleagues (2009) optimized production of biodiesel from *Madhuca longifolia* seed oil and evaluated its performance in a single-cylinder (DI) compression ignition diesel engine. They suggested that physico-chemical parameters of *Madhuca longifolia* biodiesel were close to petroleum diesel and were in range as per ASTM standards. From diesel engine analysis, they suggested that the use of *Madhuca longifolia* biodiesel and its blends (5% and 20%) results in low carbon monoxide, hydrocarbons, and smoke emissions. There was remarkable high efficiency shown by biodiesel blends as compared to pure *Maduca longifolia* biodiesel. Sankaranarayanan and Pugazhvadivu (2012) reported that the brake thermal efficiency, peak pressure, and heat release rate of a diesel engine increased with addition of hydrogen to *Madhuca longifolia* biodiesel. The engine noise, knocking, and nitrogen oxide emission increased with increase in concentration of hydrogen above 40 lpm. They concluded that the mixing of hydrogen up to 40 lpm to *Madhuca longifolia* biodiesel improves the thermal efficiency of diesel engines and reduces the emission of pollutants.

5.2.15 Azadirachta indica

Vijayaraj and Sarangan (2012) evaluated performances of *Jatropha curcus*, *Pongamia pinnata*, and *Azadirachta indica* neat biodiesel in a direct injection diesel engine. However, there was reduced performance with increased emission of nitrogen oxide as compared with petroleum diesel. Therefore, they advanced fuel injection to 3CA, and 3% volume of diethyl ether was added to the biodiesel. These increased performance of the diesel engine, and reduced emission of hydrocarbons, nitrogen oxide, and carbon monoxide, with slight increase in smoke emission as compared to petroleum diesel. The specific fuel consumption and mechanical efficiency was better in *Azadirachta indica* biodiesel as compared to *Jatropha curcus* and *Pongamia pinnata* biodiesel. There was increase in peak pressure, ignition delay, and temperature of peak cylinder with addition of diethyl ether in biodiesels and modification of the diesel engine. Awolu and Layokun (2013) also optimized biodiesel production from *Azadirachta indica* seed oil and evaluated its physico-chemical parameters. They suggested that the biodiesel yield obtained after transesterification process was 85.13% at optimum condition (1:3 oil and methanol ratio, reaction temperature 50°C, NaOH 0.70%) as catalyst and reaction time 60 minutes for transesterification). The physico-chemical parameters (moisture content 0.05%, specific gravity 0.9 at 25°C, kinematic viscosity 5.5 mm^2/s, acid value 207 mg KOH/g, iodine value 70.7 g I_2/100 g, cetane number 55.31, calorific value 39.85 MJ/Kg, pour point 4°C, cloud point 8°C, and flash point 110°C) of biodiesel were in range as per ASTM specifications. Mathiyazhagan and colleagues (2013) evaluated the combustion, emission and performance parameters of *Azadirachta indica* biodiesel blends (10% to 50%) in a single-cylinder, four-stroke diesel engine. They suggested that consumption of brake specific fuel in blends of biodiesel was low with use of *Azadirachta indica* biodiesel blends due to high oxygen and cetane number, which helps in combustion of biodiesel. The brake thermal efficiency of biodiesel was equivalent to petroleum diesel up to 50% blends, but further increase of biodiesel level results in reduction of brake thermal efficiency of diesel engines. Ramakrishnan and colleagues (2018) investigated the combustion, emission, and performance parameters of *Azadirachta indica* methyl esters and pumpkin methyl ester in the Kirloskar TV1 model (DI) four-stroke diesel engine. They suggested that the brake thermal efficiency of petroleum diesel was higher than the biodiesel samples. The consumption of brake specific fuel was lower for petroleum diesel than *Azadirachta indica* methyl esters and pumpkin methyl esters. The *Azadirachta indica* methyl esters exhibited lower brake specific fuel consumption than pumpkin methyl esters. The emission of carbon monoxide, hydrocarbons, and smoke were lower with *Azadirachta indica* methyl esters as compared to pumpkin methyl esters and petroleum diesel. However, nitrogen oxide emission was high in both biodiesel samples as compared to petroleum diesel. Datla and colleagues (2019) evaluated the combustion, emission, and performance parameters of *Azadirachta indica* biodiesel and its blends (20%, 40%, 60%, and 80%) in a variable compression ratio diesel engine. They suggested that the brake thermal efficiency of biodiesel blends were very similar to that of petroleum diesel at partial load conditions. However, the consumption of brake-specific fuel was higher than the petroleum diesel. At maximum load, consumption of brake-specific fuel for all blends were almost equivalent to that of petroleum diesel. The emission of exhaust gases such as carbon monoxide and hydrocarbons were low, but nitrogen oxide emission was higher than the petroleum diesel.

5.2.16 *Hibiscus cannabinus*

Jindal and Goyal (2012) evaluated combustion, emission and performance of *Hibiscus cannabinus* biodiesel and its blends (10%–90%) in diesel engine. They reported that the 20% blend of *Hibiscus cannabinus* biodiesel delivers high brake thermal efficiency as compared to petroleum diesel. The combustion of blended biodiesel is better than pure biodiesel with smooth engine operation. The emission of exhaust gases such as hydrocarbons and smoke were reduced by using lower proportion of biodiesel in the blends.

5.2.17 *Luffa cylindrica* and *Blighia unijugata*

Adewuyi and colleagues (2012b) reported that the seed of *Luffa cylindrica* has 40% oil content and *Blighia unijugata* has 50% oil content, which can be used for biodiesel production. The biodiesel produced from *Luffa cylindrica* seed oil has physico-chemical parameters such as density 890 kg/m^3, kinematic viscosity 4.10 mm^2/s at 40°C, iodine value 108.20 g I_2/100 g, pour point 4°C, acid value 0.06 mg KOH/g, flash point 148°C, and calorific value 39 MJ/Kg which were almost in range as per EN 14214 standard specifications. The biodiesel produced from *Blighia unijugata* seed oil also had physico-chemical parameters such as acid value 0.04 mg KOH/g, iodine value 47.60 g I$_2$/100 g, density 880 Kg/m³, kinematic viscosity 4.52 mm²/s at 40°C, calorific value 38.21 MJ/Kg, flash point 185°C and pour point 12°C which were almost in range as per EN 14214 standard specifications. However, the biodiesel produced from *Luffa cylindrica* seed oil has very low oxidative stability (0.58 h), whereas the biodiesel produced from *Blighia unijugata* seed oil has high oxidative stability (44.3 h). Both the biodiesels had better transportation safety properties. Bamgboye and Oniya (2012) evaluated the combustion, emission, and performance characteristics of *Luffa cylindrica* methyl esters and its blends (5%, 10%, 15%, 20%) in a constant speed, 2.46 kW stationary diesel engine. They suggested that the engine torque, speed, exhaust gas temperature, the consumption of brake specific fuel, the thermal efficiency of brakes, and fuel equivalent power of a diesel engine fueled with biodiesel blends was almost equivalent to petroleum diesel. There was no significant difference in mean values of exhaust temperature, speed, torque, or fuel consumption rate at all load conditions.

5.2.18 *Simarouba glauca*

Garlapati and colleagues (2013) optimized the lipase mediated transesterification process for biodiesel production from *Simarouba glauca* seed oil. From investigation, they observed that after 36 hours of reaction time at 34°C, the maximum molar conversion was 91.5% with respect to 62.23% of methyl oleate (1:1 oil:methanol). The transesterification was successfully done in presence of 10 U immobilized lipase enzyme under the solvent n-hexane. They suggested that the harmful effect of end products on immobilized lipase enzyme was overcome by using an n-hexane solvent system. The activity of lipase enzyme was retained relatively up to 95% when it was reused for six cycles. Jeyalakshmi (2019) evaluated the physico-chemical parameters and performance of *Simarouba glauca* biodiesel in a single-cylinder (DI) diesel engine. It was observed that the viscosity, density, calorific value, acid value, carbon residue, sulfur content, ash content, and copper corrosion of *Simarouba glauca* biodiesel was almost equivalent to petroleum diesel and ASTM standard biodiesel parameters. However, flash point, cloud point, fire point, pour point, and cetane number were higher as compared to petroleum diesel. Since cetane number was high, it reduced the delay in period, which means diesel engine starts rapidly and runs smoothly (Issariyakul et al., 2007). From the combustion and emission analysis of biodiesel and its blends, it was observed that the consumption of brake specific fuel at all loads was high in biodiesel and its blends as compared to petroleum diesel. Even the heat release rate of petroleum diesel was higher than biodiesel at full load. The emission of smoke and hydrocarbons was low at high load conditions; however, emission of nitrogen oxide was slightly higher in pure biodiesel and its blends as compared to petroleum diesel.

5.2.19 *Ceiba pentandra*

Sivakumar and colleagues (2013) optimized biodiesel production from *Ceiba pentandra* and observed physico-chemical parameters of biodiesel. After analysis, they suggested that except flash point (169°C), all physico-chemical parameters (specific gravity 0.876, cloud point 1°C, kinematic viscosity 4.17 mm²/s at 40°C, acid number 0.036 mg KOH/g, carbon residues 0.042%, water and sediments 0.031%, copper strip corrosion 1A, sulfated ash 0.01%, phosphorus content 0.0008%, sodium and potassium content 4.2 ppm, calcium and magnesium content 2 ppm, cetane number 47) of biodiesel were in range as per ASTM standards specifications. Norazahar and colleagues (2012) suggested that the biodiesel produced from *Ceiba pentandra* has higher oxidative stability than the standard values. Vedharaj and colleagues (2013) suggested that the blending of *Ceiba pentandra* biodiesel with petroleum diesel is a possible source of transportation fuel. From analysis of the combustion, emission, and performance parameters of the diesel engine, they concluded that blending of 25% biodiesel with petroleum diesel shows better performance than conventional diesel. Asokan and colleagues (2016) investigated the combustion and emission parameters of *Ceiba pentandra* biodiesel blends (20%, 30%, 40%, and 100%) in a single cylinder direct injection diesel engine. From their analysis, they suggested that 30% blend matched closer to diesel than that of other biodiesel blends. They suggested that the consumption of brake specific fuel and exhaust gas temperature of blended biodiesel were higher than conventional diesel. The brake thermal efficiency, rate of heat release, and combustion parameters of cylinder pressure of all blended biodiesels were similar to petroleum diesel. The pure biodiesel of *Ceiba pentandra* decreased the emission of carbon monoxide up to 23.33%, hydrocarbons up to 40%, and smoke up to 45.53% as compared to petroleum diesel. However, there was remarkable high emission of nitrogen oxide up to 3.5%. They suggested that the use of *Ceiba pentandra* biodiesel blends is a better alternative for diesel engines.

5.1.20 *Calophyllum inophyllum*

Sahoo and colleagues (2007) investigated the combustion and emission parameters of pure and blended (20%, 40%, 60%, and 80%) biodiesel produced from *Calophyllum inophyllum* in a high-speed single-cylinder diesel engine. They observed that the performance of diesel engine was better in brake thermal efficiency, consumption of brake specific fuel, opacity of smoke, and emission of exhaust gases such as hydrocarbons, nitrogen oxide, carbon monoxide, carbon dioxide, etc., as compared to petroleum diesel. There was 35% decrease in smoke emission with use of 60% blended biodiesel of *Calophyllum inophyllum* as compared to petroleum diesel. Venkanna and Reddy (2009) evaluated the performance of preheated *Calophyllum inophyllum* biodiesel in a direct injection diesel engine. They reported that the preheated *Callophyllum inophyllum* biodiesel reduces emission of carbon monoxide and hydrocarbons as compared to diesel. However, it increases exhaust gas temperature, brake thermal efficiency, and emission of nitrogen oxide. Atabani and colleagues (2013) also reported that *Calophyllum inophyllum* biodiesel blends (10% and 20%) reduce brake power up to 5%–6%. They reduce torque up to 5%–7% and increase consumption of brake-specific fuel by 6%–12%. There was 16%–34% reduction in carbon monoxide emission, 5%–19% reduction in hydrocarbons emission, and 0.5%–1% reduction in nitrogen oxide as compared to diesel. Atabani and César (2014) reported that *Calophyllum inophyllum* had higher peak pressure, maximum heat release rate, and shorter ignition delay as compared to conventional diesel, *Jatropha curcus* biodiesel, and *Pongamia pinnata* biodiesel. They suggested that a 20% blend of *Calophyllum inophyllum* decreases brake power, and increases consumption of brake-specific fuel and energy consumption, as compared to diesel. Fattah and colleagues (2014) reported that biodiesel blends of *Calophyllum inophyllum* (10% and 20%) reduces brake power 0.3%–0.7%, increases consumption of brake specific fuel 2%–3%, increases nitrogen oxide emission 2%–8%, and decreases exhaust emission of hydrocarbons 9%–17%, carbon monoxide 15%–26%, and smoke as compared to diesel in a four-cylinder indirect diesel engine (2.5 L, 55 kW). Miraculas and colleagues (2016) reported that the optimum performance with low emission characteristics of a single-cylinder, four-stroke, water-cooled diesel engine was observed in 30% blend of *Calophyllum inophyllum* biodiesel at a compression ratio of 1:9. The high compression ratio increases cylinder temperature, which enhances vaporization of fuel; hence, better performance of diesel engines can remain up to a certain extent. At higher biodiesel blends and compression ratio, the temperature of the cylinder is high during operation of the engine, which in turn increases nitrogen oxide emissions. The better combustion process during these conditions reduces the emission of carbon monoxide and unburned hydrocarbons. Ayyasamy and colleagues (2018) suggested that the use of *Calophyllum inophyllum* biodiesel blended with biogas in diesel engines shows better performance with respect to brake thermal efficiency, mechanical efficiency, and specific fuel efficiency. It reduces emission of hydrocarbons, but increases nitrogen oxide and carbon monoxide emission with increase in load. Hence, further studies should be conducted to improve *Calophyllum inophyllum* biodiesel with respect to improving engine performance, combustion, and exhaust emissions.

5.2.21 *Raphanus sativus*

Shah and colleagues (2013) optimized biodiesel production from *Raphanus sativus* seed oil and suggested that the physico-chemical parameters such as kinematic viscosity (5 mm^2/s), cloud point (10°C), pour point (−14°C), cold filter plugging point (−2°C), lubricity (201 μm), oxidative stability (1.6 h), acid value (0.08 mg KOH/g), and calorific value (38.75 MJ/kg) of *Raphanus sativus* biodiesel were similar to that of soyabean methyl esters as per ASTM and EN standards, except pour point. To improve this property, *Raphanus sativus* biodiesel blends were prepared by mixing ultra low sulfur diesel with 5% and 20% of biodiesel.

They observed that 5% and 20% blends of *Raphanus sativus* biodiesel in ultra low sulfur diesel had acceptable fuel properties as per ASTM D975 (for 5% blend) and ASTM D7467 (for 20% blend). Paturu and Vinothkanna (2018) also evaluated the combustion, performance, and emission parameters of *Raphanus sativus* biodiesel and its blends in a single-cylinder, direct-injection diesel engine which was modified by coating its cylinder head, valves, and piston crown with 100 microns of nickel chrome aluminum bond and 450 microns of partially stabilized zirconia (PSZ). They suggested that the petroleum diesel has good combustion characteristics and biodiesel has good emission characteristics. Among the biodiesel blends, 25% biodiesel blended with petroleum diesel had low brake thermal efficiency, consumption of brake specific fuel, and emission of carbon dioxide, hydrocarbons, smoke, and nitrogen oxide. They concluded that 25% blend of biodiesel has better performance, combustion, and emission characteristics in the PSZ-coated diesel engine.

5.2.22 Cleome viscosa

Kumari and colleagues (2013) evaluated physico-chemical characteristics of *Cleome viscosa* biodiesel and suggested that its properties were almost equivalent to that of *Jatropha curcus* biodiesel. They observed that all the physico-chemical parameters (kinematic viscosity at 40°C was 4.5 mm^2/s, density at 15°C was 0.89 g/cm^3, flash point 175°C, cloud point 21.2°C, oxidative stability 1 h, carbon residues 0.04%, acid value 0.1 mg KOH/g, cetane number 55.2, free glycerine 0.0045%, total glycerine 0.0240%, and calorific value 39.7 MJ/kg) of *Cleome viscosa* biodiesel were in range as per Indian and ASTM standards. Perumal and Ilangkumaran (2018) evaluated the performance, combustion, and emission characteristics of *Cleome viscosa* biodiesel and its blends (20%, 40%, 60%, and 80%) in a conventional direct injection compression ignition diesel engine. They suggested that with use of biodiesel blends, there was increase in consumption of brake-specific fuel up to 5.4% and reduction in brake thermal efficiency up to 4.7%. There was also a reduction in emission of carbon monoxide and hydrocarbons up to 30% and 20%, respectively.

5.2.23 Manilkara zapota

Kumar and Sureshkumar (2016) optimized biodiesel production from *Manilkara zapota* seed oil and tested its physico-chemical parameters. They suggested that the *Manilkara zapota* seed oil has high content of monounsaturated fatty acids and low content of saturated and polyunsaturated fatty acids. The physico-chemical parameters (density 875 kg/cm^3, kinematic viscosity 4.67 mm^2/s, iodine number 65.28 g I$_2$/100 g, acid value 3.79 mg KOH/g, pour point −6°C, flash point 174°C, calorific value 37.2 MJ/kg, cetane number 52) of *Manilkara zapota* biodiesel was in range as per EN 14214 standards, except calorific value and cetane number, which was slightly higher than standard values. Karmee (2018) also optimized biodiesel production from *Manilkara zapota* using four commercially available lipases obtained from *Porcine pancreas*, *Candida rugosa*, *Pseudomonas cepacia*, and *Candida antarctica-B*. Results suggested that immobilized forms of *Candida antarctica-B* (Novozyme-435 and CLEA) were potential biocatalysts for the transesterification of *Manilkara zapota* seed oil with 93% and 84% biodiesel yields, respectively. The Novozyme-435 can be reused up to six cycles with 21% loss of activity. The deactivated lipase can be regenerated after washing it with tert-butanol. Kumar and colleagues (2018) evaluated the performance, combustion, and exhaust emission characteristics of *Manilkara zapota* biodiesel and its blends (25%, 50%, 75%) in a compression ignition, direct injection single-cylinder diesel engine. They suggested that among the various blends, 50% blend of biodiesel had 17% high brake thermal efficiency, 14.34% low consumption of brake-specific fuel, 34.21% low carbon monoxide emission, 4.32% low unburned hydrocarbons emission, and 38.91% high carbon dioxide emission as compared to petroleum diesel. Similarly, Dewangan and colleagues (2019) evaluated the performance, emission, and combustion characteristics of 20% blend of *Manilkara zapota* biodiesel in a multi-cylinder diesel engine. They suggested that there was high consumption of brake-specific fuel and low brake power as compared to petroleum diesel. At full load conditions, the emission of carbon monoxide and hydrocarbons were lower, but nitrogen oxide emission was higher, than the petroleum diesel.

5.2.24 Aegle marmelos

Yatish and colleagues (2016) optimized biodiesel production from *Garcinia gummi-gutta*, *Terminalia bellerica*, and *Aegle marmelos*, and evaluated their physico-chemical parameters. They suggested that the yield obtained after transesterification process was 89.4% from *Garcinia gummi-gutta*, 91.6% from *Terminalia bellerica*, and 93% from *Aegle marmelos*. The physico-chemical parameters of all the three samples were almost in range as per ASTM standard specifications. They recommended that the blending of *Aegle marmelous* biodiesel up to 30% with petroleum diesel does not affect its fuel properties. Similarly, Selvan and colleagues (2018) optimized biodiesel production from *Aegle marmelos* seed oil and evaluated its physico-chemical parameters. They suggested that a yield of 98% biodiesel can be obtained after 60 minutes in presence of 1% KOH and 10:1 ratio of methanol to oil. Most of the physico-chemical parameters (kinematic viscosity 5.67 mm²/s, specific gravity 0.893, cloud point 7°C, pour point 3°C, cetane number 49, water and sediments 0.036%) of *Aegle marmelos* biodiesel were in range as per ASTM D6751 standards, except cloud point, pour point, and cetane number. These suggest that *Aegle marmelos* biodiesel cannot be used in cold climate countries. Rajak and colleagues (2019) tested the performance and emission parameters of *Aegle marmelos* biodiesel and its 20% blend in a compression ignition diesel engine. They reported that when pure *Aegle marmelos* biodiesel (100%) was used, the smoke and particulate matter emissions were low but nitrogen oxide emission was higher than the petroleum diesel. Even the cylinder pressure was high at full load conditions. When 20% blend of biodiesel was used, there was remarkable decrease in thermal efficiency up to 3.72%, peak heat release rate up to 4.2%, exhaust gas temperature up to 3.52%, pressure rise rate up to 6.68%, and ignition delay up to 17%. However, there was high specific fuel consumption up to 3.16%, cylinder peak pressure up to 1.4%, and Sauter mean diameter up to 8.5% as compared to petroleum diesel at full load conditions. There was reduction in particulate matter up to 17.2% and smoke up to 20.8%, but nitrogen oxide emission was high, up to 8.5%. The performance of 20% blend was better than the pure biodiesel at high fuel injection pressure of 220 bars and full load conditions. Paramasivam and colleagues (2018) optimized biodiesel production from *Aegel marmelos* seed cake using pyrolysis method and tested its performance in compression ignition diesel engine. They suggested that the blending of biodiesel with petroleum diesel results in decrease of brake thermal efficiency and increase of consumption of brake specific fuel. Even, increase of biodiesel ratio with petroleum diesel decreases emission of exhaust gases such as carbon monoxide and hydrocarbons, but increases nitrogen oxide emission. They concluded that the biodiesel blend of 20% can be used for better engine operation characteristics. Baranitharan and colleagues (2019) optimized the performance and emission parameters of 20% *Aegle marmelos* biodiesel blend in a variable compression ratio diesel engine by mixing it with 1000 ppm tert-butyl hydroxyl quinone antioxidant. They reported that the mixture of biodiesel blend (20%) and antioxidant gave better performance than petroleum diesel with consumption of brake-specific fuel of 0.33 kg/kWh, brake thermal efficiency of 22.01%, emission of carbon monoxide up to 0.67%, hydrocarbon up to 244 ppm, carbon dioxide up to 8.33%, and nitrogen oxide up to 351 ppm. Similarly, Kolli and colleagues (2020) optimized the performance and emission parameters of 35% *Aegle marmelos* biodiesel blend in a single-cylinder, compression ignition, four-stroke diesel engine by mixing it with graphene nanosheet and oxygenated diethyl ether. They reported after analysis that there was remarkable improvement in peak brake thermal efficiency up to 3%, consumption of brake-specific fuel up to 5.5%, carbon monoxide emission up to 41%, nitrogen oxide emission up to 13.5%, particulate matter emission up to 45%, and hydrocarbons emission up to 39% with the use of optimum mixture blends and additives. The optimum combination required to reduce dangerous exhaust gases is 50% diesel, 35% biodiesel, 100 ppm graphene nanosheets, and 15% diethyl ether.

5.2.25 Thevetia peruviana

Yadav and colleagues (2017a) evaluated performance and emissions of exhaust gases in a single-cylinder, four-stroke, water-cooled diesel engine using *Thevetia peruviana* biodiesel blends (10%, 20%, and 30%). They reported that the 20% blend of *Thevetia peruviana* biodiesel blend performs satisfactorily with diesel engines. It increases the brake thermal efficiency of diesel engine up to 2% as compared to petroleum diesel. It decreases emission of carbon monoxide and unburned hydrocarbons emission up to 41.4% and 32.3%, respectively, as compared to petroleum diesel. However, emission of nitrogen oxide was high in all biodiesel blends up to 5%–6%.

5.2.26 Aleurites moluccanus

Imdadul and colleagues (2017) evaluated the performance of *Aleurites moluccanus* biodiesel blends (10%, 20%, and 30%) in a Yanmar TF120M single cylinder, natural aspirated, water cooled direct injection diesel engine arranged with an SAJ SE-20 eddy current dynamometer. They observed that consumption of brake specific fuel was high in all biodiesel blends as compared to conventional diesel. The brake power of petroleum diesel was higher than the biodiesel blends, and there was no change in brake power with increase or decrease of biodiesel proportion in blends. The brake thermal efficiency of biodiesel blends was lower than conventional diesel, and it decreased gradually with increase of biodiesel proportion in blends. The emission of nitrogen oxide was high in all biodiesel blends as compared to petroleum diesel due to high oxygen content in biodiesel. The emission of carbon monoxide and hydrocarbons emission was low in all biodiesel blends as compared to conventional diesel due to complete combustion of biodiesel blends at lower in cylinder temperature. They suggested that the increase in the proportion of biodiesel in blends decreases the working efficiency of diesel engine.

5.2.27 Syzygium cumini

Ramalingam and colleagues (2018) evaluated performance, emission, and combustion characteristics of *Syzygium cumini* biodiesel and its blends in a direct injection diesel engine at different loads. They reported that the consumption of brake-specific fuel, exhaust gas temperature and nitrogen oxide emission of *Syzygium cumini* biodiesel and its blends was high as compared to petroleum diesel. The brake thermal efficiency, peak pressure, and emission of carbon monoxide, hydrocarbons, and smoke from *Syzygium cumini* biodiesel and its blends were low as compared to petroleum diesel. The heat release rate was low initially due to a delay in ignition, and then it increases with initiation of combustion in the chamber. The combustion duration was low for *Syzygium cumini* biodiesel and its blends as compared to petroleum diesel.

5.2.28 Hevea brasiliensis

Adam and colleagues (2018) evaluated performance, combustion, and emission of exhaust gases in a multi-cylinder, indirect injection diesel engine (IDI model XLD 418D) using biodiesel prepared from *Hevea brasiliensis* seed oil, palm seed oil, and their combined blends (mixtures of *Hevea brasiliensis* and palm seed oil biodiesel). They observed that the consumption of brake specific fuel, carbon monoxide, hydrocarbons, and smoke opacity were low and brake power was high in combined blends as compared to *Hevea brasiliensis* seed biodiesel. However, consumption of brake specific fuel, carbon monoxide, hydrocarbons, and smoke opacity was high and brake power was lower in combined blends as compared to palm seed biodiesel. The increase of biodiesel ratio decreases the volumetric heating value of blends, which results in a decrease of brake power. They suggested that the carbon double bond in *Hevea brasiliensis* seed biodiesel increases formation of free radicals and hydrocarbons that results in higher nitrogen oxide emission as compared to petroleum diesel. The biodiesel should have low density, viscosity, and unsaturated components to reduce nitrogen oxide emissions. The high peak pressure of biodiesels was due to high cetane number and short ignition delay, which increases fuel evaporation and combustion rate during the initial combustion phase.

5.2.29 Guizotia abyssinica

Jaikumar and colleagues (2019) evaluated the performance, combustion, and emission characteristics of *Guizotia abyssinica* biodiesel and its blends (10%, 20%, and 40%) in a four-stroke, multifuel variable compression research engine. They observed that among these blends, the brake thermal efficiency of 20% biodiesel blend was nearer to petroleum diesel. The consumption of brake specific fuel was high in all biodiesel blends as compared to standard diesel. The net heat release rate and cylinder pressure increased significantly with decreases in the biodiesel ratio. The 20% blend had higher cylinder pressure and net heat release rate as compared to petroleum diesel. There was notable decrease in emissions of carbon monoxide, hydrocarbons, and smoke with increase in proportion of biodiesel in blends. However, there was increase in nitrogen oxide emissions with increase in proportion of biodiesel in blends.

5.2.30 *Annona reticulata*

Yadav and colleagues (2019) reported that the use of *Annona reticulata* biodiesel blends (10%–100%) in diesel engines decreases the brake thermal efficiency and increases consumption of brake specific fuel with increase in proportion of biodiesel in blends. At higher loads, emission of smoke is reduced in all biodiesel blends as compared to petroleum diesel.

5.2.31 *Citrullus colocynthis*

Sivalingam and colleagues (2019) evaluated the performance of conventional type transesterified biodiesel blend [blend-T = 20% (vol) biodiesel, 80% (vol) mineral diesel] and enzymatic lipase immobilized transesterified biodiesel blend [blend-L = 20% (vol) biodiesel, 80% (vol) mineral diesel] in a Kirloskar Tv1 model single-cylinder, four-stroke, water-cooled diesel engine. They determined that the biodiesel blends had good combustion, performance, and reduced emission of exhaust gases. The brake thermal efficiency of blend-L (29.86%) and blend-T (28.93%) at full load was lower than petroleum diesel (31.33%). The heat release rate and peak pressure of combusted fuel was lower than petroleum diesel in blend-L. The emission of hydrocarbons and carbon monoxide was low in blend-L as compared to petroleum diesel. However, the brake-specific energy consumption at full load was high in blend-L (7.85%), blend-T (11.76%), and pure biodiesel (17.23%) as compared to petroleum diesel. The emission of nitrogen oxide and particulate matter was high in blend-L and blend-T as compared to petroleum diesel.

5.2.32 *Pistacia khinjuk*

Karthickeyan and colleagues (2019) evaluated performance of *Pistacia khinjuk* biodiesel (100%) and its blends (20%, 40%, and 50%) in a tangentially vertical (TV1) model single-cylinder direct injection diesel engine. They suggested that at full load, pure biodiesel of *Pistacia khinjuk* had lower brake thermal efficiency due to its low heating value which results in poor atomization of fuel and early start of combustion. The 20% blend of biodiesel had high brake thermal efficiency, as it exhibited calorific value almost equivalent as that of petroleum diesel. The consumption of brake specific fuel was high at full load when biodiesel and its blends were used as compared to petroleum diesel to attain the rated power of the engine. The carbon monoxide, hydrocarbon, and smoke emissions remained low at all loads in pure biodiesel and its blends as compared to petroleum diesel, due to presence of natural oxygen and high temperature in the combustion chamber. The nitrogen oxide emission was high at all loads in 100% biodiesel and its blends as compared to diesel fuel. There was remarkable decrease in cylinder pressure with use of biodiesel and its blends as compared to conventional diesel, due to poor mixing of air and biodiesel in the cylinder. The heat release rate was low in 100% biodiesel and its blends as compared to diesel fuel. To enhance biodiesel properties of 20% blend, they added antioxidants (1% volume of geraniol and pyrogallol) to it. They observed that, at full load, the brake thermal efficiency increased, consumption of brake-specific fuel decreased, carbon monoxide and hydrocarbons emissions were high, and nitrogen oxide and smoke emission were low with addition of antioxidants. There was increase in cylinder pressure with addition of antioxidants due to enhanced fuel properties. The heat release rate increased with addition of antioxidants in 20% biodiesel blend. The combination of 20% biodiesel blend and pyrogallol was almost equivalent to petroleum diesel and more efficient than 20% biodiesel blend and geraniol.

5.3 Biodiesels That Remain to Be Evaluated in Diesel Engines

5.3.1 *Terminalia catappa*

Dos Santos and colleagues (2008) reported that the seed kernel oil of *Terminalia catappa* has 49% oil content, which can be a good source for biodiesel production. It has lower content of unsaturated fatty acids, which are almost equivalent to that of EN 14214 standard specifications. The biodiesel produced after transesterification has physico-chemical parameters, such as kinematic viscosity (4.3 mm²/s), density (873 kg/m³), calorific value (36.97 MJ/kg), iodine value (83.2 g I_2/100 g), were almost in range as per ASTM D6751 and EN 14214 standards. The calorific value is lower than that of petroleum diesel (46.65 MJ/kg) (DeOliveira et al., 2006) and similar to that of soyabean, palm oil, and other biodiesels (Fröhlich and Rice, 2005; Dorado et al., 2004; Lang et al., 2001).

5.3.2 *Limnanthes alba*

Moser and colleagues (2010) optimized biodiesel production from *Limnanthes alba* seed oil using sodium methoxide as a catalyst in the transesterification process. From physico-chemical analysis of biodiesel and its blends, they suggested that the *Limnanthes alba* biodiesel had high kinematic viscosity (6.18 mm²/s), oxidative stability (41.5 h), and cetane number (66.9). The low temperature operation points, such as cold filter plugging (−9°C), cloud point (−6°C), and pour point (−10°C), also required to be improved. Other parameters (acid value, free and total glycerin content, sulfur content, phosphorus content) were low as compared to ASTM D6751 and EN 14214. The wear scar of *Limnanthes alba* biodiesel at 60°C was 119 μM, which is lower than the petrodiesel standards described by ASTM D6079, which indicates good lubricity. When 40% *Limnanthes alba* biodiesel was blended with soyabean biodiesel, its kinematic viscosity and induction period was improved. Blending of *Limnanthes alba* biodiesel with petroleum diesel meets the specifications described by ASTM D975 and D7467 standards. Blending of 20% *Limnanthes alba* biodiesel with petroleum diesel improves the lubricity of petroleum diesel without affecting its oxidative stability.

5.3.3 *Vernicia montana*

Chen and colleagues (2010) investigated the physico-chemical properties of *Vernicia montana* biodiesel and its blends. They suggested that *Vernicia montana* biodiesel has poor oxidative stability due to the instability of the conjugated carbon-carbon double bonds in the α-elaeostearic acid. The *Vernicia montana* biodiesel has a low cold filter pugging point (−11°C), low oxidative stability of 0.3 h at 110°C, low ester content of 94.9%, high density of 903 kg/m³ at 15°C, high kinematic viscosity of 7.84 mm²/s, and high iodine value of 161.1 gI2/100 g. They suggested that *Vernicia montana* biodiesel can be improved by blending it with canola and palm biodiesel to meet the ASTM, EN, and GIS standard specifications. Anantharaman and colleagues (2016) studied the effect of raw material composition of *Vernicia montana* and *Jatropha curcus* on properties of produced biodiesel. From analysis, they concluded that the properties of biodiesel are significantly affected by the fatty acid composition of the seed oil. The allylic position equivalence and bisallylic position equivalence are better parameters than iodine value to determine oxidative stability of biodiesel, as they make a distinction between structural configurations. The fatty acid profile of *Vernicia montana* and *Jatropha curcus* seed oil revealed that they are rich sources of unsaturated fatty acids. The results of physico-chemical parameters suggested that the biodiesel of *Vernicia montana* did not satisfy the standard specifications of the kinematic viscosity, oxidative stability, carbon residue deposition, relative density, and iodine value as compared to *Jatropha curcus* biodiesel, which satisfies all parameters as per ASTM D6751, EN 14214, and IS 15607 standard specifications.

5.3.4 *Annona cherimola*

Branco and colleagues (2010) optimized biodiesel production from *Annona cherimola* seed oil and evaluated its physico-chemical parameters. They suggested that the physico-chemical parameters (such as acid value 0.3 mg KOH/g, iodine number 99 g I_2/100 g, density 871 kg/m³, oxidative stability 1.17 h, kinematic viscosity 4.4 mm²/s, cold filter plugging point –5°C, cloud point 1°C, cetane number 53) of *Annona cherimola* methyl esters were in range as per EN 14214 standards. However, the high content of unsaturated fatty acids in *Annona cherimola* seed oil makes it susceptible to oxidation and reduces its oxidative stability. Hence, they suggested improving the oxidative stability by using some additives.

5.3.5 *Maclura pomifera*

Saloua and colleagues (2010) optimized biodiesel production from *Maclura pomifera* seed oil. The transesterification process was carried out in presence of sodium hydroxide as catalyst and biodiesel yield obtained after transesterification was 90%. The physico-chemical parameters (such as flash point 180°C, pour point –9°C, cloud point –5°C, kinematic viscosity 4.66 mm²/s, iodine number 125 g I_2/100 g, density 889 kg/m³, and cetane number 48) were almost in range as described by ASTM and EN biodiesel standards for petroleum diesel.

5.3.6 *Euphorbia lathyris* and *Sapium sebiferum*

Wang and colleagues (2011) optimized biodiesel production from *Euphorbia lathyris*, *Sapium sebiferum*, and *Jatropha curcus* seed oil. They suggested that the yield obtained after the transesterification process was 98.03% from *Sapium sebiferum*, 97.61% from *Euphorbia lathyris*, and 98.27% from *Jatropha curcus*. The physico-chemical parameters of all three biodiesels were in range as per ASTM and EN standard specifications. The quality of biodiesel produced from *Euphorbia lathyris* was better than *Sapium sebiferum* due to presence of high monounsaturation (82.66%), low polyunsaturation (6.49%), and appropriate proportion of saturated components (8.76%)v which improves the cetane number (59.6), oxidative stability (10.4 h)v and cold filter plugging point (–11°C) of the biodiesel. The cetane number and oxidative stability of *Sapium sebiferum* biodiesel were low due to the presence of high polyunsaturation content (72.79%).

5.3.7 *Licania rigida*

Macedo and colleagues (2011) optimized biodiesel production from *Licania rigida* seed oil using 6:1 ethanol to oil ratio in presence of 0.5% potassium hydroxide as catalyst. The transesterification process was carried out for 60 minutes at 32°C. From physico-chemical analysis, they observed that the acid value of *Licania rigida* biodiesel was 1.8 mg KOH/g. The chemical composition of biodiesel was 69.76% saturated and 28.15% unsaturated fatty acids, in which oleate, linoleate, and stereate of ethyl were predominant. The kinematic viscosity of *Licania rigida* biodiesel was 38.32 mm²/s. The *Licania rigida* biodiesel is oxidatively stable at 224°C and 179°C. It has thermal stability at 184°C and 94°C.

5.3.8 *Treculia africana*

Adewuyi and colleagues (2012a) reported that *Treculia africana* seed oil has high free fatty acids content, and should therefore be pretreated with 2% sulfuric acid as catalyst in methanol. The biodiesel yield obtained after transesterification process was 98% and physico-chemical parameters of biodiesel (kinematic viscosity at 40°C 4.6 mm²/s, density 880 kg/m³, iodine value 117.60 g I_2/100 g, acid value 0.30 mg KOH/g, and flash point 131°C) were in range as per EN 14214 standards.

5.3.9 *Dipteryx alata*

Batista and colleagues (2012) optimized biodiesel production from *Dipteryx alata* seeds using two solvents (methanol and ethanol) and suggested that seeds of *Dipteryx alata* contain 38.2% of fat and 23.9% of protein. From transesterification, they suggested that methanol was more efficient than ethanol for production of biodiesel from *Dipteryx alata* seeds, with biodiesel yield of 91% from methanol and 86% from ethanol. From physico-chemical analysis, they suggested that the acid number, relative density, iodine value, kinematic viscosity, water content, peroxide number, ester number, saponification value, and refractive index of methyl esters and ethyl esters were in range of biodiesel quality as per the ASTM D6751 and EN 14214 standards.

5.3.10 Xanthoceras sorbifolia

Li and colleagues (2012) optimized biodiesel production from *Xanthoceras sorbifolia* seed oil using ion exchange resin as heterogenous catalyst. They suggested that the yield of 96% can be obtained in 90 minutes at 60°C when microwave-assisted transesterification process is carried out in presence of high alkaline anion exchange resins. The physico-chemical parameters (kinematic viscosity at 40°C 4.4 mm^2/s, acid value 0.06 mg KOH/g, sulfated ash content 0.003%, cetane number 56.1, flash point 165°C, and carbon residues deposition 0.02%) of *Xanthoceras sorbifolia* biodiesel were in range as per ASTM D6751 standards. They observed that the alkaline anion exchange resin is reusable as catalyst for up to ten runs, but after six runs, the yield of biodiesel decreases to 70% due to fragmentation and loss of catalyst.

5.3.11 Lepidium sativum

Nehdi and colleagues (2012) optimized biodiesel production from *Lepidium sativum* seed oil. They suggested that *Lepidium sativum* seeds have 26.77% of oil content which is composed of 42.23% polyunsaturated and 39.62% monounsaturated fatty acids. The biodiesel yield obtained after transesterification process using base catalyst was 96.8%. The physico-chemical characteristics such as density (845 kg/m^3 at 15°C), pour point (−6°C), cloud point (−1°C), sulfur content (0.018%), kinematic viscosity (1.921 mm^2/s at 40°C), carbon residue deposition (0.04%), ash content (0.001), gross heat value (40.45 MJ/kg), saponification value (172 mg KOH/g), iodine value (128 g I$_2$/100 g), flash point (176°C), and cetane number (49.23) were in range as per ASTM D6751. They suggested that *Lepidium sativum* biodiesel can be a better source for biodiesel production.

5.3.12 Pentaclethra macrophylla

Oshieke and Jauro (2012) optimized biodiesel production from *Pentaclethra macrophylla* seed oil and observed various physico-chemical parameters of biodiesel. They synthesized methyl and ethyl esters from *Pentaclethra macrophylla* seed oil using transesterification process. The physico-chemical parameters (kinematic viscosity 4.08 mm^2/s and 4.16 mm^2/s, flash point 196°C and 198°C, density 0.877 g/cm^3 and 0.874 g/cm^3, cloud point −5°C and 2°C, pour point 0°C and 1°C, acid value 0.50 mg KOH/g and 0.20 mg KOH/g) of methyl esters and ethyl esters, respectively, of *Pentaclethra macrophylla* seed oil were in range as per the ASTM standards and were similar to that of the soyabean oil methyl esters.

5.3.13 Xanthium sibiricum

Chang and colleagues (2013) optimized biodiesel production from seed oil of *Xanthium sibiricum*. They observed that the oil content in seeds of *Xanthium sibiricum* was 42.34% and its acid value (1.38 mg KOH/g) was low. The fatty acids methyl esters and biodiesel yield from *Xanthium sibiricum* seed oil was 98.7% and 92%, respectively. The physico-chemical parameters (water and sediments 0.049%, sulfated ash 0.004%, kinematic viscosity 3.723 mm^2/s at 40°C, flash point 174°C, density at 20°C was 874.7, sulfur content 0.0003%, cetane number 48.8, acid number 0.07 mg KOH/g, free glycerine 0.0008%, total glycerine 0.135%, oxidative stability 1.2 h, and cold filter plugging point −3°C) of biodiesel were in range as per ASTM D6751 and EN 14214 standard specifications, except cetane number and oxidative stability.

5.3.14 Amygdalus pedunculata

Chu and colleagues (2013) reported that biodiesel can be produced from the oil of *Amygdalus pedunculata* seeds. They suggested that all the physico-chemical parameters (density 877 kg/m^3, kinematic viscosity 4.695 mm^2/s at 40°C, sulfur content 2 mg/kg, water content 475 mg/kg, copper strip corrosion

1A and acid number 0.32 mg KOH/g) of *Amygdalus pedunculata* biodiesel were in range as per ASTM D6751, EN 14214, and GB/T20828 standards, except cetane number (49.2) and oxidative stability (2.2 h), which can be improved up to biodiesel standards by adding 500 ppm of tert-butylhydroquinone. The cold filtration (−11°C) and transportation safety properties (flash point 169°C) of biodiesel were of better quality.

5.3.15 Forsythia suspense

Jiao and colleagues (2013) optimized biodiesel production from *Forsythia suspense* seed oil and observed that *Forsythia suspense* seeds contain 30.08 ± 2.35% of oil with low acid value (1.07 mg KOH/g). The oil contained 72.89% linoleic acid, 18.68% oleic acid, and 5.85% palmitic acid, which was similar with that of sunflower oil. The optimum biodiesel yield from *Forsythia suspense* seed oil was 90.74 ± 2.02%. The physico-chemical properties of *Forsythia suspense* biodiesel was in range as per ASTM D6751, except cetane number.

5.3.16 Reutealis trisperma

Holilah and colleagues (2015) optimized biodiesel production from *Reutealis trisperma* seed oil and suggested that the oil of *Reutealis trisperma* has high free fatty acids, and hence pretreatment of acid esterification is required before alkali transesterification process. The yield obtained after base catalyzed transesterification process was 95.15%. The physico-chemical parameters (such as flash point 148°C, iodine number 28.43 g I_2/100 g, acid number 0.41 mg KOH/g, density 887 kg/m^3, total glycerol 0.037%, pour point −15°C, ester content 97.18%, and cetane number 63) were in range as per ASTM D6751 standards.

5.3.17 Aleurites trisperma

Kumar and colleagues (2015) optimized biodiesel production from *Aleurites trisperma* seeds and observed that the density of biodiesel was 875 kg/m^3 at 15°C. The viscosity of biodiesel was 5.8 mm^2/s at 40°C. The copper strip corrosion test was at number 1A. The cloud point, pour point, and flash point of *Aleurites trisperma* was 3°C, −3°C, and 178°C, respectively. The acid and iodine value of biodiesel was 0.28 mg KOH/g and 97.8 g I_2/100 g, respectively. The fatty acids methyl esters content in biodiesel was 97.2%, which was within the EN and BIS standards. The alkali metals and sulfated ash concentrations in biodiesel were 4.3 ppm and 0.002%. The water and sediments content in biodiesel was 0.02%. The cetane index of biodiesel was 55.38, and calorific value of biodiesel was 35.5 MJ/kg. However, the *Aleurites trisperma* seeds biodiesel had very poor oxidative stability (1.5 h). Based on these results, it was suggested that it can be improved by adding antioxidants. From the analysis, they suggested that the physico-chemical characteristics of *Aleurites trisperma* Blanco seeds biodiesel matches with other biodiesel samples of non-edible seeds (*Pongamia pinnata*, *Jatropha curcus*, waste vegetable oil, *Azadhirachta indica*, and *Maduca longifolia*) (Kumar et al., 2013; Tanwar et al., 2013; Padhi and Singh, 2010).

5.3.18 Silybum marianum

Takase and colleagues (2014) optimized biodiesel production from *Silybum marianum* seed oil using heterogenous solid base catalyst TiO_2 customized by covering it with $C_4H_4O_6HK$ for transesterification under ultrasonication. They observed that the transesterification carried out using 16:1 ratio of methanol to oil in presence of 5% catalyst at 60°C had 90.1% yield of biodiesel. They suggested that the catalyst is reusable for five cycles of biodiesel production. The physico-chemical parameters (cetane number 52, kinematic viscosity 4.4 mm^2/s at 40°C, oxidative stability 2 h, cloud point −2°C, pour point −2°C, flash point 151°C, sulfur content 0.002%, acid value 0.46 KOH mg/kg, density 867 kg/m^3 at 20°C, total glycerin 0.15%) of biodiesel were in range as per ASTM standards, except oxidative stability of the biodiesel. They improve oxidative stability of biodiesel from 2–3.2 h by adding ascorbic acid.

5.3.19 Crotalaria juncea

Sadhukhan and Sarkar (2016) optimized biodiesel production from *Crotalaria juncea* seed oil using both homogeneous and heterogeneous catalysts. They suggested that between the two catalysts (calcium oxide and potassium hydroxide), the potassium hydroxide is an effective catalyst for production of biodiesel from *Crotalaria juncea* seed oil. The biodiesel yield obtained after transesterification process using KOH as catalyst was 90.25%. From the physico-chemical parameters, they suggested that the biodiesel produced using KOH had low viscosity (1.66 mm^2/s) as compared to standard values of viscosity for biodiesels (1.9–6.0 mm^2/s). The saponification value (180) and iodine value (80.5) of biodiesel determined the chemical stability of the biodiesel. The presence of double bond in biodiesel indicates that it can be degraded through oxidation and can polymerize further due to a long chain of fatty acids. The cetane number (58.51) and gross calorific value (37.68 MJ/kg) of biodiesel was equivalent to the standard values of biodiesel. The flash point and fire point was low as compared to standard values.

5.3.20 Gliricidia sepium

Knothe and colleagues (2015) optimized biodiesel production from *Gliricidia sepium* and observed physico-chemical properties of biodiesel. They observed that the cetane number of *Gliricidia sepium* biodiesel was 67.5, which indicates a higher content of saturated fatty acids methyl esters (Knothe, 2014). The kinematic viscosity of biodiesel was 4.38 mm^2/s. The oxidative stability of biodiesel was poor (2.20 h), which was below the specifications of biodiesel standards ASTM D6751 and EN 14214. The cloud point (21°C –22.4°C) and pour point (19°C) of biodiesel were high as compared to palm oil and *Moringa oleifera* biodiesel. The density of biodiesel was 0.8795 g/cm^3 at 15°C. The levels of sodium, potassium, calcium, sulfur, phosphorus, and magnesium were optimum as per ASTM and EN standards. On the basis of these observations, they suggested that *Gliricidia sepium* biodiesel could be a better feedstock for biodiesel production, if oxidative stability and low temperature properties of biodiesel are improved.

5.3.21 Pangium edule

Atabani and colleagues (2015) reported that the oil of *Pangium edule* had high acid value (19.62 mg KOH/g), so two-step acid-base transesterification process was carried out. The biodiesel produced from *Pangium edule* seed oil had physico-chemical properties almost in range as per ASTM D6751, D975, D7467, and EN 14214 standards. The *Pangium edule* biodiesel had poor oxidative stability (0.57 h); however, it can be improved by blending it with petroleum diesel. The most important feature of *Pangium edule* biodiesel was its cloud point (−6°C), pour point (−4°C), and cold filter plugging point (−8°C), which indicate that this biodiesel can be used in cold countries. Authors had recommended conducting further studies on *Pangium edule* biodiesel production for evaluating its diesel engine performance and emission tests.

5.3.22 Senna obtusifolia and Panicum virgatum

Armah-Agyeman and colleagues (2016) characterized triglycerides from *Senna obtusifolia* (coffeeweed) and *Panicum virgatum* (switchgrass) to know its potential as feedstock for biodiesel production. They observed that the switchgrass had a triglyceride yield of 67 g/kg, while coffeeweed had triglyceride yield of 53 g/kg, which is in range with other edible oil seeds (sunflower, corn, and soyabean). Characterization of fatty acids methyl ester suggested that both seed oil contains palmitic, stearic, oleic, and linoleic acids, which are the common methyl esters found in sunflower oil. Hence, they suggested that *Senna obtusifolia* and *Panicum virgatum* could be probable feedstocks for biodiesel production. However, further research is required to determine the feasibility of using *Senna obtusifolia* and *Panicum virgatum* seed oil for biodiesel production.

5.3.23 Cascabela thevetia

Sut and colleagues (2016) optimized biodiesel production from *Cascabela thevetia* seed oil and observed its physico-chemical parameters. From these observations, they suggested that all the physico-chemical parameters (density 882 kg/m³, kinematic viscosity 5.11 mm²/s, calorific value 39.76 MJ/kg, acid value 0.41 mg KOH/g, induction period 5.84 h, cloud point 6°C, pour point 8°C, flash point 130°C, cetane number 57, and ash content 0.003 wt.%) were in range as per ASTM and EN standards.

5.3.24 *Annona muricata*

Schroeder and colleagues (2018) optimized the production of ethyl esters and methyl esters from *Annona muricata* seed oil using 1% KOH as catalyst. They compared the yield and physico-chemical parameters of both the esters. The yield of methyl esters (95.5%) was higher than ethyl esters (95.2%). However, the physico-chemical parameters of both biodiesels were almost equivalent to each other. Among all the parameters, iodine value, kinematic viscosity, oxidative stability, specific mass, corrosion of copper, and carbon residue deposition meet the standard values determined by ASTM. However, the flash point, pour point, and cloud point of pure biodiesel were more than the standard values. Hence, they suggested that its blends (5%, 10%, 20%, and 30%) mixed with petroleum diesel meets the quality of standard biodiesel. Su and colleagues (2018) also optimized biodiesel production from *Annona muricata* seed oil and obtained biodiesel yield of 97.02%. They suggested that the physico-chemical parameters (density 868 kg/m³, kinematic viscosity 5.5 mm²/s at 40°C, Acid value <0.8 mg KOH/g, sulfur content 0.04%, water content 300 mg/kg, cetane number 53, flash point 123°C) were almost in range as per ASTM D6751 and EN 14214 standards.

5.3.25 *Leucaena leucocephala*

Hakimi and colleagues (2017) optimized biodiesel production from *Leucaena leucocephala* seed oil using alkali catalyzed transesterification process and evaluated its physico-chemical parameters. They suggested that the free fatty acids and water content is high in oil of *Leucaena leucocephala*; hence, pretreatment of acid-catalyzed transesterification was required. Most of the physico-chemical parameters (carbon residues 0.07%, cold filter plugging point −1°C, cloud point 1°C, flash point 118°C, pour point −3°C, oxidative stability 12 h, acid number 0.27 mg KOH/g, kinematic viscosity 4.8 mm²/s, iodine number 101.7 g I$_2$/100 g) of *Leucaena leucocephala* biodiesel are almost in range as per ASTM D6751 and EN 14214 standards. The oxidative stability of *Leucaena leucocephala* biodiesel is very high, and it can be stored for longer period of time.

5.3.26 *Ocimum basilicum*

Amini and colleagues (2017) optimized biodiesel production from *Ocimum basilicum* seed oil using lipase catalyzed transesterification process and suggested that 94.58% biodiesel yield can be obtained after 68 hours of reaction at 47°C in the presence of 6% lipase enzyme as catalyst. The physico-chemical parameters (kinematic viscosity 4.26 mm²/s, density 870 kg/cm³, acid value 0.53 mg KOH/g, calorific value 39.72 MJ/kg, oxidative stability 2 h) of *Ocimum basilicum* biodiesel was almost in range as per EN 14214 and ASTM D6751 standards. However, oxidative stability of biodiesel is very poor, which can be improved further by adding antioxidants such as butylated hydroxytoluene, butylated hydroxyanisole, di-tertbutylhydroquinone, or poly-1,2-dihydro-2,2,4-trimethylquinoline.

5.3.27 *Cucumis melo* var. *agrestis*

Ameen and colleagues (2018) optimized biodiesel production from *Cucumis melo* var. *agrestis* seed oil and tested its physico-chemical properties. From the analysis, they observed that the total oil content in *Cucumis melo* var. *agrestis* seeds was 29.1%. The biodiesel yield was 96% when transesterification process was carried out for 1 hour at 60°C in the presence of 30% methanol and 0.4% sodium hydroxide as catalyst. The GC-MS and FTIR analyses confirmed presence of methyl esters (linoleic acid, palmitic acid, lauric acid, tetradecanoic acid, isopropyl linolate, octadecynoic acid) and methoxy ester carbonyl group, respectively. The physico-chemical parameters of fatty acids methyl esters (flash point 91°C, density 0.873 g/cm³, viscosity 5.35 mm²/s, pour point −13°C, cloud point −10°C, total acid number 0.242 mg KOH/g, and sulfur content 0.0043 wt%) were almost in range of ASTM standards.

5.3.28 *Salvadora persica*

Budhwani and colleagues (2019) optimized biodiesel production from *Salvadora persica* using enzymatic transesterification process and suggested that the use of *Burkholderia cepacia* lipase increases the yield of biodiesel production in a short period of time as compared to chemical catalysts. It can be easily recovered and can be reused for many transesterification cycles. The physico-chemical parameters (density 839 kg/cm³, kinematic viscosity 3 mm²/s, cloud point 9.51°C, pour point 3.50°C, cetane number 67.51, iodine value 11.97 g I_2/100 g, saponification value 228.29 mg/g) of *Salvadora persica* biodiesel was in range as per ASTM D6751 standards.

5.3.29 *Koelreuteria paniculata*

Khan and colleagues (2020) optimized biodiesel production from *Koelreuteria paniculata* seeds which contained 28%–30% of oil with low fatty acid content (0.91%). They observed that the yield of biodiesel obtained from *Koelreuteria paniculata* seed oil was 95.2% and physico-chemical parameters (such as density 879 kg/cm³ at 15°C, iodine value 80.7 g I_2/100 mg, ignition point 175°C, kinematic viscosity 6.21 mm²/s at 40°C, acid value 0.07 mg KOH/g, flash point 147°C, saponification value 176.4 mg KOH/g, cetane number 51, cloud point 2°C, pour point −30°C, sulfated ash content 0.003%) of biodiesel were satisfactory in accordance with ASTM D6751 and EN 14214 standards.

5.4 Conclusion

Various researchers have evaluated the performance of different biodiesels and its blends in different models of diesel engines. On the basis of their analysis, they observed that biodiesel produced from some forest seeds (*Pongamia pinnata, Simmondsia chinensis, Linum usitatissimum, Madhuca longifolia, Azadirachta indica, Simarouba glauca, Calophyllum inophyllum, Thevetia peruviana, Aleurites moluccanus, Syzygium cumini, Hevea brasiliensis, Guizotia abyssinica, Annona reticulate, Citrullus colocynthis, Pistacia khinjuk*) when fueled in the diesel engine either as pure biodiesel (100%) or its blends had good combustion, performance, and reduced emission of exhaust gases with some limitations. The brake thermal efficiency, brake power, heat release rate, and peak pressure of combusted fuel at full load were lower than petroleum diesel. Emissions of carbon monoxide and hydrocarbons were lower than petroleum diesel. However, the brake-specific energy consumption at full load is higher than petroleum diesel, and the emissions of nitrogen oxide, matter particles, and exhaust gas temperature were greater as compared to petroleum diesel. This indicates that there is more energy consumption and incomplete fuel combustion. To improve these characteristics, antioxidants, hydrogen gas, biogas, or polar solvents (diethyl ether, euro diesel, butanol) were mixed with the some biodiesel blends (*Azadirachta indica, Pongamia pinnata, Pistacia khinjuk*) and modifications were made to diesel engines. With these modifications, they were able to improve the combustion and emission properties of diesel engines up to some extent, although there are some promising blends (5%–20%) of forest seed (*Jatropa curcus, Balanites aegyptiaca, Schleichera oleosa, Hibiscus cannabinus, Ceiba pentandra*) biodiesel which show performance, combustion, and emission characteristics similar to that of petroleum diesel without any modifications in diesel engines. Hence, they can be used for large-scale production of biodiesel in the near future. However, there are various biodiesels produced from forest seeds which show excellent physico-chemical parameters but remain to be evaluated for their performance, combustion, and emission characteristics in fueling diesel engines. These forest biodiesels and their blends should be tested in diesel engines, and excellent feedstock for biodiesel production from these seeds should be screened.

TABLE 5.1
Comparison of Physico-Chemical Parameters of Biodiesel Produced from Different Forest Seeds

S.N.	Scientific Name	Density (Kg/m³)	Kinematic Viscosity (mm²/s)	Acid Value (mg KOH/g)	Cetane No.	Oxidative stability at 110°C (h)	Flash Point (°C)	Cloud Point (°C)	Pour Point (°C)	Cold Filter Plugging Point	Calorific Value (MJ/kg)	Saponification Value	Iodine No.	Sulfated Ash (% mass)	Sulfur (% mass)	Water Content (mg/kg)	Free Glycerin (% mass)	Total Glycerin (% mass)	Specific Gravity (kg/m³)	Lubricity*	Carbon Residue	Ref.
1	*Aegle marmelos*	NA	5.67	–	49	–	–	7	3	–	–	–	–	–	–	–	–	–	0.893	–	–	Selvan et al. (2018)
2	*Aleurites trisperma Blanco*	875	5.8	–	55.5	–	178	3	–3	–	37.3	–	97.8	0.002	–	–	–	–	–	–	–	Kumar et al. (2015)
3	*Amygdalus pedunculata Pall*	877	4.695	0.32	49.2	2.2	169	–	–	–11	–	–	–	–	2	475	–	–	–	–	–	Chu et al. (2013)
4	*Annona cherimola Mill.*	871	4.4	0.3	53	1.17	–	1	–	–5	–	–	99	–	–	–	–	–	–	–	–	Branco et al. (2010)
5	*Annona muricata L.*	868	5.5	0.8	53	–	123	–	–	–	–	–	–	–	0.04	300	–	–	–	–	–	Su et al. (2018)
6	*Annona reticulata*	NA	5.25	–	–	1.93	178.5	12	5	10	–	–	101.24	–	–	–	–	–	–	–	–	Schroeder et al. (2018)
7	*Azadirachta indica*	870	5.3	–	57	–	160	5	3	–	39.4	–	–	–	–	–	–	–	–	–	–	Yadav et al. (2019)
		874	4.4	0.88	–	–	138	11	–6	–	39.25	–	–	–	0.36	–	–	–	0.876	–	–	Mathiyazhagan et al. (2013)
		862.3	4.4	–	–	–	167	–	–	–	41.3	–	–	–	–	–	–	–	–	–	–	Ramakrishnan et al. (2017)
		NA	5.5	–	55.31	–	110	8	4	–	39.85	207	70.5	–	–	–	–	–	0.9	–	–	Awolu and Layokun (2013)
8	*Balanites aegyptiaca*	890	4.2	–	56	–	131	7	3	3	–	–	100	–	5	450	–	–	–	126	0.2	Chapagain et al. (2009)
9	*Blighia unijugata*	880	4.52	0.04	–	44.3	185	–	12	–	38.21	–	47.6	–	–	–	–	–	–	–	–	Adewuyi et al. (2012b)
10	*Calophyllum inophyllum L.*	–	4.2	–	–	–	110	–	3	–	40.8	–	–	–	–	–	–	–	0.885	–	–	Miraculas et al. (2016)
		905	4.21	–	–	–	95	–	–	–3	41.45	–	–	–	–	–	–	–	–	–	–	Ayyasamy et al. (2018)
11	*Camelina sativa L. Crantz*	–	4.15	0.31	52.8	2.5	–	3	–4	–	–	–	151	–	–	–	–	–	–	122	–	Moser and Vaughn (2010)
		–	4.32	–	–	0.6	172	2.7	–	1	–	–	–	–	–	–	–	–	–	–	–	Soriano et al. (2012)
12	*Cannabis sativa L.*	884	3.48	0.25	–	–	162	–5	–	–	–	–	–	–	–	–	0.005	0.1	–	–	–	Li et al. (2010)
		891	3.86	–	51.63	4.46	180	3	–4	–10	–	194.99	125.68	–	–	–	–	–	–	–	–	Mohammed et al. (2020)

(Continued)

TABLE 5.1

Comparison of Physico-Chemical Parameters of Biodiesel Produced from Different Forest Seeds (Continued)

S. N.	Scientific Name	Density (Kg/m³)	Kinematic Viscosity (mm²/s)	Acid Value (mg KOH/g)	Cetane No.	Oxidative stability at 110°C (h)	Flash Point (°C)	Cloud Point (°C)	Pour Point (°C)	Cold Filter Plugging Point	Calorific Value (MJ/kg)	Saponification Value	Iodine No.	Sulfated Ash (% mass)	Sulfur (% mass)	Water Content (mg/kg)	Free Glycerin (% mass)	Total Glycerin (% mass)	Specific Gravity (kg/m³)	Lubricity*	Carbon Residue	Ref.
13	*Cascabela thevetia*	882	5.11	0.41	57	5.84	130	6	8	–	39.76	–	–	–	–	–	–	–	–	–	–	Sut et al. (2016)
14	*Ceiba pentandra*	–	4.17	0.036	47	–	169	1	–5	–	–	–	–	0.01	–	–	–	–	0.876	–	0.042	Sivakumar et al. (2013)
15	*Citrullus colocynthis*	878	5.96	–	–	–	140	–	–6	–	38.978	–	–	–	–	–	–	–	–	–	–	Asokan et al. (2016)
		886	3.45	0.27	66.81	–	134	2		–	40	–	–	–	–	–	–	–	–	–	–	Sivalingam et al. (2019)
16	*Cleome viscosa*	890	4.5	0.1	55.2	1	175	21.2	3	9.1	39.7	–	116.3	0.005	0.0015	–	0.0045	0.024	–	198	0.04	Kumari et al. (2013)
17	*Crotalaria juncea*	NA	1.66	–	58.51	–	60	–	–	65	37.68	180	80.5	–	–	–	–	–	0.89	296	–	Sadhukhan and Sarkar (2016)
18	*Cucumis melo var. agrestis*	873	5.35	0.242	–	–	–	–10	–13	–	–	–	–	–	0.004	–	–	–	–	–	–	Ameen et al. (2018)
19	*Dipteryx alata Vog*	878	3.2	0.42	–	–	181	–	–	–	–	173.9	50.9	–	–	420	–	–	–	–	–	Batista et al. (2012)
20	*Euphorbia lathyris* L.	876.1	4.637	0.19	59.6	10.4	–	–	–	–11	–	–	–	–	–	400	0.01	0.09	–	–	–	Wang et al. (2011)
21	*Forsythia suspense* [(Thunb.) Vahl	–	3.79	0.08	45.3	–	184	5	0	–	–	–	–	0.003	–	–	0.01	0.01	–	–	0.02	Jiao et al. (2013)
22	*Gliricidia sepium*	879.5	4.38	0	67.5	2.2	–	22.4	19	–	–	–	–	–	5.175	–	0.005	0.028	–	140	–	Knothe et al. (2015)
23	*Guizotia abyssinica*	888.6	4.14	0.3	59	–	157	3	–	–	41.25	–	–	–	–	–	–	–	–	–	–	Jaikumar et al. (2019)
24	*Hevea brasiliensis*	878	4.2	0.36	51.9	8.2	152	3.4	–2	–	39.2	–	129.8	–	–	–	–	–	–	–	–	Adam et al. (2018)
25	*Hibiscus cannabinus*	–	5.23	–	–	–	124	–	–	–	39.141	–	104.06	–	–	–	–	–	0.8575	–	–	Jindal and Goyal (2012)
26	*Jatropha curcas* L.	–	4.4	–	57.1	3.23	163	4	–	–	–	–	–	0.002	0.004	–	0.01	0.02	–	–	–	Sarin et al. (2007)
		867	4.8	0.21	–	–	173	–	–	–	–	201.13	–	–	–	–	–	–	–	–	–	Hailegiorgis et al. (2012)
		880.2	4.312	0.04	55.4	8	147	–	–	–8	–	–	–	0.01	–	–	0.02	0.14	–	–	–	Wang et al. (2011)
		886.9	5.33	–	53	–	162	8	–5	–	43,299	–	–	0.0012	0.084	–	–	–	–	–	0.65	Vijayaraj and J. Sarangan (2012)
		865	4.722	–	–	–	182.5	5	3	–	39.827	–	–	–	–	–	–	–	–	–	–	Palash et al. (2013)

S. N.	Scientific Name	Density (Kg/m³)	Kinematic Viscosity (mm²/s)	Acid Value (mg KOH/g)	Cetane No.	Oxidative stability at 110°C (h)	Flash Point (°C)	Cloud Point (°C)	Pour Point (°C)	Cold Filter Plugging Point	Calorific Value (MJ/kg)	Saponi-fication Value	Iodine No.	Sulfated Ash (% mass)	Sulfur (% mass)	Water Content (mg/kg)	Free Glycerin (% mass)	Total Glycerin (% mass)	Specific Gravity (kg/m³)	Lubr-icity*	Carbon Residue	Ref.
27	Koelreuteria paniculata	879	6.21	0.07	51	–	147	2	–30	–18	–	176.4	80.7	0.003	–	–	–	–	–	–	–	Khan et al. (2020)
28	Lepidium sativum Linn.	845	1.921	–	49.23	–	176	–1	–6	–	–	173	128	–	0.018	–	–	–	–	–	0.04	Nehdi et al. in (2012)
29	Leucaena leucocephala	–	4.8	0.27	–	12	118	1	–3	–1	–	–	101.7	–	–	–	–	–	–	–	–	Hakimi et al. (2017)
30	Licania rigida Benth	–	38.32	1.8	–	–	–	–	–	–	–	188.3	–	–	–	–	–	–	–	–	–	Macedo et al. (2011)
31	Limnanthes alba L.	–	6.18	0.02	66.9	41.5	–	–6	–10	–9	–	–	–	–	–	–	0.006	0.239	–	–	–	Moser et al. (2010)
32	Luffa cylindrical	890	4.1	0.06	–	0.58	148	–	0.4	–	39	–	108.2	–	–	–	–	–	–	–	–	Adewuyi et al. (2012b)
33	Maclura pomifera (Rafin.) Schneider	889	4.66	0.4	–	–	180	–5	–9	–	–	183	125	–	–	–	–	–	–	–	–	Saloua et al. (2010)
34	Madhuca indica	883	4.85	–	51	–	129	–	5	–	36.9	–	–	–	–	–	–	–	–	–	0.01	Kapilan et al. (2009)
35	Manilkara zapota (L.)	875	4.67	3.79	52	–	174	–	–6	–	37.2	–	65.28	–	–	–	–	–	–	–	–	Kumar and Sureshkumar (2016)
36	Moringa oleifera	–	4.83	–	67.07	3.61	–	18	17	–	–	–	–	–	–	–	–	–	–	138.5	–	Rashid et al. (2008)
37	Ocinum basilicum L.	870	4.26	0.53	–	2	–	–7	–15	–	39.72	–	–	–	–	–	–	–	–	–	–	Amini et al. (2017)
38	Pangium edule Reinw	871	5.2296	–	47	0.57	–	–6	–4	–8	40	201	119	–	–	–	–	–	–	–	–	Atabani et al. (2015)
39	Pentaclethra macrophylla	877	4.08	0.5	36.72	–	196	–5	0	–	–	–	–	–	–	–	–	–	0.904	–	–	Oshieke and Jauro (2012)
40	Pistacia khinjuk	887	4.65	–	53	–	172	–2	–5	–	38.2	–	–	–	–	–	–	–	–	–	–	Karthickeyan et al. (2019)
41	Pongamia pinnata	–	4.2	0.35	–	–	150	–	–	–	–	–	–	0	–	–	–	–	–	–	–	Agarwal & Bajaj (2009)
		–	4.8	0.62	–	–	150	–	–	–	–	–	–	0.005	–	–	–	–	–	–	–	Karmee & Chadha (2005)
		900	5.34	0.136	–	–	197	6	–	–	39.39	–	–	–	–	–	–	–	–	–	–	Kumar et al. (2011)

(Continued)

TABLE 5.1

Comparison of Physico-Chemical Parameters of Biodiesel Produced from Different Forest Seeds (Continued)

S.N.	Scientific Name	Density (Kg/m³)	Kinematic Viscosity (mm²/s)	Acid Value (mg KOH/g)	Cetane No.	Oxidative stability at 110°C (h)	Flash Point (°C)	Cloud Point (°C)	Pour Point (°C)	Cold Filter Plugging Point	Calorific Value (MJ/kg)	Saponification Value	Iodine No.	Sulfated Ash (% mass)	Sulfur (% mass)	Water Content (mg/kg)	Free Glycerin (% mass)	Total Glycerin (% mass)	Specific Gravity (kg/m³)	Lubricity*	Carbon Residue	Ref.
42	*Raphanus sativus* L.	–	5	0.08	–	1.6	–	10	–14	–2	38.75	–	111	–	–	–	–	–	–	201	–	Shah et al. (2013)
43	*Reutealis trisperma*	887	6.71	0.41	63.5	–	148	–13	–15	–	–	–	28.43	–	–	–	–	0.037	–	–	–	Holilah et al. (2015)
44	*Salvadora persica*	839	3	–	67.51	–	–	9.51	3.5	–135	–	228.29	11.97	–	–	–	–	–	–	–	–	Budhwani et al. (2019)
45	*Sapium sebiferum* L.	892	3.698	0.15	40.2	0.8	180	–	–	11	–	–	–	0.01	–	300	0.01	0.09	–	–	–	Wang et al. (2011)
46	*Schleichera oleosa*	870	4.275	–	–	–	147	–	–	–	35	–	–	–	–	–	–	–	0.904	–	–	Acharya et al. (2011)
46	*Schleichera oleosa*	870	4.7	–	–	6.5	152	–	–	–4	39.45	–	–	–	–	–	–	–	–	–	–	Yadav et al. (2017b)
47	*Silybum marianum*	867	4.4	0.46	52	2	151	–2	–2	–	–	–	–	–	0.002	–	–	0.15	–	–	–	Takase et al. (2014)
48	*Simarouba glauca*	862	3.1	0.4	56.8	–	178	17	14	–	40.325	–	–	0.01	0.01	–	–	–	–	–	0.06	Jeyalakshmi (2019)
49	*Simmondsia chinensis*	863.5	9	–	–	–	–	–	–	–	41.52	–	–	–	–	–	–	–	–	–	–	Canoira et al. (2006)
50	*Sterculia foetida*	870	5.1	0.34	–	–	162	–3	–3	–	38.39	–	–	–	–	–	–	–	–	–	–	Sambasivam and Murugavelh (2019)
51	*Terminalia belerica* Roxb.	850	4.72	–	–	3.42	179	3	3	–	–	–	72.6	–	–	–	–	–	–	–	–	Bindhu et al. (2012)
51	*Terminalia belerica* Roxb.	–	4.79	0.014	53.4	2.09	162	5	0	–	–	–	–	0.001	–	–	0.003	0.011	–	–	–	Sarin et al. (2010)
52	*Terminalia catappa* L.	873	4.3	–	–	–	–	–	–	–	36.97	–	83.2	–	–	–	–	–	–	–	–	dos Santos et al. (2008)
53	*Thevetia Peruviana*	870	4.1	0.62	72	8	174	–5	–8	–	42.69	–	76.9	–	–	–	–	–	–	–	–	Yadav et al. (2017b)
54	*Treculia africana*	880	4.6	0.3	–	–	131	–	–	–	–	–	117.6	–	–	–	0.01	–	–	–	–	Adewuyi et al. (2012a)
55	*Vernicia montana*	904	7.32	0.23	–	0.7	174	–10.5	–	–11.6	–	–	165	0.005	12.9	–	0.018	0.08	–	–	0.076	Anantharaman et al. (2016)
55	*Vernicia montana*	903	7.84	0.124	39	0.3	–	–	–	–11	–	–	161.1	–	–	–	–	0.03	–	–	–	Chen et al. (2010)
56	*Xanthium sibiricum* Patr	874.7	3.723	0.07	48.8	1.2	174	–	–	–3	–	–	–	0.004	0.0003	–	0.0008	0.135	–	–	–	Chang et al. (2013)
57	*Xanthoceras sorbifolia* Bunge	–	4.4	0.06	56.1	–	165	–	–	–	–	–	–	0.003	–	–	–	–	–	–	0.02	Li et al. (2012)
58	*Zanthoxylum bungeanum* Maxim	882.6	4	0.3	47	–	174	2	–	2	–	–	–	–	–	–	–	–	–	–	0.1	Yang et al. (2008)

* For specific time, surface size, temperature, and pressure.

REFERENCES

Abdulla, R., Chan, E.S. and Ravindra, P., 2011. Biodiesel production from *Jatropha curcas*: A critical review. *Critical Reviews in Biotechnology*, *31*(1), 53–64.

Acharya, S.K., Swain, R.K., Mohanty, M.K., Mishra, A.K. and Mahapatra, S., 2011. The performance and emission characteristics of a diesel engine using preheated Kusum oil and Kusum diesel blend. *International Journal of Energy Technology and Policy*, *7*(5–6), 503–518.

Achten, W.M., Mathijs, E., Verchot, L., Singh, V.P., Aerts, R. and Muys, B., 2007. Jatropha biodiesel fueling sustainability? *Biofuels, Bioproducts and Biorefining: Innovation for a Sustainable Economy*, *1*(4), 283–291.

Adam, I.K., Aziz, A., Rashid, A., Heikal, M.R., Yusup, S., Ahmad, A.S., Abidin, Z. and Zharif, E., 2018. Performance and emission analysis of rubber seed, palm, and their combined blend in a multi-cylinder diesel engine. *Energies*, *11*(6), 1522.

Adewuyi, A., Oderinde, R.A. and Ojo, D.F., 2012a. Biodiesel from the seed oil of *Treculia africana* with high free fatty acid content. *Biomass Conversion and Biorefinery*, *2*(4), 305–308.

Adewuyi, A., Oderinde, R.A., Rao, B.V.S.K., Prasad, R.B.N. and Anjaneyulu, B., 2012b. *Blighia unijugata* and *Luffa cylindrica* seed oils: Renewable sources of energy for sustainable development in rural Africa. *BioEnergy Research*, *5*(3), 713–718.

Afif, M.K. and Biradar, C.H., 2019. Production of biodiesel from *Cannabis sativa* (Hemp) seed oil and its performance and emission characteristics on DI engine fueled with biodiesel blends. *International Research Journal of Engineering and Technology*, *6*(8), 246–253.

Agarwal, A.K., 2005. Experimental investigations of the effect of biodiesel utilization on lubricating oil tribology in diesel engines. *Proceedings of the Institution of Mechanical Engineers, Part D: Journal of Automobile Engineering*, *219*(5), 703–713.

Agarwal, A.K. and Agarwal, D., 2004. Experimental study of linseed oil as an alternative fuel for diesel engine, Technical Session IV. In *The proceedings of national conference on biodiesel, held at Central Institute of Agriculture Engineering (CIAE), Bhopal, India*.

Agarwal, A.K. and Bajaj, T.P., 2009. Process optimisation of base catalysed transesterification of Karanja oil for biodiesel production. *International Journal of Oil, Gas and Coal Technology*, *2*(3), 297–310.

Ali, O.M., Mamat, R., Abdullah, N.R. and Abdullah, A.A., 2016. Commercial and synthesized additives for biodiesel fuel: A review. *ARPN Journal of Engineering and Applied Sciences*, *11*(6), 3650–3654.

Alptekin, E. and Canakci, M., 2008. Determination of the density and the viscosities of biodiesel—diesel fuel blends. *Renewable Energy*, *33*(12), 2623–2630.

Al-Widyan, M.I. and Al-Muhtaseb, M.A., 2010. Experimental investigation of Jojoba as a renewable energy source. *Energy Conversion and Management*, *51*(8), 1702–1707.

Ameen, M., Zafar, M., Ahmad, M., Shaheen, A. and Yaseen, G., 2018. Wild melon: A novel non-edible feedstock for bioenergy. *Petroleum Science*, *15*(2), 405–411.

Amini, Z., Ong, H.C., Harrison, M.D., Kusumo, F., Mazaheri, H. and Ilham, Z., 2017. Biodiesel production by lipase-catalyzed transesterification of *Ocimum basilicum* L. (Sweet basil) seed oil. *Energy Conversion and Management*, *132*, 82–90.

Anantharaman, G., Krishnamurthy, S. and Ramalingam, V., 2016. Effects of raw material composition of tung (*Vernicia montana*) and Jatropha (*Jatropha curcas* L.) oil methyl esters on their fuel properties: A comparative study in fuel quality perspectives. *International Journal of Oil, Gas and Coal Technology*, *12*(2), 210–230.

Armah-Agyeman, G., Gyamerah, M., Biney, P.O. and Woldesenbet, S., 2016. Extraction and characterization of triglycerides from coffeeweed and switchgrass seeds as potential feedstocks for biodiesel production. *Journal of the Science of Food and Agriculture*, *96*(13), 4390–4397.

Asokan, M.A., Vijayan, R., Prabu, S.S. and Venkatesan, N., 2016. Experimental studies on the combustion characteristics and performance of a DI diesel engine using kapok oil methyl ester/diesel blends. *International Journal of Oil, Gas and Coal Technology*, *12*(1), 105–119.

Atabani, A.E., Badruddin, I.A., Mahlia, T.M.I., Masjuki, H.H., Mofijur, M., Lee, K.T. and Chong, W.T., 2013. Fuel properties of *Croton megalocarpus*, *Calophyllum inophyllum*, and *Cocos nucifera* (coconut) methyl esters and their performance in a multicylinder diesel engine. *Energy Technology*, *1*(11), 685–694.

Atabani, A.E., Badruddin, I.A., Masjuki, H.H., Chong, W.T. and Lee, K.T., 2015. Pangium edule Reinw: A promising non-edible oil feedstock for biodiesel production. *Arabian Journal for Science and Engineering*, *40*(2), 583–594.

Atabani, A.E. and da Silva César, A., 2014. *Calophyllum inophyllum* L. A prospective non-edible biodiesel feedstock. Study of biodiesel production, properties, fatty acid composition, blending and engine performance. *Renewable and Sustainable Energy Reviews, 37*, 644–655.

Augustus, G.D.P.S., Jayabalan, M. and Seiler, G.J., 2002. Evaluation and bioinduction of energy components of *Jatropha curcas*. *Biomass and Bioenergy, 23*(3), 161–164.

Awolu, O.O. and Layokun, S.K., 2013. Optimization of two-step transesterification production of biodiesel from neem (*Azadirachta indica*) oil. *International Journal of Energy and Environmental Engineering, 4*(1), 39.

Ayyasamy, T., Balamurugan, K. and Duraisamy, S., 2018. Production, performance and emission analysis of Tamanu oil diesel blends along with biogas in a diesel engine in dual cycle mode. *International Journal of Energy Technology and Policy, 14*(1), 4–19.

Azad, A.K., Rasul, M.G. and Bhatt, C., 2019. Combustion and emission analysis of Jojoba biodiesel to assess its suitability as an alternative to diesel fuel. *Energy Procedia, 156*, 159–165.

Bamgboye, A.I. and Oniya, O.O., 2012. Fuel properties of loofah (*Luffa cylindrica* L.) biofuel blended with diesel. *African Journal of Environmental Science and Technology, 6*(9), 346–352.

Baranitharan, P., Ramesh, K. and Sakthivel, R., 2019. Measurement of performance and emission distinctiveness of *Aegle marmelos* seed cake pyrolysis oil/diesel/TBHQ opus powered in a DI diesel engine using ANN and RSM. *Measurement, 144*, 366–380.

Batista, A.C.F., de Souza Rodrigues, H., Pereira, N.R., Hernandez-Terrones, M.G., Vieira, A.T. and de Oliveira, M.F., 2012. Use of baru oil (*Dipteryx alata* Vog.) to produce biodiesel and study of the physical and chemical characteristics of biodiesel/petroleum diesel fuel blends. *Chemistry and Technology of Fuels and Oils, 48*(1), 13–16.

Bindhu, C.H., Reddy, J.R.C., Rao, B.V.S.K., Ravinder, T., Chakrabarti, P.P., Karuna, M.S.L. and Prasad, R.B.N., 2012. Preparation and evaluation of biodiesel from *Sterculia foetida* seed oil. *Journal of the American Oil Chemists' Society, 89*(5), 891–896.

Bouaid, A., Bajo, L., Martinez, M. and Aracil, J., 2007. Optimization of biodiesel production from Jojoba oil. *Process Safety and Environmental Protection, 85*(5), 378–382.

Branco, P.C., Castilho, P.C., Rosa, M.F. and Ferreira, J., 2010. Characterization of *Annona cherimola* Mill. seed oil from Madeira island: A possible biodiesel feedstock. *Journal of the American Oil Chemists' Society, 87*(4), 429–436.

Budhwani, A.A.A., Maqbool, A., Hussain, T. and Syed, M.N., 2019. Production of biodiesel by enzymatic transesterification of non-edible *Salvadora persica* (Pilu) oil and crude coconut oil in a solvent-free system. *Bioresources and Bioprocessing, 6*(1), 41.

Canakci, M. and Van Gerpen, J., 2001. Biodiesel production from oils and fats with high free fatty acids. *Transactions of the ASAE, 44*(6), 1429–1436.

Canoira, L., Alcantara, R., García-Martínez, M.J. and Carrasco, J., 2006. Biodiesel from Jojoba oil-wax: Transesterification with methanol and properties as a fuel. *Biomass and Bioenergy, 30*(1), 76–81.

Chang, F., Hanna, M.A., Zhang, D.J., Li, H., Zhou, Q., Song, B.A. and Yang, S., 2013. Production of biodiesel from non-edible herbaceous vegetable oil: *Xanthium sibiricum* Patr. *Bioresource Technology, 140*, 435–438.

Chapagain, B.P., Yehoshua, Y. and Wiesman, Z., 2009. Desert date (*Balanites aegyptiaca*) as an arid lands sustainable bioresource for biodiesel. *Bioresource Technology, 100*(3), pp. 1221–1226.

Chen, Y.H., Chen, J.H., Chang, C.Y. and Chang, C.C., 2010. Biodiesel production from tung (*Vernicia montana*) oil and its blending properties in different fatty acid compositions. *Bioresource Technology, 101*(24), 9521–9526.

Chhetri, A.B., Tango, M.S., Budge, S.M., Watts, K.C. and Islam, M.R., 2008. Non-edible plant oils as new sources for biodiesel production. *International Journal of Molecular Sciences, 9*(2), 169–180.

Chu, J., Xu, X. and Zhang, Y., 2013. Production and properties of biodiesel produced from *Amygdalus pedunculata* Pall. *Bioresource Technology, 134*, 374–376.

Datla, R., Puli, R.K., Chandramohan, V.P. and Geo, V.E., 2019. Biodiesel production process, optimization and characterization of *Azadirachta indica* biodiesel in a VCR diesel engine. *Arabian Journal for Science and Engineering, 44*(12), 10141–10154.

Demirbaş, A., 1997. Calculation of higher heating values of biomass fuels. *Fuel, 76*(5), 431–434.

Demirbaş, A., 2003. Biodiesel fuels from vegetable oils via catalytic and non-catalytic supercritical alcohol transesterifications and other methods: A survey. *Energy Conversion and Management, 44*(13), 2093–2109.

Demirbas, A., 2008. Relationships derived from physical properties of vegetable oil and biodiesel fuels. *Fuel, 87*(8–9), 1743–1748.

DeOliveira, E., Quirino, R.L., Suarez, P.A. and Prado, A.G., 2006. Heats of combustion of biofuels obtained by pyrolysis and by transesterification and of biofuel/diesel blends. *Thermochimica Acta*, *450*(1–2), 87–90.

Devan, P.K. and Mahalakshmi, N.V., 2009. Performance, emission and combustion characteristics of poon oil and its diesel blends in a DI diesel engine. *Fuel*, *88*(5), 861–867.

Dewangan, A., Yadav, A.K., Mallick, A., Pal, A. and Singh, S., 2019. Comparative study of *Manilkara zapota* and Karanja based biodiesel properties and its effect on diesel engine characteristics. *Energy Sources, Part A: Recovery, Utilization, and Environmental Effects*, 1–11.

Diaby, M., Sablier, M., Le Negrate, A., El Fassi, M. and Bocquet, J., 2009. Understanding carbonaceous deposit formation resulting from engine oil degradation. *Carbon*, *47*(2), 355–366.

Divakara, B.N., Upadhyaya, H.D., Wani, S.P. and Gowda, C.L., 2010. Biology and genetic improvement of *Jatropha curcas* L.: A review. *Applied Energy*, *87*(3), 732–742.

Dorado, M.P., Ballesteros, E., De Almeida, J.A., Schellert, C., Löhrlein, H.P. and Krause, R., 2002. An alkali-catalyzed transesterification process for high free fatty acid waste oils. *Transactions of the ASAE*, *45*(3), 525–529.

Dorado, M.P., Ballesteros, E., López, F.J. and Mittelbach, M., 2004. Optimization of alkali-catalyzed transesterification of *Brassica carinata* oil for biodiesel production. *Energy & Fuels*, *18*(1), 77–83.

dos Santos, I.C.F., De Carvalho, S.H.V., Solleti, J.I., de La Salles, W.F., de La, K.T.D.S. and Meneghetti, S.M.P., 2008. Studies of *Terminalia catappa* L. oil: Characterization and biodiesel production. *Bioresource Technology*, *99*(14), 6545–6549.

Encinar, J.M., Gonzalez, J.F., Rodriguez, J.J. and Tejedor, A., 2002. Biodiesel fuels from vegetable oils: Transesterification of *Cynara cardunculus* L. oils with ethanol. *Energy & fuels*, *16*(2), 443–450.

Fattah, I.R., Kalam, M.A., Masjuki, H.H. and Wakil, M.A., 2014. Biodiesel production, characterization, engine performance, and emission characteristics of Malaysian Alexandrian laurel oil. *RSC Advances*, *4*(34), 17787–17796.

Foidl, N., Foidl, G., Sanchez, M., Mittelbach, M. and Hackel, S., 1996. *Jatropha curcas* L. as a source for the production of biofuel in Nicaragua. *Bioresource Technology*, *58*(1), 77–82.

Fröhlich, A. and Rice, B., 2005. Evaluation of *Camelina sativa* oil as a feedstock for biodiesel production. *Industrial Crops and Products*, *21*(1), 25–31.

Garlapati, V.K., Kant, R., Kumari, A., Mahapatra, P., Das, P. and Banerjee, R., 2013. Lipase mediated transesterification of *Simarouba glauca* oil: A new feedstock for biodiesel production. *Sustainable Chemical Processes*, *1*(1), p. 11.

Ghadge, S.V. and Raheman, H., 2005. Biodiesel production from mahua (*Madhuca indica*) oil having high free fatty acids. *Biomass and Bioenergy*, *28*(6), 601–605.

Gogoi, T.K., 2013. Exergy analysis of a diesel engine operated with koroch seed oil methyl ester and its diesel fuel blends. *International Journal of Exergy*, *12*(2), 183–204.

Gogoi, T.K. and Baruah, D.C., 2011. Performance and energy analyses of a diesel engine fuelled with Koroch seed oil methyl ester and its diesel fuel blends. *International Journal of Energy Technology and Policy*, *7*(5–6), 433–454.

Hailegiorgis, S.M., Mahadzir, S. and Subbarao, D., 2012. In situ transesterification of non-edible oil in the presence of cetyltrimethylammonium bromide. *International Journal of Global Environmental Issues*, *12*(2–4), 161–170.

Hakimi, M.I., Goembira, F. and Ilham, Z., 2017. Engine-compatible biodiesel from *Leucaena leucocephala* seed oil. *Journal of the Society of Automotive Engineers Malaysia*, *1*(2).

Harrington, K.J., 1986. Chemical and physical properties of vegetable oil esters and their effect on diesel fuel performance. *Biomass*, *9*(1), 1–17.

Holilah, H., Prasetyoko, D., Oetami, T.P., Santosa, E.B., Zein, Y.M., Bahruji, H., Fansuri, H., Ediati, R. and Juwari, J., 2015. The potential of *Reutealis trisperma* seed as a new non-edible source for biodiesel production. *Biomass Conversion and Biorefinery*, *5*(4), 347–353.

Imdadul, H.K., Zulkifli, N.W.M., Masjuki, H.H., Kalam, M.A., Kamruzzaman, M., Rashed, M.M., Rashedul, H.K. and Alwi, A., 2017. Experimental assessment of non-edible candlenut biodiesel and its blend characteristics as diesel engine fuel. *Environmental Science and Pollution Research*, *24*(3), 2350–2363.

Issariyakul, T., Kulkarni, M.G., Dalai, A.K. and Bakhshi, N.N., 2007. Production of biodiesel from waste fryer grease using mixed methanol/ethanol system. *Fuel Processing Technology*, *88*(5), 429–436.

Jaichandar, S. and Annamalai, K., 2011. The status of biodiesel as an alternative fuel for diesel engine-An Overview. *Journal of Sustainable Energy & Environment*, *2*(2), 71–75.

Jaikumar, S., Bhatti, S.K. and Srinivas, V., 2019. Experimental investigations on performance, combustion, and emission characteristics of Niger (*Guizotia abyssinica*) seed oil methyl ester blends with diesel at different compression ratios. *Arabian Journal for Science and Engineering*, *44*(6), 5263–5273.

Jeyalakshmi, P., 2019. Characterization of *Simarouba glauca* seed oil biodiesel. *Journal of Thermal Analysis and Calorimetry*, 136(1), 267–280.

Jian, Z., Xuanjun, W., Qilong, H., Mingjun, H. and Shuyan, L., 2015. Physicochemical properties, combustion and emission performance of a novel *Zanthoxylum bungeanum* seed oil methylic ester biodiesel. *International Journal of Green Energy*, 12(12), 1255–1262.

Jiao, J., Gai, Q.Y., Wei, F.Y., Luo, M., Wang, W., Fu, Y.J. and Zu, Y.G., 2013. Biodiesel from *Forsythia suspense* [(Thunb.) vahl (Oleaceae)] seed oil. *Bioresource Technology*, 143, 653–656.

Jindal, S. and Goyal, K., 2012. Evaluation of performance and emissions of *Hibiscus cannabinus* (Ambadi) seed oil biodiesel. *Clean Technologies and Environmental Policy*, 14(4), 633–639.

Kapilan, N., Babu, T.A. and Reddy, R.P., 2009. Characterization and effect of using Mahua oil biodiesel as fuel in compression ignition engine. *Journal of Thermal Science*, 18(4), 382.

Karaosmanoğlu, F., Kurt, G. and Özaktaş, T., 2000. Long term CI engine test of sunflower oil. *Renewable Energy*, 19(1–2), 219–221.

Karmee, S.K., 2018. Enzymatic biodiesel production from *Manilkara zapota* (L.) seed oil. *Waste and Biomass Valorization*, 9(5), 725–730.

Karmee, S.K. and Chadha, A., 2005. Preparation of biodiesel from crude oil of *Pongamia pinnata*. *Bioresource Technology*, 96(13), 1425–1429.

Karthickeyan, V., 2019. Effect of cetane enhancer on *Moringa oleifera* biodiesel in a thermal coated direct injection diesel engine. *Fuel*, 235, 538–550.

Karthickeyan, V., Ashok, B., Thiyagarajan, S., Nanthagopal, K., Geo, V.E. and Dhinesh, B., 2019. Comparative analysis on the influence of antioxidants role with *Pistacia khinjuk* oil biodiesel to reduce emission in diesel engine. *Heat and Mass Transfer*, 1–18.

Kavitha, K.R., Beemkumar, N. and Rajasekar, R., 2019. Experimental investigation of diesel engine performance fuelled with the blends of *Jatropha curcas*, ethanol, and diesel. *Environmental Science and Pollution Research*, 26(9), 8633–8639.

Kavitha, M.S. and Murugavelh, S., 2019. Optimization and transesterification of sterculia oil: Assessment of engine performance, emission and combustion analysis. *Journal of Cleaner Production*, 234, 1192–1209.

Khan, I.U., Yan, Z. and Chen, J., 2020. Production and Characterization of Biodiesel Derived From a novel source *Koelreuteria Paniculata* seed Oil. *Energies*, 13(4), 791.

Knothe, G., 2005. Dependence of biodiesel fuel properties on the structure of fatty acid alkyl esters. *Fuel Processing Technology*, 86(10), 1059–1070.

Knothe, G., 2006. Analyzing biodiesel: Standards and other methods. *Journal of the American Oil Chemists' Society*, 83(10), 823–833.

Knothe, G., 2014. A comprehensive evaluation of the cetane numbers of fatty acid methyl esters. *Fuel*, 119, 6–13.

Knothe, G., de Castro, M.E.G. and Razon, L.F., 2015. Methyl esters (biodiesel) from and fatty acid profile of *Gliricidia sepium* seed oil. *Journal of the American Oil Chemists' Society*, 92(5), 769–775.

Knothe, G., Matheaus, A.C. and Ryan III, T.W., 2003. Cetane numbers of branched and straight-chain fatty esters determined in an ignition quality tester. *Fuel*, 82(8), 971–975.

Kolli, V., Gadepalli, S., Debbarma, J., Mandal, P. and Barathula, S., 2020. Experimental analysis on performance, combustion & emissions of a diesel engine fueled by *Aegle marmelos* seed oil biodiesel with additives: Graphene nanosheets and oxygenated diethyl ether. *Energy Sources, Part A: Recovery, Utilization, and Environmental Effects*, 1–20.

Kumar, K.R., Chandrika, K., Prasanna, K.T. and Gowda, B., 2015. Biodiesel production and characterization from non-edible oil tree species *Aleurites trisperma* Blanco. *Biomass Conversion and Biorefinery*, 5(3), 287–294.

Kumar, R., Kumar, G.R. and Chandrashekar, N., 2011. Microwave assisted alkali-catalyzed transesterification of *Pongamia pinnata* seed oil for biodiesel production. *Bioresource Technology*, 102(11), 6617–6620.

Kumar, R.K., Prasanna, K.T. and Gowda, B., 2013. Design and performance study on polypropylene biodiesel pilot plant for non-edible oils. *Biomass Conversion and Biorefinery*, 3(2), 79–86.

Kumar, R.S. and Sureshkumar, K., 2016. *Manilkara zapota* (L.) seed oil: A new third generation biodiesel resource. *Waste and Biomass Valorization*, 7(5), 1115–1121.

Kumar, R.S., Sureshkumar, K. and Velraj, R., 2018. Combustion, performance and emission characteristics of an unmodified diesel engine fueled with *Manilkara zapota* methyl ester and its diesel blends. *Applied Thermal Engineering*, 139, 196–202.

Kumari, R., Mallavarapu, G.R., Jain, V.K. and Kumar, S., 2013. Corresponding properties of fatty oils of *Cleome viscosa* and *Jatropha curcas* as resources of biodiesel. *Agricultural Research*, 2(4), 393–399.

Lang, X., Dalai, A.K., Bakhshi, N.N., Reaney, M.J. and Hertz, P.B., 2001. Preparation and characterization of bio-diesels from various bio-oils. *Bioresource Technology*, *80*(1), 53–62.

Li, J., Fu, Y.J., Qu, X.J., Wang, W., Luo, M., Zhao, C.J. and Zu, Y.G., 2012. Biodiesel production from yellow horn (*Xanthoceras sorbifolia* Bunge.) seed oil using ion exchange resin as heterogeneous catalyst. *Bioresource Technology*, *108*, 112–118.

Li, S.Y., Stuart, J.D., Li, Y. and Parnas, R.S., 2010. The feasibility of converting *Cannabis sativa* L. oil into biodiesel. *Bioresource Technology*, *101*(21), 8457–8460.

Lim, S., Hoong, S.S., Teong, L.K. and Bhatia, S., 2010. Supercritical fluid reactive extraction of *Jatropha curcas* L. seeds with methanol: A novel biodiesel production method. *Bioresource Technology*, *101*(18), 7169–7172.

Ma, F. and Hanna, M.A., 1999. Biodiesel production: A review. *Bioresource Technology*, *70*(1), pp. 1–15.

Macedo, F.L., Candeia, R.A., Sales, L.L.M., Dantas, M.B., Souza, A.G. and Conceição, M.M., 2011. Thermal characterization of oil and biodiesel from *oiticica* (*Licania rigida* Benth). *Journal of Thermal Analysis and Calorimetry*, *106*(2), 531–534.

Makkar, H.P.S., Becker, K., Sporer, F. and Wink, M., 1997. Studies on nutritive potential and toxic constituents of different provenances of *Jatropha curcas*. *Journal of Agricultural and Food Chemistry*, *45*(8), 3152–3157.

Manurung, R., Daniel, L., van de Bovenkamp, H.H., Buntara, T., Maemunah, S., Kraai, G., Makertihartha, I.G.B.N., Broekhuis, A.A. and Heeres, H.J., 2012. Chemical modifications of *Sterculia foetida* L. oil to branched ester derivatives. *European Journal of Lipid Science and Technology*, *114*(1), 31–48.

Mathiyazhagan, M., Elango, T., Senthilkumar, T. and Ganapathi, A., 2013. Assessment of fuel efficiency of neem biodiesel (*Azadirachta indica*) in a single cylinder diesel engine. *International Journal of Energy Technology and Policy*, *9*(3–4), 279–285.

Miraculas, G.A., Bose, N. and Raj, R.E., 2016. Optimization of biofuel blends and compression ratio of a diesel engine fueled with *Calophyllum inophyllum* oil methyl ester. *Arabian Journal for Science and Engineering*, *41*(5), 1723–1733.

Misra, R.D. and Murthy, M.S., 2011. Performance, emission and combustion evaluation of soapnut oil—Diesel blends in a compression ignition engine. *Fuel*, *90*(7), 2514–2518.

Mofijur, M., Masjuki, H.H., Kalam, M.A., Atabani, A.E., Arbab, M.I., Cheng, S.F. and Gouk, S.W., 2014. Properties and use of *Moringa oleifera* biodiesel and diesel fuel blends in a multi-cylinder diesel engine. *Energy Conversion and Management*, *82*, 169–176.

Mohammed, M.N., Atabani, A.E., Uguz, G., Lay, C.H., Kumar, G. and Al-Samaraae, R.R., 2020. Characterization of hemp (*Cannabis sativa* L.) biodiesel blends with euro diesel, butanol and diethyl ether using FT-IR, UV-Vis, TGA and DSC techniques. *Waste and Biomass Valorization*, *11*(3), 1097–1113.

Moser, B.R., Knothe, G. and Cermak, S.C., 2010. Biodiesel from meadowfoam (*Limnanthes alba* L.) seed oil: Oxidative stability and unusual fatty acid composition. *Energy & Environmental Science*, *3*(3), 318–327.

Moser, B.R. and Vaughn, S.F., 2010. Evaluation of alkyl esters from *Camelina sativa* oil as biodiesel and as blend components in ultra low-sulfur diesel fuel. *Bioresource Technology*, *101*(2), 646–653.

Nabi, M.N. and Hoque, S.N., 2008. Biodiesel production from linseed oil and performance study of a diesel engine with diesel bio-diesel. *Journal of Mechanical Engineering*, *39*(1), 40–44.

Nehdi, I.A., Sbihi, H., Tan, C.P. and Al-Resayes, S.I., 2012. Garden cress (*Lepidium sativum* Linn.) seed oil as a potential feedstock for biodiesel production. *Bioresource Technology*, *126*, 193–197.

Nelson, L.A., Foglia, T.A. and Marmer, W.N., 1996. Lipase-catalyzed production of biodiesel. *Journal of the American Oil Chemists' Society*, *73*(9), 1191–1195.

Norazahar, N., Ahmad, J. and Abu Bakar, S., 2012. Utilization of kapok seed as potential feedstock for biodiesel production. In W*SEAS 7th international conference of energy and environment*,- Rhodes Island, Greece: WSEAS Press

Oni, B.A. and Oluwatosin, D., 2020. Emission characteristics and performance of neem seed (*Azadirachta indica*) and Camelina (*Camelina sativa*) based biodiesel in diesel engine. *Renewable Energy*, *149*, 725–734.

Openshaw, K., 2000. A review of *Jatropha curcas*: An oil plant of unfulfilled promise. *Biomass and Bioenergy*, *19*(1), 1–15.

Oshieke, K.C. and Jauro, A., 2012. Production of biodiesel from *Pentaclethra macrophylla* seed oil. *International Journal of Renewable Energy Technology*, *3*(4), 400–409.

Padhi, S.K. and Singh, R.K., 2010. Optimization of esterification and transesterification of Mahua (*Madhuca indica*) oil for production of biodiesel. *Journal of Chemical and Pharmaceutical Research*, *2*(5), 599–608.

Padmanabhan, S., Rajasekar, S., Ganesan, S., Saravanan, S. and Chandrasekaran, M., 2017. Performance and emission analysis on CI engine using Soapnut oil as biofuel. *ARPN Journal of Engineering and Applied Sciences*, *12*(8), 2491–2495.

Palash, S.M., Kalam, M.A., Masjuki, H.H., Masum, B.M. and Sanjid, A., 2013. Impacts of *Jatropha* biodiesel blends on engine performance and emission of a multi cylinder diesel engine. In *Proceedings of the internatial conference on future trends in structural, civil, environmental and mechanical engineering (FTSCEM)*, (Vol. 2013, pp. 84–88). California, MA: Institute of Research Engineers and Doctors.

Papagiannakis, R.G., Hountalas, D.T. and Rakopoulos, C.D., 2007. Theoretical study of the effects of pilot fuel quantity and its injection timing on the performance and emissions of a dual fuel diesel engine. *Energy Conversion and Management*, *48*(11), 2951–2961.

Paramasivam, B., Kasimani, R. and Rajamohan, S., 2018. Characterization of pyrolysis bio-oil derived from intermediate pyrolysis of *Aegle marmelos* deoiled cake: Study on performance and emission characteristics of CI engine fueled with *Aegle marmelos* pyrolysis oil blends. *Environmental Science and Pollution Research*, *25*(33), 33806–33819.

Paturu, P. and Vinothkanna, I., 2018. Experimental investigation of performance and emissions characteristics on single-cylinder direct-injection diesel engine with PSZ coating using radish biodiesel. *International Journal of Ambient Energy*, *41*(7), 744–753.

Perumal, V. and Ilangkumaran, M., 2018. Experimental analysis of operating characteristics of a direct injection diesel engine fuelled with *Cleome viscosa* biodiesel. *Fuel*, *224*, 379–387.

Rajak, U., Nashine, P., Verma, T.N. and Pugazhendhi, A., 2019. Alternating the environmental benefits of Aegle-diesel blends used in compression ignition. *Fuel*, *256*, 115835.

Ramakrishnan, M., Rathinam, T.M. and Viswanathan, K., 2018. Comparative studies on the performance and emissions of a direct injection diesel engine fueled with neem oil and pumpkin seed oil biodiesel with and without fuel preheater. *Environmental Science and Pollution Research*, *25*(5), 4621–4631.

Ramalingam, S., Ganesan, R., Rajendran, S. and Ganesan, P., 2018. A novel alternative fuel for diesel engine: A comparative experimental investigation. *International Journal of Global Warming*, *14*(1), 40–60.

Raman, L.A., Deepanraj, B., Rajakumar, S. and Sivasubramanian, V., 2019. Experimental investigation on performance, combustion and emission analysis of a direct injection diesel engine fuelled with rapeseed oil biodiesel. *Fuel*, *246*, 69–74.

Rashed, M.M., Kalam, M.A., Masjuki, H.H., Mofijur, M., Rasul, M.G. and Zulkifli, N.W.M., 2016. Performance and emission characteristics of a diesel engine fueled with palm, jatropha, and moringa oil methyl ester. *Industrial Crops and Products*, *79*, 70–76.

Rashid, U., Anwar, F., Moser, B.R. and Knothe, G., 2008. *Moringa oleifera* oil: A possible source of biodiesel. *Bioresource Technology*, *99*(17), 8175–8179.

Rathore, V. and Madras, G., 2007. Synthesis of biodiesel from edible and non-edible oils in supercritical alcohols and enzymatic synthesis in supercritical carbon dioxide. *Fuel*, *86*(17–18), 2650–2659.

Rudreshaiah, O.B., Venkatesh, Y.K. and Ramappa, S., 2020. *Terminalia bellirica*: A new biodiesel for diesel engine: A comparative experimental investigation. *Environmental Science and Pollution Research*, *27*(13), 14432–14440.

Sadhukhan, S. and Sarkar, U., 2016. Production of biodiesel from *Crotalaria juncea* (Sunn-Hemp) oil using catalytic trans-esterification: Process optimisation using a factorial and Box-Behnken design. *Waste and Biomass Valorization*, *7*(2), 343–355.

Sahoo, P.K., Das, L.M., Babu, M.K.G. and Naik, S.N., 2007. Biodiesel development from high acid value polanga seed oil and performance evaluation in a CI engine. *Fuel*, *86*(3), pp. 448–454.

Saloua, F., Saber, C. and Hedi, Z., 2010. Methyl ester of *Maclura pomifera* (Rafin.) Schneider seed oil: Biodiesel production and characterization. *Bioresource Technology*, *101*(9), 3091–3096.

Sambasivam, K.M. and Murugavelh, S., 2019. Optimisation, experimental validation and thermodynamic study of the sequential oil extraction and biodiesel production processes from seeds of *Sterculia foetida*. *Environmental Science and Pollution Research*, *26*(30), 31301–31314.

Sankaranarayanan, G. and Pugazhvadivu, M., 2012. Effect of hydrogen enriched air on the performance and emissions of mahua oil fuelled diesel engine. *International Journal of Renewable Energy Technology*, *3*(1), 94–106.

Sarin, R., Sharma, M., Sinharay, S. and Malhotra, R.K., 2007. *Jatropha*–palm biodiesel blends: An optimum mix for Asia. *Fuel*, *86*(10–11), 1365–1371.

Sarin, R., Sharma, M. and Khan, A.A., 2010. *Terminalia belerica* Roxb. seed oil: A potential biodiesel resource. *Bioresource Technology*, *101*(4), 1380–1384.

Sayyar, S., Abidin, Z.Z., Yunus, R. and Muhammad, A., 2009. Extraction of oil from *Jatropha* seeds optimization and kinetics. *American Journal of Applied Sciences*, 6(7), 1390–1395.

Schroeder, P., dos Santos Barreto, M., Romeiro, G.A. and Figueiredo, M.K.K., 2018. Development of energetic alternatives to use of waste of *Annona muricata* L. *Waste and Biomass Valorization*, 9(8), 1459–1467.

Selvan, S.S., Pandian, P.S., Subathira, A. and Saravanan, S., 2018. Comparison of response surface methodology (RSM) and artificial neural network (ANN) in optimization of *Aegle marmelos* oil extraction for biodiesel production. *Arabian Journal for Science and Engineering*, 43(11), 6119–6131.

Shah, S.N., Iha, O.K., Alves, F.C., Sharma, B.K., Erhan, S.Z. and Suarez, P.A., 2013. Potential application of turnip oil (*Raphanus sativus* L.) for biodiesel production: Physical—chemical properties of neat oil, biofuels and their blends with ultra-low sulphur diesel (ULSD). *Bioenergy Research*, 6(2), 841–850.

Shah, S.N., Sharma, B.K., Moser, B.R. and Erhan, S.Z., 2010. Preparation and evaluation of jojoba oil methyl esters as biodiesel and as a blend component in ultra-low sulfur diesel fuel. *Bioenergy Research*, 3(2), 214–223.

Shehata, M.S. and Razek, S.A., 2011. Experimental investigation of diesel engine performance and emission characteristics using jojoba/diesel blend and sunflower oil. *Fuel*, 90(2), 886–897.

Silitonga, A.S., Ong, H.C., Masjuki, H.H., Mahlia, T.M.I., Chong, W.T. and Yusaf, T.F., 2013. Production of biodiesel from *Sterculia foetida* and its process optimization. *Fuel*, 111, 478–484.

Sivakumar, P., Sindhanaiselvan, S., Gandhi, N.N., Devi, S.S. and Renganathan, S., 2013. Optimization and kinetic studies on biodiesel production from underutilized *Ceiba pentandra* oil. *Fuel*, 103, 693–698.

Sivalingam, A., Kandhasamy, A., Kumar, A.S., Venkatesan, E.P., Subramani, L., Ramalingam, K., Thadhani, J.P.J. and Venu, H., 2019. *Citrullus colocynthis*: An experimental investigation with enzymatic lipase based methyl esterified biodiesel. *Heat and Mass Transfer*, 55(12), 3613–3631.

Soriano Jr, N.U. and Narani, A., 2012. Evaluation of biodiesel derived from *Camelina sativa* oil. *Journal of the American Oil Chemists' Society*, 89(5), 917–923.

Sridharan, R., Mathai, I.M., 1974. Transesterification reactions. *Journal of Scientific and Industrial Research*, 33, 178–187.

Srivastava, A. and Prasad, R., 2000. Triglycerides-based diesel fuels. *Renewable and Sustainable Energy Reviews*, 4(2), 111–133.

Su, C.H., Nguyen, H.C., Pham, U.K., Nguyen, M.L. and Juan, H.Y., 2018. Biodiesel production from a novel nonedible feedstock, soursop (*Annona muricata* L.) seed oil. *Energies*, 11(10), 2562.

Sut, D., Chutia, R.S., Bordoloi, N., Narzari, R. and Kataki, R., 2016. Complete utilization of non-edible oil seeds of *Cascabela thevetia* through a cascade of approaches for biofuel and by-products. *Bioresource Technology*, 213, 111–120.

Takase, M., Chen, Y., Liu, H., Zhao, T., Yang, L. and Wu, X., 2014. Biodiesel production from non-edible *Silybum marianum* oil using heterogeneous solid base catalyst under ultrasonication. *Ultrasonics Sonochemistry*, 21(5), 1752–1762.

Tanwar, D., Ajayta, D.S. and Mathur, Y.P., 2013. Production and characterization of neem oil methyl ester. *International Journal of Engineering Research Technology*, 2(5), 1896–1903.

Teoh, Y.H., How, H.G., Masjuki, H.H., Nguyen, H.T., Kalam, M.A. and Alabdulkarem, A., 2019. Investigation on particulate emissions and combustion characteristics of a common-rail diesel engine fueled with *Moringa oleifera* biodiesel-diesel blends. *Renewable Energy*, 136, 521–534.

Tesfa, B., Mishra, R., Gu, F. and Powles, N., 2010. Prediction models for density and viscosity of biodiesel and their effects on fuel supply system in CI engines. *Renewable Energy*, 35(12), 2752–2760.

Upadhyay, Y.P. and Sharma, R.B., 2013. Biodiesel: An alternative fuel and its emission effect. *IOSR Journal of Mechanical and Civil Engineering (IOSR-JMCE)*, 5(3), 01–04.

Van Gerpen, J., 1996, September. Cetane number testing of biodiesel. In *Proceedings, third liquid fuel conference: Liquid fuel and industrial products from renewable resources* (pp. 197–206). St. Joseph, MI: American Society of Agricultural Engineers.

Vedharaj, S., Vallinayagam, R., Yang, W.M., Chou, S.K., Chua, K.J.E. and Lee, P.S., 2013. Experimental investigation of kapok (*Ceiba pentandra*) oil biodiesel as an alternate fuel for diesel engine. *Energy Conversion and Management*, 75, 773–779.

Venkanna, B.K. and Reddy, C.V., 2009. Biodiesel production and optimization from *Calophyllum inophyllum* Linn. oil (Honne oil): A three stage method. *Bioresource Technology*, 100(21), 5122–5125.

Vijayaraj, S. and Sarangan, J., 2012. Performance and emission characteristics of methyl ester from non-edible oils in a DI diesel engine with additive and advance injection. *International Journal of Alternative Propulsion*, 2(2), 89–108.

Wang, L.B., Yu, H.Y., He, X.H. and Liu, R.Y., 2012. Influence of fatty acid composition of woody biodiesel plants on the fuel properties. *Journal of Fuel Chemistry and Technology*, *40*(4), 397–404.

Wang, R., Hanna, M.A., Zhou, W.W., Bhadury, P.S., Chen, Q., Song, B.A. and Yang, S., 2011. Production and selected fuel properties of biodiesel from promising non-edible oils: *Euphorbia lathyris* L., *Sapium sebiferum* L. and *Jatropha curcas* L. *Bioresource Technology*, *102*(2), 1194–1199.

Williamson, A.M. and Badr, O., 1998. Assessing the viability of using rape methyl ester (RME) as an alternative to mineral diesel fuel for powering road vehicles in the UK. *Applied Energy*, *59*(2–3), 187–214.

Yadav, A.K., Khan, M.E. and Pal, A., 2017a. Biodiesel production from oleander (*Thevetia Peruviana*) oil and its performance testing on a diesel engine. *Korean Journal of Chemical Engineering*, *34*(2), 340–345.

Yadav, A.K., Khan, M.E. and Pal, A., 2019. Custard apple seed oil as promising biodiesel feedstock using advanced techniques and experimental investigation on diesel engine. *International Journal of Oil, Gas and Coal Technology*, *20*(4), 473–492.

Yadav, A.K., Khan, M.E., Pal, A. and Dubey, A.M., 2017b. Experimental investigations of performance and emissions characteristics of Kusum (*Schleichera oleosa*) biodiesel in a multi-cylinder transportation diesel engine. *Waste and Biomass Valorization*, *8*(4), 1331–1341.

Yang, F.X., Su, Y.Q., Li, X.H., Zhang, Q. and Sun, R.C., 2008. Studies on the preparation of biodiesel from *Zanthoxylum bungeanum* Maxim seed oil. *Journal of Agricultural and Food Chemistry*, *56*(17), 7891–7896.

Yatish, K.V., Lalithamba, H.S., Suresh, R. and Omkaresh, B.R., 2016. Synthesis of biodiesel from *Garcinia gummi-gutta*, *Terminalia belerica* and *Aegle marmelos* seed oil and investigation of fuel properties. *Biofuels*, *9*(1), 121–128.

Zhang, J. and Jiang, L., 2008. Acid-catalyzed esterification of *Zanthoxylum bungeanum* seed oil with high free fatty acids for biodiesel production. *Bioresource Technology*, *99*(18), 8995–8998.

6

Development of Low-Cost Production Medium and Cultivation Techniques of Cyanobacteria *Arthrospira platensis* (Spirulina) for Biofuel Production

R. Dineshkumar, N. Sharmila Devi, M. Duraimurugan,
A. Ahamed Rasheeq, and P. Sampathkumar

CONTENTS

6.1 Introduction 165
    6.1.1 Biofuel Importance 166
    6.1.2 Large-Scale Biofuel Production 167
        6.1.2.1 Cultivation Methods..167
    6.1.3 Mass-Scale Production 168
    6.1.4 Commercial Biofuel Production Using Microalgae Species............................168
    6.1.5 *Spirulina* in Biofuel Production 168
6.2 Materials and Methods..169
    6.2.1 Isolation and Identification of Microalgae 169
    6.2.2 Culture Development and Maintenance 169
    6.2.3 Mass-Scale Production of *Spirulina platensis* 169
    6.2.4 Determination of Algal Growth 169
    6.2.5 Estimation of Carbon Content and Carbon Dioxide Fixation Rate 170
    6.2.6 Determination of Total Biomass 170
    6.2.7 Biochemical Composition of Dry Biomass in *Spirulina platensis* 171
6.3 Results 171
    6.3.1 Growth Factor and Biomass Production 171
    6.3.2 Carbon Content and Carbon Dioxide Fixation Rate 172
    6.3.3 Biochemical Composition 172
6.4 Discussion 173
6.5 Conclusion 174
Acknowledgment 175
References..175

6.1 Introduction

Globally, there are nearly 8,700 marine algal species described which include many microalgae and macroalgae that play a fundamental role in the marine environment as both food and for the habitats they form (Ali *et al*., 2010; Guiry, 2012). Algae grow everywhere—wherever they find suitable sources of light, water, and nutrients—and can be seen from Antarctic ice to hot springs and from the tip of the tropical rain forests to the cryptoendolithic species inside the rocks; moreover, these are responsible for the half of the photosynthesis in the world. Some algae aid in fixing of nitrogen in nutrient-poor systems, from paddy fields to coral reefs. A wide variety of essential metabolites—including proteins, lipids, carbohydrates, carotenoids, and vitamins that have a pivotal role in health, food and feed additives, antioxidants, cosmetics, and for energy production—were documented (Mata et al., 2010; Williams and Laurens, 2010; Kalidasan et al., 2015).

Blue-green algae can be easily cultivated in a wide range of environments; are able to grow in freshwater, brackish water, and sea water; and are used all over the globe as a food product and as a rich source of dietary supplement (Borowitzka, 1999; Liu et al., 2010; Sakthivel et al., 2011; Priyadarshani and Rath, 2012). High amounts of protein—between 55% and 70% dry weight—is found in *Spirulina* (Dineshkumar et al., 2015). The natural source of vitamin B_{12} are the carotenoids, xanthophylls, and phycocyanin which are extracted from the *Spirulina* biomass (Henrikson, 1989; Spolaore et al., 2006; Kemka et al., 2007).

The demand for *Spirulina* has been increasing day by day. Production of *Spirulina* with cheap costs is obligatory while considering the large-scale production for industrial purposes. The high cost of nutrients stands as a second major factor that influences the production cost of biomass, following labor charges (Vonshak, 1997). Zarrouk medium (Zarrouk, 1966) has effectively aided as the standard medium for culture of *Spirulina* for many years. Conversely, they have several disadvantages, including the higher rate of mineral costs. Zarrouk or modified Zarrouk medium or Society of Toxicology (SOT) medium are highly expensive since they require increased quantities of $NaHCO_3$, $NaNO_3$, Na_2CO_3, and some trace metals. The economics of production mainly depend on the cost of biomass production. Consequently, a low-cost medium with a minimal number of expensive synthetic chemicals will reduce the cost.

To overcome the demand of producing low-cost, nutritive, and convenient foods, studying out the feasible sources with positive environmental and economic impacts are much needed (Bolanho et al., 2014). Several epidemiological studies recommend the vitamins, minerals, and other nutrient intakes help in protecting the body against heart disease, cancer, and the aging process, and the antioxidants also have a positive effect on prevention or reduction of the severity of the diseases (Hsia et al., 2007; Namasivayam and Steele, 2015).

From this view, this study was focused on evaluation of large-scale production of algal biomass of *Spirulina platensis* in a low-cost simple nutritive medium (baking soda) and also subjecting the harvested biomass for carbon sequestration and biochemical composition estimation.

6.1.1 Biofuel Importance

With the vital demand for fuel, the scientific community is urged to look for alternative biofuel sources for various operations worldwide. The concept of biofuel demand is that they are really accepted renewable sources of energy in support of the significant properties of their non-toxicity and biodegradability. Another notable reason why biofuel is safer is the rising level of carbon dioxide emissions. Also, these fuels have very low sulfur content and have excellent lubricity. First-generation biofuels are defined as those derived from plant oils and from the fermentation of starch from corn flour or sugarcane molasses. Unfortunately, biofuel production from food crops will adversely affect the economic value of food. Related to that, non–food-based feedstock and technology are showing interest everywhere for the target of biofuel production. The current trend in biofuel production has aimed to alter the energy compounds like lipids, polysaccharides, and other hydrocarbons, which are highly stored in algae genetic patterns and metabolic pathways. The thought of using genetically engineered algae in biofuel production is not only resulting in high yields of biofuels, but it is also economically effective. The problem of making biofuel for engines is heavy investment required, and that the existing technology makes it complicated for use in engines. Various byproducts are obtained from agricultural sources and it is highly competitive for the diversion of using other kinds of resources.

Nowadays, genetic engineering techniques are really primarily focused on use for the production of biohydrogen, diesel fuel substitutes, alcohols obtained from starch, etc., from microalgae. The energy-containing compounds like lipids, polysaccharides, and other hydrocarbons stored in microalgae are manipulated by using advanced genetic engineering techniques. Thus, the genetically engineered algae species not only give higher yield of biofuel; they leave a safe environment by releasing limited carbon dioxide content. As biofuel is obtained from microalgae, which is renewable and biodegradable source, it is ecofriendly and can be the perfect alternative to the conventional fossil fuels. For the better exchange of diesel, the current study and various experiments are suggested that can use fatty acid methyl esters

(FAMEs, biodiesel), fatty alcohols, alkanes, and linear or cyclic isoprenoids. In general, biofuel engine production takes a lot of investment and it is very tedious to use biofuels directly in engines commercially.

6.1.2 Large-Scale Biofuel Production

Biofuels are highly encouraged as sustainable energy supporters with safe zones of reduced Greenhouse Gas emissions (GHGs). There are plenty of income opportunities based on this biofuel production for farmers subsisting in rural areas. In due course, improved development systems will fulfill the earth's need associated with environmental factors. Commercial producers of biofuel production have to face to many challenges in various business angles that have to be met in profitable manner for success.

Commercial-scale producers of this biofuel have to maintain the regulatory requirements for safety standards. The term biofuel is defined as sourcing renewable raw materials using solid, liquid, or gaseous fuels. In literature, agricultural crops are categorized as first-generation biofuels because of their use in food; in that way, using microalgae for biofuel production is not using such agricultural land for processing.

6.1.2.1 Cultivation Methods

The microalgae nature with oil content and their physico-chemical characteristics are similar to vegetable oils. They are more concerned with raw materials for the production of biofuels. The conventional system of microalgae cultivation is showing higher productivity; thus how many scientific studies about microalgae in biofuel production exhibit that they can meet the global demand in a promising manner. They are also considered a good source of lipid content, possessing notably 15–75% dry weight according to various studies based on the species and characteristics.

The growth characteristics of microalgae importantly depend on the cultivation methods based on phototrophic, heterotrophic, mixotrophic, and photoheterotrophic techniques. The cultivation methods in various ways are environmentally advantageous, like renewable energy and raw material conversion. Few studies reported that supplementation of carbondioxide could increase the productivity of biofuel. Additionally, some microalgae will use organic carbon in the absence of light, and that kind of cultivation method is specified as heterotrophic. The positive results in these heterotrophic methods are showing limited problems associated with light in microalgal biomass production. In another method, the mixotrophic way will produce the microalgae for better living conditions under both phototrophic and heterotrophic conditions. In mixotrophic methods, the microalgae use light as the only source, whereas in photoheterotrophic methods, the microalgae require both light and other organic resources for biomass production. The study also reported that in both the methods, the contamination is much reduced.

Microalgae hold high carbon dioxide sequestering, along with high photosynthetic levels and higher growth rates. Particularly, they use nitrogen and phosphorous from agricultural and industrial wastewater that naturally reduce the nutrient content in wastewater. Microalgae contain considerable amounts of lipids, which produce non-toxic and highly biodegradable biofuels. The algae cultivation for the production of biofuel is very simple to grow and process, due to their simple nature and growth characteristics. On the other hand, many factors will adversely affect the optimum algal growth, lipid accumulation, CO_2, temperature, pH, and light. The prime suggestion for biofuel production is to select the best species for optimization of higher productivities and biofuel production under various conditions. Performance of that species is to enhance its both biomass and biofuel production.

The selection of cultivation method of microalgae for biofuel production is very important due to its phytoremediation and yield of biofuel. Open-pond systems are comparatively less expensive, but they are weaker in controlled pattern than the closed system of microalgae cultivation. According to the literature, the most commonly used cultivation systems are raceway ponds, circular pond tanks, closed ponds, and shallow big ponds. The drawbacks to using this system are poor light penetration, unregulated temperature, evaporation loss, and CO_2. Another effective method is photobioreactors, which have higher productivity rates than open ponds due to their well-maintained controlled growth parameters which ensure good light penetration. Later, in third-generation feedstock utility, microalgae possess a huge potential for biofuel production because of their fast growth, great biomass yield, and high lipid and carbohydrate contents. In general, biodiesel, biogas, bioethanol, and biomethane are the potential biofuels produced by microalgae (Tan et al., 2020). According to the size of the microalgae, the small (μm) size has to be cultivated in a system keenly designed for that use.

6.1.3 Mass-Scale Production

Large-scale systems of microalgae cultivation were developed during the 20th century. Generally, microalgae species can be cultivated in a wide range of cultivation systems for commercial production. The production mainly requires inputs such as energy, water, CO_2, and mineral nutrients. The better development of microalgae cultivation methods is still ongoing in an effective manner to propose the economically sound large-scale production of microalgae cultivation for biomass production. In this, the quality of the biofuel to the level of international standards are undergoing various process to bring effective quality (Cruz et al., 2018).

Related to the experimental studies, there are two main approaches for cultivating microalgae on a large scale: open and closed cultivation methods. The main difference between open and closed systems is related to how they operate (e.g., cooling and gas exchange), vulnerability to outside influences (e.g., rainwater and introduction of unwanted species), and costs for building and operating the system.

6.1.4 Commercial Biofuel Production Using Microalgae Species

Microalgae are resources to be used for the production of renewable energy. In order to cultivate these microalgae in a commercial way, many methods are propagated and implemented. Due to its wide range of species with different growth and other characteristics, the method of cultivation for the biofuel production is an important concern. Apart from that, the commercial analysis of economic values of microalgal potential is a trial and basis, as each species differ in their biochemical content and capacity, and show major changes in culturing pattern and biomass quantity.

For microalgae, the development of dedicated culture systems only started in the 1950s when algae were investigated as an alternative protein source for the increasing world population. Later, algae were researched for the interesting compounds they produce, to convert CO_2 to O_2 during space travel and for remediation of wastewater. The energy crisis in the 1970s initiated research on algae as a source of renewable energy. In addition, cultivation of microalgae can be performed using wastewater such as domestic sewage water and palm oil milling effluents, which can provide aid for the purpose of bioremediation, along with biofuel production. Perfect microalgae cultivation systems work with the following factors, as clearly explained in experimental reviews: surplus light source, transfer of material across the liquid–gas barrier, simple procedures, minimal contamination rate, cheap overall building and production costs, and high land efficiency. The main advantages of using microalgae are: they do not require land like agriculture, simple cultivation systems are required, there is no competition with agriculture, there is less consumption of water compared to all terrestrial living organisms, there is no seasonal basis, and higher level of productivity with lipid content in dry weight, wastewater purifying methods, bioremediation methods, and environmentally carbon dioxide sequestration.

6.1.5 *Spirulina* in Biofuel Production

Basically, the *Arthrospira platensis* is a planktonic filamentous cyanobacterium which is composed of individual cells and grows in subtropical alkaline lakes with the temperature optimally above 35°C. In cultures, they are cultivated in shallow mixed ponds or semi-closed tubular photobioreactors in a successful method. In laboratory scale methods, the growth medium consists of inorganic salts with a high concentration of bicarbonate, which maintains a pH value between 9 and 10. Apart from this biofuel production, *Spirulina* sp. is the most cultivated photosynthetic one and is widely used in health food, feed supplements, and as a source of fine chemicals. Experimental determination of this species contains proteins, polyunsaturated fatty acids (PUFA), phycobiliproteins, carotenoids, polysaccharides, vitamins, and minerals.

According to the morphological nature, *Spirulina* (*Arthrospira platensis*) is a filamentous and multicellular blue-green algae which is effectively capable of reducing inflammation and also manifesting antioxidant effects in living organisms. They are rich in multiple sources of vitamins, especially vitamin B_{12}; minerals; proteins; and carotenoids. It is multicellular filamentous, and alkaliphilic algae referred to as *Spirulina*, a name previously given to this commercially exploited species.

The research on using the *Spirulina platensis* for biofuel production is a significant part of global demand of bioenergy. In many studies, it has been revealed that the effect of reaction variables affect the cost of biodiesel production from the non-edible *Spirulina platensis*. With the objective of the application of biodiesel production technology using an acid catalyst to the *in situ* transesterification of microalgae, *Spirulina platensis* shows a cost-effective process over all. According to the studies, these variables are the catalyst concentration, reacting alcohol volume, the temperature, the reaction time, and process stirring. Evaluation of different variables that are affecting the hydrothermal liquefaction of microalgae (*Spirulina platensis*) under subcritical water conditions are studied and revealed with favorable responses in the case of biofuel production, where they are using factorial design and response surface methodology to determine the physical and chemical properties of the product.

In another study, biocrude oil was produced by using the method of hydrothermal liquefaction of microalgae *Spirulina platensis* under subcritical water conditions. The results using *Spirulina* sp. for biofuel production expose high lipid content of algae is not the exclusive choice for future biofuel applications. The microalgae for the production of biofuel is nowadays becoming promising, beside many disadvantages (Masojídek and Torzillo, 2008).

This chapter describes our research in the field of development of low-cost production medium and cultivation techniques of cyanobacteria *Arthrospira platensis* (*Spirulina*) for biofuel production.

6.2 Materials and Methods

6.2.1 Isolation and Identification of Microalgae

Cyanobacteria, the marine microalgae, were collected from the Vellar estuary, southeast coast of India, Cuddalore district, Tamil Nadu state. *Spirulina* was isolated by serial dilution method and identified based on the morphological and microscopical characteristics. The serial dilution method was developed in the algal culture laboratory, Faculty of Marine Sciences, Annamalai University, India (Prescott, 1959).

6.2.2 Culture Development and Maintenance

Pure cultures of *Spirulina platensis* were maintained in Zarrouk medium at 4°C (Tomaselli, 1999). In order to develop the culture, a loop of *Spirulina* was inoculated in a 50ml flask containing 10 ml of Zarrouk medium under sterile conditions (Zarrouk, 1966), and the pH was adjusted to 8.8–9.0. The culture was maintained in an illuminated (4500 lux = lm/m²) culture room at 30 ± 2°C under 12/12-hour light-dark cycles. The culture was shaken manually three times daily (Ehimen *et al.*, 2010).

6.2.3 Mass-Scale Production of *Spirulina platensis*

The mass-scale production of *Spirulina platensis* was made on low-cost medium containing environmental cultivation baking soda (sodium bicarbonate) 60 g/L (simple medium), potassium nitrate 8 g/L, iron sulfate 0.05 g/L and sea salt 5 g/L. The comparative study was made with *Spirulina platensis* cultivated in Zarrouk medium under laboratory cultivation. Comparison of cost (trial grade) of the media is shown in the Table 6.1 and Table 6.2. A 250 ml Erlenmeyer flask was filled with 100ml of Zarrouk medium inoculated with pure culture of *Spirulina platensis* (Zarrouk, 1966).

6.2.4 Determination of Algal Growth

In order to measure the optical density (OD), samples were analyzed at 680 nm (OD_{680}) using a spectrophotometer (Genesys 5, Spectronic Instruments, UK).

TABLE 6.1

Composition and Approximate Cost for 1,000 Liters of Zarrouk Medium for Cultivation of *Spirulina platensis*

S. No.	Ingredients	Quantity of Chemicals in Kilograms/1,000 Liters	Price in Rupees for 1,000 Liters	Price in US Dollars for 1,000 liters (1 USD = 64.97 Rupees)
1	$NaHCO_3$	8	4032	62.06
2	$NaCl$	5	1260	19.39
3	$NaNO_3$	2.5	1500	23.09
4	K_2HPO_4	0.5	366	5.63
5	$MgSO_4 7H_2O$	0.16	504	7.76
6	$FeSO_4 7H_2O$	0.01	87.7	1.35
7	K_2SO_4	0.5	7.3	0.11
Total Cost = Rs. 7,758				Total Cost = 119.39 (USD)

TABLE 6.2

Composition and Approximate Cost for 1,000 Liters of Simple Nutrient Media (Baking Soda) Cultivation for *Spirulina platensis*

S. No.	Ingredients	Quantity of Chemicals in Kilograms/1,000 Liters	Price in Indian Rupees (1,000 Liters)	Price in US Dollars for 1,000 Liters (1 USD = 64.97 Rupees)
1	Baking soda	7	600	9.24
2	Sea salt	1	20	0.31
3	KNO3	0.2	250	3.85
4	FeS	0.20	120	1.85
Total Cost = Rs. 990				Total Cost = 15.25 (USD)

6.2.5 Estimation of Carbon Content and Carbon Dioxide Fixation Rate

In a 500 ml conical flask, 0.2 mg of dried algal samples were added and 10 ml of 1N potassium dichromate and 20 ml of concentrated H_2SO_4 mixture was diluted with 200 ml of distilled water and 10 ml of hydro phosphate (H_3PO_4), and 1 ml of diphenyl amine was added. Later, 4 N ferrous ammonium sulfate (FAS) was titrated against it with the end point of brilliant green color. The carbon content was estimated using the formula A = 3.951/g (1−T/S) where carbon content is A, weight of the sample in grams is g, FAS with blank is T, FAS with sample (ml) is S. Carbondioxide fixation rate was estimated using the formula of Yun *et al.* (1997).

$$R\ CO_2 = C_c \times \mu L\ (M_{CO_2}/M_c)$$

where $R\ CO_2$ and μL are the CO_2 fixation rate (g CO_2 m^{-3} h^{-1}) and the volumetric growth rate (g dry weight m^{-3} h^{-1}), respectively, in the linear growth phase. M_{CO_2} and M_C represented the molecular weights of CO_2 and elemental carbon, respectively, and C_C is average carbon content (algal dry weight/g).

6.2.6 Determination of Total Biomass

The biomass was harvested in the final day by alum using the flocculation technique and dried at room temperature. The empty petri plates were weighted previously to prevent errors. The harvested dried and filtered biomass was kept in the petri plate and weighted (dry weight).

The total biomass can be calculated by using the following formula:

$$\text{Total biomass} = \text{Dry weight} - \text{Initial weight (wet biomass)}$$

6.2.7 Biochemical Composition of Dry Biomass in *Spirulina platensis*

Protein content (Lowry *et al.*, 1951), carbohydrate (CHO) (Dubois *et al.*, 1956), total lipids (Bligh and Dyer, 1959), chlorophyll (Richardson, 2002), and carotenoids (Palett and Young, 1993) and various other biochemical parameters were analyzed. The cost/benefit ratio was worked out for the two media used, and is shown in Indian rupees (INR) and U.S. dollars (USD).

6.3 Results

6.3.1 Growth Factor and Biomass Production

The highest growth rate was recorded in simple nutrient medium (baking soda) at OD_{680}–3.38 and Zarrouk medium (selective media) OD_{680}–3.29 followed on the 28th day by decline phase (Figure 6.1). The first 16 days average specific growth rate was found to be 0.03, 0.052, 1.56, 2.50, and 0.948 day^{-1} for Zarrouk medium. The specific GR (growth rate) in simple nutrient medium in the first 16 days was found to be 0.03, 0.067, 1.49, 2.38, and 2.79 day^{-1}. The maximum GR was showed on the 20th day in Zarrouk medium and simple nutrient medium OD_{680} as 2.98 and 2.99, respectively.

The *Spirulina* grown in Zarrouk medium and simple nutrient medium (baking soda) were analyzed and compared. The obtained biomasses were subjected to biochemical analysis, and the chemical composition was found and calculated to find the comparison between the two media. Comparing all those parameters except protein, it is found that dry biomass from baking soda showed promising results (Table 6.2 and Table 6.3). The biomass weight grown in Zarrouk medium was found to be 2.084 g/L, and in case

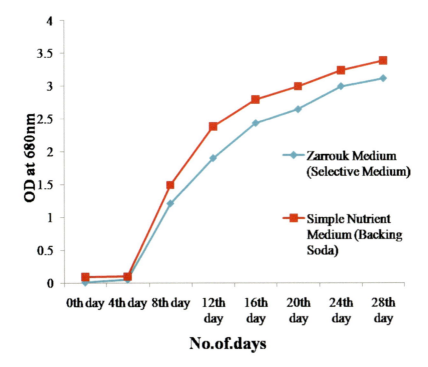

FIGURE 6.1 Algal growth curves in the two different mediums at 680 nm.

of baking soda, it was yielded to be 4.815g/L. The dry weight in Zarrouk-grown medium was 0.368 g/L and 0.602 g/L was recorded in baking soda–grown medium.

6.3.2 Carbon Content and Carbon Dioxide Fixation Rate

The algal biomass was taken and the carbon content was analyzed. The increasing carbon content enhanced the fixation of CO_2 (Table 6.4). Maximum of 19.367 g/dry weight was the carbon content of *Spirulina platensis* and 0.070 g/ml/day was the CO_2 fixation rate.

6.3.3 Biochemical Composition

The Zarrouk medium–grown biomass had a protein content of about 47.56 ± 1.33 mg/G, while in baking soda, it showed 47.73 ± 1.28 g/L; carbohydrates held 39.09 ± 0.92 mg/G in Zarrouk and in baking soda showed 39.13 ± 0.85. The lipid content from *Spirulina platensis* with the dry weight obtained from Zarrouk medium was 9.3 ± 0.12 and baking soda showed 9.8 ± 0.04. Total chlorophyll was analyzed and found to be 39.76 ± 0.78 in Zarrouk medium, while in baking soda, the chlorophyll was observed to be 35.76 ± 0.61. The total carotenoids were analyzed and recorded to be 31.63 ± 0.79 in Zarrouk and 32.14 ± 0.66 in simple nutrient medium (Figure 6.2).

TABLE 6.3

Estimation of Total Biomass in Zarrouk Medium and Baking Soda

Algae	Zarrouk Medium (g/L)		Baking Soda (g/L)	
Spirulina platensis	**Wet Weight**	**Dry Weight**	**Wet Weight**	**Dry Weight**
	2.084	0.368	4.815	0.602

TABLE 6.4

Estimation of Carbon and CO_2 Fixation Rate of *Spirulina platensis*

Algae	Carbon Content (g/Dry Weight)	CO_2 Fixation Rate (g/ml/Day)
Spirulina platensis	19.367	0.070

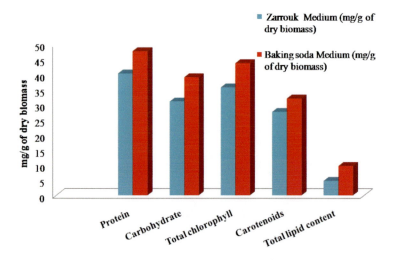

FIGURE 6.2 Biochemical and elemental analysis of cyanobacteria, *Spirulina platensis* (dry biomass basis).

6.4 Discussion

Spirulina can be grown naturally in lakes, ponds, and rivers, and artificially in photo bioreactors (PBRs), artificial ponds, or basins (Borowitzka, 1999; Huang et al., 2011). About 10–12 grams of *Spirulina* per square meter can be grown in a pond with suitable parameters (Sakthivel et al., 2011). Light intensity, nutrients, pH, and various other environmental factors affect the growth of the organisms (Danesi, 2004; Colla et al., 2007; Ogbonda et al., 2007; Hempel et al., 2012; Priyadarshani et al., 2014). In this study, the dry biomass obtained from *Spirulina platensis* from Zarrouk medium and simple nutrient medium (baking soda) were analyzed and compared, between which dry biomass obtained from the new baking soda used medium came out with astonishingly good results (Figure 6.2). The wet weight of Zarrouk medium was 2.084 g/L, and the wet weight of baking soda medium was 4.815g/L; 0.368 g/L was the dry weight of Zarrouk medium and the dry weight of baking soda medium was 0.602 g/L. It is also observed that after the 20th day, there was a gradual decreasing trend in growth rate with the unfavorable environmental conditions for microalgae culture; similar results were reported in previous studies (González-Fernández and Ballesteros, 2012; Ma et al., 2014) of *Chlorella vulgaris* and *Nannochloropsis gaditana*, respectively. Phytoplankton cells have the capacity to harvest and use the light by changing their physiological properties. The pigments which are involved in light harvest are increasing at low light regimes by maximizing the use of available energy (Finkel et al., 2004).

As low yields have been one of the major limiting factors in the production of algal biomass, the attainment of maximal productivity and its maintenance should be one of the most important operational objectives. Harting et al. (1988) studied the growth of *Spirulina platensis* using a flat plate photobioreactor with a light path of 7.5–200 nm. The results showed that microalgae had cell density in the optimal range and biomass productivity to be high. The total biomass of the bold basal medium (BBM) and the sewage water cultured *C. vulgaris* were found to be 2.034 g/L and 3.615 g/L (dry weight), and 0.268 g/L and 0.402 g/L (wet weight), respectively. The physico-chemical parameters of the sewage water were analyzed initially, and at the end of the study, to determine the chemical consumption by the microalgae. Likewise, the protein, carbohydrate, and lipid content of the BBM and the sewage water cultured *C. vulgaris* were recorded (34.56 ± 1.33 and 36.56 ± 1.28 mg/g), (41.09 ± 0.92 and 42.13 ± 0.85 mg/g), and (28.20 ± 0.89 and 28.68 ± 0.82 mg/g), respectively (Dineshkumar et al., 2017).

In the present study, the algal biomass was taken and the carbon content was analyzed. The increasing carbon content enhanced the fixation of CO_2 and maximum of 19.367 g/dw was the carbon content of *Spirulina platensis*, and 0.070 g/ml/day was the CO_2 fixation rate. *Spirulina platensis* can be suggested for CO_2 sequestration, as it has very high fixing level of CO_2. The carbon fixation rate was found to be higher, about 0.070. The bacteria uses oxygen and bio-oxidizes the organic compound in wastewater. During photosynthesis, the carbondioxide is filled in the cell carbon as an end product (Wang et al., 2009). Similar reports were made in previous studies by several authors; for instance: in the first method, it was presumed that 1g of biomass produced equal to 1.8 g of carbon dioxide recycled, and it was imprecise (Chisti, 2007; Sudhakar et al., 2012), while in the second method employed, the parameters such as the molecular weight of C and CO_2, percentage of carbon in biomass, and the productivity was a highly precise method (De Morais and Costa, 2007). Hence, the estimation of the amount of captured CO_2 depends on the carbon content present in the biomass. The higher productivity means higher rate of biomass growth, and it results in a larger amount of sequestrated carbon dioxide (CO_2) (De Morais and Costa, 2007). *Spirulina platensis* was found to grow better at higher CO_2 concentrations. Microalgae are very commonly used in carbon sequestration. Microalgae easily grow at high levels of CO_2 and their cell organism containing chlorophyll a and chlorophyll b have high photosynthetic efficiency to convert CO_2 to O_2 (Singh and Singh, 2014). Both microalgae have shown higher biomass concentration at CO_2 concentration, from 0.03%–10%, respectively. However, some researchers showed that the CO_2 concentration was found higher than 5%, and it shows some negative effects on microalgae growth (Chiu et al., 2008; De Morais and Costa, 2007; Yoo et al., 2010). The increase in CO_2 sequestration is very efficient by maneuvering chemically aided biological sequestration of CO_2; *Chlorella* sp. and *Spirulina platensis* showed 46% and 39% mean fixation efficiency, respectively, and an input CO_2 concentration of 10% (Dineshkumar et al., 2017).

The Zarrouk medium–grown biomass had the maximum protein content, carbohydrates, lipid, total chlorophyll and carotenoids. All the components were found to be higher in baking soda medium when compared to that of the same grown in Zarrouk medium.

The Zarrouk medium–grown biomass had the protein content of about 47.56 ± 1.33 mg/g, while in baking soda it showed 47.73 ± 1.28 g/L; carbohydrates hold 39.09 ± 0.92 mg/g in Zarrouk, and baking soda showed 39.13 ± 0.85. The protein production of 7.0 mgL^{-1} in Chu and 6.8 mgL^{-1} in WC media are higher than those obtained in other studies, but similar to the results of Bertoldi *et al.* (2008). The highest carbohydrates production by *Spirulina platensis* in the LC oligo cultures was 7.36 μg.mL^{-1}.

Under stressed conditions, algae produce more lipids, and the amount of algae is also being reduced (Griffiths and Harrison, 2009; Prartono *et al.*, 2010; Fulke *et al.*, 2013). More lipids are accumulated in the cell by microalgae under nutrient depletion; about 30%–50% were found to be favorable on accumulation of PUFA, total lipids, and fatty acids (Makareviciene *et al.*, 2011; Tang *et al.*, 2011). Microalgae was grown in nutrient media with optimum CO_2 for a period of 14 days in the present study. The lipid content from *Spirulina platensis* with the dry weight obtained from Zarrouk medium shows 9.3 ± 0.12 and baking soda shows 9.8 ± 0.04. There was less lipid accumulation by test algae species when compared with previous research results; this lipid content reflects microalgae as a source of biodiesel. The test algae can accumulate more lipids if they are grown under nutrient limitations (Pokoo-Aikins *et al.*, 2010).

Total chlorophyll was analyzed and found to be 39.76 ± 0.78 in Zarrouk medium, while in baking soda, the chlorophyll was observed to be 35.76 ± 0.61 (Cruz *et al.*, 2018). The *Spirulina platensis* batch cultures which were grown in the baking soda medium reporting with a maximum chlorophyll a concentration of 5 μg.mL^{-1}) within 10 days has been reported by Chinnasamy *et al.* (2009). After 28 days of the culture, the total chlorophyll content was observed to be 43.76 ± 0.61. The total chlorophyll a was also analyzed and it was found to be 40 ± 1.73 mg/g. The carotenoids were analyzed, and showed 29.18 ± 0.93 mg/g. The lipid content of *Spirulina platensis* obtained from the dry weight as 8.3 ± 0.16 mg/g (Dineshkumar *et al.*, 2017). The decrease in content of chlorophyll a and other light-harvesting pigments with increasing irradiance (Dubinsky *et al.*, 1995) is consistent with the highest pigment contents obtained at 60–105 μmol photons m^{-2}/S^{-1}. However, Finkel *et al.* (2004) reported an increase in intracellular chlorophyll a content with irradiance at very low levels of light, which shifted to an inverse relationship above "a" threshold value due to saturation effect in photo harvesting.

The simple nutrient medium baking soda has a good CHO production of about 39.13 ± 0.85. With semi-continuous grown cells, the appropriate culture conditions can be used and high productivity can be prolonged for longer times (Cruz *et al.*, 2018). The microalgal biomass is showing constant biochemical composition, which can be controlled in order to increase the nutritive value my manipulating the environmental parameters (Ferreira and Gyourko, 2009). Baking soda medium has shown very potent results, as it is much cheaper than the synthetic media, reducing the cost of production (Table 6.1 and Table 6.2).

6.5 Conclusion

Baking soda medium provides good mass-scale production in an eco-friendly and economically cheaper way. The production of value-added products enriched with protein, carbohydrates, carotenoids, and total lipids from *Spirulina* microalgal biomass can help farmers and improve their livelihoods. From the results of the experiment reported in this chapter, *S. platensis* could adopt and thrive well in both media, as there was no lag phase found. Moreover, this study shows the high-growth CO_2 fixation. Studies must be done in this field to confirm the biological activity of different fractions of *Spirulina platensis*. Above all, this study confirms that baking soda can be used as a cheaper medium to spur mass production of *Spirulina* and benefit the common people, and it has various uses in the pharmaceutical industry. The cost/benefit ratios of both the media describe that the algae growth in Zarrouk medium was found to be approximately 10 times higher than that of baking soda media. It depicted that the cultivation of algae in this alternate medium will surely increase the profit to the producers to a greater extent.

Acknowledgment

The authors are thankful to Dean and Director, CAS in Marine Biology, Faculty of Marine Sciences, Parangipettai-608502, Tamil Nadu, India, for providing the necessary lab facilities.

Conflict of interest statement: We declare that we have no conflict of interest.

REFERENCES

Ali, M., Rau, A.R.P. and Alber, G., 2010. Quantum discord for two-qubit X states. *Physical Review A*, *81*(4), p. 042105.
Bertoldi, F.C., Sant'Anna, E. and Oliveira, J.L.B., 2008. Chlorophyll content and minerals profile in the microalgae *Chlorella vulgaris* cultivated in hydroponic wastewater. *Ciência Rural*, *38*(1), 54–58.
Bligh, E.G. and Dyer, W.J., 1959. A rapid method of total lipid extraction and purification. *Canadian Journal of Biochemistry and Physiology*, *37*(8), pp. 911–917.
Bolanho, B.C., Danesi, E.D.G. and Beléia, A.D., 2014. Characterization of flours made from peach palm (*Bactrisgasipaes kunth*) by-products as a new food ingredient. *Journal of Food and Nutrition Research*, *53*(1), pp. 51–59.
Borowitzka, M.A., 1999. Commercial production of microalgae: Ponds, tanks, and fermenters. In *Progress in industrial microbiology* (Vol. 35, pp. 313–321).
Chinnasamy, S., Ramakrishnan, B., Bhatnagar, A. and Das, K., 2009. Biomass production potential of a wastewater alga *Chlorella vulgaris* ARC 1 under elevated levels of CO_2 and temperature. *International Journal of Molecular Sciences*, *10*(2), pp. 518–532.
Chisti, Y., 2007. Biodiesel from microalgae. *Biotechnology Advances*, *25*(3), pp. 294–306.
Chiu, S.Y., Kao, C.Y., Chen, C.H., Kuan, T.C., Ong, S.C. and Lin, C.S., 2008. Reduction of CO_2 by a high-density culture of *Chlorella* sp. in a semicontinuous photobioreactor. *Bioresource Technology*, *99*(9), pp. 3389–3396.
Colla, L.M., Reinehr, C.O., Reichert, C. and Costa, J.A.V., 2007. Production of biomass and nutraceutical compounds by *Spirulina platensis* under different temperature and nitrogen regimes. *Bioresource Technology*, *98*(7), pp. 1489–1493.
Cruz, Y.R., Aranda, D.A.G., Seidl, P.R., Diaz, G.C., Carliz, R.G., Fortes, M.M., da Ponte, D.A.M.P. and de Paula, R.C.V., 2018. *Cultivation systems of microalgae for the production of biofuels*. https://doi.org/10.5772/intechopen.74957.
Danesi, M., 2004. *Messages, signs, and meanings: A basic textbook in semiotics and communication* (3rd ed., Vol. 1). Studies in Linguistic and Cultural Anthropology.
De Morais, M.G. and Costa, J.A.V., 2007. Carbon dioxide fixation by *Chlorella kessleri, C. vulgaris, Scenedesmus obliquus* and *Spirulina* sp. cultivated in flasks and vertical tubular photobioreactors. *Biotechnology Letters*, *29*(9), pp. 1349–1352.
Dineshkumar, R., Narendran, R., Jayasingam, P. and Sampathkumar, P., 2017. Cultivation and chemical composition of microalgae *Chlorella vulgaris* and its antibacterial activity against human pathogens. *Journal of Aquaculture and Marine Biology*, *5*, p. 00119.
Dineshkumar, R., Umamageswari, P., Jayasingam, P. and Sampathkumar, P., 2015. Enhance the growth of *Spirulina platensis* using molasses as organic additives. *World Journal of Pharmaceutical Research*, *4*(6), pp. 1057–1066.
Dubinsky, A.J., Yammarino, F.J., Jolson, M.A. and Spangler, W.D., 1995. Transformational leadership: An initial investigation in sales management. *Journal of Personal Selling & Sales Management*, *15*(2), pp. 17–31.
Dubois, M., Gilles, K.A., Hamilton, J.K., Rebers, P.T. and Smith, F., 1956. Colorimetric method for determination of sugars and related substances. *Analytical Chemistry*, *28*(3), pp. 350–356.
Ehimen, E.A., Sun, Z.F. and Carrington, C.G., 2010. Variables affecting the in situ transesterification of microalgae lipids. *Fuel*, *89*(3), pp. 677–684.
Ferreira, F. and Gyourko, J., 2009. Do political parties matter? Evidence from US cities. *The Quarterly Journal of Economics*, *124*(1), pp. 399–422.
Finkel, J., Dingare, S., Nguyen, H., Nissim, M., Manning, C. and Sinclair, G., 2004. Exploiting context for biomedical entity recognition: From syntax to the web. In *Proceedings of the international joint workshop on natural language processing in biomedicine and its applications (NLPBA/BioNLP)* (pp. 91–94).

Fulke, A.B., Chambhare, K.Y., Sangolkar, L.N., Giripunje, M.D., Krishnamurthi, K., Juwarkar, A.A. and Chakrabarti, T., 2013. Potential of wastewater grown algae for biodiesel production and CO_2 sequestration. *African Journal of Biotechnology*, *12*(20).

González-Fernández, C. and Ballesteros, M., 2012. Linking microalgae and cyanobacteria culture conditions and key-enzymes for carbohydrate accumulation. *Biotechnology Advances*, *30*(6), pp. 1655–1661.

Griffiths, M.J. and Harrison, S.T., 2009. Lipid productivity as a key characteristic for choosing algal species for biodiesel production. *Journal of Applied Phycology*, *21*(5), pp. 493–507.

Guiry, M.D., 2012. How many species of algae are there. *Journal of Phycology*, *48*(5), pp. 1057–1063.

Harting, J.K., Huerta, M.F., Hashikawa, T., Weber, J.T. and Van Lieshout, D.P., 1988. Neuroanatomical studies of the nigrotectal projection in the cat. *Journal of Comparative Neurology*, *278*(4), pp. 615–631.

Hempel, S., Newberry, S.J., Maher, A.R., Wang, Z., Miles, J.N., Shanman, R., Johnsen, B. and Shekelle, P.G., 2012. Probiotics for the prevention and treatment of antibiotic-associated diarrhea: A systematic review and meta-analysis. *JAMA*, *307*(18), pp. 1959–1969.

Henrikson, R., 1989. *Earth food spirulina* (p. 187). Laguna Beach, CA: Ronore Enterprises, Inc.

Hsia, J., Heiss, G., Ren, H., Allison, M., Dolan, N.C., Greenland, P., Heckbert, S.R., Johnson, K.C., Manson, J.E., Sidney, S. and Trevisan, M., 2007. Calcium/vitamin D supplementation and cardiovascular events. *Circulation*, *115*(7), pp. 846–854.

Huang, H.J., Yuan, X.Z., Zeng, G.M., Wang, J.Y., Li, H., Zhou, C.F., Pei, X.K., You, Q.A. and Chen, L.A., 2011. Thermochemical liquefaction characteristics of microalgae in sub- and supercritical ethanol. *Fuel Processing Technology*, *92*(1), pp. 147–153.

Kalidasan, M., Nagarajaprakash, R., Forbes, S., Mozharivskyj, Y. and Rao, K.M., 2015. Synthesis, spectroscopic and molecular studies of half-sandwich η6-arene ruthenium, *Cp* rhodium* and *Cp* iridium* metal complexes with bidentate ligands. *Zeitschrift für anorganische und allgemeine Chemie*, *641*(3–4), pp. 715–723.

Kemka, H.O., Rebecca, A.A. and Gideon, O.A., 2007. Influence of temperature and pH bioresource and protein biosynthesis in putative *Spirulina* sp. *Bioresource Technology*, *98*, pp. 2207–2211.

Liu, J., Huang, J., Fan, K.W., Jiang, Y., Zhong, Y., Sun, Z. and Chen, F. (2010). Production potential of *Chlorella zofingiensis* as a feedstock for biodiesel. *Bioresource Technology*, *101*(22), pp. 8658–8663.

Lowry, O.H., Rosebrough, N.J., Farr, A.L. and Randall, R.J., 1951. Protein measurement with the Folin phenol reagent. *Journal of Biological Chemistry*, *193*, pp. 265–275.

Ma, X., Zhou, W., Fu, Z., Cheng, Y., Min, M., Liu, Y. and Ruan, R., 2014. Effect of wastewater-borne bacteria on algal growth and nutrients removal in wastewater-based algae cultivation system. *Bioresource Technology*, *167*, pp. 8–13.

Makareviciene, V., Andrulevičiūtė, V., Skorupskaitė, V. and Kasperovičienė, J., 2011. Cultivation of microalgae *Chlorella* sp. and *Scenedesmus* sp. as a potentional biofuel feedstock. *Environmental Research, Engineering and Management*, *57*(3), pp. 21–27.

Masojídek, J. and Torzillo, G., 2008. Mass cultivation of freshwater microalgae. In *Encyclopedia of ecology*. Amsterdam: Academic Press.

Mata, T.M., Martins, A.A. and Caetano, N.S., 2010. Microalgae for biodiesel production and other applications: A review. *Renewable and Sustainable Energy Reviews*, *14*(1), pp. 217–232.

Namasivayam, A.M. and Steele, C.M., 2015. Malnutrition and dysphagia in long-term care: A systematic review. *Journal of Nutrition in Gerontology and Geriatrics*, *34*(1), pp. 1–21.

Ogbonda, K.H., Aminigo, R.E. and Abu, G.O., 2007. Influence of temperature and pH on biomass production and protein biosynthesis in a putative *Spirulina* sp. *Bioresource Technology*, *98*(11), pp. 2207–2211.

Palett, K.E. and Young, A.J. 1993. Carotenoids. In *Antioxidants in higher plants* (Edited by R. G. Alscher and J. L. Hess) (pp. 60–89). Boca Raton, FL: CRC Press Inc.

Pokoo-Aikins, G., Nadim, A., El-Halwagi, M.M. and Mahalec, V., 2010. Design and analysis of biodiesel production from algae grown through carbon sequestration. *Clean Technologies and Environmental Policy*, *12*(3), pp. 239–254.

Prartono, T.R.I., Kawaroe, M., Sari, D.W. and Augustine, D., 2010. Fatty acid content of Indonesian aquatic microalgae. *HAYATI Journal of Biosciences*, *17*(4), pp. 196–200.

Prescott, D.M., 1959. Variations in the individual generation times of *Tetrahymenageleii* HS. *Experimental Cell Research*, *16*(2), pp. 279–284.

Priyadarshani, I. and Rath, B., 2012. Commercial and industrial applications of micro algae—A review. *Journal of Algal Biomass Utilization*, *3*(4), pp. 89–100.

Priyadarshani, I., Thajuddin, N. and Rath, B., 2014. Influence of aeration and light on biomass production and protein content of four species of marine Cyanobacteria. *International Journal of Current Microbiology and Applied Sciences*, *3*(12), pp. 173–182.

Richardson, G.E., 2002. The metatheory of resilience and resiliency. *Journal of Clinical Psychology*, *58*(3), pp. 307–321.

Sakthivel, R., Ren, Y. and Mahmudov, N.I., 2011. On the approximate controllability of semilinear fractional differential systems. *Computers & Mathematics with Applications*, *62*(3), pp. 1451–1459.

Singh, S.P. and Singh, P., 2014. Effect of CO_2 concentration on algal growth: A review. *Renewable and Sustainable Energy Reviews*, *38*, pp. 172–179.

Spolaore, P., Joannis-Cassan, C., Duran, E. and Isambert, A., 2006. Commercial applications of microalgae. *Journal of Bioscience and Bioengineering*, *101*(2), pp. 87–96.

Sudhakar, K., Rajesh, M. and Premalatha, M., 2012. A mathematical model to assess the potential of algal bio-fuels in India. *Energy Sources, Part A: Recovery, Utilization, and Environmental Effects*, *34*(12), pp. 1114–1120.

Tan, J.S., Lee, S.Y., Chew, K.W., Lam, M.K., Lim, J.W., Ho, S.-H. and Show, P.L., 2020 A review on microalgae cultivation and harvesting, and their biomass extraction processing using ionic liquids. *Bioengineered*, *11*(1), pp. 116–129. https://doi.org/10.1080/21655979.2020.1711626.

Tang, H., Chen, M., Garcia, M.E.D., Abunasser, N., Ng, K.S. and Salley, S.O., 2011. Culture of microalgae *Chlorella minutissima* for biodiesel feedstock production. *Biotechnology and Bioengineering*, *108*(10), pp. 2280–2287.

Tomaselli, L. (1997). Morphology, ultrastructure and taxonomy of *Arthrospira (Spirulina) maxima* and *Arthrospira (Spirulina) platensis*. *Spirulina platensis (Arthrospira): Physiology, Cell-Biology and Biotechnology*, pp. 1–16.

Vonshak, A., 1997. Spirulina: Growth, physiology and biochemistry. In *Spirulina platensis arthrospira* (pp. 61–84)

Wang, L., Feng, Z., Wang, X., Wang, X. and Zhang, X., 2009. DEGseq: An R package for identifying differentially expressed genes from RNA-seq data. *Bioinformatics*, *26*(1), pp. 136–138.

Williams, P.J.L.B. and Laurens, L.M., 2010. Microalgae as biodiesel & biomass feedstocks: Review & analysis of the biochemistry, energetics & economics. *Energy & Environmental Science*, *3*(5), pp. 554–590.

Yoo, Y., Henfridsson, O. and Lyytinen, K., 2010. Research commentary—The new organizing logic of digital innovation: An agenda for information systems research. *Information Systems Research*, *21*(4), pp. 724–735.

Yun, Y.S., Lee, S.B., Park, J.M., Lee, C.I. and Yang, J.W., 1997. Carbon dioxide fixation by algal cultivation using wastewater nutrients. *Journal of Chemical Technology & Biotechnology: International Research in Process, Environmental and Clean Technology*, *69*(4), pp. 451–455.

Zarrouk, C., 1966. *Contribution al'etuded'une Cyanophycee. Influence de Divers Facteurs Physiques et Chimiquessur la croissance et la photosynthese de Spirulina mixima*. Thesis. University of Paris, France.

7
Biofuel and Halophytes

Aneesha Singh and Krupali Dipakbhai Vyas

CONTENTS

7.1 Introduction 179
7.2 Opportunities and Conditions for Halophytes as a Resource for Biofuel 180
7.3 Biodiesel, Bioethanol, Biogas, Bioderived SPK, and Other Energy Source Generation from Halophytes 181
    7.3.1 Case Study: *Thespesia* sp. 183
        7.3.1.1 Fuel Properties of *Thespesia populnea* Methyl Ester (TPME) 183
7.4 Conclusion and Future Aspects 185
References 185

7.1 Introduction

The world population may reach to between 8 billion and 10.5 billion by 2050. The constant advancement in technologies may result in substantial increase in energy consumption. Fossil fuels are being depleted, and doing so has calamitous effect on the environment due to greenhouse gas emissions. To cope with increasing fuel consumption, due to growing demand, there is need to identify natural resources for more efficient use of renewable energy. Biofuel is an alternative to fossil fuels. Biofuels are environmentally friendly and renewable energy sources produced from different feedstock, like vegetable oils, animal fats, and waste cooking oil. Fuel produced from biological means like agricultural crops, marine resources, microbes, and agricultural and municipal wastes in the form of solids, liquids, or gases is termed as biofuel (Aburas and Demirbas 2015).

About 71 million hectares of agricultural land of the world is in use for biofuel production, on which 24% biodiesel and 62% bioethanol are produced (Kummamuru 2016; Huang et al. 2010). Biodiesel is generally used as conveyance fuel, and biogas for production of electricity and heat. However, pure biomethane from biogas can be used as transportation fuel. Different types of materials like domestic, industrial, and agricultural waste can be used to produce biogas. The lignocellulosic biorefinery concept incorporates the use of biomass to produce biofuel (Stocker 2008). Bioethanol, biodiesel, and biogas can be made from animals and microbes. First-generation biofuels were derived from edible crops, and second-generation biofuels from non-edible biomass like lignocellulose or woody plants, agricultural leftover or waste plant material, and non-edible oils. Third-generation biofuels are made from algal and microbial biomass (Jena et al. 2012).

In 2017, globally 9% of electricity and 96% of direct heat was generated using biomass (Kummamuru 2016). EU stands first with 36% of the world's biofuel production, followed by the United States, Indonesia, and Brazil (Figure 7.1). Renewable energy share is 17.8% of the world's fuel consumption, 13% from biomass and 5% from other sources. The United States leads the world in cellulosic biofuel production, and these fuels are still a drop in the biofuel bucket. India recently announced its National Policy on Biofuels 2018. The policy aspires to promote increased production of cellulosic biofuel and other alternative fuels produced from sustainable feedstocks, something no other country has managed to do so far.

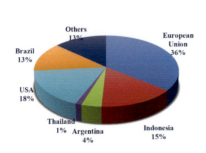
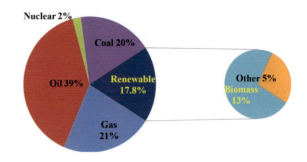

FIGURE 7.1 World biodiesel production and consumption.

7.2 Opportunities and Conditions for Halophytes as a Resource for Biofuel

Globally, forestland has been decreasing in successive years (Kummamuru 2016), due to deforestation and commercial consumption. An increase in fire incidents are also causing depletion of forest trees. There is need to explore other plants which are growing on wasteland. Halophytes cultivation have wide potential in world, as 43% land is arid and 98% of water is saline (Rozema and Flowers 2008). In India, about 8.6 million ha of land is saline and alkaline and is therefore not suitable for agricultural use. In the country, 12% area is arid and about 32 million ha of area is hot arid. About 61% of Rajasthan, 20% of Gujarat, 10% of Andhra Pradesh and Karnataka, and 9% of area in Punjab and Haryana contributes to hot arid. Jammu and Kashmir, and Himachal Pradesh, possess the remaining 7 million ha as cold arid. Halophytes are testified as one of potential sources of lignocellulose. Saline and waterlogged lands are environmental crunches that cause serious threats to food, fuel, and fiber production in the world. More than 100 countries are facing this problem (Beccles 2013). Halophytes and salt-tolerant plants grow on saline to high saline lands, as salinity is essential for their growth and development (Singh et al. 2018). True halophytes are plants that grow above 1.5% NaCl, facultative halophytes grow <200 mM NaCl, and glycophytes grow in non-saline land (Flowers and Colmer 2008). They are distributed in inland and coastal salt marshes and deserts (Fooladvand and Fazelinasab 2014). Exploration of halophytes on degraded land opens up the opportunity of using wasteland and degraded water (Singh et al. 2018) for the generation of new resource for biofuel (Glenn et al. 1998; Gul et al. 2013). In the recent past, it was known by the studies that halophytes are promising candidates for biomass generation for bioenergy. Biodiesel can be mixed with fossil fuel or can be used neat (Harrington 1986). Many salt-tolerant plants have potential to generate large biomass in saline soils (Sharma et al. 2016). Halophytes can be better choice to minimize the competition between bioenergy crops and food crops, as halophytes grow on saline soil and food crops on fertile land. Halophytic plant seeds are rich in oil content and may be explored for biodiesel. More than 350 oilseed crops are considered as sources for biodiesel (Bart et al. 2010). Several methods have been developed for biodiesel production using different oil seeds. Both lignocellulosic biomass and oil seeds from halophytes are important possible bioresource of biofuels production. Halophytes can grow in saline soil and be irrigated with saline water. They are a sustainable possible opportunity for safeguarding food security in several salt-affected regions (Wicke et al. 2011). Certainly, biofuel from halophytic biomass may signify a sustainable substitute for fossil fuels.

7.3 Biodiesel, Bioethanol, Biogas, Bioderived SPK, and Other Energy Source Generation from Halophytes

Methyl or ethyl esters of fatty acids produced by transesterification of oil is biodiesel. Several halophytes are tested for their potential for biofuel production (Table 7.1). *Salicornia* is facultative halophyte and the seeds of *Salicornia bigelovii* contain 28–36.3% oil (Glenn et al. 1991; Glenn and Brown 1999; Anwar et al. 2002; El-Araby et al. 2020), *S. brachiata* has 22% oil (Eganathan et al. 2006), and seeds of *S. fruticosa* and *A. macrostachyum* possess 25% oil. In the cases of *Halopyrum mucronatum*, *Cressa cretica*, *Haloxylon stocksii*, and *Alhagi maurorum*, the oil content was in the range of 22.7%–21.9% (Weber et al. 2007). Abideen and colleagues (2015) studied 21 halophytes that possessed 14%–30% oil, of which seven halophytes—*S. fruticosa*, *A. macrostachyum*, *A. maurorum*, *C. cretica*, *H. glomeratus*, *K. virginica*, and *A. rosea*—meet international specifications of biodiesel standards (Table 7.1).

Many halophytes are tested for their possibilities for ethanol production (Table 7.1). Enzymatic hydrolysis of *S. bigelovii* led to 90% glucose recoveries from raw biomass that corresponded to 111 kg ethanol/dry ton. Also, dry biomass of *Suaeda paradoxa*, *Atriplex nitens*, *Karelinia caspia*, and

TABLE 7.1
List of Halophytes That Have the Potential for Biofuel Production

No.	Halophyte	Biodiesel	Bioethanol	Biogas	Reference
1	*Kalidium caspicum*	–	–	✓	(Akinshina et al. 2012)
2	*Climacoptera lanata*			✓	
3	*Salicornia europaea*			✓	
4	*Aeluropus lagopoides*	–	✓	–	(Sobhanian et al. 2010)
5	*Kosteletzkya virginica*	✓	–	–	(Ruan et al. 2008)
6	*Salicornia brachiata*	✓	–	–	(Folayan et al. 2019)
7	*Salicornia bigelovii*	✓			
8	*Salicornia brachiata*	–	✓	–	(Sanandiya et al. 2010)
9	*Kosteletzkya pentacarpos*	✓	✓	–	(Moser et al. 2013)
11	*Retama retam*	–	✓	–	(Smichi et al. 2014)
12	*Salicornia sinus-persica*	–	✓	–	(Alassali et al. 2017)
13	*Juncus meritimus*	–	✓	–	(Smichi et al. 2016)
14	*Juncus rigidus* *Tamarix aphylla* *Thespesia populnea*		✓		(Singh et al. 2019)
15	*Juncus effusus* L.	–	–	✓	(Muller et al. 2020; Muller et al. 2012)
16	*Tamarix nilotica*	–	–	✓	(Kamel et al. 2019)
17	*Zygophyllum coccineum*			✓	
18	*Zygophyllum album*			✓	
19	*Tripolium pannonicum*	–	–	✓	(Turcios 2017)
20	*Salicornia* sp.			✓	
21	*Salicornia bigelovii*	–	✓	–	(Brown et al. 2014; Rashed et al. 2016; Banuelos et al. 2018; Cybulska et al. 2014)
22	*Alhagi maurorum*	✓	–	–	(Abideen et al. 2015)
23	*Allenrolfea occidentalis*	✓			

(Continued)

TABLE 7.1

List of Halophytes That Have the Potential for Biofuel Production (Continued)

No.	Halophyte	Biodiesel	Bioethanol	Biogas	Reference
24	*Arthrocnemum indicum*	✓			
25	*Atriplex heterosperma*	✓			
26	*Atriplex rosea* L.	✓			
27	*Cressa cretica* L.	✓			
28	*Halopyrum mucronatum*	✓			
29	*Haloylon stocksii*	✓			
30	*Halogeton glomeratus*	✓			
31	*Kochia scoparia*	✓			
32	*Kosteletzkya virginica*	✓			
33	*Salicornia bigelovii*	✓			
34	*Sarcobatus vermiculatus*	✓			
35	*Sarcocornia ambigua*	✓			
36	*Salicornia branchiate*	✓			
37	*Salicornia fruticosa*	✓			
38	*Salicornia europaea* L.	✓			
39	*Suaeda aralocaspica*	✓			
40	*Suaeda fruticosa*	✓			
41	*Suaeda salsa*	✓			
42	*Suaeda torreyana*	✓			
43	*Salsola baryosma*	–	✓	–	(Sharma et al. 2017)
44	*Sesuvium sesuvioides*		✓		
45	*Sporobolus helvolus*		✓		
46	*Suaeda fruticosa*		✓		
47	*Trianthema triquetra*		✓		
48	*Zygophyllum simplex*		✓		
49	*Salvadora persica*		✓		
50	*Pontederia cordata*	–	✓	–	(Lin et al. 2020)
51	*Pontederia australis*		✓		
52	*Cyprus alternifolius*		✓		
53	*Thespesia populnea*	✓	–	–	(Cyril 2017)
54	*Pongamia pinnatta*	✓			
55	*Derris trifoliata*	✓			
56	*Chenopodium quinoa*	–	–	✓	(Turcios et al. 2016)
57	*Thespesia populnea*	✓	–	–	(Rashid et al. 2015; Bhargavi et al. 2018)
58	*Suaeda salsa*	✓	–		(Wang et al. 2012)
59	*Tamarix jordanis*	–	✓	–	(Santi et al. 2014)
60	*Heleochloa setulosa*	–	✓		(Patel et al. 2019)
61	*Salvadora persica*		✓		
62	*Sesuvium portulacastrum*		✓		
63	*Atriplex griffithii*		✓		
64	*Leptochloa fusca*		✓		(Tech 2013)
65	*Sporobolus virginicus*		✓		
66	*Spartina patens*		✓		

Cynodon dactylon possessed the highest organic carbon (OC) content (243.9–396 mg/g). Low OC content (~200 mg/g) was noted in *Salicornia*, *Halostachys*, and *Climacoptera*. About 40%–60% of total dry biomass of halophytes could be decomposed into biogas. Salinity-tolerant grasses and shrubs are well studied. Halophytic woody plants (*Thespesia* and *Tamarix*) are also good sources of cellulose (Singh et al. 2019).

Bioderived synthetic paraffinic kerosene (bio-SPK) is aviation fuel that can minimize greenhouse gas emissions by aircraft. In the recent past, a task force developed to facilitate jet fuel alternatives so that aircraft can meet upcoming regulations on greenhouse gas (GHG) emissions. According to the study at Masdar Institute of Science and Technology in Abu Dhabi, ground and flight tests indicated the *Salicornia bigelovii* bio-SPK test fuels performed as well as or better than petroleum-based Jet A (www.greencarcongress.com). The biofuel of *Salicornia* oil can reduce 35%–76% GHG releases as compared to petroleum-based aviation fuel. Oil-to-jet pathway using the hydro-processed renewable jet (HRJ) process jet fuel yield from Salicornia oil was 50.2 gal/BDT (Stratton et al. 2010). Its oil heating value (0.12 MMBTU/gal) was similar to algal, *Jatropha*, and palm oil (Stratton et al. 2010; Frank et al. 2011; World 2013; Mehta and Anand 2009; Demirbas et al. 2016).

Halophytic grass (*Achnatherum splendens* L) was tested for biochar, bio-oil, and syngas. As per Irfan and colleagues (2016), *A. splendens* can be used as a possible raw material for generation of biofuel. Biochar yield was 24%, syngas yield was 54%, and bio-oil yield was 27%, with increased in temperature of pyrolysis from 300°C–700°C. Biogasification of *Kalidium caspicum*, *Salicornia europaea*, and *Climacoptera lanata* could generate about 300–500 mL CH_4 from 1 L of anaerobic digestion sludge (Akinshina et al. 2012). *A. canescens* and *A. lentiformis* biochar yields ranged from 30–39 wt.% (Sarpong et al. 2018).

7.3.1 Case Study: *Thespesia* sp.

Thespesia is the member of the family Malvaceae, commonly known as the India Tulip, Paras peepal, Poovarsan, Portia, and Malus Tree. *Thespesia* species are mostly trees, cultivated for ornamental, medicinal, and timber purposes. *T. populnea* is a useful tree for generating wood and bast fibers. *Thespesia* distributed worldwide and the plant is probably a native of Hawaii. There are about 18 species spread all over tropical regions of the South Pacific, Asia, Africa, and the Caribbean. *T. populnea* is now extensively scattered in coastal areas of the tropics and subtropics (Kabir et al. 2008). It is a fast-growing tree, attending height from 6–10 m.

CSMCRI has studied its salinity (400 mM) tolerance, water logging tolerance, and drought tolerance (Figure 7.2). In addition, it is tolerant to air pollutants with 9.99 APTI (Rupa and Venkatachalam 2018). Flowering and fruiting are seen at low to high salinity levels. It can grow at 5.5–7.5 pH, 35°C –46°C, and is an evergreen plant with some leaf shade in summer, and seed maturity in March–May. Leaves and bark are used in different medicinal and massage oils. It is also used as fodder. The timber price range in Rs. 3,000–3,500/ton. The timber is durable under water and so is used in boat making (Tamil Nadu Forest Department, tntreepedia.com). Its all plant parts are used in one or other purposes.

T. populnea seeds are good source of oil (Table 7.1). *T. populnea* seed kernels have dark color oil about 21% (w/w); it mainly contains 39.2% linoleic acid, 26.8% palmitic acid, 15.7% oleic acid, and 4.1% stearic acid. Small percentages of malvalic acid, methyl esters, and picolinyl esters are also found (Knothe et al. 2011). The 98.1% methyl ester was achieved from *Thespesia* (Knothe et al. 2011).

7.3.1.1 Fuel Properties of Thespesia populnea *Methyl Ester (TPME)*

7.3.1.1.1 Cetane Number

The cetane number (CN) reflects fuel ignition efficacy relative to cetane as standard. CN value reflects the efficiency of ignition in the engine. With increased efficiency, lag time decreases. *Thespesia populnea* methyl ester meets the ASTM and EN standards with 55.7–59.8 CN (Rashid et al. 2015; Bhargavi et al. 2018). This dissimilarity in the CN may be due to difference in the saturated fatty acid composition. Biodiesel has higher combustion efficiency than fossil diesel fuels, and thus higher CN.

FIGURE 7.2 *Thespesia populnea* plants treated with 400 mM NaCl (A); plants growing in water logged conditions (B); plants growing in clay soil under water deficit conditions.

TABLE 7.2
Thespesia Species Seed Oil and Cell Wall Content

Plant	α-Cellulose (%)	Hemicellulose (%)	Lignin (%)	Seed Oil (%)	
Thespesia lampus	64.12	20.8	14.98	–	(Reddy et al. 2014)
Thespesia populnea	45.22	26.61	26.43	–	(Senthilkumar et al. 2015)
	30	–	–	–	(Singh et al. 2019)
	–	–	–	21	(Rashid et al. 2015; Rashid et al. 2011)
	–	–	–	20.7	(Bhargavi et al. 2018)

7.3.1.1.2 Kinematic Viscosity

Kinematic viscosity (KV) of TPME at 40°C meets the standards specification of EN 14214 and ASTM D6751. High proportion of linoleic acid in the oil attributes to low 4.12–4.25 mm^2/s^{-1} viscosity of TPME. Low viscosity leads low carbon emission. High viscosity causes black smoke discharge due to improper combustion of fuel. KV of TPME was similar to other feedstocks like sesame, rapeseed, and soybean oil (Saydut et al. 2008; Canakci and Sanli 2008).

7.3.1.1.3 Cold Flow Properties and Flash Point

Cloud point (CP) is the temperature at which biodiesel forms cloud wax crystals that are visibly solid. The CP of TPME was in the range of 7.9°C–9°C (Rashid et al. 2015; Bhargavi et al. 2018). The CP of TPME was lower than palm oil (15°C), a widely used raw material for biodiesel production (Moser

2008). The flash point (FP) of biodiesel is higher than conventional diesel. Higher FP increases possibilities for safer transportation. The FP value of TPME (176°C) meets EN 14214 and ASTM D6751 standards, and it is comparative higher than palm, soybean, cottonseed, Pongamia, okra, and *Jatropha* biodiesel (Rashid et al. 2008; Anwar et al. 2010; Sarin et al. 2007), but it was less than that of sunflower oil biodiesel (Sarin et al. 2007). However, varying alcohol content likely caused differences in FP values. TPME also possessed excellent lubricity (136 mm).

7.4 Conclusion and Future Aspects

Biofuel is a renewable energy source and a very important alternate energy for fulfilling future needs of a growing population with depleting fossil fuels. Halophytes are studied for biofuel production, and extensive studies are required as halophytes can minimize competition with arable land and fresh water for cultivation. Identification of elite germplasm and breeding programs will lead to sustainable varieties for biofuel production. As compared to herbaceous halophytic plants, woody halophytes should be studied more for biofuel production. The perennial plants will ensure multiple harvests, resulting in increasing yields with plant age. Identification of crops for intercropping to increase intermediate earnings of farmers will encourage halophyte cultivation among coastal peoples. Developing biosaline agriculture more intensively will encourage halophyte cultivation in coastal areas. *Thespesia* was tested for biodiesel production and found to be a good source of both oil and cellulose. Tolerance to salinity, water logging, drought, and pollution make it potential candidate for biofuel biomass generation.

REFERENCES

Abideen Z, Qasim M, Rizvi RF, Gul B, Ansari R, Khan MA (2015) Oilseed halophytes: a potential source of biodiesel using saline degraded lands. *Biofuels*, 6:241–248.

Aburas H, Demirbas A (2015) Evaluation of beech for production of bio-char, bio-oil and gaseous materials. *Process Safety and Environmental Protection*, 94:29–36.

Akinshina N, Naka D, Toderich K, Azizov A, Yasui H (2012) Anaerobic degradation of halophyte biomass for biogas production. *Journal of Arid Land Studies*, 22:227–230.

Alassali A, Cybulska I, Galvan AR, Thomsen MH (2017) Wet fractionation of the succulent halophyte *Salicornia* sinus-persica, with the aim of low input (water saving) biorefining into bioethanol. *Applied Microbiology and Biotechnology*, 101:1769–1779.

Anwar F, Bhanger MI, Nasir MKA, Ismail S, 2002. Analytical characterization of *Salicornia bigelovii* seed oil cultivated in Pakistan. *Journal of Agricultural and Food Chemistry*, 50:4210–4214.

Anwar F, Rashid U, Ashraf M, Nadeem M (2010) Okra (*Hibiscus esculentus*) seed oil for biodiesel production. *Applied Energy*, 87:779–785.

Banuelos JA, Velazquez-Hernandez I, Guerra-Balcazar M, Arjona N (2018) Production, characterization and evaluation of the energetic capability of bioethanol from *Salicornia bigelovii* as a renewable energy source. *Renewable Energy*, 123:125–134.

Bart JC, Palmeri N, Cavallaro S (2010) *Biodiesel science and technology: from soil to oil*. Boca Raton, FL: CRC Press.

Beccles R (2013) Biofuel as the solution of alternative energy production? *Future of Food: Journal on Food, Agriculture and Society*, 1:22–26.

Bhargavi G, Rao PN, Renganathan S (2018) Production of biodiesel from *Thespesia populnea* seed oil through rapid in situ transesterification-an optimization study and assay of fuel properties. *MS&E*, 330:012046.

Brown JJ, Cybulska I, Chaturvedi T, Thomsen MH (2014) Halophytes for the production of liquid biofuels. In *Sabkha ecosystems* (pp. 67–72). Dordrecht: Springer.

Canakci M, Sanli H (2008) Biodiesel production from various feedstocks and their effects on the fuel properties. *Journal of Industrial Microbiology and Biotechnology*, 35:431–441.

Cybulska I, Chaturvedi T, Brudecki GP, Kadar Z, Meyer AS, Baldwin RM, Thomsen MH (2014) Chemical characterization and hydrothermal pretreatment of *Salicornia bigelovii* straw for enhanced enzymatic hydrolysis and bioethanol potential. *Bioresource Technology*, 153:165–172.

Cyril N (2017) *Investigation on the potential of some selected wetland plants for the biodiesel production.* Research Project, Kerala.

Demirbas A, Bafail A, Ahmad W, Sheikh M (2016) Biodiesel production from non-edible plant oils. *Energy Exploration and Exploitation*, 34:290–318.

Eganathan P, Subramanian HMSR, Latha R, Srinivasan RC (2006) Oil analysis in seeds of *Salicornia brachiata*. *Industrial Crops and Products* 23:177–179.

El-Araby, R, Rezk, AI, El-Enin, SAA et al. (2020) Comparative evaluation of *Salicornia bigelovii* oil planted under different treatments. *Bulletin of the National Research Centre*, 44:133.

Flowers TJ, Colmer TD (2008) Salinity tolerance in halophytes. *New Phytologist*, 179:945–963.

Folayan AJ, Anawe PAL, Ayeni, AO (2019) Synthesis and characterization of *Salicornia bigelovii* and *Salicornia brachiata* halophytic plants oil extracted by supercritical CO_2 modified with ethanol for biodiesel production via enzymatic transesterification reaction using immobilized *Candida antarctica* lipase catalyst in tert-butyl alcohol (TBA) solvent. *Cogent Engineering*, 6:1625847.

Fooladvand Z, Fazelinasab B (2014) Evaluate the potential halophyte plants to produce biofuels. *European Journal of Biotechnology and Bioscience*, 2:1–3.

Frank ED, Han J, Palou-Rivera I, Elgowainy A, Wang MQ (2011) *User manual for algae life cycle analysis with GREET: Version 0.0.* Argonne National Laboratory.

Glenn EP, Brown JJ (1999) Salt tolerance and crop potential of halophytes. *Critical Reviews in Plant Sciences* 18(2):227–255.

Glenn EP, Brown JJ, O'Leary JW (1998) Irrigating crops with seawater. *Scientific American*, 279:76–81.

Glenn EP, O'Leary JW, Watson MC, Thompson TL, Kuehl RO (1991) *Salicornia bigelovii* Torr.: an oilseed halophyte for seawater irrigation. Science, 251(4997):1065–1067.

Gul B, Abideen Z, Ansari R, Khan MA (2013) Halophytic biofuels revisited. *Biofuels*, 4(6):575–577.

Harrington KJ (1986) Chemical and physical properties of vegetable oil esters and their effect on diesel fuel performance. *Biomass*, 9:1–17.

Huang GH, Chen F, Wei D, Zhang XW, Chen G (2010) Biodiesel production by microalgal biotechnology. *Applied Energy*, 87(1):38–46.

Irfan M, Chen Q, Yue Y, Pang R, Lin Q, Zhao X, Chen H (2016) Co-production of biochar, bio-oil and syngas from halophyte grass (*Achnatherum splendens* L.) under three different pyrolysis temperatures. *Bioresource Technology*, 211:457–463.

Jena J, Nayak M, Panda HS, Pradhan N, Sarika C, Panda PK, Sukla, LB (2012) Microalgae of odisha coast as a potential source for biodiesel production. *World Environment*, 2(1):11–16.

Kabir M, Iqbal MZ, Shafiq M, Farooqi ZR (2008) Reduction in germination and seedling growth of *Thespesia populnea L.*, caused by lead and cadmium treatments. *Pakistan Journal of Botany*, 40:2419–2426.

Kamel M, Hammad S, Khalaphallah R, Abd Elazeem M (2019) Halophytes and salt tolerant wild plants as a feedstock for biogas production. *Journal of BioScience and Biotechnology*, 8:151–159.

Knothe G, Rashid U, Yusup S, Anwar F (2011) Fatty acid profile of *Thespesia populnea*. Mass spectrometry of the picolinyl esters of cyclopropene fatty acids. *European Journal of Lipid Science and Technology*, 113:980–984.

Kummamuru B (2016) *WBA global bioenergy statistics 2016.* Stockholm, Sweden: World Bioenergy Association. www.worldbioenergy.org.

Lin Y, Zhao Y, Ruan X, Barzee TJ, Zhang Z, Kong H, Zhang X (2020) The potential of constructed wetland plants for bioethanol production. *BioEnergy Research*, 13:43–49.

Mehta PS, Anand K (2009) Estimation of a lower heating value of vegetable oil and biodiesel fuel. *Energy and Fuels*, 23:3893–3898.

Moser BR (2008) Influence of blending canola, palm, soybean, and sunflower oil methyl esters on fuel properties of biodiesel. *Energy Fuels*, 22:4301–4306.

Moser BR, Dien BS, Seliskar DM, Gallagher JL (2013) Seashore mallow (*Kosteletzkya pentacarpos*) as a salt-tolerant feedstock for production of biodiesel and ethanol. *Renewable Energy*, 50:833–839.

Muller J, Jantzen C, Kayser M (2012) *The biogas potential of Juncus effusus L. using solid phase fermentation technique.* Grassland—A European Resource?

Muller J, Jantzen C, Wiedow D (2020) The energy potential of soft rush (*Juncus effusus* L.) in different conversion routes. *Energy, Sustainability and Society*, 10:1–13.

Patel MK, Pandey S, Brahmbhatt HR, Mishra A, Jha B (2019) Lipid content and fatty acid profile of selected halophytic plants reveal a promising source of renewable energy. *Biomass and Bioenergy*, 124:25–32.

Rashed SA, Ibrahim MM, Hatata MM, El-Gaaly GA (2016) Biodiesel production and antioxidant capability from seeds of *Salicornia bigelovii* collected from Al Jubail, Eastern province, Saudi Arabia. *Pakistan Journal of Botany*, 48:2527–2533.

Rashid U, Anwar F, Knothe G (2011) Biodiesel from Milo (*Thespesia populnea* L.) seed oil. *Biomass and Bioenergy*, 35:4034–4039.

Rashid U, Anwar F, Moser BR, Knothe G (2008) *Moringa oleifera* oil: a possible source of biodiesel. *Bioresource Technology*, 99:8175–8179.

Rashid U, Anwar F, Yunus R, Al-Muhtaseb AH (2015) Transesterification for biodiesel production using *Thespesia populnea* seed oil: an optimization study. *International Journal of Green Energy*, 12:479–484.

Reddy KO, Ashok B, Reddy KRN, Feng YE, Zhang J, Rajulu AV (2014) Extraction and characterization of novel lignocellulosic fibers from *Thespesia lampas* plant. *International Journal of Polymer Analysis and Characterization*, 19(1):48–61.

Rozema J, Flowers, T (2008) Crops for a salinized world. *Science*, 1478–1480.

Ruan CJ, Li H, Guo YQ, Qin P, Gallagher JL, Seliskar DM, Mahy G (2008) *Kosteletzkya virginica*, an agro-ecoengineering halophytic species for alternative agricultural production in China's east coast: ecological adaptation and benefits, seed yield, oil content, fatty acid and biodiesel properties. *Ecological Engineering*, 32:320–328.

Rupa P, Venkatachalam T (2018) Studies on air pollution tolerance index of native plant species to enhance greenery in industrial area. *Indian Journal of Ecology*, 45:1–5.

Sanandiya ND, Prasad K, Meena R, Siddhanta AK (2010) Cellulose of *Salicornia brachiate*. *Natural Product Communications*, 5(4):603–606.

Santi G, D'Annibale A, Eshel A, et al. (2014) Bioethanol production from xerophilic and salt-resistant *Tamarix jordanis* biomass. *Biomass and Bioenergy*, 61:73–81.

Sarin R, Sharma M, Sinharay S, Malhotra RK (2007) *Jatropha*-palm biodiesel blends: an optimum mix for Asia. *Fuel*, 86:1365–1371.

Sarpong K, Brewer C, Idowu OJ (2018) *Biochar for desalination concentrate management*. Research project report, U.S. Department of the Interior Denver Federal Center.

Saydut A, Duz MZ, Kaya C, Kafadar AB, Hamamci C (2008) Transesterified sesame (*Sesamum indicum* L.) seed oil as a biodiesel fuel. *Bioresource Technology*, 99:6656–6660.

Senthilkumar N, Murugesan S, Sumathi R, Babu DS (2015) Compositional analysis of lignocellulosic biomass from certain fast growing tree species in India. *Der Chemica Sinica*, 6:25–28.

Sharma R, Wungrampha S, Singh V, Pareek A, Sharma MK (2016) Halophytes as bioenergy crops. *Frontiers in Plant Science*, 7:1372.

Sharma V, Joshi A, Ramawat KG, Arora J (2017) Bioethanol production from halophytes of Thar desert: A "Green Gold". In: Basu SK, Zandi P, Chalaras SK (eds.) *Environment at crossroads: Challenges, dynamics and solutions* (pp. 219–235). Tehran, Iran: Haghshenass Publishing.

Singh A, Ranawat B, Meena R (2019) Extraction and characterization of cellulose from halophytes: next generation source of cellulose fibre. *SN Applied Sciences*, 1:1311.

Singh A, Sharma S, Shah MT (2018) Successful cultivation of *Salicornia brachiata*—A sea asparagus utilizing RO reject water: a sustainable solution. *International Journal of Waste Resource*, 8:332–337.

Smichi N, Messaoudi Y, Ksouri R, Abdelly C, Gargouri M (2014) Pretreatment and enzymatic saccharification of new phytoresource for bioethanol production from halophyte species. *Renewable Energy*, 63:544–549.

Smichi N, Messaoudi Y, Moujahed N, Gargouri M (2016) Ethanol production from halophyte *Juncus maritimus* using freezing and thawing biomass pretreatment. *Renewable Energy*, 85:1357–1361.

Sobhanian H, Motamed N, Jazii FR, Razavi K, Niknam V, Komatsu S (2010) Salt stress responses of a halophytic grass *Aeluropus lagopoides* and subsequent recovery. *Russian Journal of Plant Physiology*, 57:784–791.

Stocker M (2008) Biofuels and biomass to liquid fuels in the biorefinery: catalytic conversion of lignocellulosic biomass using porous materials. *Angewandte Chemie International Edition*, 47:9200–9211.

Stratton RW, Wong HM, Hileman JI (2010) Life cycle greenhouse gas emissions from alternative jet fuels. *PARTNER Project*, 28:133.

Tamil Nadu Forest Department, www.tntreepedia.com

Tech ET (2013) Testing of some halophytic plants for forage, biofuel production and soil bioremediation. *Journal of Environmental Treatment Techniques*, 1:183–189.

Turcios PAE (2017) *Use of halophytes as biofilter to decrease organic and inorganic contaminants in water and their further use for biogas productio*n (Doctoral dissertation). Hannover: Gottfried Wilhelm Leibniz Universitat.

Turcios PAE, Weichgrebe D, Papenbrock J (2016) Potential use of the facultative halophyte *Chenopodium quinoa* Willd. as substrate for biogas production cultivated with different concentrations of sodium chloride under hydroponic conditions. *Bioresource Technology*, 203:272–279.

Wang L, Zhao ZY, Zhang K, Tian CY (2012) Oil content and fatty acid composition of dimorphic seeds of desert halophyte *Suaeda aralocaspica*. *African Journal of Agricultural Research*, 7:1910–1914.

Weber DJ, Ansari R, Gul B, Khan MA (2007) Potential of halophytes as source of edible oil. *Journal of Arid Environments*, 68:315–321.

Wicke B, Smeets E, Dornburg V, Vashev B, Gaiser T, Turkenburg W, Faaij A (2011) The global technical and economic potential of bioenergy from salt-affected soils. *Energy & Environmental Science*, 4:2669–2681.

World J (2013) *Jatropha biodiesel making*. www.nrel.gov/publications.

8

Eco-Friendly Applications of Natural Secondary Metabolites and Status of Siderophores

Pratika Singh, Azmi Khan, Micky Anand, Hemant Kumar,
Shivpujan Kumar, and Amrita Srivastava

CONTENTS

8.1	Secondary Metabolite: A Potential Resource	189
8.2	Biosynthesis of Secondary Metabolites	190
	8.2.1 Terpenoid Biosynthesis	190
	8.2.2 Alkaloid Biosynthesis	191
	8.2.3 Polyketide Biosynthesis	192
	8.2.4 Non-Ribosomal Peptide Synthesis Pathway	193
	8.2.5 Hybrid PKS-NRPS Pathway	193
	8.2.6 Lignin Biosynthesis	194
	8.2.7 Shikimate Pathway	194
8.3	Applications of Secondary Metabolite	195
	8.3.1 Biofuel Production	197
	8.3.2 Bioremediation	198
	8.3.3 Edible Dyes	198
8.4	Strategies for Improving Secondary Metabolite Production	199
	8.4.1 Traditional Strategies	199
	8.4.2 Metabolic Engineering	200
	8.4.3 Bioreactors	201
8.5	Siderophores: Promising Secondary Metabolites	201
	8.5.1 Medical Applications of Siderophores (Iron Chelation)	201
	8.5.2 Siderophore-Mediated Bioremediation	202
8.6	Conclusion	202
Acknowledgment		202
References		203

8.1 Secondary Metabolite: A Potential Resource

Metabolism is the sum of all life-sustaining biochemical reactions occurring inside cells of an organism. The chemical transformations result in production of small-molecule intermediates known as metabolites. These metabolites can be categorized as primary metabolites or central metabolites and secondary metabolites (SMs). Primary metabolites are the products formed during metabolic reactions that are crucial for growth and development of organisms and are derived from processes like glycolysis, tricarboxylic acid cycle, etc. Central metabolites thus play an important role in optimum physiological activities and so their formation occurs during exponential growth phase. Secondary metabolites, on the other hand, are not directly involved in growth and reproduction of organisms, rather taking part in functions like pigment formation, stress response, and ecological interactions. These are synthesized during late growth phase or stationary phase. These specialized molecules are economically beneficial for their diverse applications as antibiotics, biofuels, and agents of bioremediation, biodegradation, and other industrial processes (Kang and Lee 2016; Thirumurugan et al. 2018; Khan et al. 2019). It is noteworthy that unlike primary metabolism, which encompasses reactions that are highly conserved, huge diversity occurs during secondary metabolism reactions at the tissue level and at different developmental stages (Wink 2010). This is due to catalytic promiscuity as a result of poor selection pressure occurring in enzymes encoding metabolite (Tokuriki et al. 2012). Evolutionarily, it is considered that secondary metabolism involves enzymes of primary metabolism indicating the origin of secondary metabolic pathways from other nodes of the central metabolic pathway (Carrington et al. 2018). This may have occurred in order to produce compounds for adapting to a changing environment. On the basis of chemical structure, diverse biosynthetic pathways, and their origins, SMs are categorized broadly as alkaloids, polyketides, cytochalasins, terpenoids, and non-ribosomal peptides (Thirumurugan et al. 2018).

Secondary metabolite production is widespread in all domains of life, i.e. archaea, bacteria, and eukarya. Abundance of SM production can be observed primarily in ascomycota (31%), actinobacteria (23%), euryarchaeota (12%), proteobacteria (9%), and firmicutes (8%) (Yadav et al. 2019). Various species belonging to *Streptomyces* sp., *Actinomadura* sp., *Micromonospora* sp., *Kitasatospora* sp., *Agrococcus* sp., *Moorea* sp., *Muscodor* sp., and *Oceanicola* sp. produce a range of secondary metabolites that are now explored for their eco-friendly applications (Kusaka et al. 1968; Higgens and Kastner 1971; Kawamoto et al. 1983; Kudalkar et al. 2012).

In the last few decades, several chemical compounds have been synthesized and commercialized to meet human demands. This has, however, led to deterioration of the ecosystem, provoking researchers to look for an alternative which is effective, economic, eco-friendly, and aims at sustainable development. SMs seem to be a promising product in this regard. This chapter provides a detailed account of secondary metabolite biosynthesis and strategies to enhance their production. An elaborate account of their eco-friendly application is also provided, along with a specific mention of a universal metal chelating secondary metabolite, siderophore.

8.2 Biosynthesis of Secondary Metabolites

Biosynthesis is the production of substances by reactions and processes taking place in the living organism (Bentley 1999). The major biosynthetic pathways to secondary metabolite synthesis include the isoprenoid pathway giving rise to terpenoids, polyketide pathway, shikimate pathway, alkaloid biosynthesis pathway, lignin biosynthesis pathway, and hybrid polyketide–non-ribosomal polypeptide synthesis (PKS-NRPS) pathway (Bentley 1999; Croteau et al. 2000; Liu et al. 2018).

8.2.1 Terpenoid Biosynthesis

All terpenoid compounds such as artemisinin, azadirachtin A, and ginkgolides are synthesized by fusion of the basic building block called isoprene which was obtained on decomposition of many terpenoids (Ruzicka 1953). As a result, terpenoids are also referred to as "isoprenoids." But curiously, the precursor for terpenoid biosynthesis was found to be dimethylallyl pyrophosphate (DMAPP) or isopentenyl diphosphate (IPP), an isomer of the former and not the isoprene units (Bentley 1999; Croteau et al. 2000; Chan et al. 2019). Precursors are synthesized either via the 2-C-methyl-d-erythritol 4-phosphate (MEP) pathway or the mevalonic acid (MVA) pathway (Chan et al. 2019). Briefly, the terpenoid synthesis pathway includes synthesis of IPP, polymerization of IPP forming prenyl diphosphate precursors, and further extension of the precursors by terpenoid synthases, followed by specific modifications giving rise to the specific terpenoid metabolite. Archaebacteria, eubacteria, and higher plants use the MVA pathway (Kuzuyama 2002). In the first step, two acetyl-CoAs condense together and produce acetoacetyl-CoA which is catalyzed by acetyl-CoA acetyltransferase. Another acetyl-CoA molecule undergoes condensation with the acetoacetyl-CoA catalyzed by hydroxymethylglutaryl (HMG)-CoA synthase that gives rise to HMG-CoA. HMG-CoA reductase further reduces it in two coupled reactions using two molecules of NADPH and produces MVA. Mevalonate kinase catalyzes ATP-dependent phosphorylation of MVA, producing mevalonic acid 5-phosphate (MVAP) which undergoes another ATP-dependent phosphorylation catalyzed by MVAP kinase producing mevalonic acid 5-diphosphate (MVAPP). IPP is then produced by ATP-dependent phosphorylation mediated decarboxylation of MVAPP catalyzed by MVAPP decarboxylase. Enzyme IPP isomerase further isomerizes IPP to form DMAPP in the presence of FMN and NAD(P)H. IPP and DMAPP then serve as precursors to isoprenoid synthesis (Croteau et al. 2000; Kuzuyama 2002; Kuzuyama and Seto 2012). The MEP pathway is used by chloroplasts of higher plants, green algae, and many eubacteria (Kuzuyama and Seto 2012). In the first step, 1-deoxy-D-xylulose 5-phosphate (DXP) is synthesized. The latter is formed in a condensing reaction involving d-glyceraldehyde 3-phospahte and pyruvate catalyzed via DXP synthase. DXP reductoisomerase causes DXP to gain electrons and be converted to MEP. NADPH provides the reducing power for the reaction. MEP is converted by MEP cytidylyltransferase into 4-(cytidine 5′-diphospho)-2-C-methyl-d-erythritol (CDP-ME) by using CTP. CDP-ME kinase catalyzes an ATP-driven conversion of CDP-ME to 2-phospho-4-(cytidine 5′-diphospho)-2-C-methyl-d-erythritol (CDP-ME2P). CMP is now eliminated from CDP-ME2P. This elimination generates MECDP. Hydroxymethylbutenyl diphosphate (HMBDP) synthase then converts MECDP into HMBDP. Completion of this step requires a NADPH/flavodoxin reductase system. HMBDP is further reduced by HMBDP reductase in the presence of NADPH and ferrodoxin, yielding a 1:5 mixture of DMAPP and IPP (Kuzuyama 2002; Kuzuyama and Seto 2012; Chan et al. 2019). The DMAPP thus generated contains an allylic double bond which can be ionized, generating a carbocation stabilized by resonance. This carbocation can now undergo condensation with IPP to form a ten-carbon intermediate geranyl pyrophosphate (GPP). Repetition of the cycle multiple times will subsequently produce farnesyl diphosphate (FPP) and geranylgeranyl diphosphate (GGPP). Each of these elongations is carried out by specific "prenyltransferases" named for their respective products. Monoterpenes are formed from GPP, sesquiterpenes from FPP, and diterpenes from GGPP through similar reactions. Monoterpenes, sesquiterpenes, and diterpenes are synthesized by monoterpene synthase, sesquiterpene synthase, and diterpene synthase, respectively (Croteau et al. 2000).

8.2.2 Alkaloid Biosynthesis

A significant number of the alkaloids such as morphine, colchicine, nicotine, and quinine are derived from l-amino acids such as tyrosine, lysine, phenylalanine, tryptophan, and arginine or from a combination of these amino acids with a terpenoid-type or a steroidal product (Bentley 1999; Croteau et al. 2000). The monoterpenoid indole alkaloids (MIA) pathway involves generation of tryptamine by conversion of tryptophan. This step is catalyzed by tryptophan decarboxylase (TDC). All MIAs are derived from strictosidine through Pictet-Spengler condensation of tryptamine and secologanin. Strictosidine synthase (STR) catalyzes the reaction. Strictosidine β-d-glucosidase removes the glucose moiety of strictosidine. The resulting aglycone is converted to dehydrogeissoschizine, which then leads to several different MIA pathways (Croteau et al. 2000; Ziegler and Facchini 2008). Benzylisoquinoline alkaloids (BIA) synthesis is initiated with the production of tyramine and dopamine by decarboxylation of tyrosine and dihydroxyphenylalanine, respectively. This reaction is catalyzed by tyrosine decarboxylase (TYDC). Deamination of tyramine produces 4-hydroxyphenylacetaldehyde which undergoes another Pictet-Spengler–type condensation with dopamine top produce (S)-norcoclaurine. This reaction is catalyzed by norcoclaurine synthase (NCS). Further, O-methylation of (S)-norcoclaurine leads to formation of (S)-reticuline. (S)-reticuline then acts as the central intermediate for the BIA synthesis pathway and is aptly called the chemical chameleon (Croteau et al. 2000). One notable exception is the synthesis of bisbenzylisoquinolone alkaloids, whereby the enzyme Cyp80A1 catalyzes conversion of N-methylcoclaurine to berbamunine. (S)-scoulerine is an alternative product derived from (S)-reticuline that is acted on by scoulerine 9-O-methyltransferase, (S)-tetrahydroprotoberberine oxidase, and canadine synthase (in the presence of NADPH) to produce berberine (Croteau et al. 2000; Ziegler and Facchini 2008). Tropane alkaloids are another class of plant-derived alkaloids. In this pathway, putrescine is first synthesized by ornithine decarboxylase (ODC) from ornithine. Arginine is transformed to N-carbamoylputrescine via agmatine in the presence of arginine decarboxylase (ADC). N-methylputrescine is produced by putrescine N-methyltransferase (PMT) by methylation of putrescine. N-methylputrescine undergoes oxidative deamination to produce 4-methylaminobutanal. Deamination is catalyzed by the enzyme N-methylputrescine oxidase (MPO). MPO requires copper as a cofactor. The 4-methylaminobutanal cyclizes immediately to yield N-methyl-Δ^1-pyrrolium cation, which acts as the central intermediate for the synthesis of alkaloids with tropane ring and nicotine. N-methyl-Δ^1-pyrrolium cation is converted to tropinone by a yet unknown enzyme, although the role of a 2-carboxytropinone intermediate has been observed (Ziegler and Facchini 2008). Tropinone reductase (TR) catalyzes the next step. Tropinone is reduced by TR-I to tropine with a 3α-hydroxy group while TR-II reduces it to pseudotropine with 3β-hydroxy group. Tropine condenses with (R)-phenylacetate producing (R)-littorine. Oxidation and rearrangement of (R)-littorine is catalyzed by Cyp80F1, generating hyoscyamine. Hyoscyamine 6β-hydroxylase (H6H) is a bifunctional enzyme which catalyzes the epoxidation and hydroxylation of hyoscyamine to scopolamine (Croteau et al. 2000; Ziegler and Facchini 2008; Georgiev et al. 2013). Synthesis of purine alkaloids occurs with xanthosine 7-N-methyltanferase (XMT) catalyzed N-methylation of xanthosine. The ribose residue from 7-methylxanthosine is removed by 7-methylxanthosine nucleosidase, yielding 7-methylxanthine. Two methyltransferases, theobromine synthase (TS) and caffeine synthase (CS), methylate 7-methylxanthine at the 3-N position, generating 3,7-dimethylxanthine (theobromine), which undergoes another methylation at 1-N position, producing 1,3,7-trimethylxanthine (caffeine), respectively. The methyl group donor at each step is S-adenosyl-l-methionine (SAM). Pyrrolizidine alkaloids (PA) are known to participate in herbivore–plant interactions. Polyamines putrescine and spermidine are the building blocks of PAs, both of which are derived from arginine. Spermidine synthase (SPDS) catalyzes the formation of spermidine. An aminopropyl moiety is transferred from decarboxylated SAM by SPDS to putrescine. Apart from arginine, putrescine is also synthesized from ornithine by the action of ornithine decarboxylase (in animals). Homospermidine synthase (HSS) transfers the aminobutyl moiety of spermidine to putrescine in the presence of NAD+ producing homospermidine. Oxidative deamination of homospermidine converts it into the necine base moiety senecionine. Further components of this metabolic pathway are not well understood (Ziegler and Facchini 2008; Dreger et al. 2009).

8.2.3 Polyketide Biosynthesis

Polyketides such as erythromycin, geldanamycin, and aflatoxin B1 are synthesized by polyketide synthases (PKS), which are similar to fatty acid synthase. PKS are prevalent of three types: I, II, and III (Weissman 2009). PKS Type I comprises several domains (Shen 2003). The acyltransferase (AT) domain can recognize a range of acyl-CoAs as starting molecules. The extender unit is bound to the ketosynthase (KS) domain. KS domain then decarboxylates the extender unit and is followed by a Claisen-like condensation between the cysteine thiol attached intermediates and the two carbons of the extender unit. Except the first AT domain, the remaining are involved in selecting the appropriate extender unit and the chiral center for commencing elongation. The first AT domain is tasked with selecting a range of acyl-CoA to serve as a starter molecule. The phosphopantethiene region of acyl carrier protein (ACP) forms a thioester (TE) linkage with the growing chain. This phosphopantethiene region also serves the role of presenting the elongated chain to a reductase domain. The last module of the enzyme contains a TE domain releasing the synthesized polyketide (PK) by a hydrolysis reaction. The synthesized PK can then undergo specific modifications by cyclases, dehydratases, and aromatases, giving rise to a range of secondary metabolites (Gokulan et al. 2014). Type II PKS form several SMs using malonyl-CoA (MCoA) as the extender unit for chain elongation. They are composed of two domains. The synthesis of aromatic PKs requires the presence of ACP domain and two units of the KS domain (KSα and KSβ). KSα and KSβ are identical, with the exception that KSβ lacks an active cysteine residue. KSα catalyzes the Claisen-like C-C condensation between acyl intermediates and decarboxylated MCoA. MCoA binds to ACP with the help of KSβ, which determines the length of the carbon chain. Synthesis of aromatic PKs begins with acetate, although other starter molecules—such as malonate or butyrate—can be used (Gokulan et al. 2014). Flavonoids and phenylpropanoids are synthesized by Type III PKS. Type III PKS do not possess ACP domain and a single active site carries out substrate priming, repetitive decarboxylation, extension and cyclization. For initiation, these enzymes can use anywhere from a C_2 to a C_{20} acyl-CoA but uses MCoA as the extender. Some Type III PKS have also been reported to use methyl-MCoA along with MCoA as the extender unit. The initiator acyl-CoA binds to one substrate binding pocket, followed by a nucleophilic attack by the site's cysteine residue, forming a TE linkage. Extender units bind in the second pocket, where they undergo subsequent decarboxylation and transfer of two carbon units to the growing chain by a Claisen-like condensation. Variations at this step are responsible for the diversity of products synthesized from the Type III PKS (Shen 2003; Weissman 2009; Gokulan et al. 2014).

8.2.4 Non-Ribosomal Peptide Synthesis Pathway

The non-ribosomal peptide synthesis (NRPS) pathway is the source of compounds such as bacillibactin, enterobactin, peptidolactones, and many more diverse compounds from even non-proteinogenic and d-amino acids (Marahiel et al. 1997; Mootz et al. 2002). The multi-enzyme complexes involved in NRPS are modular peptide synthetases, with each module capable of acting independently of each other (Marahiel et al. 1997). Each modular peptide synthetase has functional domains for carrying out specific reaction steps (Marahiel et al. 1997; Mootz et al. 2002). The adenylation (A) domain located at the N-terminal region catalyzes two reactions. In the first step, amino acid gets activated to aminoacyl-adenylate in presence of Mg^{2+}-ATP. In the second step, aminoacyl thioester is obtained with AMP as the leaving group. This step requires the presence of the thiolation (T) domain. The 4′-phosphopantetheine (4′PP) cofactor binds to the peptidyl carrier protein domain. Additionally, this domain also performs substrate acylation. Aminoacyl-adenylate intermediate is transferred from the A domain to the sulfhydryl group of the 4′-PP cofactor via a thioester linkage, which then attacks the aminoacyl-adenylate. Between each pair of A and T domain lies the condensation (C) domain (Mootz et al. 2002). C domains catalyze formation of peptide bonds between aminoacyl-S-PCP or peptidyl-S-PCP on the flanking A and T domains. This results in the transfer of the growing peptide chain from one module to another. To ensure proper directionality of synthesis, the C domain has separate "donor" and "acceptor" sites, with acceptor sites binding the aminoacyl-S-PCP tightly until completion of condensation (Mootz et al. 2002; Bloudoff and Schmeing 2017). The thioesterase (TE) domain has been found in many NRPS toward the C-terminal end (Mootz et al. 2002). The peptidyl intermediate is transferred from the T domain to the catalytic serine (cysteine, in some cases) residue in the TE domain, which then leads to the release of the polypeptide by mechanisms such as oligomerization, hydrolysis, and cyclization (Mootz et al. 2002). Interestingly, eukaryotic NRPS have been found to lack a TE domain and thus produce cyclic products, mediated by intramolecular attacks within the linear intermediate (Marahiel et al. 1997). Some NRPS possess additional modifying or tailoring domains, which modify the intermediate during its synthesis. They include domains for epimerization, methyltransferase, oxidase, formylation, reductase, and halogenase activity (Marahiel et al. 1997; Bloudoff and Schmeing 2017).

8.2.5 Hybrid PKS-NRPS Pathway

PKS synthesize polyketides, while NRPS produce a diverse class of short-chain peptides including non-proteinogenic and d-amino acids (Mootz et al. 2002). Both NRPS and PKS are modular enzymes and use carrier proteins PCP and ACP (Lambalot et al. 1996; Walsh et al. 1997). The hybrid PKS-NRPS pathway merges the isolated PKS and NRPS pathways and leads to the synthesis of secondary metabolites such as bleomycin and cytochalasans (Boettger and Hertweck 2013). The pathway begins with the elongation of an acetyl-CoA initiator unit by KS domain. Only malonyl-CoA is used as an initiator in fungi, although bacteria have been shown to use branched extenders (Hertweck 2009). The remaining steps of the PKS section of the hybrid are similar to the biosynthetic pathway of conventional polyketides (Staunton and Weissman 2001). The only marked difference hybrid PKS-NRPS and conventional PKS pathway is the absence of the ER domain in the PKS module of the hybrid (Halo et al. 2008). The ER domain might be either non-essential or might be present elsewhere in any biosynthetic pathway and acting in trans (Boettger and Hertweck 2013). C-methyltransferase domains are responsible for introduction of chain branches by α-methylation, with the methyl group being donated by S-adenosylmethionine. NRPS domain catalyzes the same reactions as the conventional NRPS pathway (Boettger and Hertweck 2013). Once the polyketide portion of the product has been synthesized, it is fused to an activated amino acid by the C domain, producing an amide (Cox 2007). Release of the fusion product has been proposed through two routes. The first route uses a reduction mechanism generating an aldehyde which is further converted into a pyrrolinone by Knoevenagel condensation (Sims and Schmidt 2008). The second route involves a Dieckmann cyclization releasing a tetramic acid derivative catalyzed by the terminal domain (Liu and Walsh 2009). The released product can undergo further tailoring reactions to yield the diverse products (Boettger and Hertweck 2013).

8.2.6 Lignin Biosynthesis

Lignins are another class of plant-based secondary metabolites derived from the phenylalanine/tyrosine pathway (Whetten and Sederoff 1995; Boerjan et al. 2003; Liu et al. 2018). In the first step, lignin monomers are formed followed by their transport and polymerization (Liu et al. 2018). The pathway starts with the formation of cinnamate from phenylalanine by phenylalanine ammonia-lyase (PAL) catalyzed deamination. The released ammonia is believed to be reused by the plant mediated by glutamine synthase (Whetten and Sederoff 1995). In the next step, cinnamate 4-hydroxylase (C4H) produces *p*-coumaric acid by hydroxylation of cinnamte. C4H cleaves an oxygen molecule and transfers it to the aromatic ring of cinnamte. p-coumaric acid can also be derived from tyrosine using tyrosine ammonia-lyase (TAL; Whetten and Sederoff 1995; Liu et al. 2018). *P*-coumarate is hydroxylated to caffeate by the enzyme coumarate 3-hydroxylase (C3H). Coumaryl-CoA 3-hydroxylase then hydroxylates *p*-coumaryl-CoA to yield caffeoyl-CoA. Caffeic acid-O-methyltransferase (C-OMT) methylates caffeate to form ferulic acid. S-adenosyl methionine acts as the donor of the methyl group. This methylation restricts the atoms that can take part in polymerization (Whetten and Sederoff 1995). Ferulate is converted to 5-hydroxyferulate by the action of ferulate 5-hydroxylase, which is another cytochrome P-450–linked monooxygenase. Additionally, C-OMT methylates 5-hydroxyferulate to sinapate further down in the pathway by a similar mechanism. The 5-hydroxyferuloyl-CoA leads to formation of sinapoyl-CoA, and similarly, caffeoyl-CoA is formed from feruloyl-CoA. Both the reactions are catalyzed by caffeoyl-CoA O-methyltransferase. Feruloyl-CoA and 5-hydroxyferuloyl-CoA can also be obtained from ferulate and 5-hydroxyferulate, respectively, by the action of 4CL. All the synthesized CoA thioesters are reduced to their respective aldehydes by the action of cinnamoyl-CoA reductase (CCR). CCR is the marker enzyme for monolignol synthesis. Cinnamyl alcohol dehydrogenase further reduces the aldehydes, producing their respective monolignols (Whetten and Sederoff 1995; Boerjan et al. 2003; Liu et al. 2018). Polymerization of monolignols occurs after their channelized transport to cell walls and occurs via free radical intermediates generated through oxidation. Two enzymes, laccase and peroxidase, catalyze this process. Laccases are oxidases dependent on oxidation, and peroxidases are H_2O_2-dependent oxidases. The free radicals generated on the free phenolic groups of the monolignols undergo cross-coupling, which leads to polymerization. It is at this step that multiple combinations of the monomers can take place, leading to the great diversity of lignins (Whetten and Sederoff 1995; Boerjan et al. 2003; Tobimatsu and Schuetz 2019).

8.2.7 Shikimate Pathway

Aromatic acids of plants and microorganisms are synthesized via the shikimate pathway, which has never been found in animals (Herrmann and Weaver 1999). Microorganisms, due to their very high metabolic rates, consume most of these amino acids for protein synthesis as soon as they are formed, and thus, they do not produce secondary metabolites through this pathway (Herrmann 1983; Pittard 1987). Meanwhile, plants use these amino acids as precursors for secondary metabolites in large quantities as a defensive measure (Herrmann 1995; Herrmann and Weaver 1999). Erythrose 4-phosphate (E4P) and phosphoenolpyruvate (PEP) condense in the first step of the shikimate pathway which ends with chorismate synthesis in seven metabolic steps (Herrmann and Weaver 1999). The first intermediate of this pathway synthesized by the DAHP synthase catalyzed condensation of PEP and E4P is 3-deoxy-d-arabino-heptulosonate7-phosphate (DAHP). Inorganic phosphate is also produced in this step. Microbial DAHPs are effectively inhibited by aromatic amino acids, unlike their plant counterparts (Herrmann 1995; Herrmann and Weaver 1999). In the next step, DHQ is synthesized in a 3-dehydroquinate (DHQ)

Eco-Friendly Applications

synthase catalyzed reaction by removal of the phosphate from DAHP. It was found that DAHP binds to DHQ synthase in its pyran form (Herrmann and Weaver 1999). NAD+ and a divalent cation (Zn^{2+} and Co^{2+} in bacteria) are required by DHQ synthase for its catalytic activity. Inorganic phosphate activates DHQ synthase to catalyze the redox neutral reaction (Herrmann and Weaver 1999). The complicated reaction involves transfer of the oxygen from DAHP ring to the carbon at the seventh position following cleavage of the phosphoester bond and elimination of inorganic phosphate. Dehydration of DHQ by DHQ dehydratase gives 3-dehydroshikimate (DHS). Syn and anti-elimination of water is catalyzed by Type I and Type II DHQ dehydratase, respectively, although both enzymes show no sequence similarity. Type I DHQ reductase is a dimer, while Type II DHQ reductase is a dodecamer. In microorganisms, an NADP-dependent shikimate dehydrogenase or a pyrrolo-quinoline quinine–dependent shikimate dehydrogenase reduces DHS to shikimate. In contrast, plants have a bifunctional DHQ dehydratase-shikimate dehydrogenase which catalyzes the conversion of DHQ to shikimate with the N-terminal of the enzyme possessing the dehydratase activity (Herrmann and Weaver 1999). The next step is catalyzed by shikimate kinase, which yields shikimate 3-phosphate (S3P) by phosphorylation of shikimate. The second double bond is formed by removal of this phosphate in the last step of this pathway. The sixth step is catalyzed by 5-enolpyruvylshikimate 3-phosphate (EPSP) synthase that catalyzes the condensation of S3P with another molecule of PEP, yielding EPSP and inorganic phosphate. Chorismate synthase catalyzes the last reaction of the pathway using $FMNH_2$ as a cofactor. Chorismate is formed as a result of loss of phosphate group from EPSP. All the products from the seven enzymatic reactions can serve as precursors to secondary metabolites (Herrmann 1995; Herrmann and Weaver 1999).

8.3 Applications of Secondary Metabolite

Secondary metabolites from different origins find enormous application in diverse fields. Their applications range from eco-friendly usage in bioremediation, biofuel production, etc., to commercial ones such as medicine, food additives, and cosmetics ingredients (Table 8.1).

TABLE 8.1

Economically Important Secondary Metabolites and Their Applications

S. No.	Name of Secondary Metabolite	Application	Source	Reference
1	Glycyrrhizin	Natural sweetener	Licorice	(Seki et al. 2011)
2	Lignan podophyllotoxin	Antitumor (anticancer drug-etoposide)	*Podophyllum peltatum*	(Ardalani et al. 2017)
3	Maklamicin	Antimicrobial against gram-positive bacteria	Endophytic *Micromonospora* sp.	(Igarashi et al. 2011)
4	Lipopeptide surfactin	Surfactant	*Bacillus subtilis*	(Kakinuma et al. 1969)
5	Erythromycin A	Macrolide antibiotic (gram-positive bacteria)	*Saccharopolyspora erythraea*	(Vara et al. 1989)
6	Rapamycin	Antifungal, antitumor, and immunosuppressant activity	*Streptomyces hygroscopicus*	(Kuscer et al. 2007)
7	Morphine	Analgesic	*Papaver bracteatum*	(Brochmann-Hanssen and Wunderly 1978)
8	Petrobactin	Protects against oxidative stress	*Bacillus anthracis*	(Hagan et al. 2018)
9	Scytonemin	Sunscreen	*Lyngbya aestuarii*	(Rastogi and Incharoensakdi 2014)
10	Curcumin	Active ingredient in blemish balm or bases cream (bb cream)	*Curcuma longa*	(Arct et al. 2014)

(Continued)

TABLE 8.1

Economically Important Secondary Metabolites and Their Applications (Continued)

S. No.	Name of Secondary Metabolite	Application	Source	Reference
11	Desferrioxamine	Desferal, iron overload diseases	*Streptomyces pilosus*	(Bannerman et al. 1962)
12	Lignin	Biofuel	Plants including herbaceous plants, grasses, etc.	(Gellerstedt et al. 2008)
13	Sabinene	Biofuel	Bioengineered strain of *Escherichia coli*	(Zhang et al. 2014)
14	Limonene and Myrcene	Fuel additive	Coniferous plants and citrus fruits	(Tracy et al. 2009)
15	Riboflavin	Food additives	*Ashbya gossypii*	(Dharmaraj et al. 2009)
16	Prodigiosin	Anticancer against leukemia cells	*Pseudoalteromonas* sp. 1020R	(Wang et al. 2012)
17	Melanin	Anticancer against skin cancer cells	*Streptomyces glaucescens* NEAE-H	(El-Naggar and El-Ewasy 2017)
18	Anthraquinone derivatives	Anticancer against breast cancer	*Alternaria* sp. ZJ9–6B	(Huang et al. 2011)
19	Astaxanthin	Food additives	*Xanthophyllomyces dendrorhous*	(Dharmaraj et al. 2009)
20	Anthraquinones	Antioxidants	*Stemphylium lycopersici*	(Li et al. 2017)
21	Carotenoid	Antioxidants	*Pedobacter*	(Correa Llanten et al. 2012)
22	Arpink red™	Food additives	*Penicillium oxalicum*	(Dharmaraj et al. 2009)
23	Lycopene	Food additives	*Erwinia uredovora, Fusarium sporotrichioides*	(Dharmaraj et al. 2009)
24	Monascorubramine, Rubropunctamine	Food additives	*Monascus* sp.	(Chen et al. 1969; Sweeny et al. 1981)
25	Prodigiosin	Textile industry	*Vibrio* spp.	(D'Aoust and Gerber 1974)
26	Anthraquinone	Textile industry	*Fusarium* sp.	(Nagia and El-Mohamedy 2007)
27	Pravastatin	Anticholesterolemic	*Penicillium citrinum*	(Alvarez et al. 2010)
28	Capsaicin	Chiles	*Capsicum annuum*	(Pandhair et al. 2006)
29	Stevioside	Saffron	*Stevia rebaudiana*	(Pandey et al. 2016)
30	Isoprenoids	Biofuel	Metabolically engineered *E. coli* and *S. cerevisiae*	(Ajikumar et al. 2010)
31	Enterobactin	Reference standard for HPLC and tryptophan quenching experiments	*E. coli*	(Gehring et al. 1997)
32	Azadiractin	Act as feeding inhibitor in desert locust	Neem (*Azadiracta indica*)	(Butterworth and Morgan 1968)
33	Agistatine D	Insecticidal agent	Endophytic *Xylaria* sp.; isolated from *Vitis labrusca*	(Ibrahim et al. 2014)
34	Ajmalicine	α1-adrenoceptor antagonist	*Catharanthus roseus*	(Vázquez-Flota et al. 1994)

S. No.	Name of Secondary Metabolite	Application	Source	Reference
35	Psoralen	Used to treat hyperproliferative skin disorder like psoriasis	*Ficus carica*; originates from cumarin in shikimate pathway	(Zaynoun et al. 1984)
36	Avermectin	Ivermectin, anti-malarial/ anthelmintic	*Streptomyces avermitilis*	(Pemberton et al. 2001)
37	Hydroxamate	Metalloid detoxification	*Aspergillus nidulans*	(Kumari et al. 2019)
38	Catecholate	Bioaccumulation/ bioremediation of heavy metal	*Bacillus subtilis*	(Khan et al. 2019)
39	Ellagitannin	Antioxidant, skin whitening	*Fragaria vesca*	(Zhentian et al. 1999)
40	Taxifolin	Skin whitening	*Polygonum hydropiper*	(Ge et al. 2018)
41	Flavonoids	Antioxidant	*Allium sativum*	(Singh and Kumar 2017)
42	Pelargonidin	Skin firming and whitening	*Dendrobium*	(Kanlayavattanakul et al. 2018)
43	Nicotine	Stimulant effect and improve attention and memory	*Nicotiana tabacum*	(Shitan et al. 2015)
44	Codeine	Antitussive, opioid, analgesic, narcotic	*Papaver somniferum*	(Parker et al. 1972)
45	Tetrahydrocannabinols	Stimulates dopamine release	*Cannabis sativa*	(Marks et al. 2009)
46	Hyoscyamine	Anticholinergics/ antispasmodics	*Datura stramonium*	(Underhill and Youngken Jr. 1962)

8.3.1 Biofuel Production

The existing resources are depleting at an alarming rate while the demand is increasing uncontrollably, which is a major global threat. Thus, there is a need to look for an alternative that aims for sustainable development by fulfilling the needs of today and keeping in mind the needs of future generations. Non-conventional energy sources, i.e. fossil fuels, are expected to be exhausted in a couple of decades. Biofuels, the products of cellular conversion of biomass into fuels, are considered an eco-friendly alternative. Microbes—plant sources like forests, agriculture products, etc.—form biomass, and fuels in different states (solid, liquid, and gas) can be obtained from these (Dufey 2006). Although microbes are potential producers of biofuel, limited information about genetic regulation of secondary metabolism causes hindrance in its production (Khan et al. 2020). Countries like Brazil, the United States, India, China, and the nations of Europe pose ample opportunities for biofuel production (Spiess 2011).

Secondary metabolites can either be directly used as biofuel or can be used to enhance production of biofuel from different sources. Isoprenoids, a component of various plant-derived and microbial secondary metabolites, have been suggested to be used as biofuel because of their cyclized structure of hydrocarbon alkene with methyl branches. Isoprenoids have been procured using different approaches, including use of metabolically engineered microbes such as *E. coli* and *S. cerevisiae* (Kang and Lee 2016; Wang et al. 2017). In engineered *E. coli*, for example, heterologous expression of mevalonic acid (MVA) pathway or the 2-C-methyl-d-erythritol 4-phosphate (MEP) pathway enhances production of basic units of isoprenoids, i.e. isopentenyl pyrophosphate (IPP) and dimethylallyl pyrophosphate (DMAPP), respectively (Ajikumar et al. 2010).

A number of different biofuels like methane and alcohol are derived from lignin, a plant-derived organic polymer. Lignin valorization, or conversion to various products, including fuel, can be carried out by oxidation, hydrogenation, gasification, pyrolysis, etc., yielding phenols, methane, cyclohexanes, fuel additives, ethane, propane, etc. (Abdullah et al. 2017). Hydrogenation of lignin that depolymerizes lignin, using hydrogen as reductant, under pressure and high temperature, is widely suggested. Terpenes such as camphene, limonene, etc., on the other hand, have been regarded as "specialty biofuels" that can either be used as is or mixed with other fuels like gasoline, diesel, etc. (Tracy et al. 2009; Meylemans et al. 2012). Increased production of sabinene, a bicyclic monoterpene, was also achieved by metabolic engineering of MVA pathway, similar to the case of isoprenoid (Zhang et al. 2014).

Iron is an essential nutrient for growth and metabolism of almost all living organisms, including microalgae, used for biofuel production. However, iron is not present in a soluble form in the environment, and its uptake needs specific mechanisms—one of which is the use of iron-chelating molecules. In an indirect approach, a microbial secondary metabolite called siderophore has been suggested to enhance biofuel production from microalgae by facilitating iron uptake. These secondary metabolites are known to chelate iron with high affinity; thus, it enhances microalgal growth that might in turn increase biofuel yield (Amin et al. 2009). This happens because microalgae do not possess any biosynthetic pathway to produce siderophores of their own, but are known to take up foreign siderophores complexed with iron (Hopkinson and Morel 2009). Such symbiotic interactions were observed in cases of bacteria like *Marinobacter* and *Holomonas*, enhancing growth of microalgae, e.g. *Dunaliella* sp. (Amin et al. 2009; Baggesen et al. 2014).

8.3.2 Bioremediation

Several plant-based SMs are responsible for bioremediation by stimulating plant growth-promoting microbes found in the vicinity of the rhizosphere (Miya and Firestone 2001; Isidorov and Jdanova 2002). These are known for remediating xenobiotic compounds. Compounds like salicylate induce the biodegradation of polycyclic aromatic hydrocarbon (PAH) in plants through systemic acquired resistance in association with microbes (De Meyer et al. 1999; Bogan et al. 2010). Bioremediation of compounds like pyrene, chrysene, etc., can be achieved through salicylate. These plant–microbe consortia effectively degrade polychlorinated biphenyls (PCBs; Singer et al. 2003).

PCBs are congeners and are enormously used in industries. However, due to severe toxicity and carcinogenic properties, their use was restricted in 1970 (Hornbuckle and Robertson 2010). It was then prioritized as a chemical that needs to be removed by 2025, owing to its persistence in the soil (Xu et al. 2010). Some microbes have the ability to degrade PCBs, but their interaction with plant exudates enhances the degradation ability. It is evident that plants' secondary metabolites (PSMs) have the ability to stimulate microbes, thereby enhancing PCB remediation termed as rhizo-remediation (Uhlik et al. 2013). Pino and colleagues (2016) formed a plant (*Avena sativa*, *Brachiaria decumbers*, *Medicago sativa*, *Brassica juncea*)–microbe (*Pseudomonas* sp., *Stenotrophomonas* sp.) consortium. It was observed that percentage of degradation of PCBs increased by acting as the consortium acted as inducer of PCB catabolic pathway. Flavone and its derivatives were identified as active PSMs. Other SMs like terpenes can be used in oxidation of PCB and other xenobiotic compounds (Focht 1995). Natural substrates like limonene, carvone, and thymolete can enhance biodegradation in case of *Arthrobacter* sp. (Pino et al. 2016).

Similarly, SMs also degrade hydrocarbons, which are the major cause of pollution in marine environments. Microbes use pollutants as carbon or energy sources, thereby releasing water and carbon dioxide as byproducts. Microorganisms belonging to *Pseudomonas* sp., *Bacillus* sp., *Nocardiopsis* sp., and *Micrococcus* sp. are involved in bioremediation of hydrocarbon from soil and marine contaminated regions (Sharma et al. 2019). It is quite interesting that PSMs not only remediate xenobiotic compounds but also stimulate microbes to carry out the same process. However, the mechanism requires further research.

8.3.3 Edible Dyes

The application of pigments as food coloring agents, paints, dyes, etc., has been in practice for ages. In India, it can be traced back to Indus Valley civilization (Aberoumand 2011). The use of color from natural extracts started in Egypt and later flourished in Japan in the 8th century (Nara period). In due course of time, use of synthetic color was practiced in the 18th century and rapidly extended as its production became technically and commercially feasible. Later on, its toxic and deleterious effects were observed in the environment and human health; thereby, some of them were withdrawn by the U.S. Food and Drug Administration. This led to an alternative option of natural pigments in order to meet worldwide demand.

Plants and microbes produce several beneficial pigments with commercial applications, despite few limitations like pigment instability and non-availability throughout all seasons. Bacteria and fungi are known to produce several pigments such as lycopene, canthaxanthin, riboflavin, etc. (Narsing Rao et al. 2017). Pigment from azaphilone, which is a group of *Monascus* sp., is extensively used as coloring agent in Asian countries, France, Germany, the United States, etc. The biosynthesis of azaphilone occurs via polyketide biosynthetic pathway (Tallapragada and Dikshit 2017). Recently, approximately 50 *Monascus* pigments have been isolated and extensively studied and a minimum 50 patents have been issued focusing its use as food. *Monascus* pigments widely used in the food industry include monascorubramine and rubropunctamine (purple/red colored pigment) used as sauces, ankaflavin, monascine (yellow), and orange-colored pigments rubtopunctatin and monascorubrine (Salomon and Karrer 1932; Chen et al. 1969; Sweeny et al. 1981). Steps involved in synthesis of *Monascus* pigment are as follows: (a) polyketide synthase condense acetyl-CoA and malonyl-CoA to form hexaketide chromophore; (b) beta-keto acid (obtained through fatty acid biosynthesis pathway) and hexaketide compound joined to form orange pigment via transesterification; and (c) reduction and amination in orange pigments give rise to yellow and red pigments, respectively (Sumathi 2009).

Besides *Monascus*, several other pigments are used as food additives, viz. astaxanthin, arpink red, riboflavin, beta-carotene, and lycopene isolated from *Xanthophyllomyces* sp., *Penicillium* sp., *Ashbya* sp., *Blakeslea* sp., and *Erwinia* sp., respectively (Dharmaraj et al. 2009).

Apart from the food industry, tons of natural extracts are used as dyes in cases of wool, silk, cotton, nylon, etc., as they are environment friendly. However, more than 1 million tons of synthetic dyes are being used that remain persistent in the environment due to their stability and are affecting the ecosystem adversely. Thus, microbial pigments like prodigiosin (*Vibrio* sp.), anthraquinone (*Fusarium* sp.), etc., are considered to be an eco-friendly option to replace synthetic dyes (Narsing Rao et al. 2017).

8.4 Strategies for Improving Secondary Metabolite Production

For commercial purposes, it is essential to increase the yield of secondary metabolites. Recently, there has been progress to develop approaches for high-yield production of SMs from plants and microbes. With the arrival of genetic engineering and advanced molecular tools, its regulation can be maintained efficiently. There are various stimuli that regulate SMs production, like nutrients, light intensity, temperature, pH, inter/intraspecies communication, etc. Moreover, the enhanced production is quite dependent on strain and fermentation process. Multiple strategies can be employed in order to achieve enhanced production (Figure 8.1)

8.4.1 Traditional Strategies

Scientists have used several approaches, and strain improvement is one such typical method within plant cells which depends on the parent body containing required products. This is done for callus induction and obtaining better cell lines (Buitelaar and Tramper 1992). Selection and screening is an important procedure for cultivation, as the major drawback in cells is epigenetic instability. There are various factors that influence SMs production, viz. strain improvement (selection and screening), media variations (nutrients and phytohormones), culture condition (pH, light, temperature), and specialized techniques such as immobilization, elicitation, two-phase, and two-stage systems, etc. (Murthy et al. 2014; Isah et al. 2018).

It is a known fact that SMs are produced in response to changes in environmental conditions such as biotic or abiotic stress. Microbial attacks lead to stimulation of secondary metabolism, and such stimulators are referred to as elicitors, which are either endogenous or exogenous (West 1981; Darvill and Albersheim 1984). These include salicylic acid, methyl jasmonate (Giri and Zaheer 2016), chitosan, heavy metals, salts, UV radiation, etc. (Narayani and Srivastava 2017). The reactions during elicitation

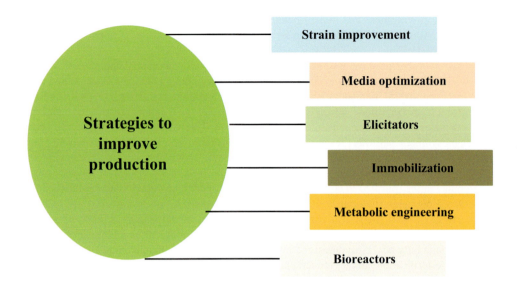

FIGURE 8.1 Strategies for improvement of secondary metabolite production.

include recognition and binding of elicitors to receptor proteins present on plasma membranes, reduction of proton gradient by inhibiting ATPase activity, upregulation of genes responsible for SMs production, stimulation of defense response leading to enhancement in biosynthesis, and accumulation of SM (Halder et al. 2019).

Low shear resistance is a common problem encountered in commercial application of SMs which can be overcome by immobilization, a technique using gel entrapment. In this strategy, cells get entrapped in gel made up of specific matrices like calcium alginate, agar, polyacrylamide, agarose, etc., thereby helping in easy separation of SM from medium (Felix et al. 1981; Murthy et al. 2014). The advantages include reuse of biocatalyst due to entrapment, maintenance of stable biocatalyst, matrix protected cells that prevent cells against damage, continuous process by removing inhibitors, increased cell densities, and productivity (Gonçalves and Romano 2018).

8.4.2 Metabolic Engineering

Metabolic engineering provides an advanced approach to uplift SMs production through genetic manipulation in biosynthetic pathways (O'Connor 2015). This involves enzymatic reaction studies, overexpression of biosynthetic pathways, competitive pathway inhibition, etc. Different metabolic engineering tools have been introduced in recent years to enhance SMs production—for example, yeast oligo-mediated genome engineering, gateway cloning, Gibson assembly, DNA assembler, and the CRISPR-Cas system which modify the pathways at pre-transcriptional level; synthetic promoters modify at transcriptional level; genetically encoded biosensors act at post transcriptional level thereby improving and optimizing SMs production (Gibson et al. 2009; DiCarlo et al. 2013; Vogl and Glieder 2013). Such strategies are being used in yeast like *Saccharomyces cerevisiae* and *Pichia pastoris*, and show enhanced production of various SMs like resveratrol, taxol/taxadiene, forskolin, amorphadiene, dihydrochalcones, gurmarin, etc. (Rahmat and Kang 2020). To study modulation in gene expression, the CRISPR-Cas9 module has become the most emerging strategy to engineer SMs production (De Frias et al. 2018). In order to accumulate specific preceding compounds or intermediates, certain steps involved in biosynthetic pathways can be inhibited. This has a huge role in large-scale production. However, clear understanding of pathways is still in the experimental stage, which is the main hindrance for its practical use. Thus, to carry out mass-scale production, knowledge about rate limiting steps, etc., is a prerequisite.

8.4.3 Bioreactors

It is noteworthy that mass-scale production of SM products is hindered by various limitations like genetic instability, shear sensitivity, oxygen transfer, slow rate of growth, etc. Scaling up production involves the use of bioreactors possessing several advantages over others. Metabolites can be isolated in a much efficient, simple, and predictable manner from media or biomass. Taxol, shikonin, and ginseng are some of the examples obtained by this strategy (Tabata and Fujita 1985; Hibino and Ushiyama 1999; Wink et al. 2005). Various reactors are used, viz. mechanically agitated and pneumatically agitated (bubble column reactor, airlift reactor), where airlift comparatively has a better role in submerged conditions and in less shear-sensitive cells (Murthy et al. 2014). Production of shikonin in the root culture of *Lithospermum erythrorhizon* was carried out *in situ* by using n-hexadecane layer in a two-phase bubble column reactor, and 95% product recovery was observed. However, due to the aggregation of hairy root culture, oxygen limitation occurred that can be overcome by using other bioreactors (Sim and Chang 1993). Enhanced production of shikonin isolated from *Arnebia* sp. was carried out using a two-stage culture system where a growth medium was employed for cell biomass and pigment formation using modified M9 medium in an airlift bioreactor (Gupta et al. 2014).

8.5 Siderophores: Promising Secondary Metabolites

Siderophores (Greek. *Sidero*: iron; *phores*: carrier) are specifically known for iron (Fe^{3+}) chelation and have molecular weight less than 10k Da (Devireddy et al. 2010). Iron is an important metal content in all living organisms for their development and metabolic growth (Khan et al. 2018). Despite its abundance in the lithosphere, it is not easily accessible to microorganisms. This is because oxidation of iron to insoluble oxyhydroxide polymer under aerobic conditions and biological pH makes it biologically unavailable (Paul and Dubey 2015). During biofilm formation, iron plays a crucial role in motility and maintenance of surface matrix consisting of polysaccharides (Weinberg 2004). Chelation of metals other than iron have recently been noticed, but in these cases, metal ion uptake by bacteria decreases the siderophore chelation activity (O'Brien et al. 2014). Rather than iron (Fe^{+3}), siderophores can bind with other toxic heavy metals and metalloids, although with less efficiency than iron (Braud et al. 2009). Unlike iron-siderophore complex, siderophores binding with other heavy metals is not always meant for internalization, and sometimes it chelates metals and prevents them from entering inside organisms (Kumar et al. 2010). Thus, siderophores in this context may be used for bioremediation (O'Brien et al. 2014).

8.5.1 Medical Applications of Siderophores (Iron Chelation)

Iron content in the human body is most essential to sustain major parts of biological processes. In the human body, the iron content is about 3–4 grams, a majority of which takes part to form porphyrin rings that are present in hemoglobin, myoglobin, and ferritin (Kurth et al. 2016). Iron in excess may be toxic to the human body. Thus, the body maintains its iron content by striking a balance between iron intake and excretion (Ganz 2013). Excess iron leads to unregulated rise in level of reactive oxygen species leading to cellular damage and organ failure, which can be cured by removal of excess plasma iron by iron chelation therapy (Kontoghiorghe and Kontoghiorghes 2016).

Desferrioxamine B (DFO), a hydroxamate siderophore isolated from *Streptomyces pilosus*, was the first agent used to treat iron overload, in the early 1960s (Bickel et al. 1960). Later, *Deferiprone* (DFP) and *Deferasirox* (DFX), siderophores like artificially synthesized chelators with improved half-life in blood plasma, proved highly efficient (Bickel et al. 1960). This is possible due to the synergistic effect of DFP and DFX, and its shuttle mechanism in which DFP helps in removal of cellular iron and DFX helps to excrete the iron from the body (Breuer et al. 2001).

Sideromycins are antibiotics conjugated with siderophores. Griseins were the first sideromycins, introduced by Reynolds et al. in 1947. Some of the effectively applicable sideromycins include danomycin, salmycin, and ferrimycin (Sackmann et al. 1962; Tsukiura et al. 1964; Vértesy et al. 1995). Sideromycins get entry in bacterial cells through conventional siderophore transport systems. After their entry inside the cells, hydrolysis of ester/amide bonds occurs, cleaving conjugated antibiotics from siderophore moiety (Wencewicz et al. 2009). The smart delivery system of azithromycin—also referred to as a "Trojan horse"—enhances the delivery and uptake efficiency of the drug (Braun et al. 2009; Zheng and Nolan 2014).

8.5.2 Siderophore-Mediated Bioremediation

Siderophores have great bioremediation potential in the ecosystem. They help in removing heavy toxic metals from affected areas and enhance the degradation ability of microbes. Siderophores are also used in recovery of rare earth elements.

Siderophores have the ability to chelate several metals in addition to iron; therefore, siderophore synthesis in microbes during metal exposure is an intricate balance between metal chelation requirements and metal-induced toxicity (Schalk et al. 2011; Fones and Preston 2013). Siderophores help to solubilize the heavy metals and make them bioavailable. These solubilized heavy metals are concentrated by siderophores to separate from the soil matrix, and for these, different types of bioreactor have been developed (Diels et al. 2009). Such application was aptly observed in soil purification by *Pseudomonas azotoformans* (Nair et al. 2008).

The addition of a large amount of hydrocarbons in the oceans and its bioremediation is a serious concern in petroleum-polluted ocean areas (Head and Swannell 1999). Here, siderophores can be used to degrade hydrocarbons by applying siderophore-producing microorganisms. Sabirova and colleagues (2006) observed that iron uptake and oxidation of alkanes is possible due to expression of relevant genes in *Alkanivorax borkumensis* during use of alkanes instead of pyruvate as a carbon source. Iron demand is increased during activity of alkane oxidizing enzymes required for hydrocarbon degradation (Austin and Groves 2011).

Alkanivorax borkmensis, most abundantly found in oil-polluted oceans, produces two types of siderophore, apmphibactine and pseudomonines. The latter help to maintain the hydrocarbon degradation efficiency of bacteria under iron-limiting conditions (Sabirova et al. 2006; Denaro et al. 2014). Amphibactine is also used by some hydrocarbon-degrading *Vibrio* sp. Siderophores aid in alkane degradation in two ways: (a) by adding iron uptake to alkane degrading microbes, which enhance their degradation potential; and (b) by emulsifying the alkane so that the solubility of oil increases to bacterial surface (Kem et al. 2014). *Marinobactor hydrocarbonoclasticus* is another commonly used oil degrading microbe which produces petrobactin and petrobactin sulfonate (Barbeau et al. 2002; Hickford et al. 2004).

8.6 Conclusion

Human population is increasing exponentially, and thus the demand to meet the requirement needs to be channelized without deteriorating the environment. The ecosystem is overloaded with synthetic and chemically synthesized compounds which, due to their harmful effects, need replacement. The multifaceted role of secondary metabolites is an excellent example to deliver required results aiming at sustainable development due to their varied chemical structure and origin. In the last few decades, besides antibiotics development, eco-friendly uses were explored in industries, agriculture, bioremediation, etc. Although extensive research is being carried out to bring these compounds to market, its optimization and production require further research. Strain improvement, genetic engineering, and advanced biotechnological tools need to happen to achieve the desirable results.

Acknowledgment

Azmi Khan and Pratika Singh are thankful to UGC, New Delhi for providing financial support as fellowship.

REFERENCES

Abdullah, B., Muhammad, S., Mahmood, N., 2017. Production of biofuel via hydrogenation of lignin from biomass. In *New Advances in Hydrogenation Processes—Fundamentals and Applications*. London: IntechOpen.

Aberoumand, A., 2011. A review article on edible pigments properties and sources as natural biocolorants in foodstuff and food industry. *World J. Dairy Food Sci.* 6, 71–78.

Ajikumar, P. K., Xiao, W. H., Tyo, K. E., Wang, Y., Simeon, F., Leonard, E., Mucha, O., Phon, T. H., Pfeifer, B., Stephanopoulos, G., 2010. Isoprenoid pathway optimization for Taxol precursor overproduction in *Escherichia coli*. *Science* 330, 70–74.

Alvarez, E., Rodiño-Janeiro, B. K., Somoza, R., González-Juanatey, J. R., 2010. Pravastatin counteracts angiotensin II-induced upregulation and activation of NADPH oxidase at plasma membrane of human endothelial cells. *J. Cardiovasc. Pharmacol.* 55(2), 203–212.

Amin, S. A., Green, D. H., Hart, M. C., Kupper, F. C., Sunda, W. G., Carrano, C. J., 2009. Photolysis of iron-siderophore chelates promotes bacterial-algal mutualism. *PNAS* 106, 17071–17076.

Arct, J., Ratz-Łyko, A., Mieloch, M., Witulska, M., 2014. Evaluation of skin colouring properties of curcuma longa extract. *Indian J. Pharm. Sci.* 76(4), 374–378.

Ardalani, H., Avan, A., Ghayour-Mobarhan, M., 2017. Podophyllotoxin: a novel potential natural anticancer agent. *Avicenna J. Phytomed.* 7(4), 285–294.

Austin, R. N., Groves, J. T., 2011. Alkane-oxidizing metalloenzymes in the carbon cycle. *Metallomics* 3, 775.

Baggesen, C., Gjermansen, C., Brandt, A. B., 2014. *A Study of Microalgal Symbiotic Communities with the Aim to Increase Biomass and Biodiesel Production*. Technical University of Denmark, 92pp.

Bannerman, R. M., Callender, S. T., Williams, D. L., 1962. Effect of desferrioxamine and D.T.P.A. in iron overload. *Br. Med. J.* 2(5319), 1573–1577.

Barbeau, K., Zhang, G. P., Live, D. H., Butler, A., 2002. Petrobactin, a photoreactive siderophore produced by the oil-degrading marine bacterium *Marinobacter hydrocarbonoclasticus*. *J. Am. Chem. Soc.* 124, 378–379.

Bentley, R., 1999. Secondary metabolite biosynthesis: the first century. *Crit. Rev. Biotechnol.* 19(1), 1–40.

Bickel, H., Bosshardt, R., Gäumann, E., Reusser, P., Vischer, E., Voser, W., Wettstein, A., Zähner, H., 1960. Antibiotic. *Helv. Chim. Acta.* 43, 2118.

Bloudoff, K., Schmeing, T. M., 2017. Structural and functional aspects of the nonribosomal peptide synthetase condensation domain superfamily: discovery, dissection and diversity. *Biochim. Biophys. Acta.* 1865(11 Pt. B), 1587–1604.

Boerjan, W., Ralph, J., Baucher, M., 2003. Lignin biosynthesis. *Annu. Rev. Plant Biol.* 54, 519–546.

Boettger, D., Hertweck, C., 2013. Molecular diversity sculpted by fungal PKS-NRPS hybrids. *ChemBioChem* 14(1), 28–42.

Bogan, B. W., Lahner, L. M., Paterek, J. R., 2010. Limited roles for salicylate and phthalate in bacterial PAH bioremediation. *Bioremediat. J.* 5(2), 93–100.

Braud, A., Hoegy, F., Jezequel, K., Lebeau, T., Schalk, I. J., 2009. New insights into the metal specificity of the *Pseudomonas aeruginosa pyoverdine*-iron uptake pathway. *Environ. Microbiol.* 11, 1079–1091.

Braun, V., Pramanik, A., Gwinner, T., Köberle, M., Bohn, E., 2009. *Sideromycins*: tools and antibiotics. *Biometals* 22, 3.

Breuer, W., Ermers, M. J., Pootrakul, P., Abramov, A., Hershko, C., Cabantchik, Z. I., 2001. Desferrioxamine-chelatable iron, a component of serum non-transferrin-bound iron, used for assessing chelation therapy. *Blood* 97, 792.

Brochmann-Hanssen, E., Wunderly, S. W., 1978. Biosynthesis of morphine alkaloids in *Papaver bracteatum*. Lindl. *J. Pharm. Sci.* 67(1), 103–106.

Buitelaar, R. M., Tramper, J., 1992. Strategies to improve the production of secondary metabolites with plant cell cultures: a literature review. *J. Biotechnol.* 23(2), 111–141.

Butterworth, J. H., Morgan, E. D., 1968. Isolation of a substance that suppresses feeding in locusts. *J. Chem. Soc. Chem. Commun.* 23–24.

Carrington, Y., Guo, J., Fillo, A., Kwon, J., Tran, L. T., Ehlting, J., 2018. Evolution of a secondary metabolic pathway from primary metabolism: shikimate and quinate biosynthesis in plants. *Plant J.* 95, 823–833.

Chan, J. Y., Nguyen, A. D., Lee, E. Y., 2019. Bioproduction of isoprenoids and other secondary metabolites using methanotrophic bacteria as an alternative microbial cell factory option: current stage and future aspects. *Catalysts* 9, 883.

Chen, F. C., Manchard, P. S., Whalley, W. B., 1969. The structure of monascin. *J. Chem. Soc. D* 3, 130–131.

Correa Llanten, D. N., Amenabar, M. J., Blamey, J. M., 2012. Antioxidant capacity of novel pigments from an Antarctic bacterium. *J. Microbiol.* 50, 374–379.

Cox, R. J., 2007. Polyketides, proteins and genes in fungi: programmed nano-machines begin to reveal their secrets. *Org. Biomol. Chem.* 5(13), 2010–2026.

Croteau, R., Kutchan, T. M., Lewis, N. G., 2000. Natural products (secondary metabolites). In Buchanan, B., Gruissem, W., Jones, R. (eds.), *Biochemistry and Molecular Biology of Plants* (Vol. 24, pp. 1250–1319). Rockville, MD: American Society of Plant Physiologists.

D'Aoust, J. Y., Gerber, N. N., 1974. Isolation and purification of prodigiosin from *Vibrio psychroerythrus*. *J. Bacteriol.* 118(2), 756–757.

Darvill, A. G., Albersheim, P., 1984. Phytoalexins and their elicitors: a defense against microbial infection in plants. *Ann. Rev. Plant Physiol.* 35, 243–275.

De Frias, U. A., Pereira, G. K. B., Guazzaroni, M.-E., Silva-Rocha, R., 2018. Boosting secondary metabolite production and discovery through the engineering of novel microbial biosensors. *BioMed Res. Int.* 1–11.

De Meyer, G., Capieau, K., Audenaert, K., Buchala, A., Métraux, J. P., Höfte, M., 1999. Nanogram amounts of salicylic acid produced by the rhizobacterium *Pseudomonas aeruginosa* 7NSK2 activate the systemic acquired resistance pathway in bean. *Mol. Plant Microbe Interact.* 12, 450–458.

Denaro, R., Crisafi, F., Russo, D., Genovese, M., Messina, E., Genovese, L., Carbone, M., Ciavatta, M. L., Ferrer, M., Golyshin, P., Yakimov, M. M., 2014. *Alcanivorax borkumensis* produces an extracellular siderophore in iron-limitation condition maintaining the hydrocarbon-degradation efficiency. *Mar. Genomics.* 17, 43.

Devireddy, L., Hart, D., Goetz, D., Green, M., 2010. A mammalian siderophore synthesized by an enzyme with a bacterial homolog involved in enterobactin production? *Cell* 141(6), 1006–1017.

Dharmaraj, S., Kumar, A. B., Dhevendaran, K., 2009. Food-grade pigments from *Streptomyces* sp. isolated from the marine sponge *Callyspongia diffusa*. *Food. Res. Int.* 42, 487–492.

DiCarlo, J. E., Conley, A. J., Penttila, M., Jantti, J., Wang, H. H., Church, G. M., 2013. Yeast oligo-mediated genome engineering (YOGE). *ACS Synth. Biol.* 2, 741–749.

Diels, L., Van, R. S., Taghavi, S., Van, H. S., 2009. From industrial sites to environmental applications with *Cupriavidus metallidurans*. *Antonie Van Leeuwenhoek* 96, 247.

Dreger, M., Stanislawska, M., Krajewska-Patan, A., Mielcarek, S., Mikolajczak, P. L., Buchwald, W., 2009. Pyrrolizidine alkaloids—chemistry, biosynthesis, pathway, toxicity, safety and perspectives of medicinal usage. *Herba. Pol.* 55. 127–147.

Dufey, A., 2006. *Biofuels Production, Trade and Sustainable Development: Emerging Issues*. London: International Institute for Environment and Development.

El-Naggar, N. E., El-Ewasy, S. M., 2017. Bioproduction, characterization, anticancer and antioxidant activities of extracellular melanin pigment produced by newly isolated microbial cell factories Streptomyces glaucescens NEAE-H. *Sci. Rep.* 14, 42129.

Felix, H. R., Brodelius, P., Mosbach, K., 1981. Enzyme activities of the primary and secondary metabolism in simultaneously permeabilized and immobilized plant cells. *Anal. Biochem.* 116, 462–470.

Focht, D.D., 1995. Strategies for the improvement of aerobic metabolism of polychlorinated-biphenyls. *Curr. Opin. Biotechnol.* 6, 341–346.

Fones, H., Preston, G. M., 2013. The impact of transition metals on bacterial plant disease. *FEMS Microbiol. Rev.* 37(4), 495–519.

Ganz, T., 2013. Systemic iron homeostasis. *Physiol. Rev.* 93, 1721.

Ge, F., Tian, E., Wang, L., Li, L., Zhu, Q., Wang, Y., Zhong, Y., Ge, R., 2018. Taxifolin suppresses rat and human testicular androgen biosynthetic enzymes. *Fitoterapia* 125.

Gehring, A., Bradley, K. and Walsh, C., 1997. Enterobactin biosynthesis in *Escherichia coli*: Isochorismate Lyase (EntB) is a bifunctional enzyme that is phosphopantetheinylated by EntD and then acylated by EntE using ATP and 2,3-dihydroxybenzoate. *Biochemistry* 36(28), 8495–8503.

Gellerstedt, G., Li, J., Eide, I., Kleinert, M. and Barth, T., 2008. Chemical structures present in biofuel obtained from lignin. *Energy Fuels* 22(6), 4240–4244.

Georgiev, V., Marchev, A. S., Berkov, S., Pavlov, A., 2013. Plant in vitro systems as sources of tropane alkaloids. In Ramawat, K.G., Mérillon, J. (eds.), *Natural Products*. Berlin & Heidelberg: Springer.

Gibson, D. G., Young, L., Chuang, R. Y., Venter, J. C., Hutchison, C. A., Smith, H. O., 2009. Enzymatic assembly of DNA molecules up to several hundred kilobases. *Nat. Methods* 6, 343–345.

Giri, C. C., Zaheer, M., 2016. Chemical elicitors versus secondary metabolite production in vitro using plant cell, tissue and organ cultures: recent trends and a sky eye view appraisal. *Plant Cell Tissue Organ Cult.* 126, 1–18.

Gokulan, K., Khare, S., Cerniglia, C., 2014. Metabolic pathways: production of secondary metabolites of bacteria. In Batt, C. A. (ed.), *Encyclopedia of Food Microbiology* (2nd ed., pp. 561–569). Amsterdam: Elsevier Ltd, Academic Press.

Gonçalves, S., Romano, A., 2018. Production of plant secondary metabolites by using biotechnological tools. In Vijayakumar, R., Raja, S.S.S. (eds.), *Secondary Metabolites—Sources and Applications*. London: IntechOpen. https://doi.org/10.5772/intechopen.76414

Gupta, K., Garg, S., Singh, J., Kumar, M., 2014. Enhanced production of napthoquinone metabolite (shikonin) from cell suspension culture of *Arnebia* sp. and its up-scaling through bioreactor. *3 Biotech* 4(3), 263–273.

Hagan, A. K., Plotnick, Y. M., Dingle, R. E., Mendel, Z. I., Cendrowski, S. R., Sherman, D. H., Tripathi, A., Hanna, P. C., 2018. Petrobactin protects against oxidative stress and enhances sporulation efficiency in *Bacillus anthracis* Sterne. *mBio* 9, e02079-18.

Halder, M., Sarkar, S., Jha, S., 2019. Elicitation: a biotechnological tool for enhanced production of secondary metabolites in hairy root cultures. *Eng. Life Sci*. elsc.201900058. https://doi.org/10.1002/elsc.201900058.

Halo, L. M., Marshall, J. W., Yakasai, A. A., Song, Z., Butts, C. P., Crump, M. P., Heneghan, M., Bailey, A. M., Simpson, T. J., Lazarus, C. M., Cox, R. J., 2008. Authentic heterologous expression of the tenellin iterative polyketide synthase nonribosomal peptide synthetase requires coexpression with an enoyl reductase. *ChemBioChem* 9(4), 585–594.

Head, I. M., Swannell, R. P., 1999. Bioremediation of petroleum hydrocarbon contaminants in marine habitats. *Curr Opin Biotechnol*. 10, 234.

Herrmann, K. M., 1983. The common aromatic biosynthetic pathway. In Herrmann, K.M., Somerville, R.L., (eds.), *Amino Acids: Biosynthesis and Genetic Regulation* (pp. 301–322). Reading, MA: Addison Wesley.

Herrmann, K. M., 1995. The shikimate pathway: early steps in the biosynthesis of aromatic compounds. *Plant Cell* 7(7), 907–919.

Herrmann, K. M., Weaver, L. M., 1999. The shikimate pathway. *Annu. Rev. Plant. Physiol. Plant. Mol. Biol.* 50, 473–503.

Hertweck, C., 2009. The biosynthetic logic of polyketide diversity. *Angew. Chem. (Int. Ed. Engl.)*. 48(26), 4688–4716.

Hibino, K., Ushiyama, K., 1999. Commercial production of ginseng by the plant tissue culture technology. In Fu, T. J., Singh, G., Curtis, W. R. (eds.), *Plant Cell & Tissue Culture for Production of Food Ingredients* (pp. 215–224). New York: Springer.

Hickford, S. J. H., Küpper, F. C., Zhang, G., Carrano, C. J., Blunt, J. W., Butler, A., 2004. Petrobactin sulfonate, a new siderophore produced by the marine bacterium *Marinobacter hydrocarbonoclasticus*. *J. Nat. Prod.* 67, 1897.

Higgens, C., Kastner, R., 1971. *Streptomyces clavuligerus* sp. nov., a β-lactam antibiotic producer. *Int. J. Syst. Evol. Microbiol.* 21, 326–331.

Hopkinson, B. M., Morel, F. M. M., 2009. The role of siderophores in iron acquisition by photosynthetic marine microorganisms. *Biometals* 22, 659–669.

Hornbuckle, K., Robertson, L., 2010. Polychlorinated biphenyls (PCBs): sources, exposures, toxicities. *Environ. Sci. Technol.* 44, 2749–2751.

Huang, C. H., Pan, J. H., Chen, B., Yu, M., Huang, H. B., Zhu, X., Lu, Y.-J., She, Z.-G., Lin, Y-C., 2011. Three bianthraquinone derivatives from the mangrove endophytic fungus *Alternaria* sp. ZJ9–6B from the South China Sea. *Mar. Drugs* 9, 832–843.

Ibrahim, A., Sorensen, D., Jenkins, H. A., McCarry, B. E., Sumarah, M. W., 2014. New diplosporin and agistatine derivatives produced by the fungal endophyte *Xylaria* sp. isolated from *Vitis labrusca*. *Phytochem. Lett.* 9, 179–183.

Igarashi, Y., Ogura, H., Furihata, K., Oku, N., Indananda, C., Thamchaipenet, A., 2011. Maklamicin, an antibacterial polyketide from an endophytic *Micromonospora* sp. *J. Nat. Prod*. 14(4), 670–674.

Isah, T., Umar, S., Mujib, A., Sharma, M. P., Rajasekharan, P. E., Zafar, N., Frukh, A., 2018. Secondary metabolism of pharmaceuticals in the plant in vitro cultures: strategies, approaches, and limitations to achieving higher yield. *Plant Cell Tissue Organ Cult.* 132, 239–265.

Isidorov, V., Jdanova, M., 2002. Volatile organic compounds from leaves litter. *Chemosphere* 48, 975–979.

Kakinuma, A., Oachida, A., Shima, T., Sugino, H., Isono, M., Tamura, G., Arima, K., 1969. Confirmation of the structure of surfactin by mass spectrometry. *Agric. Bioi. Chem.* 33, 1669–1671.

Kang, A., Lee, T. S., 2016. Secondary metabolism for isoprenoid-based biofuels. In *Biotechnology for Biofuel Production and Optimization* (pp. 35–71). Amsterdam: Elsevier.

Kanlayavattanakul, M., Lourith, N., Chaikul, N., 2018. Biological activity and phytochemical profiles of *Dendrobium*: a new source for specialty cosmetic materials. *Ind. Crop. Prod.* 120, 61–70.

Kawamoto, I., Yamamoto, M., Nara, T., 1983. *Micromonospora olivasterospora* sp. nov. *Int. J. Syst. Evol. Microbiol.* 33, 107–112.

Kem, M. P., Zane, H. K., Springer, S. D., Gauglitz, J. M., Butler, A., 2014. Amphiphilic siderophore production by oil-associating microbes. *Metallomics* 6, 1150.

Khan, A., Gupta, A., Singh, P., Mishra, A. K., Ranjan, R. K., Srivastava, A., 2019. Siderophore-assisted cadmium hyperaccumulation in *Bacillus subtilis*. *Int. Microbiol.* 23, 277–286.

Khan, A., Singh, P., Srivastava, A., 2018. Synthesis, nature and utility of universal iron chelator—Siderophore: a review. *Microbiol. Res.* 212–213, 103–111.

Khan, A., Singh, P., Srivastava, A., 2020. Microbial biofuels: an economic and eco-friendly approach. In Kumar, N. (ed.), *Biotechnology for Biofuels: A Sustainable Green Energy Solution* (pp. 165–196). Singapore: Springer Nature.

Kontoghiorghe, C. N., Kontoghiorghes, G. J., 2016. Efficacy and safety of iron-chelation therapy with deferoxamine, deferiprone, and deferasirox for the treatment of iron-loaded patients with non-transfusion-dependent thalassemia syndromes. *Drug Des. Devel. Ther.* 10, 465–481.

Kudalkar, P., Strobel, G., Riyaz-Ul-Hassan, S., Geary, B., Sears, J., 2012. *Muscodor sutura*, a novel endophytic fungus with volatile antibiotic activities. *Mycoscience* 53, 319–325.

Kumar, H., Bajpai, V. K., Dubey, R. C., Maheshwari, D. K., Kang, S. C., 2010. Effect of plant growth promoting rhizobia on seed germination, growth promotion and suppression of *Fusarium* wilt of fenugreek (Trigonella foenum-graecum L.). *Crop Prot.* 29, 591.

Kumari, S., Khan, A., Singh, P., Dwivedi, S., Ojha, K., Srivastava, A., 2019. Mitigation of As toxicity in wheat by exogenous application of hydroxamate siderophore of *Aspergillus* origin. *Acta Physiol. Plant.* 41(7).

Kurth, C., Kage, H., Nett, M., 2016. Siderophores as molecular tools in medical and environmental applications. *Org. Biomol. Chem.* 14(35), 8212–8227.

Kusaka, T., Yamamoto, H., Shibata, M., Muroi, M., Kishi, T., Mizuno, K., 1968. *Streptomyces citricolor* nov. sp. and a new antibiotic, aristeromycin. *J. Antibiot.* 21, 255–263.

Kuscer, E., Coates, N., Challis, I., Gregory, M., Wilkinson, B., Sheridan, R., Petković, H., 2007. Roles of rapH and rapG in positive regulation of rapamycin biosynthesis in *Streptomyces hygroscopicus*. *J. Bacteriol.* 189(13), 4756–4763.

Kuzuyama, T., 2002. Mevalonate and nonmevalonate pathways for the biosynthesis of isoprene units. *Biosci. Biotechnol. Biochem.* 66(8), 1619–1627.

Kuzuyama, T., Seto, H., 2012. Two distinct pathways for essential metabolic precursors for isoprenoid biosynthesis. *Proc. Jpn. Acad., Ser. B, Phys. Biol. Sci.* 88(3), 41–52.

Lambalot, R. H., Gehring, A. M., Flugel, R. S., Zuber, P., LaCelle, M., Marahiel, M. A., Reid, R., Khosla, C., Walsh, C. T., 1996. A new enzyme superfamily: the phosphopantetheinyl transferases. *Chem. Bio.* 3(11), 923–936.

Li, F., Xue, F., Yu, X., 2017. GC-MS, FTIR and Raman analysis of antioxidant components of red pigments from *Stemphylium lycopersici*. *Curr. Microbiol.* 74, 532–539.

Liu, Q., Luo, L., Zheng, L., 2018. Lignins: biosynthesis and biological functions in plants. *Int. J. Mol. Sci.* 19(2), 335.

Liu, X., Walsh, C. T., 2009. Cyclopiazonic acid biosynthesis in *Aspergillus* sp.: characterization of a reductase-like R* domain in cyclopiazonate synthetase that forms and releases cyclo-acetoacetyl-L-tryptophan. *Biochemistry* 48(36), 8746–8757.

Marahiel, M. A., Stachelhaus, T., Mootz, H. D., 1997. Modular peptide synthetases involved in nonribosomal peptide synthesis. *Chem. Rev.* 97(7), 2651–2674.

Marks, M. D., Tian, L., Wenger, J. P., Omburo, S. N., Soto-Fuentes, W., He, J., Gang, D. R., Weiblen, G. D., Dixon, R. A., 2009. Identification of candidate genes affecting Delta9-tetrahydrocannabinol biosynthesis in *Cannabis sativa*. *J. Exp. Bot.* 60(13), 3715–3726.

Meylemans, H., Quintana, R., Harvey, B., 2012. Efficient conversion of pure and mixed terpene feedstocks to high density fuels. *Fuel* 97, 560–568.

Miya, R. K., Firestone, M. K., 2001. Enhanced phenanthrene biodegradation in soil by slender oat root exudates and root debris. *J. Environ. Qual.* 30, 1911–1918.

Mootz, H. D., Schwarzer, D., Marahiel, M. A., 2002. Ways of assembling complex natural products on modular nonribosomal peptide synthetases. *ChemBioChem* 3(6), 490–504.

Murthy, H. N., Lee, E. J., Paek, K. Y., 2014. Production of secondary metabolites from cell and organ cultures: strategies and approaches for biomass improvement and metabolite accumulation. *Plant Cell Tissue Organ Cult.* 118, 1–16.

Nagia, F. A., El-Mohamedy, R. S. R., 2007. Dyeing of wool with natural anthraquinone dyes from *Fusarium oxysporum*. *Dyes Pigments* 75, 550–555.

Nair, A., Juwarkar, A. A., Devotta, S., 2008. Study of speciation of metals in an industrial sludge and evaluation of metal chelators for their removal. *J. Hazard Mater.* 152, 545.

Narayani, M., Srivastava, S., 2017. Elicitation: a stimulation of stress in in vitro plant cell/tissue cultures for enhancement of secondary metabolite production. *Phytochem. Rev.* 16, 1227–1252.

Narsing Rao, M. P., Xiao, M., Li, W. J., 2017. Fungal and bacterial pigments: secondary metabolites with wide applications. *Front. Microbiol.* 8, 1113.

O'Brien, S., Hodgson, D. J., Buckling, A., 2014. Social evolution of toxic metal bioremediation in *Pseudomonas aeruginosa*. *Proc. Biol. Sci.* (1787), 20140858.

O'Connor, S. E., 2015. Engineering of secondary metabolism. *Annu. Rev. Genet.* 49, 71–94.

Pandey, H., Pandey, P., Pandey, S. S., Singh, S., Banerjee, S., 2016. Meeting the challenge of stevioside production in the hairy roots of *Stevia rebaudiana* by probing the underlying process. *Plant Cell Tissue Organ Cult.* 126, 511–521.

Pandhair, V., Vinayak, V., Gosal, S. S., 2006. Biosynthesis of capsaicin in callus cultures derived from fruit explants of *Capsicum annuum* L. *Plant Cell Biotechnol. Mol. Biol.* 7. 35–40.

Parker, H. I., Blaschke, G., Rapoport, H., 1972. Biosynthetic conversion of thebaine to codeine. *J. Am. Chem. Soc.* 94, 4, 1276–1282.

Paul, A., Dubey, R., 2015. Characterization of protein involved in nitrogen fixation and estimation of co-factor. *Appl. J. Curr. Res. Biosci. Plant Biol.* 2(1), 89–97.

Pemberton, D. J., Franks, C. J., Walker, R. J., Holden-Dye, L. (2001). Characterization of glutamate-gated chloride channels in the pharynx of wild-type and mutant *Caenorhabditis elegans* delineates the role of the subunit GluCl-alpha2 in the function of the native receptor. *Mol. Pharmacol.* 59, 1037–1043.

Pino, N. J., Muñera, L. M., Peñuela, G. A., 2016. Root exudates and plant secondary metabolites of different plants enhance polychlorinated biphenyl degradation by rhizobacteria. *Bioremediat. J.* 20(2), 108–116.

Pittard, A. J., 1987. Biosynthesis of the aromatic amino acids. In *Escherichia coli and Salmonella typhimurium: Cellular and Molecular Biology* (pp. 368–394). Washington, DC: American Society for Microbiology.

Rahmat, E., Kang, Y., 2020. Yeast metabolic engineering for the production of pharmaceutically important secondary metabolites. *Appl. Microbiol. Biotechnol.* https://doi.org/10.1007/s00253-020-10587-y

Rastogi, R., Incharoensakdi, A., 2014. Characterization of UV-screening compounds, mycosporine-like amino acids, and scytonemin in the cyanobacterium *Lyngbya* sp. CU2555. *FEMS Microbiol. Ecol.* 87(1), 244–256.

Reynolds, D. M., Schatz, A., Waksman, S. A., 1947. Grisein, a new antibiotic produced by a strain of *Streptomyces griseus*. *Proc. Soc. Exp. Biol. Med.* 64(1), 50–54.

Ruzicka, L., 1953. The isoprene rule and the biogenesis of terpenic compounds. *Experientia* 9, 357–367.

Sabirova, J. S., Ferrer, M., Regenhardt, D., Timmis, K. N., Golyshin, P. N., 2006. Proteomic insights into metabolic adaptations in *Alcanivorax borkumensis* induced by alkane utilization. *J. Bacteriol.* 188, 3763.

Sackmann, W., Reusser, P., Neipp, L., Kradolfer, F., Gross, F., 1962. Ferrimycin A, a new iron-containing antibiotic. *Antibiot. Chemother.* 12, 34.

Salomon, H., Karrer, P., 1932. Pflanzenfarbstoffe XXXVIII. Ein farbstoff aus "rotem" reis monascin. *Helv Chim Acta*. 15, 18–22.

Schalk, I. J., Hannauer, M., Braud, A., 2011. New roles for bacterial siderophores in metal transport and tolerance. *Environ. Microbiol.* 13(11), 2844–2854.

Seki, H., Sawai, S., Ohyama, K., Mizutani, M., Ohnishi, T., Sudo, H., Fukushima, O., Akashi, T., Aoki, T., Saito, T., Muranakaa, T., 2011. Triterpene functional genomics in licorice for identification of CYP72A154 involved in the biosynthesis of glycyrrhizin. *Plant Cell*, 23, 4112–4123.

Sharma, N., Lavania, M., Lal, B., 2019. Microbes and their secondary metabolites: agents in bioremediation of hydrocarbon contaminated site. *Arch. Pet. Environ. Biotechnol.* 4, 151.

Shen, B., 2003. Polyketide biosynthesis beyond the type I, II and III polyketide synthase paradigms. *Curr. Opin. Chem. Biol.* 7(2), 285–295.

Shitan, N., Hayashida, M., Yazaki, K., 2015. Translocation and accumulation of nicotine via distinct spatiotemporal regulation of nicotine transporters in *Nicotiana tabacum*. *Plant Signal. Behav.* 10(7), e1035852

Sim, S. J., Chang, H. N., 1993. Increased shikonin production by hairy roots of *Lithospermum erythrorhizon* in two phase bubble column reactor. *Biotechnol. Lett.* 15(2), 145–150.

Sims, J. W., Schmidt, E. W., 2008. Thioesterase-like role for fungal PKS-NRPS hybrid reductive domains. *J. Am. Chem. Soc.* 130(33), 11149–11155.

Singer, A. C., Crowley, D. E., Thompson, I. P., 2003. Secondary plant metabolites in phytoremediation and biotransformation. *Trends Biotechnol.* 21(3), 123–130.

Singh, V., Kumar, R., 2017. Study of phytochemical analysis of Antioxidant activity of *Allium sativum* of Bundelkhand region. *Int. J. Life Sci. Scienti. Res.* 3(6), 1451–1458.

Spiess, W. E. L., 2011. Does biofuel production threaten food security. *IUFoST Food Security TaskForce.* www.iufost.org/iufostftp/Does%20Biofuel%20Production%20Threaten%20Food%20Security.pdf. Accessed 12 Nov 2016.

Staunton, J., Weissman, K. J., 2001. Polyketide biosynthesis: a millennium review. *Nat. Prod. Rep.* 18(4), 380–416.

Sumathi, B., 2009. Microbial pigments. In Singh nee' Nigam, P., Pandey, A. (eds.), *Biotechnology for Agro-Industrial Residues Utilisation* (pp. 147–162). The Netherlands: Springer.

Sweeny, J. G., Valdes, M. C. E., Guillermo, I. A., Sato, H., Sakamura, S., 1981. Photoprotection of the red pigments of *Monascus anka* in aqueous media by 1,4,6-trihydroxynaphthalene. *J. Agric. Food Chem.* 29(6), 1189–1193.

Tabata, M., Fujita, Y., 1985. Production of shikonin by the plant cell cultures. In Zatlin, M., Day, P., Hollaender, A. (eds.), *Biotechnology in the Plant Science* (pp. 207–218). Cambridge: Academic Press.

Tallapragada, P., Dikshit, R., 2017. Microbial production of secondary metabolites as food ingredients. In Holban, A.M., Grumezescu, A.M., (eds.), *Microbial Production of Food Ingredients and Additives* (pp. 317–345). Cambridge, CA: Academic Press.

Thirumurugan, D., Cholarajan, A., Raja, S. S. S., Vijayakumar, R., 2018. An introductory chapter: secondary metabolites. In Vijaykumar, R., Raja, S. S. S. (eds.), *Secondary Metabolites—Sources and Applications.* London: Intech Open. https://doi.org/10.5772/intechopen.79766.

Tobimatsu, Y., Schuetz, M., 2019. Lignin polymerization: how do plants manage the chemistry so well? *Curr. Opin. Biotechnol.* 56. 75–81.

Tokuriki, N., Jackson, C. J., Afriat-Jurnou, L., Wyganowski, K. T., Tang, R., Tawfik, D. S., 2012. Diminishing returns and tradeoffs constrain the laboratory optimization of an enzyme. *Nat. Commun.* 3, 1257–1259.

Tracy, N., Chen, D., Crunkleton, D., Price, G., 2009. Hydrogenated monoterpenes as diesel fuel additives. *Fuel*, 88(11), 2238–2240.

Tsukiura, H., Okanishi, M., Ohmori, T., koshiyama, H., Miyaki, T., Kitazima, H., Kawaguchi, H., 1964. Danomycin: A new antibiotic. *J. Antibiotic.* 17, 39.

Uhlik, O., Musilova, L., Ridl, J., Hroudova, M., Vlcek, C., Koubek, J., Holeckova, M., Mackova, M., Macek, T., 2013. Plant secondary metabolite-induced shifts in bacterial community structure and degradative ability in contaminated soil. *Appl. Microbiol. Biotechnol.* 97, 9245–9256.

Underhill, E. W., Youngken, H. W., Jr., 1962. Biosynthesis of hyoscyamine and scopolamine in *Datura stramonium. J. Pharm.* Sci. 51, 121–125.

Vara, J. E. S. U. S., Lewandowska-Skarbek, M., Wang, Y. G., Donadio, S., Hutchinson, C. R., 1989. Cloning of genes governing the deoxysugar portion of the erythromycin biosynthesis pathway in *Saccharopolyspora erythraea* (*Streptomyces erythreus*). *J. Bacteriol.* 171(11), 5872–5881.

Vázquez-Flota, F., Moreno-Valenzuela, O., Miranda-Ham, M. L., Coello-Coello, J., Loyola-Vargas, V. M., 1994. Catharanthine and ajmalicine synthesis in *Catharanthus roseus* hairy root cultures. *Plant Cell Tissue Organ Cult.* 38, 273–279.

Vértesy, L., Aretz, W., Fehlhaber, H. W., Kogler, H., 1995. Salmycin A—D, antibiotics from *Strep tomy ces violaceus*, DSM 8286, having a sidierophor-aminoglycoside structure. *Helv. Chim. Acta.* 78, 46.

Vogl, T., Glieder, A., 2013. Regulation of *Pichia pastoris* promoters and its consequences for protein production. *New Biotechnol.* 30, 385–404.

Walsh, C. T., Gehring, A. M., Weinreb, P. H., Quadri, L. E., Flugel, R. S., 1997. Post-translational modification of polyketide and nonribosomal peptide synthases. *Curr. Opin. Chem. Biol.* 1(3), 309–315.

Wang, C., Zada, B., Wei, G., Kim, S. W., 2017. Metabolic engineering and synthetic biology approaches driving isoprenoid production in *Escherichia coli. Bioresour. Technol.* 241, 430–438.

Wang, Y., Nakajima, A., Hosokawa, K., Soliev, A. B., Osaka, I., Arakawa, R., Enomoto, K., 2012. Cytotoxic prodigiosin family pigments from *Pseudoalteromonas* sp. 1020R isolated from the Pacific coast of Japan. *Biosci. Biotechnol. Biochem.* 76, 1229–1232.

Weinberg, E. D., 2004. Suppression of bacterial biofilm formation by iron limitation. *Med. Hypotheses* 63: 863–865.

Weissman, K. J., 2009. Introduction to polyketide biosynthesis. *Methods. Enzymol.* 459, 3–16.

Wencewicz, T. A., Möllmann, U., Long, T. E., Miller, M. J., 2009. Syntheses and biological studies of the naturally occurring salmycin "Trojan Horse" antibiotics and synthetic desferridanoxamine-antibiotic conjugates. *Biometals* 22, 633.

West, C. A., 1981. Fungal elicitors of the phytoalexin response of higher plants. *Narurwissenschaften* 68, 447–457.

Whetten, R., Sederoff, R., 1995. Lignin biosynthesis. *Plant Cell* 7(7), 1001–1013.

Wink, M., 2010. Introduction: biochemistry, physiology and ecological functions of secondary metabolites. In Wink, M. (Ed.), *Annual Plant Reviews: Biochemistry of Plant Secondary Metabolism* (pp. 1–19). New York: Wiley-Blackwell.

Wink, M., Alfermann, A. W., Franke, R., Wetterauer, B., Distl, M., Windhövel, J., Krohn, O., Fuss, E., Garden, H., Mohagheghzadeh, A., Wildi, E., Ripplinger, P., 2005. Sustainable bioproduction of phytochemicals by plant in vitro cultures: anticancer agents. *Plant Genet. Resour. C.* 3, 90–100.

Xu, L., Teng, Y., Li, Z., Norton, J., Luo, Y., 2010. Enhanced removal of polychlorinated biphenyls from alfalfa rhizosphere soil in a field study: the impact of a rhizobial inoculum. *Sci. Total Environ.* 408, 1007–1013.

Yadav, A. N., Kour, D., Rana, K. L., Yadav, N., Singh, B., Chauhan, V. S., Rastegari, A. A., Hesham, A. E.-L., Gupta, V. K., 2019. Metabolic engineering to synthetic biology of secondary metabolites production. In *New and Future Developments in Microbial Biotechnology and Bioengineering* (pp. 279–320). Amsterdam: Elsevier.

Zaynoun, S. T., Aftimos, B. G., Abi Ali, L., Tenekjian, K. K., Khalidi, U., Kurban, A. K., 1984. Ficus carica; isolation and quantification of the photoactive components. *Contact Derm.* 11(1), 21–25.

Zhang, H., Liu, Q., Cao, Y., Feng, X., Zheng, Y., Zou, H., Liu, H., Yang, J., Xian, M., 2014. Microbial production of sabinene—a new terpene-based precursor of advanced biofuel. *Microb. Cell Fact.* 13, 20.

Zheng, T., Nolan, E. M., 2014. Enterobactin-mediated delivery of β-lactam antibiotics enhances antibacterial activity against pathogenic *Escherichia coli*. *J. Am. Chem. Soc.* 136, 9677.

Zhentian, L., Jervis, J., Helm, R. F., 1999. C-Glycosidic ellagitannins from white oak heartwood and callus tissues. *Phytochemistry* 51(6), 751–756.

Ziegler, J., Facchini, P. J., 2008. Alkaloid biosynthesis: metabolism and trafficking. *Annu. Rev. Plant. Biol.* 59, 735–769.

9

Genetic Diversity Analysis of *Jatropha curcas*: A Biofuel Plant

Nitish Kumar

CONTENTS

9.1	Introduction	211
9.2	Conventional Method for Genetic Diversity Assessment	212
9.3	Genetic Diversity and Phylogenetics of *Jatropha* Species	215
	9.3.1 Intraspecific Genetic Diversity	217
	9.3.2 Markers for Toxic and Non-Toxic Varieties of *J. curcas*	219
9.4	Conclusion	219
References		220

9.1 Introduction

We are aware of the importance of black gold that we dig out from the earth as we often get shocks of inflation in prices of fuels like petrol-diesel obtained from this crude oil—but we also know that invention is the mother of necessity, and energy technology is currently moving from "below the earth" to "above the earth." Researchers have explored the possibility of obtaining biodiesel from plants like *Jatropha*, and its cultivation is in progress all over on gigantic scale. This versatile plant can be labeled as green gold, as it will not only provide biofuel in abundance, but also will help us to protect our environment in the future, besides its medical and other applications.

Worldwide introduction of *J. curcas* for varied purposes had met with limited success due to unreliable and low seed set, as well as oil yields resulting in low economic returns. *J. curcas* is a wild species and no varieties with desirable traits for specific growing conditions are available, which makes growing *Jatropha* a risky business (Jongschaap et al. 2007). The crop is also characterized by variable and unpredictable yield for reasons that have not been identified (Ginwal et al. 2004). This limits the large-scale cultivation and warrants the need for genetic improvement and breeding of superior genotypes of the species for which establishing genetic distances through DNA fingerprinting methods is required.

The major constraints in achieving higher-quality oil yield of this crop are lack of information about its genetic variability, oil composition, and absence of suitable ideotypes for different cropping systems. Knowledge of genetic relationship and variation in the species is a prerequisite in any breeding program because it permits the organization of germplasm, including elite lines, and provides for more efficient parental selection. Despite its ecological and economic importance, the taxonomy and genetic structure of the *Jatropha* genus is not entirely clarified due to the occurrence of natural hybridization among species (Airy Shaw 1972). Furthermore, the available germplasm lacks information on the genetic base. Hence, assessment of genetic diversity and its characterization becomes imperative, for which molecular tools have rendered their assistance in the recent past. In the present chapter, we discuss the studies conducted at the molecular level to characterize genetic diversity in *J. curcas*.

DOI: 10.1201/9780429262975-9

9.2 Conventional Method for Genetic Diversity Assessment

A wide variety of techniques has been used in the studies of forest tree relationship and variation. Studies have been conducted to identify genetic variation in populations, provenances and clones of *J. curcas* through traditional morphometric and biochemical marker techniques (Prabakaran and Sujatha 1999; Ginwal et al. 2004; Pant et al. 2006; Sunil et al. 2008). Although phenotypic traits cannot be reliable measures of genetic differences because of the influence of the environment on gene expression, we believe that it is appropriate to mention salient traditional approaches employed for assessment of variability in *J. curcas*. These studies have enriched the scientific understanding of nature of existing variation in the species (Table 9.1).

Sukarin and colleagues (1987) did not observe any differences in vegetative development and first seed yields among 42 clones originating from different locations in Thailand and planted in a provenance trial at the Khon Kaen Field Crops Research Center. Heller (1992; cited in Heller 1996) tested a collection of 13 provenances in multi-location field trials in two countries of the Sahel region of Africa: Senegal and Cape Verde. Significant differences in vegetative development were detected among the various provenances at all locations. However, plants of various provenances appeared very uniform in morphological characters such as leaf shape. In a small subset of 10 *J. curcas* accessions from central India, seed oil content was significantly correlated with seed weight, stem diameter, and total leaf area (Ginwal et al. 2004). In a range from 400–100 m elevation, Pant and colleagues (2006) found a significant positive effect of altitude on various oil yield components, including number of branches per tree, number of fruits per branch, and number of seeds per tree, but a significant reduction was observed in kernel oil content (43.1% at lower elevations vs. 30.7% at higher elevations).

The predominance of environmental factors over genetic factors has been reported by Kaushik and colleagues (2007) within the small genetic resource base of *J. curcas* accessions from Haryana state in India, although seed size and oil content and seed weight could be genetically clustered and significantly differentiated. Rao and colleagues (2008) evaluated genetic association and variability in seed and growth characters in 32 high-yielding candidates plus trees (CPTs; phenotypes judged, but not proven by test, to be unusually superior in some quality or qualities such as exceptional growth rate, desirable growth habit, high wood density, etc.) of *J. curcas* from different locations spread over 150,000 km^2 in India. Significant trait differences were observed in all the seed characters, viz., seed morphology and oil content as well as in growth characters, viz., plant height, female-to-male flower ratio and seed yield in the progeny trial. Broad sense heritability was high in general and exceeded 80% for all the seed traits studied. Sunil and colleagues (2008) recorded the phenotypic traits of *J. curcas* plants *in situ* at four different eco-geographic regions of India. They noticed pronounced differences in the nine characteristics they assessed for a total of 162 accessions in the four zones. For example, the plant height of 80% accessions in one zone was less than 1.5 m, while in another zone, 60% of the accessions were larger than 1.5 m. Similar differences were noticed in number of fruits and seed oil content and composition. Like most earlier studies, Sunil and colleagues (2008) did not undertake genetic characterization of the accessions; thus, the reason for the variability was not clear.

Only a limited number of studies involving biochemical markers for assessment of genetic diversity in *J. curcas* have been reported, which mainly involved isozyme markers. For example, Prabakaran and Sujatha (1999) used the isozymes peroxidase and superoxide dismutase to ascertain the phylogeny of *J. curcas* with other species, while Yunus (2007) was able to differentiate among *Jatropha* species from different regions in central Java and Indonesia on the basis of enzyme-like sorbitol dehydrogenase, shikimate dehydrogenase, alcohol dehydrogenase, and isocitrate dehydrogenase. Most of the studies to evaluate germplasm have been done using materials collected from CPTs of different regions, different aged plants (3–20 years), and propagated through seeds or vegetative cuttings (Anonymous 2006). Comparison of yield-contributing traits based on such accessions results in erroneous conclusions about the superiority of the identified clone, as it is strongly influenced by the mode of propagation, soil type, climatic conditions, age of the plant, and plant density (Heller 1996).

Diversity Analysis of Jatropha curcas

TABLE 9.1
A Consolidated List of Intraspecies and Interspecies Genetic Diversity Studies Conducted in *J. curcas*

Sr. No.	Reference	Marker Type Used for Analysis	Location of Collection	Number of Accessions/Species
1	(Sujatha et al. 2005)	RAPD	Andhra Pradesh	2
2	(Basha and Sujatha 2007)	RAPD, ISSR, SCAR	Tamil Nadu, Rajasthan, Kerala, Andhra Pradesh, Haryana, Madhya Pradesh	42
3	(Ranade et al. 2008)	RAPD, DAMD	Uttar Pradesh, Uttaranchal, Sikkim, Arunachal Pradesh, Meghalaya, Orissa	12
4	(Gupta et al. 2008)	ISSR, RAPD	Uttaranchal, Rajasthan, Orissa, Uttar Pradesh	13
5	(Subramanyam et al. 2009)	RAPD	Andhra Pradesh, Orissa, Tamil Nadu, Karnataka, Jarkhand, Bhihar, Kerala, Madhya Pradesh, Uttaranchal, Uttar Pradesh, Haryana, Punjab, Assam, West Bengal, New Delhi, Gujarat, Meghalaya, Tripura, Chhattisgarh, Maharashtra	40
6	(Pamidimarri et al. 2009c)	RAPD, AFLP, SSR	Gujarat	7
7	(Tatikonda et al. 2009)	AFLP	Uttar Pradesh, Gujarat, Rajasthan, Andhra Pradesh, Chhattisgarh, Madhya Pradesh	48
8	(Jubera et al. 2009)	RAPD	Tamil Nadu, Madhya Pradesh, Maharashtra, Karnataka	7
9	(Umamaheshwari et al. 2010)	ISSR	Tamil Nadu	17
10	(Ikbal et al. 2010)	RAPD	Rajasthan, Punjab, Haryana, Madhya Pradesh, Gujarat	40
11	(Rao et al. 2009)	Morphological traits	Karnataka	-
12	(Pamidimarri et al. 2009a)	nrDNA ITS sequence	Gujarat	7
13	(Pamidimarri et al. 2009b)	RAPD, AFLP	Gujarat	7
14	(Ganesh Ram et al. 2008)	RAPD	Tamil Nadu, Andhra Pradesh	12
15	(Dhillon et al. 2009)	RAPD	Haryana	23 hybrids from cross of two species
16	(Kumar et al. 2009)	RAPD	Tamil Nadu	9
17	(Vijayanand et al. 2009)	ISSR, Morphological traits	Tamil Nadu	9

(Continued)

TABLE 9.1

A Consolidated List of Intraspecies and Interspecies Genetic Diversity Studies Conducted in *J. curcas* (Continued)

Sr. No.	Reference	Marker Type Used for Analysis	Location of Collection	Number of Accessions/ Species
18	(Basha and Sujatha 2009)	RAPD, ISSR, SSR	Andhra Pradesh, Tamil Nadu	8
19	(Sujatha and Prabakaran 2003)	RAPD	Andhra Pradesh	Hybrids of two species
20	(Prabakaran and Sujatha 1999)	Isozyme	Tamil Nadu	3
21	(Basha *et al.* 2009)	RAPD, ISSR, SSR	India, Cape Verde, Mexico, Madagascar, Uganda, Africa, El Salvador, Egypt, Vietnam, China, Malaysia, Phillippines, Thailand	72
22	(Sun *et al.* 2008)	SSR, AFLP	South China Botanical Garden	58
23	(Wen *et al.* 2010)	EST-SSR, genomic SSR	Indonesia, Grenada, South America, Hainan (China), Yunnan (China)	45
24	(Ambrosi *et al.* 2010)	RAPD, ISSR	Mexico, Brazil, South America, Africa, Western Africa, Sri Lanka, Jordan, Southeast Asia, India	27
25	(Shen *et al.* 2010)	AFLP	Hainan (China)	38
26	(Xiang *et al.* 2007)	ISSR	Yunnan (China)	158
27	(Pamidimarri *et al.* 2009d)	SSR	Different parts of the globe	6 species of *Jatropha*
28	(Cai *et al.* 2010)	ISSR	China, Myanmar	224
29	(Tanya *et al.* 2011)	ISSR	Mexico, China, Vietnam, Thailand	39 accessions and four species
30	(Rosado *et al.* 2010)	RAPD, SSR	Brazil	192
31	(Subramanyam *et al.* 2010)	RAPD	India	10
32	(Pamidimarri *et al.* 2011)	SSR	Gujarat, India	6 species of *Jatropha*
33	(Zhang *et al.* 2011)	AFLP	China and Southeast Asia	240
34	(Pamidimarri *et al.* 2010)	RAPD, AFLP	India	28
35	(Mastan *et al.* 2011)	RAPD, AFLP, SSR	India	15

RAPD: Random amplification of polymorphic DNA
ISSR: Inter simple sequence repeats
SCAR: Sequence characterized amplified region
DAMD: Directed amplification of minisatellite-region DNA
AFLP: Amplified fragment length polymorphisms
SSR: Simple-sequence repeats
nrDNA ITS: Nuclear ribosomal DNA internal transcribed spacer
EST-SSR: Expressed sequence tag-simple-sequence repeats

9.3 Genetic Diversity and Phylogenetics of *Jatropha* Species

Assessment of genetic diversity using molecular markers is crucial for the efficient management and biodiversity conservation of plant genetic resources in gene banks. Assessment of diversity and phylogenetic relationships has traditionally been studied through morphological characteristics and isozyme analysis. However, such analyses have inherent disadvantages, such as limited numbers of markers, and are often less effective due to their inconsistency and sensitivity to short-term environmental fluctuations (Crawford *et al.* 1995; Essilman *et al.* 1997; Lesica *et al.* 1998; Lowrey and Crawford 1985; Soltis *et al.* 1992). Advances in the field of molecular biology have provided many tools for studying genetic diversity at the genome level in order to investigate phylogenetic relationships among different species. DNA-based molecular analysis tools are ideal for germplasm characterization and phylogenetic studies. Among the currently available DNA fingerprinting techniques, restriction fragment length polymorphism (RFLP), amplified fragment-length polymorphism (AFLP), randomly amplified polymorphic DNA (RAPD), microsatellites markers/SSR, sequence characterized amplified regions (SCAR), sequence tagged sites (STS), and nuclear ribosomal DNA internal transcribed spacer (nrDNA ITS) regions have been used to study genetic diversity and phylogenetics, and in generation of molecular markers for efficient use in breeding and genetic resource management. Limited studies have been carried out on the genetic diversity and phylogenetics of genus *Jatropha*. Puangpaka and Thaya (2003) studied the karyology of five *Jatropha* species by staining chromosomes of the microsporocyte, and reported that in most species, the chromosomes paired as bivalents at first metaphase and separated to 11:11 at first anaphase. The bivalent length ranged from 1–3.67 µm, and most species had chromosome numbers of $2n = 22$. This study reported that *J. curcas* and *J. multifida* were chromosomally similar. *J. integerrima*'s "red" and "pink" flowers had the same meiotic configuration of six ring II + five rod II, while the meiotic configuration of *J. podagrica* was eight ring II + three rod II. The karyology of *J. gossypifolia* was determined from first anaphase cells and the chromosomes separated to 11:11. Thus, according to Puangpaka and Thaya (2003), *J. curcas*, *J. multifida*, and *J. gossypifolia* appear to be related closely to each other, based on their meiotic configuration and morphological similarity. Carvalho and colleagues (2008) studied the genome size, base composition, and karyotype of *J. curcas*. The results showed that the 2C value of the *J. curcas* genome was 0.85 pg, with an average GC base composition of 38.7%. The karyotype of *J. curcas* is made up of 22 relatively small metacentric and submetacentric chromosomes whose size ranges from 1.24 to 1.71 µm. Based on cytological and peroxidase isozyme studies, Prabakaran and Sujatha (1999) reported *J. tanjorensis* as a natural interspecific hybrid of *J. curcas* and *J. gossypifolia*.

Efforts have been directed toward establishment of phylogenetic relationship of *J. curcas* with other related *Jatropha* species and natural hybrids (e.g. *J. tanjorensis J. L. Ellis et Saroja*) through RAPD, AFLP, ISSR, and other microsatellite markers, in addition to some novel molecular techniques. Fortunately, the material used in these studies sometimes incorporated accessions from many countries (Costa Rica, India, Mexico, Nigeria, Thailand, etc.), which assisted in clear comparative analyses. Ganesh Ram and colleagues (2008) investigated the genetic diversity of 12 *Jatropha* species (including five accessions of *J. curcas*) using RAPD markers. Of the 26 RAPD primers used, 18 gave reproducible amplification banding patterns of 112 polymorphic bands out of 134 bands scored, accounting for 80.2% polymorphism across the genotypes. Three primers—viz., OPA 4, OPF 11, and OPD 14—generated 100% polymorphic patterns. However, the study with few data points (18 primers) resulted in several ambiguities in establishment of genetic relationships among *Jatropha* species.

Pamidimarri and colleagues (2009b) studied the extent of genetic variability to establish phylogenetic relationship among seven species of *Jatropha*—viz., *J. curcas*, *J. glandulifera Roxb.*, *J. gossypifolia L.*, *J. integerrima Jacq.*, *J. multifida L.*, *J. podagrica Hook*, and *J. tanjorensis J. L. Ellis et Saroja* — using RAPD and AFLP markers. The percentage of loci that were polymorphic among the species was found to be 97.74% by RAPD and 97.25% by AFLP. The mean percentage of polymorphism was found to be 68.48% by RAPD and 71.33% by AFLP. They recorded maximum relatedness between *J. curcas* and *J. integerrima* which may be the reason for the success of interhybrid crosses between these two species. However, neither RAPD nor AFLP data generated in this study supported the view of *J. tanjorensis*, a natural interspecific hybrid between *J. curcas* and *J. gossypifolia*, as suggested by Prabakaran and Sujatha (1999) and emphasized that both RAPD and AFLP techniques are comparable in divergence studies of *Jatropha s*pecies.

Pamidimarri and colleagues (2009c) further studied phylogenetic relationships among these seven species using nuclear ribosomal DNA internal transcribed spacer (ITS) sequencing (nrDNA ITS) and compared the results with multilocus marker analysis systems reported earlier (Pamidimarri *et al.* 2009b) for the same genus. The size variation obtained among sequenced nrDNA ITS regions was narrow and ranged from 647–654 bp. The overall mean genetic distance (GD) of genus *Jatropha* was found to be 0.385. The present study also strongly supports high phylogenetic closeness of *J. curcas* and *J. integerrima*, while *J. podagrica* was also found clustered with *J. curcas*.

Senthil and colleagues (2009) studied genetic diversity among eight *Jatropha* species and three *J. curcas* accessions analyzed using ISSR markers. Nine ISSR primers generated reproducible amplification banding pattern of 61 polymorphic bands out of 64 scored, accounting for 98.14% polymorphism across the species. The ISSR primers—viz., I1, I2, I3, I4, I5, I6, I7 and I10—generated 100% polymorphic patterns. Jaccard's coefficient of similarity varied from 0.346–0.807, indicative of high level of genetic variation among the genotypes studied. The UPGMA cluster analysis indicated three distinct clusters, one comprised of all accessions of *J. curcas* L. (TNMJ1, TNMJ 22, and TNMJ 23); the second including four species—viz., *J. tanjorensis* J. L. Ellis et Saroja, *J. gossypifolia* L., *J. podagrica* Hook and *J. maheshwarii* Subrum and M.P. Nayer; and the third cluster including another four species—viz., *J. villosa* Wight, *J. multifida* L., *J. integerrima* Jacq and *J. glandulifera* Roxb. The overall grouping pattern of clustering corresponds well with PCA confirming patterns of genetic diversity observed among the species.

Using nuclear and organelle specific primers for supporting interspecific gene transfer, Basha and Sujatha (2009) attempted characterization of *Jatropha* species occurring in India. DNA from 34 accessions comprising eight agronomically important species—*J. curcas, J. gossypifolia, J. glandulifera, J. integerrima, J. podagrica, J. multifida, J. villosa, J. villosa* var. ramnadensis and *J. maheshwarii*—and a natural hybrid, *J. tanjorensis*, were subjected to molecular analysis using 200 RAPD, 100 ISSR, and 50 organelle-specific microsatellite primers from other angiosperms. The nuclear marker systems revealed high interspecific genetic variation (98.5% polymorphism) corroborating with the morphological differentiation of the species. Ten organelle-specific microsatellite primers resulted in single, discrete bands of which three were functional disclosing polymorphism among *Jatropha* species. PCR products from two consensus chloroplast microsatellite primer pairs (ccmp6 and 10) revealed variable number of T and A residues in the intergenic regions of ORF 77–ORF 82 and rp12–rps19 regions, respectively, in *Jatropha*. Artificial hybrids were produced between *J. curcas* and all *Jatropha* species used in this study, with the exception of *J. podagrica*. Characterization of F1 hybrids using polymorphic primers specific to the respective parental species confirmed the hybridity of the interspecific hybrids. Characterization of both natural and artificially produced hybrids using chloroplast-specific markers revealed maternal inheritance of the markers. While the RAPD and ISSR markers confirmed *J. tanjorensis* as a natural hybrid between *J. gossypifolia* and *J. curcas*, the ccmp primers (ccmp6 and 10) unequivocally established *J. gossypifolia* as the maternal parent.

As compared to multilocus markers like RAPD and AFLP, microsatellites have advantages like locus specificity, codominant nature, high reproducibility, and substantial size polymorphism (Powell et al. 1996). Generation of novel molecular markers like microsatellites provides better tools to assess the amount and distribution of molecular diversity and for population genetic studies. Pamidimarri and colleagues (2009d) isolated 12 microsatellites from *J. curcas* and characterized them in 32 accessions collected from a natural population in the Junagadh Gir forest region of Gujarat state, India, and the library was constructed following the FIASCO procedure (Zane et al. 2002), with minor modifications. Their cross-amplification was also checked in six common species of *Jatropha* (*J. glandulifera, J. gossypifolia, J. integerrima, J. multifida, J. podagrica,* and *J. tanjorensis*).

A total of 12 polymorphic loci were identified, with highest number of alleles (11) given by marker jcds24 and lowest (two) by jcps1 and jcms30. The observed and expected heterozygosities ranged from 0.94 to 0.54 and from 0.95 to 0.56, respectively. Tests for Hardy–Weinberg equilibrium showed that loci jcds58, jcds66, jcps1, jcps6 and jcms30 were not in Hardy–Weinberg equilibrium. These deviations may be due to presence of null alleles or disturbances in natural dispersal of the race in the population by anthropogenic activity (Basha and Sujatha 2007; Pamidimarri *et al.* 2009b). No significant linkage disequilibrium was detected between any pair of loci after correcting for multiple comparisons.

Assessing genetic variation by RAPD, AFLP, and combinatorial tubulin–based polymorphism (cTBP) in 38 *J. curcas* accessions from 13 countries on three continents revealed narrow genetic diversity, while the six *Jatropha* species from India exhibited pronounced genetic diversity, indicating higher possibilities of improving *J. curcas* by interspecific breeding (Popluechai et al. 2009). The samples were initially examined using 10 RAPD primers. One cluster contained all of the 17 *J. curcas* accessions, and the second contained the out-group *J. podagrica*, which showed an overall similarity of 52% with *J. curcas*. Among the *J. curcas* accessions, the similarity coefficient was high (0.78), indicating a narrow genetic base. The two Indian accessions clustered separately, while the Nigerian accession clustered with the remaining 14 Thai accessions. The six provenances of Thai accessions could not be clearly differentiated, reinforcing the narrow genetic base between provenances. Popluechai and colleagues (2009) also used a novel, relatively unexploited technique of combinatorial tubulin–based polymorphism (cTBP; Breviario et al. 2007) which uses variation in the length of the first and second intron of members of the plant tubulin gene family. The approach was successfully used earlier to detect intra- and interspecies polymorphism in diverse plants including oilseed plants, rapeseed, peanut (Breviario et al. 2007) and palm. Results showed that the four accessions from Costa Rica were clearly different than those from other parts of the world, and they also exhibited intraspecific polymorphism in both intron I and II. In all these studies, *J. curcas* accessions from different eco-geographic regions of India were 60%–80% similar. Results from the studies above suggested the importance of testing accessions from wider eco-geographic regions of the world. However, the analysis of 38 accessions from 13 countries around the world, along with six different species of *Jatropha* from India, again indicated 75% similarity among the global *J. curcas* accessions (Popluechai et al. 2009).

9.3.1 Intraspecific Genetic Diversity

Assessment of within-species genetic variation involving accessions/populations from different agroclimatic zones and/or geographical areas in India and China (with the exception of some studies in which accessions from Asia, Latin America, and Africa were collectively evaluated) have been carried out, employing a wide variety of molecular marker systems to reveal low to moderate levels of genetic diversity. The extent of genetic diversity was assessed in a representative set of 42 accessions of *J. curcas* encompassing different agroclimatic zones of India, along with a non-toxic genotype from Mexico. Molecular polymorphism was 42.0% with 400 RAPD primers and 33.5% with 100 ISSR primers between accessions, indicating modest levels of genetic variation in the Indian germplasm. The within-population variation based on RAPD polymorphism was 64%, at par with interpopulation variation. Population-specific bands have been identified for accessions from Kerala (2 RAPD markers), Neemuch-1 from Rajasthan (1 each of RAPD and ISSR markers) and Mexican genotype (17 RAPD and 4 ISSR markers), which serve as diagnostic markers in genotyping (Basha and Sujatha 2007). However, among the 23 selected provenances from 300 collected provenances from all over India, Reddy and colleagues (2007) reported relatively lower polymorphism using RAPD (14%–16%) and AFLP (8%–10%) techniques.

Ranade and colleagues (2008) employed two single-primer amplification reaction (SPAR) methods to assess the diversity among the accessions of *J. curcas*, both among already held collections, as well as from a few locations in the wild. They concluded that this relatively recently introduced plant species shows adequate genetic diversity in India and the accessions from the North East Region of India were most distant from all other accessions in UPGMA analysis. The phylogenetic relationships of 13 *J. curcas* genotypes from different parts of India (Rajasthan, Uttaranchal, Uttar Pradesh, and Orissa) were analyzed by Gupta and colleagues (2008) using 34 PCR markers (20 RAPDs and 14 ISSRs). Amplification of genomic DNA of the 13 genotypes, using RAPD analysis, yielded 107 fragments that could be scored, of which 91 were polymorphic, with an average of 4.55 polymorphic fragments per primer. The number of amplified fragments ranged from one (OPA20, OPB19, OPD13) to nine (OPA18), and they varied in size from 200–2,500 bp; percentage of polymorphism ranged from 40% (OPB18) to a maximum of 100% (14 primers); and resolution power ranged from a minimum of 0.153 (OPA20, OPB19) to a maximum of

11.23 (OPB15). The genotypes from Orissa (Orissa 6 and Orissa 7) appeared to be distinct from other genotypes.

Out of 25 ISSR primers used in the same study, only 14 were able to give rise to reproducible amplification products. These primers produced 81 bands across 13 genotypes, of which 62 were polymorphic with an average of 4.42 polymorphic fragments per primer. The number of amplified fragments ranged from two (ISSR 7, ISSR 8, ISSR 16) to nine (ISSR 12), and varied in size from 200–2,500 bp. Percentage of polymorphism ranged from 37.5% (ISSR 2, ISSR10) to a maximum of 100% (seven primers). The primers based on poly (GA) produced the maximum number of bands (nine) while, poly (AT) and many other motifs gave no amplification at all with any of these 13 genotypes (Gupta et al. 2008). RAPD markers were more efficient than the ISSR assay with regard to polymorphism detection, as they detected 84.26% polymorphism as compared to 76.54% for ISSR markers. However, resolving power (Rp), average bands per primer, Nei's genetic diversity (h), Shannon's Information Index (I), total genotype diversity among population (Ht), within-population diversity (Hs), and gene flow (Nm) estimates were more for ISSR (7.098, 5.79, 0.245, 0.374, 0.244, 0.137 and 0.635, respectively) as compared to RAPD markers (5.669, 5.35, 0.225, 0.359, 0.225, 0.115 and 0.518, respectively). The regression test between the two Nei's genetic diversity indices gave r2 = 0.3318, showing low regression between RAPD- and ISSR-based similarities. Regression value for ISSR and ISSR + RAPD combined data was moderate (0.6027), while it is maximum for RAPD- and ISSR + RAPD–based similarities (0.9125). Clustering of genotypes within groups was not similar when RAPD and ISSR derived dendrograms were compared, whereas the pattern of clustering of the genotypes remained more or less the same in RAPD and combined data of RAPD + ISSR (Gupta et al. 2008).

In China, Sun and colleagues (2008) assessed genetic relationships of 58 *J. curcas* accessions from different geographic locations based on simple sequence repeat (SSR) and AFLP analyses. Seventeen SSR markers were developed using the FIASCO (Fast Isolation by AFLP of Sequences Containing repeats) protocol; only one SSR primer was polymorphic with two alleles. The seven AFLP primer combinations amplified 70 polymorphic loci in total, 14.3% of which were polymorphic. The clustering of genotypes based on the AFLP markers shows that the genetic diversity of *J. curcas* in Guizhou region of China was notably different than the other samples. AFLP was also employed to assess the diversity in the elite germplasm collection of *J. curcas* from six different states of India. Forty-eight accessions were used, with seven AFLP primer combinations that generated a total of 770 fragments with an average of 110 fragments per primer combination. A total of 680 (88%) fragments showed polymorphism in the germplasm analyzed, of which 59 (8.7%) fragments were unique (accession specific) and 108 (15.9%) fragments were rare (present in fewer than 10% of accessions). In order to assess the discriminatory power of seven primer combinations used, a variety of marker attributes like polymorphism information content (PIC), marker index (MI), and resolving power (RP) values were calculated. Although the PIC values ranged from 0.20 (E-ACA/M-CAA) to 0.34 (E-ACT/M-CTT), with an average of 0.26 per primer combination, and the MI values were observed in the range of 17.60 (E-ACA/M-CAA) to 32.30 (E-ACT/M-CTT), with an average of 25.13 per primer combination. Genotyping data obtained for all 680 polymorphic fragments were used to group the accessions analyzed using UPGMA-phenogram and principal component analysis (PCA). Accessions coming from Andhra Pradesh were found to be diverse as these were scattered in different groups, whereas accessions coming from Chhattisgarh showed an occurrence of a higher number of unique/rare fragments (Tatikonda et al. 2009).

Recently, 225 accessions of *J. curcas* collected from over 30 countries in Latin America, Africa, and Asia were studied. Samples were analyzed (AFLP) at San Carlos University in Guatemala by nucleotide binding site (NBS)-profiling (conserved sequence based on NBS-gene family), and in the Netherlands at Wageningen University and Research Centre—Plant Research International. Genetic variability was low in African and Indian *J. curcas* accessions, but high genetic variability was found in Guatemalan and other Latin American accessions. These studies have proved that molecular markers provide an efficient and quick tool in characterization of genetic diversity among the clones, accessions and populations of *J. curcas*. Several authors (e.g. Reddy et al. 2007; Basha and Sujatha 2007; Gupta et al. 2008; Sun et al. 2008) have also recorded relative effectiveness of different marker systems, as well as low to moderate levels of within-species genetic diversity, in *J. curcas*. However, considerable variability in morphometric features and oil yield in promising genotypes (CPTs) of *J. curcas* identified after rigorous selection in the whole of eastern India comprising the states of Bihar, Jharkhand, Orissa, and West Bengal (62 CPTs identified in an area of more than 4 million hectares) during the first phase of our research under the aegis of the Indian Council of Forestry Research and Education has been confirmed through 20 RAPD markers in 28 CPTs/their clones (Bhatia et al. 2009; Singh et al. 2009).

9.3.2 Markers for Toxic and Non-Toxic Varieties of *J. curcas*

A non-toxic *J. curcas* variety from Mexico has been reported with seeds that can be used for human consumption after roasting (Makkar *et al.* 1998). Cultivation of non-toxic varieties could provide oil for biodiesel and de-oiled cake as livestock feed, and thus add value to the crop (Becker and Makkar 1998). No significant morphological, qualitative, or quantitative differences were observed between toxic and non-toxic varieties, except for phorbol ester content (Makkar *et al.* 1998; Becker and Makkar 1997). Development of markers that can help distinguish non-toxic from toxic varieties will not only add value to the product, but will also help in the selective cultivation of non-toxic varieties. Molecular markers specific to toxic and non-toxic varieties were identified using selective primers in a single PCR reaction using RAPD and AFLP techniques (Pamidimarri et al. 2009c). The primers IDT E-18 and OPL-14, and the AFLP selective primer combination E-ACC/M-CAC, resulted in polymorphic markers for both toxic and non-toxic varieties. Novel microsatellite markers were also tested for their polymorphism in toxic and non-toxic varieties, and seven out of 12 markers succeeded in showing polymorphism (Pamidimarri et al. 2009c). The isolated SCAR markers confirmed their specificity and reproducibility in discriminating toxic and non-toxic genotypes (33). In total, 371 RAPD and 1441 AFLP were analyzed, and 56 (15.09%) RAPD and 238 (16.49%) AFLP markers were found specific to either of the varieties. Genetic similarity between non-toxic and toxic varieties was found to be 0.92 by RAPD and 0.90 by AFLP fingerprinting, and demonstrated that both techniques were equally competitive in identifying polymorphic markers and differentiating varieties.

Basha and colleagues (2009) elucidated that genetic background of 72 *J. curcas* accessions representing 13 countries has been under taken using molecular analysis and biochemical traits. Seed kernel protein, oil content, ash content, and phorbol esters revealed variation, with accessions from Mexico containing low levels of phorbol esters. Molecular characterization disclosed polymorphism of 61.8% and 35.5% with RAPD and ISSR primers, respectively, and a Mantel test (Mantel 1967) revealed positive correlation between the two marker systems. A dendrogram based on pairwise genetic similarities and three-dimensional principal coordinate analysis (PCA) using data from RAPD and ISSR marker systems showed close clustering of accessions from all countries and grouped the Mexican accessions separately in clusters III–VI.

Presence of the toxic phorbol esters is major concern, and analysis of 28 Mexican accessions resulted in identification of molecular markers associated with high and low phorbol ester content. The identified RAPD and ISSR markers were converted to SCARs for increasing their reliability and use in marker-assisted programs aimed at development of accessions with reduced toxicity. Twelve microsatellite primers differentiated the non-toxic Mexican accessions and disclosed novel alleles in Mexican germplasm. Amplification with primers specific to the curcin coding sequence and promoter region of ribosome-inactivating protein (RIP) revealed polymorphism with one primer specific to RIP promoter region specifically in accessions with low phorbol ester levels. Narrow genetic variation among accessions from different regions of the world and rich diversity among Mexican genotypes in terms of phorbol ester content and distinct molecular profiles indicates the need for exploitation of germplasm from Mexico in *J. curcas* breeding programs (Basha *et al.* 2009). The specific markers generated will be useful in distinguishing non-toxic from toxic varieties of *J. curcas*, which could be used further in marker-assisted selection (MAS), quantitative trait loci (QTL) analysis, and further molecular breeding studies.

9.4 Conclusion

With advancement of scientific research in the arena of plant molecular biology, a number of molecular techniques have been developed in recent years. Most of these have been efficiently employed for characterization of the genetic diversity of *J. curcas* L. to produce similar notions of low level of genetic diversity in the species despite wide phenotypic variability and significant differences in oil content in accessions from different geographical regions. *J. curcas*, an undomesticated plant species, uniquely exhibits naturally widespread genetic monomorphism, as revealed in most of the molecular studies. The reasons for the global low genetic variability seen in *J. curcas* remain unclear. Most likely, the anthropogenic and environmental influences in generating genetic variability are missing because: (a) it is not a crop; (b) it is a well-surviving, undomesticated plant; (c) it is highly stress tolerant due to adaptive genomic characters probably acquired before its global distribution; and (d) a limited stock has been vegetatively and apomictically propagated, since *J. curcas* is known to exhibit apomixis (Bhattacharya et al. 2005). Furthermore, Richards and colleagues (2006) observed that a pronounced phenotypic plasticity is in itself a genotypic trait that allows the plant to respond to different environments through morphological and physiological changes for its survival. Another reason behind this seems to be introduced by the nature of *J. curcas* in countries (like India) where these molecular studies have been conducted with limited numbers of accessions in most of the cases. Further, variable seed yield and oil content in different accessions not commensurate with genetic differences indicates toward overriding influence of prevailing environmental conditions for seed oil production. However, incorporation of genotypes from a wide geographical area—including Central America, especially Mexico and Guatemala, and the Caribbean, in addition to Africa and Asia—for evaluation at the molecular level will present a clear picture. Therefore, a collaborative global *Jatropha* genetic diversity evaluation effort is immediately needed for better use of this valuable species in breeding programs, considering its potential of biodiesel and motor fuel production.

REFERENCES

Airy Shaw HK (1972) The Euphorbiaceae of Siam. *Kew Bulletin* 26:191–363.
Ambrosi DG, Galla G, Purelli M, Barbi T, Fabbri A, Lucretti S *et al* (2010) DNA markers and FCSS analyses shed light on the genetic diversity and reproductive strategy of *Jatropha curcas* L. *Diversity* 2:810–836.
Basha SD, Francis G, Makkar HPS, Becker K, Sujatha M (2009) A comparative study of biochemical traits and molecular markers for assessment of genetic relationships between *Jatropha curcas* L. germplasm from different countries. *Plant Sci* 176:812–823.
Basha SD, Sujatha M (2009) Genetic analysis of *Jatropha* species and interspecific hybrids of *Jatropha curcas* using nuclear and organelle specific markers. *Euphytica* 168:197–214.
Basha SD, Sujatha M. (2007) Inter- and intra-population variability of *Jatropha curcas* (L.) characterized by RAPD and ISSR markers and development of population-specific SCAR markers. *Euphytica* 56:375–386.
Becker K, Makkar HPS (1997) Potential of *Jatropha* seed cake as protein supplement in livestock feed and constraints to its utilization. In: *Proceedings of Jatropha 97: International Symposium on Biofuel and Industrial Products from Jatropha curcas and Other Tropical Oilseed Plants*, 23–27 February 1997, Managua/Nicaragua Mexico.
Becker K, Makkar HPS (1998) Toxic effects of phorbolesters in carp (*Cyprinus carpio* L.). *Vet Human Toxicol* 40:82–86.
Bhatia SK, Singh P, Singh S (2009) Genetic evaluation and molecular characterization of *Jatropha curcas* L. of eastern India. In: *National Conference on Biofuel: Potential and Challenges*, 25–26 February, 2009, Jabalpur, India, p 89 (Abstract).
Bhattacharya A, Datta K, Datta SK (2005) Floral biology, floral resource constraints and pollination limitation in *Jatropha curcas* L. *Pakistan J Biol Sci* 8:456–460.
Breviario D, Baird WmV, Sangoi S, Hilu K, Blumetti P, Giani S (2007) High polymorphism and resolution in targeted fingerprinting with combined beta-tubulin introns. *Molecular Breeding* 20:249–259.
Cai Y, Wu G, Peng J (2010) ISSR-based genetic diversity of *Jatropha curcas* germplasm in China. *Biomass Bioenergy* 34(12):1739–1750. https://doi.org/10.1016/ j.biombioe.2010.07.001
Carvalho CR, Clarindo WR, Praca MM, et al. (2008) Genome size, base composition and karyotype of *Jatropha curcas* L., an important biofuel plant. *Plant Sci* 174:613–617.
Crawford AM, Dodds KG, Ede AJ, et al. (1995) An autosomal genetic linkage map of the sheep genome. *Genetics* 140:703–724.
Dhillon RS, Hooda MS, Jattan M, Chawla V, Bhardwaj M, Goyal SC (2009) Development and molecular characterization of interspecific hybrids of *Jatropha curcas 9 J. integerrima*. *Indian J Biotech* 8:384–390.

Essilman EJ, Crawford DJ, Brauner S, et al. (1997) RAPD marker diversity within and divergence among species of *Dendroseris* (Asteraceae: Lactuceae). *Am J Bot* 4:591–596.

Ganesh Ram S, Parthiban KT, Kumar RS, Thiruvengadam V, Paramathma M (2008) Genetic diversity among *Jatropha* species as revealed by RAPD markers. *Genet Resour Crop Evol* 55:803–809.

Ginwal HS, Rawat PS, Srivastava RL (2004) Seed source variation in growth performance and oil yield of *Jatropha curcas* Linn. in central India. *Silvae Genet* 53:186–192.

Gupta S, Srivastava M, Mishra GP, Naik PK, Chauhan RS, Tiwari SK *et al* (2008) Analogy of ISSR and RAPD markers for comparative analysis of genetic diversity among different *Jatropha curcas* genotypes. *African J Biotech* 7:4230–4243.

Heller J (1992) Untersuchungen über genotypische Eigenschaften und Vermehrungsund Anbauverfahren bei der Purgiernuß (*Jatropha curcas* L.) [Studies on genotypic characteristics and propagation and cultivation methods for physic nuts (*Jatropha curcas* L.)]. Dr. Kovac, Hamburg.

Heller J (1996) *Physic nut—Jatropha curcas L. Promoting the conservation and use of underutilized and neglected crops*. International Plant Genetic Resources Institute, Rome, Italy (www.ipgri.cgiar.org/publications/pdf/161.pdf).

Ikbal, Boora KS, Dhillon RS (2010) Evaluation of genetic diversity in *Jatropha curcas* L. using RAPD markers. *Indian J Biotech* 9:50–57.

Jongschaap REE, Corre WJ, Bindraban PS, Brandenburg WA (2007) *Claims and Facts on Jatropha curcas L. Global Jatropha curcas Evaluation, Breeding and Propagation Programme*. Plant Research International B.V., Wageningen, The Netherlands, pp. 1–4.

Jubera MA, Janagoudar BS, Biradar DP, Ravikumar RL, Koti RV, Patil SJ (2009) Genetic diversity analysis of elite *Jatropha curcas* (L.) genotypes using randomly amplified polymorphic DNA markers. *Karnataka J Agric Sci* 22:293–295.

Kaushik N, Kumar K, Kumar S, Kaushik N, Roy S (2007) Genetic variability and divergence studies in seed traits and oil content of Jatropha (*Jatropha curcas* L.) accessions. *Biomass Bioenergy* 31:497–502.

Kumar RV, Tripathi YK, Shukla P, Ahlawat SP, Gupta VK (2009) Genetic diversity and relationships among germplasm of *Jatropha curcas* L. revealed by RAPDs. *Trees-Struct Funct* 23:1075–1079.

Lesica P, Leary RF, Allendort FR, Bilderbecl DE (1998) Lack of genetic diversity within and among populations of an endangered plant, *Hawellia aquatilis*. *Conserv Biol* 2:275–282.

Lowrey TK, Crawford DJ (1985) Allozyme divergence and evolution in *Tertramolopium* (Compositae: Astereae) on the Hawaiian Islands. *Syst Bot* 10:64–72.

Makkar HPS, Becker K, Schmook B (1998) Edible provenances of *Jatropha curcas* from Quintana Roo state of Mexico and effect of roasting on antinutrient and toxic factors in seeds. *Plant Foods Human Nutr* 52:31–36.

Mantel NA (1967) The detection of disease clustering and a generalized regression approach. *Cancer Res* 27:209–220.

Mastan SG, Sudheer PDVN, Rahman H, Ghosh A, Rathore MS, Ravi Prakash C, Chikara J (2011) Molecular characterization of intra-population variability of *Jatropha curcas* L. using DNA based molecular markers. *Mol Biol Rep* 39(4):4383–4390. https://doi.org/10.1007/s11033-011-1226-z

Pamidimarri DVNS, Chattopadhyay B, Reddy MP (2009a) Genetic divergence and phylogenetic analysis of genus *Jatropha* based on nuclear ribosomal DNA ITS sequence. *Mol Biol Rep* 36:1929–1935.

Pamidimarri DVNS, Mastan G, Rahman H, Reddy MP (2010) Molecular characterization and genetic diversity analysis of *Jatropha curcas* L. in India using RAPD and AFLP analysis. *Mol Biol Rep* 37:2249–2257. https://doi.org/10.1007/s11033-009-9712-2

Pamidimarri DVNS, Mastan SG, Rahman H, Ravi Prakash C, Singh S, Reddy MP (2011) Cross species amplification ability of novel microsatellites isolated from *Jatropha curcas* and genetic relationship with sister taxa cross species amplification and genetic relationship of Jatropha using novel microsatellites. *Mol Biol Rep* 38:1383–1388. https://doi.org/10.1007/s11033-010-0241-9

Pamidimarri DVNS, Pandya N, Reddy MP, Radhakrishnan T (2009b) Comparative study of interspecific genetic divergence and phylogenic analysis of genus *Jatropha* by RAPD and AFLP. *Mol Biol Rep* 36:901–907.

Pamidimarri DVNS, Singh S, Mastan SG, Patel J, Reddy MP (2009c) Molecular characterization and identification of markers for toxic and non-toxic varieties of *Jatropha curcas* L. using RAPD, AFLP and SSR markers. *Mol Biol Rep* 36:1357–1364.

Pamidimarri DVNS, Sinha R, Kothari P, Reddy MP (2009d) Isolation of novel microsatellites from *Jatropha curcas* L. and their cross-species amplification. *Mol Eco Resour* 9:431–433.

Pant KS, Khosla V, Kumar D, Gairola S (2006) Seed oil content variation in *Jatropha curcas* Linn. in different altitudinal ranges and site conditions in H.P. India. *Lyonia* 11(2):31–34.

Popluechai S, Breviario D, Sujatha M, Makkar HPS, Raorane M, Reddy AR et al (2009) Narrow genetic and apparent phonetic diversity in *Jatropha curcas*: initial success with generating low phorbol ester interspecific hybrids. *Nat Proc* 3:1–44.

Powell W, Morgante M, Andre C, Hanafey M, Vogel J, Tingey S, Rafalski A (1996) The comparison of RFLP, RAPD, AFLP and SSR (microsatellite) markers for germplasm analysis. *Molecular Breeding* 2(3):225–238.

Prabakaran AJ, Sujatha M (1999) *Jatropha tanjorensis Ellis & Saroja*, a natural interspecific hybrid occurring in Tamil Nadu, India. *Genet Resour Crop Evol* 46:213–218.

Puangpaka SS, Thaya JJ (2003) Karyology of *Jatropha* (*Euphorbiaceae*) in Thailand. *Thai For Bull (Bot)* 3:105–112.

Ranade SA, Srivastava AP, Rana TS, Srivastava J, Tuli R (2008) Easy assessment of diversity in *Jatropha curcas* L. plants using two single-primer amplification reaction (SPAR) methods. *Biomass Bioenergy* 32:533–540.

Rao GR, Korwar GR, Shanker AK, Ramakrishna YS (2008) Genetic associations variability and diversity in seed characters, growth, reproductive phenology and yield in *Jatropha curcas* (L.) accessions. *Trees* 22:697–709.

Rao MRG, Ramesh S, Prabuddha HR, Rao AM, Gangappa E (2009) Genetic diversity in *Jatropha* [*Jatropha curcas* L.]. *Indian J Crop Sci* 4(1&2).

Reddy MP, Chikara J, Patolia JS, Ghosh A (2007) Genetic improvement of *Jatropha curcas* adaptability and oil yield. In: *FACT SEMINAR on Jatropha curcas L.: agronomy and genetics*. Wageningen, The Netherland, 26-28 March 2007, FACT Foundation. Availablae at: http://www.jatropha-alliance.org/fileadmin/documents/ knowledgepool/ReddyChikara_Genetic_Improvement.pdf (accessed 25/10/10)

Richards CL, Bossdorf O, Muth NZ, Gurevitch J, Pigliucci M (2006) Jack of all trades, master of some? On the role of phenotypic plasticity in plant invasions. *Ecol Lett* 9:981–993.

Rosado TB, Laviola BG, Faria DA, Pappas MR, Bhering LL, Quirini B, Grattapaglia D (2010) Molecular markers reveal limited genetic diversity in a large germplasm collection of a biofuel crop *Jatropha curcas* L. in Brazil. *Crop Sci* 50:2372–2382.

Senthil KR, Parthiban KT, Govinda RM (2009) Molecular char acterization of *Jatropha* genetic resources through inter-simple sequence repeat (ISSR) markers. *Mol Biol Rep*. https://doi.org/10.1007/s11033-008-9404-3

Shen Jl, Jia XN, Ni HQ, Sun PG, Niu SH, Chen XY (2010) AFLP analysis of genetic diversity of *Jatropha curcas* grown in Hainan, China. *Tree Structure Funct* 24:455–462.

Singh P, Bhatia SK, Singh S (2009) Promising genotypes of *Jatropha curcas* L. in Eastern India and their molecular characterization. In: *International Conference on Recent Trends in Life Science Researches vis-à-vis Natural Resource Management, Sustainable Development and Human Welfare*, 27–29 June 2009, Hazaribagh, India, p 109 (Abstract)

Soltis PS, Soltis DE, Tucker TL, Lang A (1992) Allozyme variability is absent in the narrow endemic *Bensoniella oregona* (Saifragacear). *Conserv Biol* 6:131–134.

Subramanyam K, Muralidhararao D, Devanna N (2009) Genetic diversity assessment of wild and cultivated varieties of *Jatropha curcas* (L.) in India by RAPD analysis. *African J Biotech* 8:1900–1910.

Subramanyam K, Rao DM, Devanna N, Aravinda A, Pandurangadu V (2010) Evaluation of genetic diversity among *Jatropha curcas* (L.) by RAPD analysis. *Indian J Biotechnol* 9:283–288.

Sujatha M, Makkar HPS, Becker K (2005) Shoot bud proliferation from axillary nodes and leaf sections of non-toxic *Jatropha curcas* L. *Plant Growth Reg* 47:83–90.

Sujatha, M, Prabakaran, AJ (2003) New ornamental *Jatropha* hybrids though interspecific hybridization. *Genet Resour Crop Evol* 50:75–82.

Sukarin W, Yamada Y, Sakaguchi S (1987) Characteristics of physic nut *Jatropha curcas* L. as a new biomass crop in the Tropics (Japan). *Japan Agric Res Q* 20:302–303.

Sun QB, Li LF, Li Y, Wu GJ, Ge XJ (2008) SSR and AFLP markers reveal low genetic diversity in the biofuel plant *Jatropha curcas* in China. *Crop Sci* 48:1865–1871.

Sunil N, Varaprasad KS, Sivaraj N, Kumar TS, Abraham B, Prasad RBN (2008) Accessing *Jatropha curcas* L. germplasm in-situ: A case study. *Biomass and Bioenergy* 32:198–202.

Tanya P, Taeprayoon P, Hadkam Y, Srinives P (2011) Genetic diversity among Jatropha and Jatropha-related species based on ISSR markers. *Plant Mol Biol Rep* 29:252–264.

Tatikonda L, Wani SP, Kannan S, Beerelli N, Sreedevi TK, Hoisington DA *et al* (2009) AFLP-based molecular characterization of an elite germplasm collection of *Jatropha curcas* L., a biofuel plant. *Plant Sci* 176:505–513.

Umamaheshwari D, Paramathma M, Manivannan N (2010) Molecular genetic diversity analysis in seed sources of Jatropha (*Jatropha curcas* L.) using ISSR markers. *Electronic J Plant Breed* 1:268–278.

Vijayanand V, Senthil N, Vellaikumar S, Paramathama M (2009) Genetic diversity of Indian *Jatropha* species as revealed by morphological and ISSR markers. *J Crop Sci Biotech* 12:115–120.

Wen M, Wang H, Xia Z, Zou M, Lu C, Wang W (2010) Development of EST-SSR and genomic-SSR markers to assess genetic diversity in *Jatropha curcas* L. *BMC Res Notes* 3:42.

Xiang ZY, Song SQ, Wang GJ, Chen MS, Yang CY, Long CL (2007) Genetic diversity of *Jatropha curcas* (*Euphorbiaceae*) collected from Southern Yunnan, detected by inter-simple sequence repeat (ISSR). *Acta Bot Yunnanica* 29:6.

Yunus A (2007) Indentifikasi keragaman genetik jarak pagar (*Jatropha curcas* L.) di Jawa Tengah berdasarkan penanda Isoenzym. *Biodiversitas* 80:249–252.

Zane L, Bargelloni L, Patarnello T (2002) Strategies for microsatellite isolation: A review. *Mol Ecol* 11:1–16.

Zhang Z, Guo X, Liu B, Tang L, Chen F (2011) Genetic diversity and genetic relationship of *Jatropha curcas* between China and Southeast Asian revealed by amplified fragment length polymorphisms. *African J Biotechnol* 10(15):2825–2832.

10

Cellulase Immobilization on Magnetic Nanoparticles for Bioconversion of Lignocellulosic Biomass to Ethanol

Prabhpreet Kaur and Monica Sachdeva Taggar

CONTENTS

10.1 Introduction 225
10.2 Potential of Lignocellulosic Substrate for Bioethanol Production 227
    10.2.1 Pretreatment 227
        10.2.1.1 Physical Pretreatment 228
        10.2.1.2 Chemical Pretreatment 228
        10.2.1.3 Biological Pretreatment 229
10.3 Cellulase 229
    10.3.1 Cellulose and Its Degradation by Cellulases 229
    10.3.2 Cellulase Production 231
        10.3.2.1 Fungal Cellulases 231
        10.3.2.2 Bacterial Cellulases 232
10.4 Enzyme Immobilization 232
    10.4.1 Various Techniques of Immobilization 232
        10.4.1.1 Entrapment Method 232
        10.4.1.2 Adsorption 234
        10.4.1.3 Cross-Linking 234
10.5 Immobilization of Cellulase on Magnetic Nanoparticles and Their Characterization 235
10.6 Saccharification of Biomass Using Immobilized Cellulases 237
References 238

10.1 Introduction

The increasing global dependence on fossil fuel, combined with their gradual depletion and speedy rise in prices, is a growing subject of concern that has tempted us to look for alternative fuels which are economically viable and environmentally safe. This search has resulted in increasing interest in the production of bioethanol. Bioethanol is the fuel suitable for future: its production from sucrose (sugarcane, sugar beet) and starch (wheat, corn) containing feedstocks is a potential substitute to the fossil fuels. These are called first-generation biofuels. Currently, the substrates for production of bioethanol are mainly the traditional food crops such as corn (United States), sugarcane (Brazil), wheat (France, England, Germany, and Spain), and sorghum (India), i.e. the feedstocks used as raw material generally depend upon the dominant agricultural product of the region (Mojović et al. 2006). The United States and Brazil are the primary producers of ethanol, producing 94 billion liters per year from corn grain and sugarcane, respectively, which accounts for about 85% of production worldwide (Bertrand et al. 2016).

    Although the production of first-generation bioethanol is predicted to rise to more than 100 billion liters by 2022 (Goldemberg and Guardabasi 2009), these are insufficient to fulfill the increasing demand for fuels. Also, the application of food crops for production of bioethanol might lead to food scarcity and an increase in the cost of food grains and related products (Searchinger et al. 2015). Additionally, it has adverse impact on biodiversity and may even result in deforestation to acquire more farmland (Hahn-Hägerdal et al. 2006). The cumulative effect of these concerns limits the food crop's functionality for bioethanol production. However, the lignocellulosic biomass—such as agricultural residues, wood, paper, and municipal solid waste used for production of second-generation biofuels—are feasible options for sustainable production of bioethanol, as it does not present a competition for food and fodder (Kumar et al. 2008; Rubin 2008). Agricultural residues like sugarcane bagasse, wheat straw, corn stover, rice straw, etc., are rich in cellulose, abundantly available, cheap, and renewable. Moreover, providing residues as raw material for bioethanol production can solve the problems related to disposal of agricultural residues and offer income opportunities to small-scale farmers.

The high amounts of cellulose and hemicellulose present in lignocellulosic biomass can be readily hydrolyzed into fermentable sugars. Fermentable sugar production is a prerequisite for bioethanol production, and enzymatic saccharification is an important step for bioconversion of cellulose and hemicellulose into glucose and xylose. For this cellulase, a complex enzyme system is required that acts synergistically to hydrolyze cellulose to monomers. The standard model for cellulose breakdown involves the sequential, cooperative action of endoglucanases (EC 3.2.1.4) which randomly hydrolyze internal bonds in the cellulose chain, the exoglucanases or cellobiohydrolases (EC 3.2.1.19) which hydrolyze the reducing (cellobiohydrolase I) or non-reducing end (cellobiohydrolase II) of the cellulose chains, and β-glucosidases (EC 3.2.1.37) which convert cellobiose into glucose (de Souza Moreira et al. 2016). Cellulose breakdown yields glucose units, whereas hemicellulose produces several hexoses and pentoses. These simple sugars produced as a result of saccharification are further used by several microorganisms for fermentation to ethanol.

The major challenge in the commercial realization of bioethanol production from lignocellulosic feedstock is the exceptionally expensive cellulase enzyme required for the saccharification step, which accounts for approximately 40% of the overall production cost (Ahamed and Vermette 2008). Some of the commercially important cellulase cocktails—namely Accelerase, Cellulocast, Novozyme 88, and Cellic CTec 2 and 3—are available, but the high cost limits their application for an economically feasible process (Kumar et al. 2018). Under the processing conditions, the enzyme lacks long-term stability, so its recovery from reaction mixture and repeated use is difficult (Zhang et al. 2013). Moreover, during saccharification, the enzyme gets inactivated due to its denaturation and unspecific binding with lignin (Chundawat et al. 2017).

These hindrances can be overcome by immobilization of cellulase onto different carriers and supports, which do not interfere with the biological functionality of the enzyme (Mitchell et al. 2002). Several natural polymers like collagen, chitosan, alginate, chitin (Datta et al. 2013), and inorganic supports like alumina, silica, and zeolites are commonly available matrices for enzyme immobilization (Hudson et al. 2008; Hartmann and Kostrov 2013). But recently, the application of nanoparticles has become widely used as compared to bulk materials for immobilization of enzymes. This is due to the large surface area–to-volume ratio that increases the enzyme loading capacity and other advantages such as their robustness and easy dispersion (Verma et al. 2013). Covalent immobilization of enzymes on nanoparticles may also lead to certain structural and conformational changes in the enzyme structure, thus inducing greater thermal stability and wider temperature and pH ranges than the free enzyme (Abraham et al. 2014).

Exceptional properties of magnetic nanoparticles—such as high surface area–to-volume ratio, low toxicity, easy mobility, and most importantly, easy recovery from reaction mixture by applying an external magnetic field—make them important tools for immobilization of enzymes (Bilal et al. 2018). Moreover, magnetic nanoparticle-immobilized enzymes can retain the activity even after being recycled several times (Song et al. 2016); immobilized cellulase hydrolyzed 83% of untreated synthetic substrate, i.e. carboxymethylcellulose, while remaining stable for up to seven hydrolysis cycles (Abraham et al. 2014). Easy recovery makes the enzyme compatible for reuse, thus reducing enzyme consumption to improve economic viability of the entire process at commercial scale. The half-life of Fe_3O_4@SiO_2-graphene oxide (GO) composite immobilized cellulase increased 3.34-fold (from 216 minutes to 722 minutes) at 50°C upon immobilization (Li et al. 2015). The free and immobilized β-glucosidase was used for cellobiose hydrolysis and conversion rate of 90% was achieved after 16 and 5 hours of incubation, respectively (Verma et al. 2013). This demonstrates that immobilized enzymes shows better activity than free enzymes.

The majority of the research related to cellulase immobilized nanoparticles involves the use of single and pure substrates such as microcrystalline cellulose, carboxymethyl cellulose, filter paper, avicel, or xylan to assess the retained activity of immobilized enzyme (Liang and Cao 2012; Zhang et al. 2012; Yu et al. 2013). The lignocellulosic biomass is, however, a complex mixture of crystalline cellulose, amorphous cellulose, hemicellulose, and lignin. The hydrolytic behavior of immobilized cellulase on lignocellulosic biomass is quite different than that on pure model substrates. This chapter focuses on recent advances in the hydrolysis of lignocellulosic biomass using magnetic nanoparticle-immobilized cellulases.

10.2 Potential of Lignocellulosic Substrate for Bioethanol Production

The main components of lignocellulosic biomass are cellulose, hemicellulose, and lignin. Cellulose and hemicellulose polymers are tightly cross-linked to lignin via hydrogen bonds and covalent bonds which form a robust and recalcitrant structure (Limayem and Ricke 2012). The complexity of lignocellulosic biomass is dependent on these structural and compositional factors, which further determine the digestibility of biomass for bioethanol production (Mosier et al. 2005). Bioethanol production from lignocellulosic biomass involves three major steps: (a) pretreatment; (b) saccharification (enzymatic hydrolysis); and (c) fermentation of hydrolyzed biomass to bioethanol (Sukumaran et al. 2009).

Pretreatment is a major step in bioethanol production, as it helps to overcome the lignin barrier so that the sequential enzymatic hydrolysis can be used to achieve maximum sugar production rapidly (Mosier et al. 2005). In the next step, i.e. saccharification, complex carbohydrates are converted into simple sugars. The enzymatic approach for cellulosic biomass degradation involves introducing enzyme onto the substrate at particular temperature and pH levels optimum for enzyme activity. Further, the fermentable sugars produced are fermented into ethanol by suitable microorganisms (Azhar et al. 2017).

10.2.1 Pretreatment

Compared to the other dominant fossil fuels, certain properties of biomass render it inferior for energy and heat production. Some of these particularly include its fibrous nature, high silica content, and poor handling and transportation due to bulkiness. The lignin present in these cellulosic substrates hinders its saccharification into monomeric sugars. Certainly, various pretreatment methods are a prerequisite to overcome the lignin barrier so that the sequential enzymatic hydrolysis can be used to achieve maximum sugar production rapidly. Without pretreatment, enzymatic hydrolysis of native biomass generally gives below 20% of the maximum theoretical yield of ethanol (Mosier et al. 2005).

Pretreatment is the first and the most challenging step in lignocellulosic biomass conversion to bioethanol. This step is needed to modify size and structure of biomass at macroscopic and microscopic scale, in addition to its chemical composition at submicroscopic scale so that digestion of the carbohydrate part can be carried out more rapidly and with greater sugar yields (Sindhu et al. 2016). An ideal pretreatment procedure should help to avoid size reduction of biomass, make lignocellulosic biomass vulnerable to fast hydrolysis with improved sugar yields, and minimize formation of the inhibitory compounds (Gupta and Verma 2015). Apart from these, other criteria that should be given due consideration include the recovery of high-value–added co-catalysts and pretreatment catalyst, recycling of catalyst, and waste treatment. Also, the methods must be considered on basis of ease of separation of products and minimizing the energy and capital demands for lower operational costs.

Various pretreatment techniques have been extensively studied which challenge the complexity of biomass structure and are based on the application and type of pretreatment catalyst. These techniques are generally classified into three categories: physical, chemical, and biological. Combination of methods from same or different categories is also practiced, as no one method can be designated as best; each has its merits and demerits. The selection of pretreatment procedure is determined by various factors, such as the type of lignocellulosic biomass, effect on environment, economic feasibility, etc.

10.2.1.1 Physical Pretreatment

Physical pretreatment is uncatalyzed, typically including grinding and milling, irradiation, steam explosion, liquid hot water treatment, etc. Chipping (reducing size to 10–20 mm), grinding, and milling (reducing size to 0.2–2 mm) are practiced to reduce the cellulose crystallinity and degree of polymerization. The small particle size reduces the energy input required for comminution. Reduction of particle size is mostly needed to make handling of the material easier and to increase surface area–to-volume ratio, thereby increasing the surface available for enzymatic attack. Hideno and colleagues (2009) found that wet disk milling of rice straw was a more practical pretreatment than ball milling in terms of both glucose recovery and energy savings.

Uncatalyzed steam explosion, referred to as as autohydrolysis, includes the use of only steam water. It is a cost-effective method, which has been demonstrated from pilot scale to commercial application. In commercially available steam-explosion equipment, high-pressure saturated steam rapidly heats up the biomass to promote the hemicellulose breakdown, then the biomass undergoes an explosive decompression with a sudden release of pressure from the equipment. Grous and colleagues (1986) showed that the enzymatic hydrolysis of steam-pretreated poplar chips reached 90% as compared to only 15% for untreated chips.

Steam pretreatment loosens the cellulose-hemicellulose-lignin complex, removes the pentoses, and increases the surface area. However, the limitation of this process is the production of certain inhibitors of cellulase enzyme which might later interfere with the hydrolysis process.

Microwave irradiation alters the ultra-structure of cellulose, degrades hemicelluloses and lignin, and increases the susceptibility of lignocellulosic materials to enzymatic hydrolysis. The reaction rate of the subsequent enzymatic hydrolysis of irradiated cellulose was increased by approximately 200% (Imai *et al.* 2004). Microwave pretreatment can disrupt the silicified waxy layer and degrade the lignin-hemicellulose complex for efficient hydrolysis (Ma *et al.* 2009). However, such methods involving high-energy radiation are often energy intensive, expensive, and slow. Based on present approximation of overall cost of such methods, they are usually less fascinating at commercial scale.

10.2.1.2 Chemical Pretreatment

The lignocellulosic biomass cannot be converted to sugars effectively without prior chemical pretreatment. The most likely chemicals for pretreatment of rice straw include alkali, ammonia, and acid. Table 10.1 shows the mode of action, merits, and demits of most commonly used chemical pretreatment methods.

TABLE 10.1

Comparison of Different Pretreatment Methods

Chemical	Mode of Action	Merits	Demerits
Dilute acid	Removal of hemicellulose	High reaction rates, increased accessibility to cellulose	Formation of reaction inhibitors, corrosive to reactor
Conc. Acid	Solubilization of hemicellulose, hydrolysis of cellulose to glucose	Suitable for all types of biomass	Uncontrolled hydrolysis, corrosive to reactor
Alkaline hydrolysis	Removal of lignin (mainly), removal of hemicellulose	Low formation of fermentation inhibitors	High cost of chemicals
Organosolv	Extraction of lignin and solubilization of hemicellulose	Recovery and reuse of used solvent, high recovery of pure lignin	High cost and high energy consumption for solvent recovery
Hydrogen peroxide	Solubilization of hemicellulose and lignin, bleaching effect	Efficient removal of lignin	High costs of chemical

Concentrated acid pretreatment, such as concentrated H_2SO_4 and HCl hydrolysis, has become a major technology for lignocellulosic biomass hydrolysis for production of fermentable sugars (Hsu *et al.* 2010; Zhu *et al.* 2015; Kapoor *et al.* 2017; de Assis Castro *et al.* 2017), as it is very effective for cellulose hydrolysis. However, the concentrated acid is hazardous and corrosive; also, the construction of corrosion resistant reactors is expensive. Furthermore, the concentrated acid must be recovered after hydrolysis and reused to increase the economic feasibility of the entire process.

Alkali pretreatment is primarily a delignification process, with solubilization of a notable amount of hemicellulose. It lowers the degree of crystallinity and polymerization, and increases the internal surface area and disruption of lignin structure (Fan *et al.* 1987). As a result of this, the relative cellulose content is increased, thereby enhancing the rate of saccharification (Jeya *et al.* 2009) and ethanol production (Ko *et al.* 2009; Salehi *et al.* 2012).

By using sequential acid followed alkali pretreatment processes, the pretreatment efficiency can be increased by allowing stepwise separation of components from lignocellulosic biomass (Weerasai *et al.* 2014).

10.2.1.3 Biological Pretreatment

Of the several pretreatment methods available, biological pretreatment seems to be an attractive choice, being eco-friendly and economically feasible. It involves low input of energy and capital, and does not generate any inhibitor compounds or polluting byproducts. It employs the use of microorganisms such as white, brown, and soft-rot fungi and bacteria that can degrade the structure of lignocellulosic biomass and modify the chemical composition for easy enzyme digestibility.

Generally, brown and soft-rot fungi break down the cellulose part while causing minor changes in lignin structure, and white-rot fungi degrades the lignin part effectively (Zheng *et al.* 2009). The reduction in crystallinity index and microstructural changes were observed in rice straw upon treatment with *Dichomitus squalens*. This pretreated rice straw could be further used for ethanol production with 54.2% of the theoretical maximum yield.

Major limitations of biological pretreatment include long treatment time for effective delignification, breakdown of residual sugars, and heat generation during the process. Identification of competent lignin degrading microbes using advanced molecular techniques still needs to be explored. Mostly, the extraction methods are energy intensive, costly, and require chemicals which need special handling, disposal, or production techniques. The extraction methods developed in the laboratory rarely find application at the industrial level. Breaking this technology barrier is the key for cellulose conversion into bioethanol on large scale.

10.3 Cellulase

10.3.1 Cellulose and Its Degradation by Cellulases

Cellulose is the most abundant polysaccharide on earth. It represents approximately 1.5×10^{12} tons of the total yearly biomass production through photosynthesis (Klemm *et al.* 2005). It is the main structural constituent of the primary cell wall of green plants and many forms of algae. Likewise, the principal component of lignocellulosic biomass is also cellulose. Cellulose is an unbranched linear homopolysaccharide consisting of glucose (D-glucopyranose) units linked together by β-(1–4) glycosidic bonds (β-D-glucan) (Crawford 1981) with degree of polymerization ranging from 800–10,000 (Klemm *et al.* 2005). The repeating unit of the cellulose is the disaccharide cellobiose. Cellulose is a crystalline biopolymer which exists as overlying sheets of glucopyranose rings stacked in parallel on the top of each other to form a three-dimensional particle. The adjacent cellulose molecules are linked by hydrogen bonds and van der Waal's forces (Zhang and Lynd 2004). This crystalline array is relatively impermeable to not just large enzyme molecules, but also sometimes to water. So, in order to hydrolyze this polysaccharide, a member of the glycoside hydrolase family of enzymes is responsible: cellulase.

Cellulase is an inducible multi-enzyme complex mechanistically consisting of consortium of three classes of enzymes: endo-(1,4)-β-D-glucanase (EC 3.2.1.4), exo-(1,4)-β-D-glucanase (EC 3.2.1.91), and β-glucosidases (EC 3.2.1.21) (Deswal et al. 2011). Cellulase has simple architecture comprised of independently folding, structurally and functionally distinct units called domains or modules (Sinnott 2007). A typical free cellulase has cellulose binding domain at C-terminal that binds to substrate and catalytic domain at T-terminal that performs the catalytic function. The binding domain is functionally divided into three types: Type A, which binds to crystalline regions of polysaccharides and shows no or weak affinity for soluble carbohydrates; Type B, which binds to polysaccharide chains either as soluble oligosaccharides or as amorphous regions of insoluble polysaccharides; and Type C, which binds to small sugars (Sinnott 2007). The binding domain is not involved in catalysis, but its removal lowers the enzyme activity significantly (Bayer et al. 1998).

Prior to undergoing an attack by cellulase and forming reducing sugars, the native cellulose undergoes some changes. Such changes include swelling, fragmentation, transverse cracking, loss in tensile strength, and reduction in the degree of polymerization (Lee and Fan 1980). The mode of action of different components of cellulase enzyme (Singh 1999) is as follows. The first component—endo-β-1,4 glucanase—acts randomly on internal O-glycosidic bonds of crystalline cellulose to release mainly cellobiose and cellotriose. It cannot further hydrolyze cellobiose. It mainly acts on carboxymethylcellulose, phosphoric acid, swollen cellulose and cellodextrin. The second component—exo-β-1,4-glucanase—removes single glucose from non-reducing end of the chain of 4–7 units produced by the action of the endo-glucanase. The third component—β-glycosidase—acts specifically on the cellobiose disaccharides and short-chain cellooligosaccharides to produce glucose, but has no effect on the cellulose. While it rapidly hydrolyzes cellobiose and cellotriose, its rate of attack decreases markedly with an increasing degree of polymerization (Lee and Fan 1980). Therefore, to accomplish the complete cellulolysis, a complex reaction involving these three enzymes is needed. Figure 10.1 shows the mode of action of each component of cellulase enzyme system.

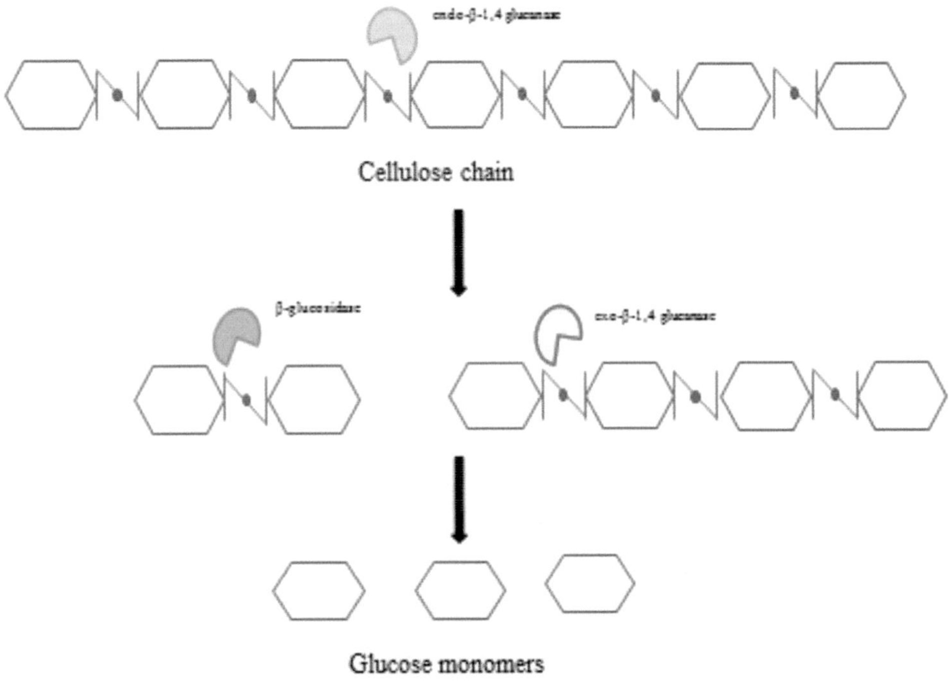

FIGURE 10.1 Mode of action of cellulase enzyme.

10.3.2 Cellulase Production

Traditionally, cellulase is produced by submerged fermentation (SmF), in which microorganisms are cultivated in the aqueous solution containing nutrients. These days, however, solid-state fermentation (SSF) is mostly practiced, as it involves the growth of microorganisms on solid substrates rather than on free liquids (Cannel and Young 1980). SSF uses less water and energy than does SmF (Zhuang *et al.* 2007), and it is well suited to fermentation methods involving fungi and microorganisms that require less moisture (Babu and Satyanarayana 1996). The crude enzyme prepared by SSF is concentrated and can be further used in agrobiotechnological applications and lignocellulosic hydrolysis. SSF uses solid substrates like bran, paddy straw, bagasse, other agricultural wastes, and paper waste (Panday *et al.* 1999: Subramaniyam and Vimala 2012). The use of such low-cost, nutrient-rich, readily available agricultural waste is advantageous to lower the overall cellulase production cost.

Cellulases are produced by a wide range of microorganisms, generally growing on dead and decaying organic matter. Screening and isolating such microbes from nature is the most important source of cellulases (Juturu and Wu 2014). Microbes that outdo industrial conditions like high cellulase production, secreting the complete spectrum of cellulase for depolymerization of biomass into simple glucose units, are generally selected. Aerobic and anaerobic bacteria are known to produce cellulose degrading enzymes. However, the most commercially available enzymes are fungal cellulases, especially from *Trichoderma* and *Aspergillus* species. Producing high yields of enzyme and using secretory pathways for its easy recovery and purification make fungi more advantageous than bacteria (Behera and Ray 2016).

10.3.2.1 Fungal Cellulases

Fungal cellulases are in the light of research for biofuel production due to many reasons. They have more activity than bacterial cellulases. Besides, the fungi can grow on cheaper substrates such as lignocellulosic wastes, and downstream processing for fungal cellulase extraction is also easier. This lowers the overall cost of enzyme production, thus making them the preferred source as compared to bacteria (Panchapakesan and Shankar 2016). Some of the commonly used fungal cellulase sources are from the genus *Trichoderma*, titled as "King of Cellulytic Fungi" (Gusakov 2011). Some notable species of this family included *T. reesei, T. viride*, and *T. atroviride*. Different agricultural residues like cassava bagasse, sugarcane bagasse, rice straw, and wheat bran have been used for cellulose production from *T. reesei* with supplementation of NH_4NO_3 as nitrogen source and cellobiose as an inducer (Singhania *et al.* 2006). Likewise, other residues like apple pomace (Sun *et al.* 2010), corncob (Pandey *et al.* 2015), *Kallar* grass (Kalsoom *et al.* 2019), etc., have been used as carbon sources for cellulose production from *Trichoderma* spp.

Cellulases from *Aspergillus* species have been widely studied to explore their ability to degrade lignocellulosic substrates and in comparison to the *Trichoderma* species, and produce very good amounts of β-glucosidases (Damisa *et al.* 2011). A number of characteristics—such as good fermentation capacity, high protein secretion, high sporulation capacity, and ability to suppress other microorganisms—make them ideal candidates for use at industrial scale. Several species—like *A. niger* (Kang *et al.* 2004; Acharya *et al.* 2008; Lee *et al.* 2011; Reddy *et al.* 2015), *A. fumigatus* (Sherief *et al.* 2010; Ang *et al.* 2013), *A. terreus* (Jahromi *et al.* 2011), *A. saccharolyticus*, etc.—produce industrially important extracellular enzymes like β-galactosidases, β-mannanases, xylanases, and cellulases (de Vries and Visser 2001).

Another notable cellulase producing species is *Penicillia* (*P. brasilianum, P. echinulatum, P. chrysogenum, P. funiculosum, P. pinophilum, P. decumbens*, and *P. purpurogenum*). The cellulose degrading ability of some of these cellulases was reported to be even better than that of *T. reesei*. The major problem of cellulase production from *Trichoderma* was low levels of β-glucosidases, which led to only partial hydrolysis of lignocellulosic biomass. Martins and colleagues (2008) used *P. echinulatum* to produce filter paper activity of 0.27 U mL^{-1}, carboxymethylcellulase 1.53 U mL^{-1}, and birchwood xylanase 3.16 U mL^{-1} on various cellulosic substrates. The substrate hydrolysis was then compared with hydrolysis by *T. reesei* cellulases (Celluclast 1.5L FG®, Novozymes), and the results demonstrated that *P. echinulatum* enzymes had higher β-glucosidase activity than Celluclast 1.5L FG®. It has been reported by researchers worldwide that the glucose yield from *Penicillium* hydrolysis was nearly twofold compared to that from *Trichoderma*, making it an attractive source of enzyme (Gusakov 2011). However, the major challenges of enzyme production from *Penicillium* include low protein content, leading to higher production costs.

10.3.2.2 Bacterial Cellulases

Application of bacteria for cellulase production is not widely used, even though bacteria have a higher growth rate than fungi. The ability of bacteria to withstand extreme conditions enables the screening of novel cellulases to overcome various stress conditions during the biorefining process (Maki *et al.* 2009). Many cellulase producing bacteria—such as *Bacillus subtilis*, *B. pumilus* (Ariffin *et al.* 2006), *B. coagulans*, and *B. licheniformis* (Acharya and Chaudhary 2012)—have been isolated from various sources and selected due to their xylan degrading properties (Juturu and Wu 2014; El-Bakry *et al.* 2015; Rasul *et al.* 2015; Khatiwada *et al.* 2016).

10.4 Enzyme Immobilization

The large-scale industrial application of enzymes is obstructed due to lower stability and higher sensitivity to process conditions, inhibition by high substrate concentrations, and extensive cost of separation of enzyme (Guisan 2006). The lack of technology for efficient recovery and reuse challenges the practical application, considering the economic issues related to synthesis of enzymes. Such problems can be overcome by immobilization of enzyme for easy recovery and further recycling of enzyme (Liu and Chen 2016). Immobilization of enzyme refers to physically confining or localizing the enzyme in a fixed region with the retention of catalytic properties to be used repeatedly and continuously. It refers to the attachment or incorporation of enzyme molecules onto or into large structures, via binding to the support, cross-linking, and encapsulation (Ansari and Husain 2012). The major components of an immobilized enzyme system are enzyme, support, and the mode of attachment (Mohamad *et al.* 2015).

Immobilization of enzyme often enhances operational and storage stability. So, the denaturation of enzyme by organic solvents, heat, or autolysis is minimized. The enhanced stability improves the overall performance of entire process. Also, the immobilized enzyme is mostly designed in such a way that it can be easily and stably recovered from the reaction mixture and reused; such an approach provides a solution to high production costs and makes the whole process of biorefining economically viable at industrial scale (Cantone *et al.* 2012). Due to easy recovery, the protein contamination of product is minimized. In this way, an ideally immobilized enzyme would help to overcome the challenges of free enzymes such as lack of long-term operational stability, complex downstream processing, low productivity, and risk of contaminations.

10.4.1 Various Techniques of Immobilization

Fundamentally, the enzyme immobilization technologies can be grouped into three traditional categories: (a) entrapment or encapsulation within an inert polymer network or membrane; (b) carrier binding by physical adsorption or covalent interactions; and (c) cross-linking, the formation of insoluble proteins (Bezerra *et al.* 2015). Figure 10.2 illustrates the different immobilization techniques.

10.4.1.1 Entrapment Method

Physical entrapment involves encapsulation of isolated enzyme into organic or inorganic polymer networks. The reactants and products can easily diffuse through the polymeric network, but the biocatalyst cannot migrate into the bulk media. The protein molecules may be entrapped by two methods: (a) inside a matrix; or (b) inside a membrane (Liu *et al.* 2018). Commonly used matrices for enzyme entrapment include natural gel matrices like Ca-alginate, collagen, gelatin, κ-carrageenin, and agar. Endo-β-glucanase (EC 3.2.1.4), isolated from *T. reesei*, was immobilized in calcium alginate beads. The enzyme retained 75% of its original activity and showed good hydrolysis efficiency, 250–720 mg glucose/g straw during saccharification of wheat straw (Busto *et al.* 1998). Likewise, cellulase enzyme

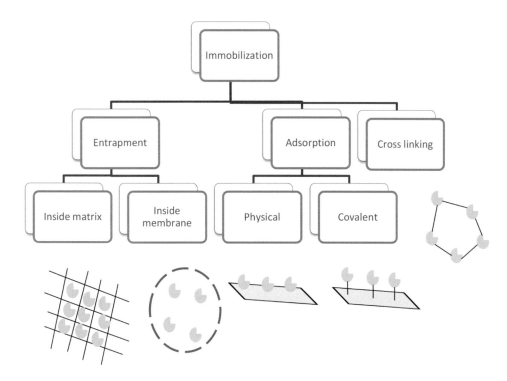

FIGURE 10.2 Different methods of enzyme immobilization.

was immobilized in calcium alginate bead with immobilization efficiency of 83.65%. The immobilized enzyme could tolerate higher temperature and pH conditions, and it could be reused many times for CMC hydrolysis, with 69.2% activity even after five recycles (Viet et al. 2013). The advantages of using natural matrix are low toxicity, mild synthesis reactions, biocompatibility, etc. However, these have low mechanical strength.

The synthetic gels such as polyacrylamide, polyvinyl alcohol, resin, etc., have good mechanical strength, but the monomers of synthetic matrices like acrylamide are highly toxic to enzymes and cells (Liu 2017). Sometimes, solid matrices such as porous ceramic, activated carbon, and diatomaceous earth can also be used for entrapment purposes. To immobilize the enzyme in a matrix, enzyme molecules are either added to a monomer solution prior to polymerization or added to a pre-formed matrix, where the enzyme molecules are "breathed in" the matrix.

Semipermeable membranes of nylon, polysulfone, and polyacrylate are commonly used for the entrapment of enzyme. Another method of entrapment using membrane is microencapsulation. Here, microscopic hollow spheres of diameter ranging from few microns to 1 mm called microcapsules are formed, with thin and semipermeable membrane (Zhu 2007). The enzyme solution is contained within the spheres, which are further enclosed within a porous membrane.

The challenges related to enzyme entrapment, such as significant diffusion limitations, enzyme leakage into solution, reduced enzyme activity, and stability and lack of control of microenvironmental conditions, need to be overcome for a successful enzyme entrapment system. Xu and colleagues (2015) developed a compatible ionic liquid (IL)-cellulase system for biorefining through an efficient encapsulation method. The cellulase from *Trichoderma aureoviride* strain HS was encapsulated in alginate beads with greatly enhanced stability in various concentrations of 1-ethyl-3-methylimidazolium dimethylphosphate ([Emim][DMP]). After treating rice straw with (Emim)(DMP) at 100°C and dilution to a final IL concentration of 40% (v/v), the pretreatment slurry was *in situ* hydrolyzed using encapsulated cellulase. Under the one-pot process, a maximum saccharification rate of 84% was obtained, increased by 65% compared to buffer-based free cellulase.

10.4.1.2 Adsorption

10.4.1.2.1 Physical

Physical adsorption of enzyme is achieved by the formation of weak, temporary bonds such as hydrogen bonds, Van der Waals force, and ionic bonds between the enzyme and support (Zang *et al*. 2014). Support with large surface area, porous structure, and many attachment points is preferred for immobilization. In a study conducted by Mo and colleagues (2020), polyporous biochar with large specific surface area was combined with magnetic particle γ-Fe_2O_3 and used for physical immobilization of cellulase enzyme. The relative activity of immobilized cellulase was 73.6% in comparison to that of free enzyme, and it could be reused for three cycles of CMC hydrolysis.

Apart from the ease of preparation and economic aspects of physical immobilization, another very important feature is that the structure of enzyme—and therefore, its activity—is rarely distorted in physical immobilization (Teofil *et al*. 2014). Recent studies show the physical adsorption of cellulase by ionic binding on supramagnetic nanoparticles (Khoshnevisan *et al*. 2011), TiO_2 nanoparticles (Ahmad and Sardar 2014), and mesoporous silica nanoparticles (Chang *et al*. 2011), and by metal affinity immobilization on Fe_3O_4 nanoparticles (Abbaszadeh and Hejazi 2019).

Physical adsorption is simple, economic and results in high retention of enzyme activity, as in covalent immobilization (Zhang et al. 2015). However, due to weaker bonds, the enzyme molecules may easily get desorbed from the surface of the matrix; therefore, the reusability of physically immobilized cellulase is rarely reported.

10.4.1.2.2 Covalent

Covalent immobilization of enzyme involves the binding of enzyme molecules onto the support by formation of covalent bonds. Due to the formation of strong covalent bonds, the chances of protein desorption are limited as compared to physical adsorption. Various cross-linking agents are used for bonding, such as glutaraldehyde. Glutaraldehyde acts as good cross-linking agent as it aims for the α-amino groups of cellulase without disrupting the active site of enzyme, and provides good binding efficiency (Tata *et al*. 2015). Cellulase produced by *A. fumigatus* JCF was covalently immobilized onto MnO_2 nanoparticles using glutaraldehyde solution (1 mol/L) to provide strong binding efficiency of 75% (Cherian *et al*. 2015). Covalently immobilized cellulases exhibit excellent stability and reusability. Roth and colleagues (2016) demonstrated excellent cellulase enzyme stability and reusability by covalently immobilizing cellulase on low-price silica coated magnetic nanoparticles (MNP @SiO). Jordan and colleagues (2011) covalently immobilized cellulase enzyme on carbodiimide activated magnetic nanoparticles. The weight ratio (mg bound enzyme/mg nanoparticles) was optimized to 0.021 for maximum binding efficiency of 65% and at the same time retaining maximum activity of 62.7 IU mg^{-1}. However, sometimes, covalent binding may lead to partial inactivation of enzyme due to conformational restrictions caused by the linking of amino acids to the support.

10.4.1.3 Cross-Linking

Cross-linked enzyme aggregates (CLEAs) are prominent technique of enzyme immobilization, without any requirement of a solid support (Sheldon *et al*. 2005). It is the technique of protein cross-linking using a cross-linker, for instance gluteraldehyde cross-linking with reactive amine groups, on the surface of protein. It was originally developed in the 1960s (Doscher and Richards 1963). CLEAs are a carrier-free immobilization technique; thereby, the unwanted negative consequences of carrier use can be avoided (Hanefeld *et al*. 2009). Generally, the methods for CLEAs preparation include two main steps: first, the soluble enzyme is precipitated with appropriate precipitant; and then second, cross-linked using a suitable cross-linker, during which the particle size increases. Commonly, glutaraldehyde is used as a cross-linking agent because it is inexpensive and easily accessible in abundant quantities. Cross-linking is an attractive method due to its simple procedure, with no necessity for highly pure enzyme and operational stability. Additionally, CLEAs are easily recoverable and recyclable, with high activity retention. CLEAs take up a roughly spherical shape, which provides higher surface area, and therefore, more exposed catalytic sites, which is beneficial as a biocatalyst (Podrepšek *et al*. 2012).

The molecular weight of the cross-linking agent is negligible as compared to that of the enzyme, so the resulting biocatalyst is essentially 100% active enzyme. Therefore, it clearly offers more enzymatic activity, and also the production costs are reduced due to elimination of additional cost of carrier agent (Vaghari et al. 2016).

Interestingly, the activity of aggregated lipase, and subsequently other enzymes, was observed to exceed that of native enzyme; this hyperactivation is thought to have originated due to protein structural changes induced by the aggregated state (López-Serrano et al. 2002). Schoevaart and colleagues (2004) demonstrated that the wide application of cross-linking of enzyme aggregates is influenced by certain factors dependent on the properties of the resultant CLEAs such as precipitant, precipitant concentration, cross-linker concentration, and protein concentration.

In conclusion, conformational enzyme changes resulting in reduced activity, mass diffusion limitations, and lowered efficacy against insoluble substrates are among the disadvantages of immobilized enzymes (Singh et al. 2013) which must be considered before choosing the immobilization technique for desired results.

10.5 Immobilization of Cellulase on Magnetic Nanoparticles and Their Characterization

Enzyme immobilization has become a popular strategy in various applications due to the ease of separation, product purity, and continuous operation (Kim et al. 2006). Improvement in the performance and reusability of enzyme determines the costs of enzyme per kilogram of product (Sheldon and van Pelt 2013). Therefore, immobilization of enzymes is especially beneficial at commercial scale due to convenience in handling and cost-effective recovery and reuse (Ansari and Husain 2012).

Reduction in the size of enzyme support used for immobilization can generally improve the efficiency of immobilized enzymes, because small-scale particles can provide an increased surface area for the binding of enzymes, causing higher enzyme loading per unit mass of particles (Jia et al. 2003). As such, nanostructures have gained large research momentum, owing to the increase in functional surface area and decrease in the diffusion limitations (Xie et al. 2009). Various studies have shown the application of nanoscale structures such as spheres, fibers, and tubes for immobilizing enzyme. Recently, the research interest in application of nanoparticles as carriers for immobilization of enzyme is growing (Daubresse et al. 1996; Martins et al. 1996; Caruso and Schüler 2000; Liao and Chen 2001; Jia et al. 2003). Consequently, nanoparticles have gained major importance in the field of enzyme immobilization.

More recently, the novel immobilization approaches involve binding the enzyme onto smart polymers that are stimulus responsive as they undergo conformational changes in response to some environmental changes (temperature, pH, and ionic strength) which makes their separation easier from reaction media (Galaev and Mattiasson 2001). Considering this aspect, magnetic particles can be easily removed from the reaction mixture by the application of a magnetic field (Yiu and Keane 2012). Compared to magnetite, nanoscale magnetic particles have unique properties such as superparamagnetism, easy separation under external magnetic fields, large surface area–to-volume ratio, favoring higher binding capacity and loading capacity, and reducing the diffusion limitation (Xu et al. 2014).

Therefore, magnetic nanoparticles (MNPs) are a particularly important area of interest. In comparison to the polymeric carriers, non-porous magnetic nanoparticles have no problem of external diffusion, thus making them attractive alternatives especially for industrial scale usage (Xu et al. 2014).

Easy retention and recovery of these enzyme-immobilized nanoparticles helps to overcome the need for an additional step of catalyst purification before each hydrolysis cycle (Vaghari et al. 2016). The reusability of magnetic nanoparticle-immobilized cellulases can prove to be of high importance for practical applications in saccharification and economic production of biofuels. Jordan and colleagues (2011) covalently immobilized cellulase enzyme on carbodiimide-activated magnetic nanoparticles with maximum binding efficiency of 65% and at the same time retaining maximum activity of 62.7 IU mg^{-1}. The immobilized enzyme was recycled six times for CMC hydrolysis before its residual activity fell to 10%. Likewise, another study conducted by Verma and colleagues

(2013) showed that magnetic iron oxide nanoparticle–immobilized β-glucosidase retained more than 50% enzyme activity up to the 16th hydrolysis cycle, which provided a cost-effective advantage for its commercial viability. Similarly, the immobilized β-glucosidase on Fe_3O_4 nanoparticles could also be reused, as it retained 86% of initial activity, even after ten consecutive cycles (Zhou et al. 2013). Several other studies have reported the recovery and reusability of cellulase immobilized on magnetic nanostructures (Li et al. 2014; Zhang et al. 2015; Tao et al. 2016; Kumar et al. 2018; Han et al. 2018).

Apart from reusability, the immobilized enzyme exhibits various other characteristics that are not usually seen in free enzyme. For instance, the storage capacity of immobilized enzyme is usually prolonged in comparison to free enzyme. Zhou and colleagues (2013) showed that the immobilized enzyme retained about 73% of activity after six weeks of storage, whereas the free enzyme retained only about 20%. Furthermore, immobilized enzymes often exhibit better thermal stability (Verma et al. 2013; Li et al. 2014), pH stability and higher activity as compared to free enzymes (Hola et al. 2015). The binding of enzyme onto nanoparticles might induce certain conformational and structural rearrangements that lead to better stability and activity.

Khoshnevisan and colleagues (2011) showed that the pH and temperature range of cellulase immobilized on supramagnetic nanoparticles was broader, thus facilitating long-term storage of the enzyme. At higher pH 9, the carboxymethylcellulase activity of immobilized enzyme was 0.07 $\mu moles^{-1}$ $min^{-1}ml$ in comparison to 0.04 $\mu moles^{-1}min^{-1}ml$ for free enzyme. Other studies also reported similar increase in optimum pH values upon immobilization of cellulase on magnetic nanoparticles (Jordan et al. 2011; Verma et al. 2013; Kaur et al. 2020). Furthermore, the immobilized enzyme also depicted relatively higher activity at broader pH range as compared to free enzyme. The stability of immobilized enzyme over broader pH range might be due to interactions between the charged groups present the enzyme molecules and charge on carrier molecule involved in covalent binding (Kumar et al. 2017).

Likewise, the variation in optimum temperature for free and immobilized enzyme has also been observed (Zhou et al. 2013; Li et al. 2014; Kumar et al. 2018). However, some studies reported that the temperature optima were same, even after immobilization (Jordan et al. 2011; Verma et al. 2013; Tao et al. 2016). In either case, the immobilized enzyme showed broader operational temperature range. The possible explanation could be the stabilization of weak hydrogen bonds and ionic forces with increased range of operating temperature of the immobilized enzyme (Jordan et al. 2011). In addition to this, the K_m value of immobilized enzyme (1.77 mmol L^{-1}) was less than that of free enzyme (3.12 mmol L^{-1}), which showed that substrate was more accessible to immobilized enzyme (Khoshnevisan et al. 2011).

Certain modifications in magnetic nanoparticles can be induced to further enhance the loading capacity, stability, and recovery. Most commonly, these issues have been addressed by surface functionalization of magnetic nanoparticles. In a recent study, the modification of polyethylene glycol dendrimers on magnetic nanomaterials effectively increased the enzyme loading capacity and the water solubility of the carrier, which further improved the catalytic activity and stability of immobilized enzyme. The hydrolysis of microcrystalline cellulose and filter paper with immobilized cellulase showed increase in the catalytic efficiency of enzyme by about 76% upon immobilization. Meanwhile, the activity of the immobilized cellulase remained 50% compared to its initial value after reusing it for six times (Han et al. 2018).

Surface modification using organic structures such as alkoxysilanes can be applied for surface functionalization of nanoparticles by introducing surface charges. Such modifications can prevent the particles from aggregating in liquids and enhance their biocompatibility. Zhang and colleagues (2015) studied cellulase immobilization on 3-(2-aminoethylaminopropyl)-trimethoxysilane (AEAPTMES)–functionalized magnetic nanospheres. The results exhibited that maximum binding of immobilized cellulase was 112 mg g^{-1} support with 87% activity recovery. Another study conducted by Li and colleagues (2014) showed immobilization of cellulase on supermagnetic silica–coated Fe_3O_4 nanoparticles. A high binding efficiency of 95% was achieved. The half-life of free and immobilized cellulase was 223 and 777 minutes, respectively, at 70°C. Reusability is another important factor, and the immobilized cellulase lost only 30% activity even after ten cycles of CMC hydrolysis.

Fe_3O_4 nanoparticles (NPs) were successfully fabricated with a layer of SiO_2 to form a core shell structure, which was further coated onto graphene oxide (GO) to form $Fe_3O_4@SiO_2$-graphene oxide (GO) composites. These prepared composites were used as the carrier for covalent binding of cellulase with above 90% immobilization efficiency. Remarkably, the thermal stability was enhanced and the half-life of cellulase upon immobilization (722 minutes) was 3.34 times greater than that of free enzyme (216 minutes) at 50°C (Li et al. 2015).

Another kinetically beneficial concept of co-immobilization can also be used for binding of two or more enzymes acting in cascade reaction, onto same support. The co-immobilization of β-glucosidase A (BglA) and cellobiohydrolase D (CelD) on superparamagnetic nanoparticles was studied by Song and colleagues (2016) using 5% gluteraldehyde. Both enzymes demonstrated 100% binding efficiency with 67.1% and 41.5% enzyme activities, respectively, compared to that of the free cellulases. The immobilized BglA showed more temperature stability than free BglA over most of the temperature points from 55°C–65°C. However, the temperature stability of immobilized CelD was not superior to free CelD. The immobilized BglA and CelD retained 85% and 43% of initial activity after being recycled for three and ten times, respectively. The co-immobilization, thus, exhibited benefits of synergistic action of cellulases during sugar production, and has a wide scope to reduce enzyme use.

10.6 Saccharification of Biomass Using Immobilized Cellulases

The cellulase enzyme might get denatured due to feedback inhibition and high temperature during the saccharification step. To get rid of these undesirable effects, immobilization of cellulase on a solid support can be done (Kumar et al. 2017). Also, previous studies suggest that the immobilization of enzymes onto nanomaterials has the potential to improve the economic viability of the entire bioenergy production process (Puri et al. 2013).

Xu and colleagues (2011) covalently immobilized crude cellulase (mixture of endocellulase, exocellulase, and a small amount of β-glucosidase) on iron oxide nanoparticles using gluteraldehyde as coupling agent. The optimum temperature for the free enzyme was 50°C, whereas it was 60°C for the immobilized enzyme. The K_m value for the immobilized enzyme decreased, indicating greater affinity for cellulosic substrates. The immobilized enzyme was further used for hydrolysis of steam-exploded corn stalk and bagasse. The glucose production increased rapidly during the initial 12 hours of hydrolysis, but leveled off after that. In bagasse hydrolysis, this turning point appeared after 24 hours of hydrolysis. The immobilized cellulase could be easily separated from hydrolysates using permanent magnet and more than 60% of activity was retained after six cycles.

Similarly, the enzymatic saccharification of pretreated hemp hurd biomass using free and magnetic nanoparticles immobilized cellulase from *T. reesei* was studied by Abraham and colleagues (2014). The optimum temperature for hydrolysis for free and immobilized enzyme was 50°C and 60°C, respectively. After 48 hours of hydrolysis of pretreated hemp hurd biomass at 2:1 enzyme-to-substrate ratio, free and immobilized cellulase resulted in 89% and 93% hydrolysis, respectively. The immobilized enzyme showed greater thermal stability at 80°C, retaining activity up to 6 hours, compared to the free enzyme that lost its activity in the first half-hour of incubation.

However, sometimes, the immobilized cellulase may not as effective as free cellulase for biomass hydrolysis. The precipitated cross-linked cellulase aggregates immobilized on 3-aminopropyltriethoxysilane modified magnetic iron oxide nanoparticles hydrolyzed bamboo biomass with 21% hydrolysis yield as compared to 44% yield using the free enzyme (Jia et al. 2017). The enzymatic saccharification of pretreated sunn hemp using free and immobilized cellulase showed hydrolysis yields of 72% and 53%, respectively (Manasa et al. 2017). The reusability of cellulase immobilized onto magnetic nanoparticles can, however, make up for the lowered activity. The glucose yield was 1.5 times higher using free cellulase as compared to immobilized cellulase for microcrystalline hydrolysis. The immobilized cellulase was easily recovered from the reaction mixture by applying an external magnetic field and preserved about 80% of its activity after 15 cycles of carboxymethylcellulose hydrolysis and 50% of activity after four cycles of *Agave* fiber hydrolysis. Immobilization, thus, could lead to reduction in the enzyme consumption during lignocellulosic biomass saccharification for bioethanol production (Sánchez-Ramírez et al. 2017).

Another approach was followed by Huang and colleagues (2015), where the difference in hydrolysis rate was compensated by the addition of ionic liquid to the reaction. Cellulase from *A. niger* was immobilized on 3-aminopropyltriethoxysilane–coated magnetic iron oxide nanoparticles. The sugar yield increased from 3.58 to 5.3 g $l^{-1}h^{-1}$ and initial rate of hydrolysis increased from 1.6 to 2.7 g l^{-1} h^{-1} on an average when ionic liquid was added to the immobilized cellulase reaction mixture. Also, the immobilized enzyme retained activity up to 16 cycles of hydrolysis. The total concentration of sugars was 208.92 g l^{-1}, which was higher than 10.5 g l^{-1} sugar concentration from the free enzyme.

The activity of cellulase immobilized on (3-aminopropyl) triethoxysilane–modified Fe_3O_4 nanoparticles was retained up to 99.1% of the free enzyme. The immobilized cellulase was used to hydrolyze pretreated corncob. The maximum decomposition rate of 61.94% was obtained after five hours of hydrolysis at 5 mg ml^{-1} initial corncob concentration. Even after 15 cycles, the immobilized enzyme retained about 91.1% of its activity (Zhang et al. 2016).

Kumar and colleagues (2018) extracted crude cellulase enzyme from *A. niger* SH3 using wheat bran and wheat straw as substrate. It was then immobilized on five different nanoparticles: magnesium oxide (MgO), silicon dioxide (SiO_2), zinc oxide (ZnO), iron oxide (Fe_2O_3), and silver oxide (Ag_2O) nanoparticles *via* physical adsorption and covalent binding. Iron oxide nanoparticles showed 60%–65% immobilization efficiency for the physically adsorbed enzyme, and the covalently bound enzyme showed 75%–80% immobilization efficiency. Further, this immobilized cellulase was used for saccharification of pretreated rice straw. Under the optimum conditions (i.e. 60°C, pH 5.0 and 6% substrate loading), 52% saccharification efficiency (375.4 mg g^{-1} sugar yield) was obtained using the immobilized cellulase compared to 47% saccharification efficiency (339.99 mg g^{-1} sugar yield) using the free cellulase. The immobilized cellulase was recovered after first and second saccharification cycles with 72.5% and 55% retained enzyme activity, respectively.

REFERENCES

Abbaszadeh, M., Hejazi, P., 2019. Metal affinity immobilization of cellulase on Fe_3O_4 nanoparticles with copper as ligand for biocatalytic applications. *Food Chem.* 290, 47–55.

Abraham, R. E., Verma, M. L., Barrow, C. J., Puri, M., 2014. Suitability of magnetic nanoparticle immobilised cellulases in enhancing enzymatic saccharification of pretreated hemp biomass. *Biotechnol. Biofuels* 7, 90.

Acharya, P. B., Acharya, D. K., Modi, H. A., 2008. Optimization for cellulase production by *Aspergillus niger* using saw dust as substrate. *Afr. J. Biotechnol.* 7, 4147–4152.

Acharya, S., Chaudhary, A., 2012. Optimization of fermentation conditions for cellulases production by *Bacillus licheniformis* MVS1 and *Bacillus* sp. MVS3 isolated from Indian hot spring. *Braz. Arch. Biol. Technol.* 55, 497–503.

Ahamed, A., Vermette, P., 2008. Culture based strategies to enhance cellulase enzyme production from *Trichoderma reesei* RUT-30 in bioreactor culture conditions. *Biochem. Eng.* 140, 399–407.

Ahmad, R., Sardar, M., 2014. Immobilization of cellulase on TiO_2 nanoparticles by physical and covalent methods: a comparative study. *Indian J. Biochem. Biophys.* 51(4), 314–320.

Ang, S. K., Shaza, E. M., Adibah, Y., Suraini, A., Madihah, M. S., 2013. Production of cellulases and xylanase by *Aspergillus fumigatus* SK1 using untreated oil palm trunk through solid state fermentation. *Process Biochem.* 48, 1293–1302.

Ansari, S. A., Husain, Q., 2012. Potential applications of enzymes immobilized on/in nano materials: a review. *Biotechnol. Adv.* 30, 512–523.

Ariffin, H., Abdullah, N., Umi Kalsom, M. S., Shirai, Y., Hassan, M. A., 2006. Production and characterization of cellulase by *Bacillus pumilus* EB3. *Int. J. Eng. Technol.* 3, 47–53.

Azhar, S. H. M., Abdulla, R., Jambo, S. A., Marbawi, H., Gansau, J. A., Faik, A. A. M., Rodrigues, K. F., 2017. Yeasts in sustainable bioethanol production: a review. *Biochem. Biophys. Rep.* 10, 52–61.

Babu, K. R., Satyanarayana, T., 1996. Production of bacterial enzymes by solid state fermentation. *J. Sci. Ind. Res.* 55, 464–467.

Bayer, E. A., Chanzy, H., Lamed, R., Shoham, Y., 1998. Cellulose, cellulases and cellulosomes. *Curr. Opin. Struct. Biol.* 8, 548–557.

Behera, S. S., Ray, R. C., 2016. Solid state fermentation for production of microbial cellulases: recent advances and improvement strategies. *Int. J. Biol. Macromol.* 86, 656–669.

Bertrand, E., Vandenberghe, L. P., Soccol, C. R., Sigoillot, J. C., Faulds, C., 2016. First generation bioethanol. In *Green fuels technology* (pp. 175–212). Cham: Springer.

Bezerra, C. S., de Farias Lemos, C. M. G., de Sousa, M., Gonçalves, L. R. B., 2015. Enzyme immobilization onto renewable polymeric matrixes: past, present, and future trends. *J. Appl. Polym. Sci.* 132(26).

Bilal, M., Zhao, Y., Rasheed, T., Iqbal, H. M., 2018. Magnetic nanoparticles as versatile carriers for enzymes immobilization: a review. *Int. J. Biolog. Macromol.* 120, 2530–2544.

Busto, M. D., Ortega, N., Perez-Mateos, M., 1998. Characterization of microbial endo-β-glucanase immobilized in alginate beads. *Acta Biotechnol.* 18(3), 189–200.

Cannel, E., Young, M. M., 1980. Solid-State cultivation systems. *Process. Biochem.* 15, 2–7.

Cantone, S., Ferrario, V., Corici, L., Ebert, C., Fattor, D., Spizzoa, P., Gardossi, L., 2012. Efficient immobilisation of industrial biocatalysts: criteria and constraints for the selection of organic polymeric carriers and immobilisation methods. *Chem. Soc. Rev.* 42, 6262–6276.

Caruso, F., Schüler, C., 2000. Enzyme multilayers on colloid particles: assembly, stability, and enzymatic activity. *Langmuir*, 16(24), 9595–9603.

Chang, R. H. Y., Jang, J., Wu, K. C. W., 2011. Cellulase immobilized mesoporous silica nanocatalysts for efficient cellulose-to-glucose conversion. *Green Chem.* 13(10), 2844–2850.

Cherian, E., Dharmendirakumar, M., Baskar, G., 2015. Immobilization of cellulase onto MnO_2 nanoparticles for bioethanol production by enhanced hydrolysis of agricultural waste. *Chin. J. Catal.* 36(8), 1223–1229.

Chundawat, S., Sousa, L. D. C., Cheh, A. M., Balan, V., Dale, B., 2017. *Methods for producing extracted and digested products from pretreated lignocellulosic biomass*. U.S. Patent, 9650657.

Crawford, R. L., 1981. *Lignin biodegradation and transformation* (p. 114). New York: John Wiley & Sons.

Damisa, D., Ameh, J. B. Egbe, N. E. L., 2011. Cellulase production by native *Aspergillus niger* obtained from soil environments. *Ferment. Technol. Bioeng.* 1, 62–70.

Datta, S., Christena, L. R., Rajaram, Y. R. S., 2013. Enzyme immobilization: an overview on techniques and support materials. *3 Biotech* 3, 1–9.

Daubresse, C., Grandfils, C., Jérôme, R., Teyssié, P., 1996. Enzyme immobilization in reactive nanoparticles produced by inverse microemulsion polymerization. *Colloid Polym. Sci.* 274(5), 482–489.

de Assis Castro, R. C., Fonseca, B. G., dos Santos, H. T. L., Ferreira, I. S., Mussatto, S. I., Roberto, I. C., 2017. Alkaline deacetylation as a strategy to improve sugars recovery and ethanol production from rice straw hemicellulose and cellulose. *Ind. Crops Prod.* 106, 65–73.

de Souza Moreira, L. R., Sciuto, D. L., Ferreira Filho, E. X., 2016. An overview of cellulose-degrading enzymes and their applications in textile industry. In Gupta, V. K. (ed.), *New and future developments in microbial biotechnology and bioengineering* (pp. 165–175). Amsterdam, Netherlands: Elsevier.

Deswal, D., Khasa, Y. P., Kuhad, R. C., 2011. Optimization of cellulase production by a brown rot fungus *Fomitopsis* sp. RCK2010 under solid state fermentation. *Bioresor. Technol.* 102, 6065–6072.

de Vries, R. P., Visser, J. A. A. P., 2001. *Aspergillus* enzymes involved in degradation of plant cell wall polysaccharides. *Microbiol. Mol. Biol. Rev.* 65, 497–522.

Doscher, M. S., Richards, F. M., 1963. The activity of an enzyme in the crystalline state: ribonuclease S. *J. Biol. Chem.* 238(7), 2399–2406.

El-Bakry, M., Abraham, J., Cerda, A., Barrena, R., Ponsa, S., Gea, T., Sanchez, A., 2015. From wastes to high value added products: novel aspects of SSF in the production of enzymes. *Crit. Rev. Environ. Sci. Tecnol.* 45, 1999–2042.

Fan, L. T., Gharpuray, M. M., Lee, Y. H., 1987. Design and economic evaluation of cellulose hydrolysis processes. In *Cellulose hydrolysis* (pp. 149–187). Berlin & Heidelberg: Springer.

Galaev, I., Mattiasson, B. (Eds.) 2001. *Smart polymers for bioseparation and bioprocessing*. London and New York: CRC Press, Taylor & Francis Group.

Goldemberg, J., Guardabasi, P., 2009. Are biofuels a feasible option? *Energy Policy* 37, 10–14.

Grous, W. R., Converse, A. O., Grethlein, H. E., 1986. Effect of steam explosion pretreatment on pore size and enzymatic hydrolysis of poplar. *Enz. Microbial Technol.* 8(5), 274–280.

Guisan, J. M., 2006. Immobilization of enzymes as the 21st century begins. In *Immobilization of enzymes and cells* (pp. 1–13). Totowa, NJ: Humana Press.

Gupta, A., Verma, J. P., 2015. Sustainable bio-ethanol production from agro-residues: a review. *Renew. Sust. Energ. Rev.* 41, 550–567.

Gusakov, A. V., 2011. Alternatives to *Trichoderma reesei* in biofuel production. *Trends. Biotechnol.* 29, 419–425.

Hahn-Hägerdal, B., Galbe, M., Gorwa-Grauslund, M. F., Lidén, G., Zacchi, G., 2006. Bio-ethanol—the fuel of tomorrow from the residues of today. *Trends Biotechnol.* 24(12), 549–556.

Han, J., Wang, L., Wang, Y., Dong, J., Tang, X., Ni, L., Wang, L., 2018. Preparation and characterization of Fe_3O_4-NH_2@4-arm-PEG-NH_2, a novel magnetic four-arm polymer-nanoparticle composite for cellulase immobilization. *Biochem. Eng. J.* 130, 90–98.

Hanefeld, U., Gardossi, L., Magner, E., 2009. Understanding enzyme immobilisation. *Chem. Soc. Rev.* 38(2), 453–468.

Hartmann, M., Kostrov, X., 2013. Immobilization of enzymes on porous silicas- benefits and challenges. *Chem. Soc. Rev.* 42, 6277–6289.

Hideno, A., Inoue, H., Tsukahara, K., Fujimoto, S., Minowa, T., Inoue, S., . . . Sawayama, S., 2009. Wet disk milling pretreatment without sulfuric acid for enzymatic hydrolysis of rice straw. *Bioresour. Technol.* 100(10), 2706–2711.

Hola, K., Markova, Z., Zoppellaro, G., Tucek, J., Zboril, R., 2015. Tailored functionalization of iron oxide nanoparticles for MRI, drug delivery, magnetic separation and immobilization of biosubstances. *Biotechnol. Adv.* 33(6), 1162–1176.

Hsu, T. C., Guo, G. L., Chen, W. H., Hwang, W. S., 2010. Effect of dilute acid pretreatment of rice straw on structural properties and enzymatic hydrolysis. *Bioresour. Technol.* 101, 4907–4913.

Huang, P. J., Chang, K. L., Hsieh, J. F., Chen, S. T., 2015. Catalysis of rice straw hydrolysis by the combination of immobilized cellulase from *Aspergillus niger* on β-cyclodextrin-Fe_3O_4 nanoparticles and ionic liquid. *BioMed. Res. Int.* 2015, 1–9.

Hudson, S., Cooney, J., Magner, E., 2008. Proteins in mesoporous silicates. *Angew. Chem. Int. Ed. Engl.* 47, 8582–8594.

Imai, M., Ikari, K., Suzuki, I., 2004. High-performance hydrolysis of cellulose using mixed cellulase species and ultrasonication pretreatment. *Biochem. Eng. J.* 17(2), 79–83.

Jahromi, M. F., Liang, J. B., Rosfarizan, M., Goh, Y. M., Shokryazdan, P., Ho, Y. W., 2011. Efficiency of rice straw lignocelluloses degradability by *Aspergillus terreus* ATCC 74135 in solid state fermentation. *Afr. J. Biotechnol.* 10, 4428–4435.

Jeya, M., Zhang, Y. W., Kim, I. W., Lee, J. K., 2009. Enhanced saccharification of alkali-treated rice straw by cellulase from *Trametes hirsuta* and statistical optimization of hydrolysis conditions by RSM. *Bioresour. Technol.* 100(21), 5155–5161.

Jia, H., Zhu, G., Wang, P., 2003. Catalytic behaviors of enzymes attached to nanoparticles: the effect of particle mobility. *Biotechnol. Bioeng.* 84(4), 406–414.

Jia, J., Zhang, W., Yang, Z., Yang, X., Wang, N., Yu, X., 2017. Novel magnetic cross-linked cellulase aggregates with a potential application in lignocellulosic biomass bioconversion. *Molecules* 22, 269.

Jordan, J., Kumar, C. S., Theegala, C., 2011. Preparation and characterization of cellulase-bound magnetite nanoparticles. *J. Mol. Catal. B. Enzym.* 68, 139–146.

Juturu, V., Wu, J. C., 2014. Microbial cellulases: engineering, production and applications. *Renew. Sust. Energ. Rev.* 33, 188–203.

Kalsoom, R., Ahmed, S., Nadeem, M., Chohan, S., Abid, M., 2019. Biosynthesis and extraction of cellulase produced by *Trichoderma* on agro-wastes. *Int. J. Environ. Sci. Technol.* 16, 921–928.

Kang, S. W., Park, Y. S., Lee, J. S., Hong, S. I., Kim, S. W., 2004. Production of cellulases and hemicellulases by *Aspergillus niger* KK2 from lignocellulosic biomass. *Bioresour. Technol.* 91, 153–156.

Kapoor, M., Soam, S., Agrawal, R., Gupta, R. P., Tuli, D. K., Kumar, R., 2017. Pilot scale dilute acid pretreatment of rice straw and fermentable sugar recovery at high solid loadings. *Bioresour. Technol.* 224, 688–693.

Kaur, P., Taggar, M. S., Kalia, A., 2020. Characterization of magnetic nanoparticle—immobilized cellulases for enzymatic saccharification of rice straw. *Biomass Convers. Biorefin.* 1–15.

Khatiwada, P., Ahmed, J., Sohag, M. H., Islam, K., Azad, A. K., 2016. Isolation, screening and characterization of cellulase producing bacterial isolates from municipal solid wastes and rice straw wastes. *J. Bioproc. Biotech.* 6, 280–285.

Khoshnevisan, K., Bordbar, A. K., Zare, D., Davoodi, D., Noruzi, M., Barkhi, M., Tabatabaei, M., 2011. Immobilization of cellulase enzyme on superparamagnetic nanoparticles and determination of its activity and stability. *Chem. Eng. J.* 171, 669–673.

Kim, J., Grate, J. W., Wang, P., 2006. Nanostructures for enzyme stabilization. *Chem. Eng. Sci.* 61(3), 1017–1026.

Klemm, D., Heublein, B., Fink, H. P., Bohn, A., 2005. Cellulose: fascinating biopolymer and sustainable raw material. *Angew. Chem. Int. Ed*. 44, 3358–3393.

Ko, J. K., Bak, J. S., Jung, M. W., Lee, H. J., Choi, I. G., Kim, T. H., Kim, K. H., 2009. Ethanol production from rice straw using optimized aqueous-ammonia soaking pretreatment and simultaneous saccharification and fermentation processes. *Bioresour. Technol*. 100(19), 4374–4380.

Kumar, A., Singh, S., Nain, L., 2018. Magnetic nanoparticle immobilized cellulase enzyme for saccharification of paddy straw. *Int. J. Curr. Microbiol. Appl. Sci*. 7, 881–893.

Kumar, A., Singh, S., Tiwari, R., Goel, R., Nain, L., 2017. Immobilization of indigenous holocellulase on iron oxide (Fe_2O_3) nanoparticles enhanced hydrolysis of alkali pretreated paddy straw. *Int. J. Biol. Macromol*. 96, 538–549.

Kumar, R., Singh, S., Singh, O. V., 2008. Bioconversion of lignocellulosic biomass: biochemical and molecular perspectives. *J. Ind. Microbiol. Biotechnol*. 35, 377–391.

Lee, C. K., Darah, I., Ibrahim, C. O., 2011. Production and optimization of cellulase enzyme using *Aspergillus niger* USM AI 1 and comparison with *Trichoderma reesei* via solid state fermentation system. *Biotechnol. Res. Int*. 2011: 1–6.

Lee, Y. H., Fan, L. T., 1980. Properties and mode of action of cellulase. *Adv. Biochem. Eng*. 17, 101–129.

Li, Y., Wang, X. Y., Jiang, X. P., Ye, J. J., Zhang, Y. W., Zhang, X. Y., 2015. Fabrication of graphene oxide decorated with $Fe_3O_4@SiO_2$ for immobilization of cellulase. *J. Nanopart. Res*. 17, 8.

Li, Y., Wang, X. Y., Zhang, R. Z., Zhang, X. Y., Liu, W., Xu, X. M., Zhang, Y. W., 2014. Molecular imprinting and immobilization of cellulase onto magnetic $Fe_3O_4@SiO_2$ nanoparticles. *J. Nanosci. Nanotechnol*. 14, 2931–2936.

Liu, S., 2017. Chapter 7: Enzymes. In Liu, S. (ed.) *Bioprocess engineering (second edition): Kinetics, sustainability, and reactor design* (pp. 297–373). Amsterdam: Elsevier.

Liang, W., Cao, X., 2012. Preparation of a pH-sensitive polyacrylate amohiphilic copolymer and its application in cellulase immobilization. *Bioresour. Technol*. 116, 140–146.

Liao, M. H., Chen, D. H., 2001. Immobilization of yeast alcohol dehydrogenase on magnetic nanoparticles for improving its stability. *Biotechnol. Lett*. 23(20), 1723–1727.

Limayem, A., Ricke, S. C., 2012. Lignocellulosic biomass for bioethanol production: current perspectives, potential issues and future prospects. *Prog. Energ. Combust. Sci*. 38(4), 449–467.

Liu, D. M., Chen, J., Shi, Y. P., 2018. Advances on methods and easy separated support materials for enzymes immobilization. *TrAC Trend. Anal. Chem*. 102, 332–342.

Liu, Y., Chen, J. Y., 2016. Enzyme immobilization on cellulose matrixes. *J. Bioact. Compat. Pol*. 31, 553–567.

López-Serrano, P., Cao, L., Van Rantwijk, F., Sheldon, R. A., 2002. Cross-linked enzyme aggregates with enhanced activity: application to lipases. *Biotechnol. Lett*. 24(16), 1379–1383.

Ma, H., Liu, W. W., Chen, X., Wu, Y. J., Yu, Z. L., 2009. Enhanced enzymatic saccharification of rice straw by microwave pretreatment. *Bioresour. Technol*. 100(3), 1279–1284.

Maki, M., Leung, K. T., Qin, W., 2009. The prospects of cellulase-producing bacteria for the bioconversion of lignocellulosic biomass. *Int. J. Biol. Sci*. 5, 500.

Manasa, P., Saroj, P., Korrapati, N., 2017. Immobilization of cellulase enzyme on zinc ferrite nanoparticles in increasing enzymatic hydrolysis on ultrasound assisted alkaline pretreated *Crotalaria juncea* biomass. *Indian J. Sci. Technol*. 10, 1–7.

Martins, L. F., Kolling, D., Camassola, M., Dillon, A. J. P., Ramos, L. P., 2008. Comparison of *Penicillium echinulatum* and *Trichoderma reesei* cellulases in relation to their activity against various cellulosic substrates. *Bioresour. Technol*. 99, 1417–1424.

Martins, M. B. F., Simoes, S. I. D., Cruz, M. E. M., Gaspar, R., 1996. Development of enzyme-loaded nanoparticles: Effect of pH. *J. Mater. Sci. Mater. Med*. 7(7), 413–414.

Mitchell, D. T., Lee, S. B., Trofin, L., Li, N., Nevanen, T. K., Soderlund, H., Martin, C. R., 2002. Smart nanotubes for bioseparations and biocatalysis. *J. Am. Chem. Soc*. 124, 11864–11865.

Mo, H., Qiu, J., Yang, C., Zang, L., Sakai, E., 2020. Preparation and characterization of magnetic polyporous biochar for cellulase immobilization by physical adsorption. *Cellulose* 27(9), 4963–4973.

Mohamad, N. R., Marzuki, N. H. C., Buang, N. A., Huyop, F., Wahab, R. A., 2015. An overview of technologies for immobilization of enzymes and surface analysis techniques for immobilized enzymes. *Biotechnol. Biotechnol. Eq*. 29, 205–220.

Mojović, L., Nikolić, S., Rakin, M., Vukasinović, M., 2006. Production of bioethanol from corn meal hydrolyzates. *Fuel* 85(12–13), 1750–1755.

Mosier, N., Wyman, C., Dale, B., Elander, R., Lee, Y. Y., Holtzapple, M., Ladisch, M., 2005. Features of promising technologies for pretreatment of lignocellulosic biomass. *Bioresour. Technol.* 96(6), 673–686.

Panchapakesan, A., Shankar, N., 2016. Fungal cellulases: an overview. In Gupta, V. (ed.), *New and future developments in microbial biotechnology and bioengineering* (pp. 9–18). Amsterdam, Netherlands: Elsevier.

Panday, A., Selvakumar, P., Soccol, C. R., Nigam, P., 1999. Solid state fermentation for the production of industrial enzymes. *Curr. Sci.* 77, 149–162.

Pandey, S., Srivastava, M., Shahid, M., Kumar, V., Singh, A., Trivedi, S., Srivastava, Y. K., 2015. *Trichoderma* species cellulases produced by solid state fermentation. *J Data Mining Genomics Proteomics* 6, 1.

Podrepšek, G. H., Primožič, M., Knez, Ž., Habulin, M., 2012. Immobilization of cellulase for industrial production. *Chem. Eng.* 27.

Puri, M., Barrow, C. J., Verma, M. L., 2013. Enzyme immobilization on nanomaterials for biofuel production. *Trends Biotechnol.* 31, 215–216.

Rasul, F., Afroz, A., Rashid, U., Mehmood, S., Sughra, K., Zeeshan, N., 2015. Screening and characterization of cellulase producing bacteria from soil and waste (molasses) of sugar industry. *Int. J. Biosci.* 6, 230–236.

Reddy, G. P. K., Narasimha, G., Kumar, K. D., Ramanjaneyulu, G., Ramya, A., Kumari, B. S., Reddy, B. R., 2015. Cellulase production by *Aspergillus niger* on different natural lignocellulosic substrates. *Int. J. Curr. Microbiol. Appl. Sci.* 4, 835–845.

Roth, H. C., Schwaminger, S. P., Peng, F., Berensmeier, S., 2016. Immobilization of cellulase on magnetic nanocarriers. *ChemistryOpen* 5(3), 183.

Rubin, E. M., 2008. Genomics of cellulosic biofuels. *Nature* 454, 841–845.

Salehi, S. A., Karimi, K., Behzad, T., Poornejad, N., 2012. Efficient conversion of rice straw to bioethanol using sodium carbonate pretreatment. *Energ. Fuel.* 26(12), 7354–7361.

Sánchez-Ramírez, J., Martínez-Hernández, J. L., Segura-Ceniceros, P., López, G., Saade, H., Medina-Morales, M. A., Ramos-Gonzalez, R., Aguilar, C. N., Ilyina, A., 2017. Cellulases immobilization on chitosan-coated magnetic nanoparticles: application for *Agave atrovirens* lignocellulosic biomass hydrolysis. *Bioprocess. Biosyst. Eng.* 40, 9–22.

Schoevaart, R., Wolbers, M. W., Golubovic, M., Ottens, M., Kieboom, A. P. G., Van Rantwijk, F., ... Sheldon, R. A., 2004. Preparation, optimization, and structures of cross-linked enzyme aggregates (CLEAs). *Biotechnol. Bioeng.* 87(6), 754–762.

Searchinger, T., Edwards, R., Mulligan, D., Heimlich, R., Plevin, R., 2015. Do biofuel policies seek to cut emissions by cutting food? *Science* 347(6229), 1420–1422.

Sheldon, R. A., Schoevaart, R., Van Langen, L. M., 2005. Cross-linked enzyme aggregates (CLEAs): a novel and versatile method for enzyme immobilization (a review). *Biocatal. Biotransfor.* 23(3–4), 141–147.

Sheldon, R. A., van Pelt, S., 2013. Enzyme immobilisation in biocatalysis: why, what and how. *Chem. Soc. Rev.* 42(15), 6223–6235.

Sherief, A. A., El-Tanash, A. B., Atia, N., 2010. Cellulase production by *Aspergillus fumigatus* grown on mixed substrate of rice straw and wheat bran. *Res. J. Microbiol.* 5, 199–211.

Sindhu, R., Binod, P., Pandey, A., 2016. Biological pretreatment of lignocellulosic biomass-An overview. *Bioresour. Technol.* 199, 76–82.

Singh, A., 1999. Engineering enzyme properties. *Indian J. Microbiol.* 39, 65–77.

Singh, R. K., Tiwari, M. K., Singh, R., Lee, J. K., 2013. From protein engineering to immobilization: promising strategies for the upgrade of industrial enzymes. *Int. J. Mol. Sci.* 14(1), 1232–1277.

Singhania, R. R., Sukumaran, R. K., Pillai, A., Prema, P., Szakacs, G., Pandey, A., 2006. Solid-state fermentation of lignocellulosic substrates for cellulase production by *Trichoderma reesei* NRRL 11460. *Indian J. Biotechnol.* 5, 332–336.

Sinnott, M., 2007. Enzyme-catalysed glycosyl transfer. In *Carbohydrate chemistry and biochemistry: Structure and mechanism* (pp. 299–477). Cambridge: Royal Society of Chemistry.

Song, Q., Mao, Y., Wilkins, M., Segato, F., Prade, R., 2016. Cellulase immobilization on superparamagnetic nanoparticles for reuse in cellulosic biomass conversion. *AIMS Bioeng.* 3, 264–276.

Subramaniyam, R., Vimala, R., 2012. Solid state and submerged fermentation for the production of bioactive substances: a comparative study. *Int. J. Sci. Nat.* 3, 480–486.

Sukumaran, R. K., Singhania, R. R., Mathew, G. M., Pandey, A., 2009. Cellulase production using biomass feed stock and its application in lignocellulose saccharification for bio-ethanol production. *Renew. Energ.* 34, 421–424.

Sun, H., Ge, X., Hao, Z., Peng, M., 2010. Cellulase production by *Trichoderma* sp. on apple pomace under solid state fermentation. *Afr. J. Biotechnol.* 9, 163–165.

Tao, Q. L., Li, Y., Shi, Y., Liu, R. J., Zhang, Y. W., Guo, J., 2016. Application of molecular imprinted magnetic $Fe_3O_4@SiO_2$ nanoparticles for selective immobilization of cellulase. *J. Nanosci. Nanotechnol.* 16, 6055–6060.

Tata, A., Sokolowska, K., Swider, J., Konieczna-Molenda, A., Proniewicz, E., Witek, E., 2015. Study of cellulolytic enzyme immobilization on copolymers of N-vinylformamide. *Spectrochim. Acta. Part A Mol. Biomol. Spectrosc.* 149, 494–504.

Teofil, J., Jakub, Z., Krajewska, B., 2014. Enzyme immobilization by adsorption: a review. *Adsorption* 20(5–6), 301–321.

Vaghari, H., Jafarizadeh-Malmiri, H., Mohammadlou, M., Berenjian, A., Anarjan, N., Jafari, N., Nasiri, S., 2016. Application of magnetic nanoparticles in smart enzyme immobilization. *Biotechnol. Lett.* 38(2), 223–233.

Verma, M. L., Barrow, C. J., Puri, M., 2013. Nanobiotechnology as a novel paradigm for enzyme immobilisation and stabilisation with potential applications in biodiesel production. *Appl. Microbiol. Biotechnol.* 97, 23–39.

Viet, T. Q., Minh, N. P., Dao, D. T. A., 2013. Immobilization of cellulase enzyme in calcium alginate gel and its immobilized stability. *Am. J. Res. Commun.* 1(12), 254–267.

Weerasai, K., Suriyachai, N., Poonsrisawat, A., Arnthong, J., Unrean, P., Laosiripojana, N., Champreda, V., 2014. Sequential acid and alkaline pretreatment of rice straw for bioethanol fermentation. *BioResources*, 9(4), 5988–6001.

Xie, T., Wang, A., Huang, L., Li, H., Chen, Z., Wang, Q., Yin, X., 2009. Recent advance in the support and technology used in enzyme immobilization. *Afr. J. Biotechnol.* 8(19).

Xu, J., Huo, S., Yuan, Z., Zhang, Y., Xu, H., Guo, Y., Liang, C., Zhuang, X., 2011. Characterization of direct cellulase immobilization with superparamagnetic nanoparticles. *Biocatal. Biotransfor.* 29, 71–76.

Xu, J., Liu, X., He, J., Hu, L., Dai, B., Wu, B., 2015. Enzymatic in situ saccharification of rice straw in aqueous-ionic liquid media using encapsulated *Trichoderma aureoviride* cellulase. *J. Chem. Technol. Biotechnol.* 90(1), 57–63.

Xu, J., Sun, J., Wang, Y., Sheng, J., Wang, F., Sun, M., 2014. Application of iron magnetic nanoparticles in protein immobilization. *Molecules*, 19(8), 11465–11486.

Yiu, H. H., Keane, M. A., 2012. Enzyme-magnetic nanoparticles hybrids: new effective catalysts for the production of high value chemicals. *J. Chem. Technol. Biotechnol.* 87, 583–594.

Yu, Y., Yuan, J., Wang, Q., Fan, X., Wang, P., 2013. Immobilization of cellulases on the reversibly soluble polymer Eufragit S-100 for cotton treatment. *Eng. Life. Sci.* 13, 194–200.

Zang, L., Qiu, J., Wu, X. et al., 2014. Preparation of magnetic chitosan nanoparticles as support for cellulase immobilization. *Ind. Eng. Chem. Res.* 53, 3448–3454.

Zhang, Q., Han, X., Tang, B., 2013. Preparation of a magnetically recoverable biocatalyst support on monodisperse Fe_3O_4 nanoparticles. *RSC Adv.* 3, 9924–9931.

Zhang, Q., Kang, J., Yang, B., Zhao, L., Hou, Z., Tang, B., 2016. Immobilized cellulase on Fe_3O_4 nanoparticles as a magnetically recoverable biocatalyst for the decomposition of corncob. *Chin. J. Catal.* 37, 389–397.

Zhang, W., Qiu, J., Feng, H., Zang, L., Sakai, E., 2015. Increase in stability of cellulase immobilized on functionalized magnetic nanospheres. *J. Magn. Magn. Mater.* 375, 117–123.

Zhang, Y. H. P., Lynd, L. R., 2004. Towards an aggregated understanding of enzymatic hydrolysis of cellulose: noncomplexed cellulase systems. *Biotechnol. Bioeng.* 88, 797–824.

Zhang, Y., Xu, J. L., Yuan, Z. H., Qi, W., Liu, Y. Y., He, M. C., 2012. Artificial intelligence techniques to optimize the EDC/NHS-mediated immobilization of cellulose on Eudragit L-100. *Int. J. Mol. Sci.* 13, 7952–7962.

Zheng, Y., Pan, Z., Zhang, R., 2009. Overview of biomass pretreatment for cellulosic ethanol production. *Int. J. Agric. Biol. Eng.* 2(3), 51–68.

Zhou, Y., Pan, S., Wei, X., Wang, L., Liu, Y., 2013. Immobilization of β-glucosidase onto magnetic nanoparticles and evaluation of the enzymatic properties. *BioResources* 8, 2605–2619.

Zhu, J. Q., Qin, L., Li, W. C., Zhang, J., Bao, J., Huang, Y. D., . . . Yuan, Y. J., 2015. Simultaneous saccharification and co-fermentation of dry diluted acid pretreated corn stover at high dry matter loading: overcoming the inhibitors by non-tolerant yeast. *Bioresour. Technol.* 198, 39–46.

Zhu, Y., 2007. Immobilized cell fermentation for production of chemicals and fuels. In *Bioprocessing for value-added products from renewable resources* (pp. 373–396). Amsterdam: Elsevier.

Zhuang, J., Marchant, M. A., Nokes, S. E., Strobel, H. J., 2007. Economic analysis of cellulase production methods for bio-ethanol. *Appl. Eng. Agric.* 23, 679–687.

11

Biogas Production in a Biorefinery Context: Analysis of the Scale Based on Different Raw Materials

J.A. Poveda-Giraldo, M. Ortiz-Sánchez, S. Piedrahita-Rodríguez,
J.C. Solarte-Toro, A.M. Zetty-Arenas, and C.A. Cardona Alzate

CONTENTS

11.1 Introduction 245
11.2 Raw Materials Applied in Biogas Production 248
11.3 Stand-Alone Production of Biogas Using Different Raw Materials 249
    11.3.1 Chemical Characterization 249
    11.3.2 Biochemical Methane Potential (BMP) 250
    11.3.3 Anaerobic Digestion of Plantain Peel 251
    11.3.4 Anaerobic Digestion of Marigold Residues 251
    11.3.5 Anaerobic Digestion of Wood Chips 252
    11.3.6 Comparison of Anaerobic Digestions 253
    11.3.7 Simulation of Biogas Production 256
11.4 Biogas Integrated in Biorefineries 258
    11.4.1 Process Description 259
        11.4.1.1 Biorefinery of Marigold Residues and Wood Chips 259
        11.4.1.2 Biorefinery of Plantain Peel 260
    11.4.2 Comparison of Biorefinery Results 260
11.5 Conclusion 264
Acknowledgments 264
References 264

11.1 Introduction

The excessive use of fossil fuels has led to several environmental problems due to greenhouse gas emissions (Lelieveld et al., 2019). Therefore, research on renewable energy source alternatives has increased in the last years (Gielen et al., 2019), among which the thermochemical and biochemical routes are the most widely used. The biochemical pathway does not generate large amounts of gas emissions when compared to the thermochemical route. In this sense, biogas has emerged as an important renewable energy source to supply a share of the worldwide energy demand (Scarlat et al., 2018).

Anaerobic digestion (AD) is a biological process that breaks down organic matter in the absence of oxygen, obtaining biogas as the main product. This biogas is a mixture of approximately 45%–75% CH_4 and 25%–55% CO_2, although it can also contain small fractions of other gases such as H_2, N_2, O_2, S, NH_3, and H_2S (Deublein and Steinhauser, 2010). AD is a biological process that involves many different consortia of microorganisms that present specific functions in each stage (Deublein and Steinhauser, 2010). Figure 11.1 shows the metabolic pathway of organic compounds present in the biomass. AD initiates with the hydrolysis stage, where complex organic polymers (i.e., carbohydrates, proteins, lipids, fats) are converted into reduced compounds (e.g., sugars, amino acids). Then, acetogenesis occurs when the reduced compounds are converted into volatile short-chain fatty acids and byproducts such as carbon dioxide, hydrogen, and alcohol. Finally, during the methanogenesis stage, acetate is divided to produce

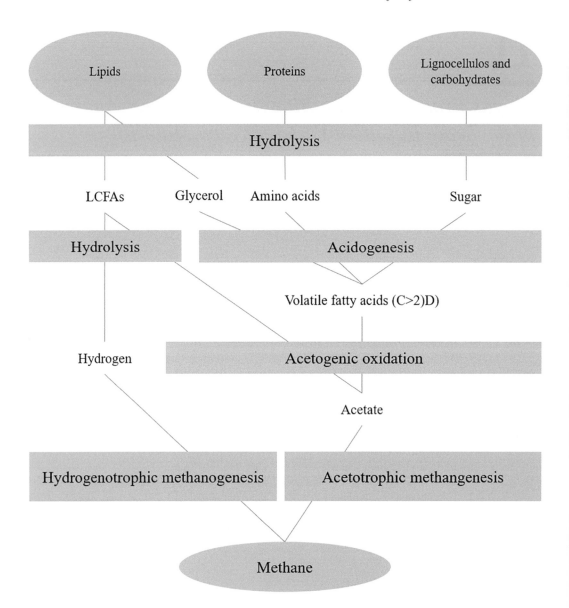

FIGURE 11.1 The metabolism pathway of organic compounds in biomass to produce biomethane.

methane and carbon dioxide; hydrogen and carbon dioxide are used as electron donors and acceptors, respectively (Li et al., 2011; Ofoefule et al., 2009).

Different types of digesters are used for anaerobic treatments. In high-rate systems with hydraulic retention time for biomass less than the sludge, digesters use anaerobic filters, fluidized bed, up-flow anaerobic sludge bed (UASB), and expanded granular mud bed (EGSB). In contrast, accumulative, plug flow, continuous stirred tank reactor (CSTR), and batch-type digesters are used in low-rate systems, where there is no biomass retention. Therefore, its hydraulic retention time is long and equal to the retention time of the sludge. According to the content of total solids (TS) in the raw material, AD can be classified into wet fermentation (low solids content) with values between 15–25%, and dry fermentation (high solids content) with TS above 30% (Mes et al., 2003). Consequently, systems with high TS content require a higher amount of inoculum (Rocamora et al., 2020). Microorganisms capable of decomposing organic material are susceptible to temperature, which is optimal between 30°C–38°C and 44°C–57°C for mesophilic and thermophilic conditions, respectively (Hilkiah Igoni et al., 2008).

In Europe, the biogas production reached 13.5 million tons in 2014 (EurObserv'ER consortium, 2019), Germany being the pioneer country (25% of world biogas production). According to the European Environment Agency (EEA), there are around 15,000 biogas plants in Europe (Wiesenthal et al., 2006). From 2010–2014, an increase of 49.6% of installation of biogas plants with a total capacity of more than 8,000 MW was estimated. Germany has more than 8,000 operating biogas plants, which generate 4 TWh. In the last decade, the biogas sector has increased in Europe due to different parameters, such as feed rates in Germany, the mandatory certification for energy renewability in the United Kingdom, and the fiscal policy (economic exemptions) in Sweden. In Germany, most of the electrical energy comes from biogas due to government initiatives that promote energy generation from waste. By 2030, a more significant amount of biogas (224 TWh) is estimated from wet manure, landfills, undigested sewage sludge, and food processing waste. The United States, China, and India have studied lignocellulosic residues to produce biogas. These countries are emerging as future biogas producers. Although lignocellulosic residues are promising raw materials for biogas production (mainly due to their high production scale), their complex composition of cellulose, hemicellulose, and lignin requires a pretreatment to break down their tight structure, creating an economic and technical barrier in the biogas production. Stages of lignocellulosic material pretreatment are milling, extrusion, treatment with steam or steam explosion, liquid hot water, microwave, dilute or strong acids, or other liquids. Subsequently, lignocellulose is hydrolyzed by using enzymes. These pretreatment technologies are primarily aimed at accelerating AD and increasing biogas production. Figure 11.2 depicts the increase in the digestion rate by including a pretreatment stage.

For facilitating the microbial degradation, the pretreatment stages in a biogas plant must overcome the structural barriers of lignocellulose. In turn, the streams obtained from the chemical pretreatment stages can be used for different products that improve the economic viability of the plant. Therefore, the biorefinery concept gains strength by making full use of lignocellulosic materials.

The biochemical methane potential (BMP) tests and the degradation rates are an estimation tool for biogas production, helping to design and size digesters, and evaluate the performance of the process at high scales. Furthermore, the BMP measures the efficiency of the process through the consumed organic

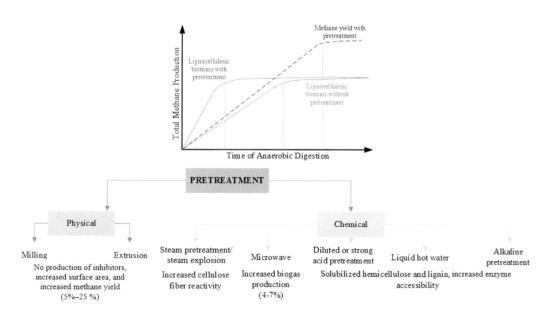

FIGURE 11.2 Influence in the digestion rate during the pretreatment stages.

matter, which allows evaluating the reuse of solid waste in the production of biofertilizers (Filer et al., 2019). Currently, there are mathematical and computer models to predict methane production from any substrate; for example, the Buswell model and novel process simulation model (PSM). The first method predicts the amount of methane, carbon dioxide, and ammonia concerning the elemental chemical composition of the substrate by balanced redox equations. Simultaneously, the PSM divides the phases of hydrolysis, acidogenesis, acetogenesis, and methanogenesis into two separate groups of reaction sets. The first set contains the hydrolysis phase based on the fractional conversion in a range of 0.0–1.0 of the reactants to products, and the second set is made up of the remaining phases, which can be simulated as first-order reactions (Rajendran et al., 2014; Cárdenas Cleves et al., 2016).

11.2 Raw Materials Applied in Biogas Production

Biomass is a versatile type of feedstock to use as a substrate in the AD process. The composition of this raw material is the main reason to define whether or not the process can reach different yields. The content of carbohydrates, cellulose, proteins, fats, and hemicellulose are crucial biomass values. The quality of the biogas obtained, and the methane yield, will depend on the substrate composition, digestion system, and retention time (Weiland, 2010). Biogas yields can be varied according to the content of the components previously mentioned. Table 11.1 shows the variation of the CO_2 and CH_4 yields based on the substrate. Fats directly affect the process retention time due to the lower accessibility. In contrast, the use of carbohydrates and proteins allows short conversion times; however, it achieves low biogas production yields. Additionally, lignin compounds are not desirable in these processes because they could affect the accessibility to smaller molecular weight components and, consequently, decrease the yields (Weiland, 2010).

Several raw materials have been studied to obtain biogas; mainly animal manure and domestic waste, and co-substrates from crops. These residues can be fruit peels, waste fruits, or grass residues, stems, or leaves (Chynoweth et al., 2001). These feedstocks offer different components that significantly benefit methane performance. They can be used either as sole or mix raw material in the biogas production processes. Many authors have carried out anaerobic co-digestion experiments. These combinations of raw materials are proposed to improve biogas production yields, the potential of each raw material, supplement nutrients, or supply the scales required to carry out the process. For example, the AD of a mixture of vegetable and fruit residues with cow manure was studied (Deressa et al., 2015). As a byproduct of the AD, digestate can be applied as a rich nitrogen and potassium (fertilizer) source in these crops, and ensures the sustainability of the process. Table 11.2 shows different types of raw materials used in anaerobic digestion processes, and biogas production yields.

Biodegradability is a factor related to the anaerobic digestion process. This term is associated with the amount of TS and volatile solids (VS) consumed in the process, and their content is substrate dependent. Therefore, the biodegradability determination is a factor susceptible to evaluation and analysis to compare the most suitable raw material for AD (Solarte Toro et al., 2017). Table 11.3 shows different compositions of TS and VS of several raw materials and the biodegradability degree.

TABLE 11.1

Theoretical Yields of CH_4 and CO_2, According to the Component Present in the Raw Material

Component	Methane Yield	Carbon Dioxide Yield
Carbohydrates	50	50
Fats	68	32
Protein	70	30
Cellulose and hemicellulose	50	42
Lignin	0	0

Source: Weiland, 2010.

TABLE 11.2

Raw Materials and Biogas Yields Reported in Literature

Raw Material	Biogas Yield L/kg VS	Reference
Banana peel	227.1	(Zheng et al., 2013)
Orange peel	276.8	
Soybean	443.9	
Potato	336.6	
Bamboo branch	65.3	
Food waste	450	(Solarte Toro et al., 2017)
Grass residues	340	
Sugarcane	230	(Chynoweth et al., 2001)
Poplar	230	
Pine chips	143	(Shafiei et al., 2014)
Eucalyptus	134	(Nakamura and Mtui, 2003)
Maize	650	(Weiland, 2010)
Wheat	700	
Sunflower	540	

TABLE 11.3

Biodegradability Index of Different Biomass

Raw Material	TS%	VS%	Biodegradability (%)	Reference
Plantain peel	95.80	53.10	55.43	(Joseph Igwe, 2014)
Banana pseudostem	5.53	4.47	80.83	(Li et al., 2016)
Potato stem	92.63	90.06	97.23	(Martínez-Ruano et al., 2019)
Milk whey	5.69	5.38	94.55	
Wheat straw	89.01	74.73	83.96	(Sun et al., 2018)
Olive cake	90.10	85.14	94.50	(Sarker, 2020)
Pumpkin stem	15.11	12.68	83.92	(Mama et al., 2019)

As mentioned previously, biomass is highly versatile. This quality provides benefits in obtaining biogas through anaerobic digestion. However, the diverse composition of biomass makes it a strong candidate susceptible to the transformation toward obtaining high-value–added products (biochemicals, biomaterials, pharmaceutical products, food products, and bioenergy). The most widely applied system for these purposes is the biorefinery.

11.3 Stand-Alone Production of Biogas Using Different Raw Materials

11.3.1 Chemical Characterization

Raw material characterization is crucial for understanding the feasibility of substrates to be converted into products. The advantages and limitations of the materials can be seen through their chemical composition. Table 11.4 illustrates the lignocellulosic characterization of plantain peel (PP), the mixture of leaves and stem of marigold (MLS), and eucalyptus chips (EC). The high carbohydrate

TABLE 11.4

Physico-Chemical Characterization of Plantain Peel, Marigold Residues, and Wood Chips

Parameter	Weight Percentage (% wt.) Dry Basis		
	Plantain Peel (Piedrahita-Rodriguez., 2020)	**Stem and Leaves of Marigold (Poveda G., 2019)**	**Wood Chips (Poveda G., 2019)**
Moisture content	87.2 ± 0.1	73.2 ± 0.7	54.7 ± 0.4
Lignocellulosic			
Cellulose	14.4 ± 0.7	20.8 ± 0.6	40.7 ± 0.2
Hemicellulose	12.6 ± 1.7	14.8 ± 1.3	18.7 ± 0.4
Total lignin	**9.7 ± 0.4**	**24.6 ± 1.5**	**36.6 ± 1.0**
Insoluble	N.R	13.9 ± 1.5	31.5 ± 1.0
Soluble	N.R	10.6 ± 0.1	5.1 ± 0.1
Extractives	41.3 ± 1.1	29.1 ± 1.2	3.5 ± 0.1
Ash	14.6 ± 1.1	10.7 ± 0.2	0.6 ± 0.1
Solids content			
Total solids	**13.2 ± 0.3**	**99.5 ± 0.2**	**96.1 ± 0.1**
Volatile solids	**10.6 ± 0.2**	**86.2 ± 0.3**	**95.6 ± 0.2**

N.R: not reported.

content for all substrates can be observed, which can be valorized into energy products through biochemical conversions. Wood chips show the highest content of carbon source, becoming optimal for biogas production. Nevertheless, it is also the raw material with the highest lignin content, limiting its application in bioconversion processes. The recalcitrance of the material is a barrier to the substrate accessibility. The biodegradability index (ratio of volatile solids to total solids) for PP, MLS, and EC are 80.3%, 86.6%, and 99.5%, respectively. Therefore, wood proves to be the best substrate for biogas production, however, it is limited by the hardness of the lignin content. Thus, PP seems to be an optimal substrate for biogas production, as it has the lowest amount of lignin, limited only by the ash content. Ashes are inorganic mineral salts non-degradable by microorganisms, whose content can restrict the AD (Lauka et al., 2015). Therefore, the ideal substrates for biogas production should have the lowest amount of lignin and ash content. Another possible limitation of PP is its extract content. Some researchers have reported that the extracts hinder AD's proper performance. The removal of extracts with dioxane has increased conversion by 60% (Oi et al., 1981). Other authors have reported that their removal negatively affects the biogas yield, decreasing the conversion from 49.5% to 7.1% (Lane, 1983). To better understand the relationship between composition and biogas production, this chapter will refer mainly to three raw materials as examples: plantain peel (PP), stem and leaves of marigold (MLS), and wood chips (EC).

11.3.2 Biochemical Methane Potential (BMP)

Technological advances in AD have enabled its application in various waste treatment schemes. These advances have allowed both minimization of the organic load of effluents and production of biofuel to supply the industrial or household sector energy demand. Different AD's of lignocellulosic waste are discussed in this section, including biomass from industrial and crop processing stages. Therefore, the performance of the BMP test for PP, MLS, and EC residues will be reviewed. The limited information on the comparative anaerobic digestion of agroindustry wastes (e.g., food, forestry, and aromatic plant processing) allowed the development of this section. It will be useful for the design and assessment of industrial-scale digesters.

11.3.3 Anaerobic Digestion of Plantain Peel

Piedrahita-Rodriguez (2020) performed anaerobic digestions of mashed PP (*Musa Paradisiaca AAB Simonds*) at 37°C in digesters of 100 mL capacity. An anaerobic sludge from a UASB reactor was used as inoculum. The characterization of the substrate in terms of solids content is presented in Table 11.4. The authors obtained maximum biogas yields of 959.3 mL/gVS (83.3% of the theoretical) and methane yields of 360.9 mL/gVS (53.6% of the theoretical), with an average of the specific biogas rate of 62.2 mL gVS^{-1}d^{-1} and a maximum methane composition of 62.1% (vol.) during 25 days of digestion. Figure 11.3 depicts the cumulative yield of anaerobic digestion for PP. Pisutpaisal and colleagues (2014) analyzed the biogas productivity at different ground banana peel loads in terms of TS (2.5%–10.0% w/v) using a microbial seed from a UASB inoculum. The authors found that the highest productivity (451 mL/gVS) was achieved by performing a substrate load of 10%. These results show differences compared to the work of Piedrahita-Rodriguez (2020) due to the higher amount of VS in the raw material, which leads to more significant biodegradation of the material. Makinde and Odokuma (2015) carried out different anaerobic digestion schemes at 37°C from the banana peel, using cow dung as inoculum. The researchers found that the maximum biogas production (456.8 mL/gVS) was achieved for a feed rate of 25% plantain peels and 75% cow dung in 150 mL of water. This difference in productivity compared to Piedrahita-Rodriguez (2020) can be explained by the type of inoculum used. The innoculum type plays a crucial role in biogas production, since it provides the initial microbial population for digestions. It has been reported that manure has a lower production rate and productivity of biogas than a slurry adapted to anaerobic digestion for a period below 30 days (Rajput and Sheikh, 2019). Likewise, other factors affecting productivity are the inoculum preparation, pre-incubation, degassing, and the amount and type of water added during the collection and digestion of the sludge (Møller et al., 2004; Raposo et al., 2012).

11.3.4 Anaerobic Digestion of Marigold Residues

Poveda G. (2019) performed an AD for the stem and leaves of marigold (*Calendula officinalis*) pretreated with 8% NaOH at 130°C for 60 min. This alkaline pretreatment removed 4.1, 63.5, and 68.7% of the cellulose, hemicellulose, and lignin fractions, respectively, leading to a material loss of 35.3%. Furthermore,

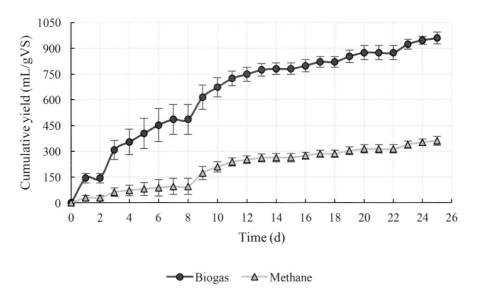

FIGURE 11.3 Experimental results of anaerobic digestions of plantain peel.

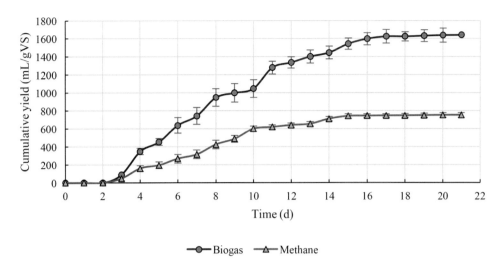

FIGURE 11.4 Experimental results of anaerobic digestions of marigold residues.

the content of VS and TS varied compared with the initial characterization, providing the solids content illustrated in Table 11.4. After pretreatment, the remaining solids were used as a substrate in anaerobic digestion conducted at 37°C during 21 days of digestion in 100 mL glass flasks. Figure 11.4 presents the accumulated yield during digestion time. The author obtained maximum biogas yields of 1637.3 mL/gVS (70.4% of the theoretical) and 751.5 mL of methane/gVS (55.9% of the theoretical). The maximum composition of methane was 64.2% (vol.), and the average of the specific biogas rate was 97.3 mL gVS^{-1}d^{-1}. Swapnavahini and colleagues (2010) studied the potential for biogas production from basil leaves (*Ocimum sanctum*) in digesters of 1.5 L at 27°C–30°C, using cow manure as an active inoculum. The authors succeed in a biogas production yield of 87.7 mL/gVS during 30 days of digestion. This low productivity can be explained because no pretreatment of the raw material was performed to improve biodegradation accessibility. Drying, mechanical, thermal, and chemical pretreatments are essential for AD (Jin et al., 2014). Kulkarni and Ghanegaonkar (2019) carried out AD of floral waste consisting of aster flowers, marigold flowers, and basil leaves, which were pretreated at different concentrations of three types of alkaline agents (NaOH, Na_2CO_3, and $NaHCO_3$) at 60°C for 24 hours. These ADs were analyzed at two temperature conditions during 31 days of digestion: heating by solar irradiation and controlled temperature of 30°C. The authors determined that the best biogas productivity (149.3 mL/g$_{total}$) was obtained by pretreating the marigold flowers with 8% and 10.66% NaOH at a controlled temperature of 30°C during digestion. The results show a lower biogas production yield than those presented by Poveda G. (2019) (423 mL/g$_{total}$), explained by the operating conditions (e.g., temperature, alkali concentration, reaction time) of the alkaline pretreatment. By decreasing the temperature in alkaline pretreatments, the carbohydrates removal is lower, being favorable for bioconversion in digestion. However, the removal of inhibitory compounds, such as the lignin fraction, is also low (Lehto and Alén, 2013). Flower residues offer a faster rate of biogas production per substrate load than plant residues (Ranjitha et al., 2014). Nevertheless, aromatic plants possess bactericidal and phytochemical characteristics, given to some compounds such as linalool, scopoletin, α-cadinol, and rosmarinic acid, among others, which inhibit the metabolism of several bacteria in the AD process (Dasgupta et al., 2012; Efstratiou et al., 2012).

11.3.5 Anaerobic Digestion of Wood Chips

Poveda G. (2019) evaluated anaerobic digestion of groundwood chips (*Eucalyptus grandis*) pretreated with 8% NaOH at 130°C for 60 minutes. During this pretreatment, removal of 58.6%, 5.1%, and 55.7%, for hemicellulose, cellulose, and lignin was obtained, respectively. Poveda G. (2019) performed AD of

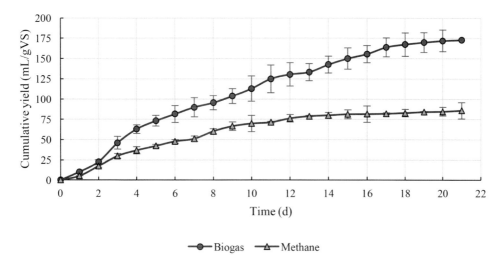

FIGURE 11.5 Experimental results of anaerobic digestions of wood chips.

pretreated wood at mesophilic conditions of 37°C for 21 days, using an anaerobic sludge from the coffee industry. Figure 11.5 illustrates the accumulated yield of anaerobic digestion of wood chips. As main results, the maximum biogas and methane yields were 174.9 mL/gVS (9.7% of the theoretical) and 85.9 mL/gVS (7.7% of the theoretical), respectively. The average of the specific rate was 11.2 mL $gVS^{-1}d^{-1}$, and the maximum methane composition was 62.9% (vol.). Li and colleagues (2019) evaluated the AD performance from wood chips (*Eucalyptus* spp.) pretreated with 2% NaOH for four hours at 90°C. For the AD, pig manure was used as inoculum for 49 days at 35°C. The methane yield for wood chips without pretreatment was 234.8 mL/gVS. In contrast, after pretreatment, the yield increased by 31.6% (309.1 mL/gVS). Significant differences can be observed in comparison to the work done by Poveda G. (2019), explained by the concentrations of the cellulose, hemicellulose, and lignin fractions after pretreatment. Li and colleagues (2019) report a higher concentration of C6 and C5 carbohydrates (41.2 and 50.9% wt., respectively) and a lower concentration of lignin (3.3% wt.). Therefore, there is a more significant carbon source for the metabolism of anaerobic bacteria, and a lower resistance of the substrate to be biodegraded due to recalcitrance of lignin. Adaptation of the inoculum to the substrate, retention time, and temperature are also essential to achieve optimal yields (Corro et al., 2013). Salehian and colleagues (2013) studied the effect of alkali pretreatment times for pine wood biogas productivity. The pretreatment was carried out using NaOH (8% w/w) and temperatures of 0°C and 100°C. Furthermore, the AD was carried out at 37°C for 45 days using an anaerobic bioreactor sludge. The authors reported that the methane yield of wood without pretreatment was 65 mL/gVS. Regarding biogas productivity after pretreatment, the methane yield was higher at shorter alkaline digestion times and temperatures of 100°C, improving by 181% (175 mL/gVS). In contrast, methane yields increased by 118% for high alkaline pretreatment times at 0°C, reaching values of 140 mL/gVS. The higher yields reported by Salehian and colleagues (2013) compared to Poveda G. (2019) can also be explained by the high carbon source content of the substrate. Although the pretreatment with NaOH increases the material porosity through the lignin removal, facilitating its bioconversion, the temperature increase produces the fractionation of the hemicellulose into soluble oligosaccharides, the decomposition of the dissolved polysaccharides, and the formation of stable alkaline end groups (detachment reactions) that negatively affect the bioconversion yield (Gupta and Tuohy, 2014; Lehto and Alén, 2013; Mirahmadi et al., 2010).

11.3.6 Comparison of Anaerobic Digestions

Figure 11.6 shows the daily biogas production for PP, MSL, and EC waste. It can be seen that the productivity of the MSL started after the third day, indicating a stage of hydrolysis. In contrast, the PP

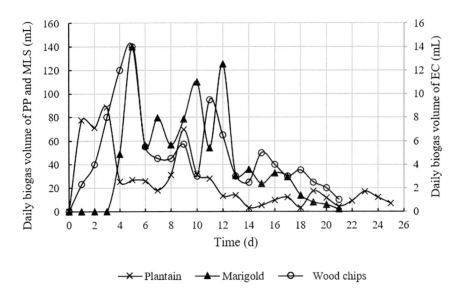

FIGURE 11.6 Daily biogas production of plantain peel, marigold residues, and wood chips.

and EC start the biogas production from day one, although at a much lower rate for PP. Even though the alkaline pretreatment increases the biodegradability of the material through the cleavage of lignin-hemicellulose bonds, much of the cellulose fraction remains intact. The amorphous composition of cellulose is affected by alkaline pretreatment (Kumar et al., 2009). It has been reported that after alkaline pretreatment, cellulose Type I can swell and recrystallize into cellulose Type II due to the formation of alkali-cellulose complexes, being more susceptible to biological conversions after water rinse (Gupta and Tuohy, 2014). However, the crystallinity index of MLS may be higher than the EC, preventing its easy bioconversion. Some authors have reported that the hydrolysis stage of agricultural residues usually limits the overall degradation rate (Azman et al., 2015; Liu et al., 2017). After these days, the biogas productivity for MLS increased due to the crystallinity breakage exerted by the cellulose, increasing the specific surface area for microbial degradation (Angelidaki and Ahring, 2000). In Figure 11.6, it is possible to observe that during the first five days of digestion, the highest rates of biogas production were obtained, being 77.6 mL/d during days 1–3 for the PP, 12.3 mL/d during days 3–5 for the MSL, and 2.6 mL/d during days 1–5 for the PP. These high rates at the beginning of AD can be explained by the conversion rates of volatile solids and nutrient degradation due to rapid microbial growth.

Following the first five days of digestion, the rate of biogas production decreased for all substrates, explained by two possible facts: (a) most of the easily degradable compounds, such as soluble carbohydrates, were consumed, and thereafter, an increase in productivity was observed due to the consumption of more complex compounds, reflected in the new peaks of productivity during days 7–9, 6–12 and 9–11 for PP, MLS and EC, respectively (Figure 11.6); or (b) partial inhibition by undesirable compounds. During the initial stage of AD, the accumulation of inhibitory volatile fatty acids (VFAs), such as propionic, acetic, and butyric acids, among others (produced during hydrolysis and acidogenesis), is promoted, leading to a decrease in the pH of the system and preventing the growth of methanogenic bacteria (Wainaina et al., 2019). Simple and soluble sugars, oligosaccharides of hemicellulose and pectin, are rapidly hydrolyzed and converted into methanogenic intermediates (mainly VFAs) which partially inhibit biogas productivity (Chanakya et al., 1995). Keeping the pH between 5.5 and 6.5 is optimal for the hydrolysis and acidogenesis stage, while a neutral pH between 6.8 and 7.2 is suitable for the methanogenic activity (methanogenesis) (Lemmer et al., 2017; Li et al., 2019). Gram-negative microorganisms are less vulnerable to VFAs inhibition than gram-positive and methanogenic microorganisms (Salminen and Rintala, 2002). In addition to the microbial type, the inoculum shape in the digester also influences the inhibition, since suspended and flocculent sludge is more vulnerable to granular sludge due to its greater specific surface area (Hwu et al., 1996). In order to increase biogas productivity and avoid this methanogenic inhibition, high inoculum-to-substrate ratios are suggested (Parawira et al., 2004). Some authors have concluded that the degradation of long-chain VFAs is the limiting step during digestion due to the low growth rate of VFAs consumer bacteria, and because the decomposition of these compounds is performed at partial hydrogen pressures (Salminen and Rintala, 2002). To prevent inhibition by VFAs, bentonite has been studied for its flocculant capacity and for the formation of precipitates that increase surface tension (Angelidaki et al., 1990). Likewise, some substances such as starch, albumin, and bile acids reduce the VFAs toxicity due to the production of complex compounds or competitive adsorption in the microbial cell for its metabolism (Salminen and Rintala, 2002).

Besides the VFAs, ammonia may also have been the cause of the partial inhibition presented after the first days of digestion (Figure 11.6). Ammonia can be inhibitory if it exceeds a specific concentration (around 0.1–2 g-N/L) that varies according to the digestion conditions and the substrate type (Parawira et al., 2004; Salminen and Rintala, 2002). It has been reported that at concentrations above the threshold level of 1.7 g-N/L, ammoniacal nitrogen has a stronger effect on acetoclastic methanogenic bacteria than on hydrogenic methanogenic bacteria (Koster and Lettinga, 1984). During AD, the balance between the production of acetoclastic methanogenic bacteria and hydrogenic bacteria is crucial. The latter helps to preserve a low hydrogen concentration to ensure a favorable thermodynamic conversion of propionic acid to hydrogen and acetic acid (Fujishima et al., 2000). Unionized ammonia (NH_3-N), produced during protein consumption, inhibits the metabolism of methanogenic bacteria due to its easy diffusion (unlike ammonia) through the cell membranes (Kadam and Boone, 1996). Furthermore, some studies have found that sulfur concentrations of 23 mg S^2/L can contribute to increased ammonia inhibition (Salminen and Rintala, 2002). Ammonia accumulation has also been found to inhibit the glycolytic pathway via glucose degradation (Fujishima et al., 2000). Other researchers have reported that similar to VFAs, ammonia inhibition is partial since methane production occurs after an adaptation period, as a result of the growth of new methanogenic bacteria rather than metabolic changes (Angelidaki and Ahring, 1993). After days 10, 13, and 15, the rates of biogas production for PP, MLS, and EC tend to decrease, explained by the possible depletion of carbon source and/or a low amount of essential bacterial nutrients for the development of AD. The amount of biogas that can be produced in an AD is proportional to the quantity and quality (referred to the chemical structure of the substrate) of the VS. Therefore, the volumetric production of methane will help to estimate the volumetric size of the digester and, thereby, the facility of digestion through a global economy.

Several treatment methods for lignocellulosic materials have been investigated to improve biodegradability (Abraham et al., 2020). The main objective is to increase the substrate accessibility to bacteria by decreasing the association of lignin with the degradable part of the biofibers. Therefore, the efficient use of lignocellulosic material requires that carbohydrates must be accessible to the hydrolytic enzymes produced by the bacteria that are capable of degrading cellulose and hemicellulose (Angelidaki and Ahring, 2000). These pretreatments have also improved the initial C/N conditions that are essential for the performance of AD. For both wood and non-wood materials, alkaline pretreatment has succeeded in decreasing the C/N levels to the optimal range of 25–35 (Li et al., 2019). Nevertheless, a low rate of biogas production for EC compared to other substrates was observed in Figure 11.6. This can be explained by the possible low bacterial growth rates due to the new conditions of the substrate and its inhibitory components, such as the lignin recalcitrance and the formation of stable alkaline end groups. Likewise, the formation of sodium ions (Na^+) during pretreatment has been reported as an inhibition factor (Abraham et al., 2020). The recalcitrance of lignocellulosic samples has been attributed to several factors, such as crystalline structures of the cellulose that are difficult to degrade, binding and content of lignin, composition and cross-linking of hemicellulose, and the overall porosity of the structure (Jin et al., 2014; Pengyu et al., 2017). Therefore, a small surface area is available where bacteria can attack the fibers and rupture their structure easily. Under these stress conditions, a bacterial adaptation could have occurred and resulted in a stationary mutation phase, preventing high rates of biogas production (Corro et al., 2013).

11.3.7 Simulation of Biogas Production

Process simulation is a highly studied topic. It is continuously growing in terms of new ways of simulating, new models, and new equations that allow representing the physical, chemical, biochemical, and thermochemical phenomena that are required. Regarding AD, several advances have been reached. Currently, there is a wide range of models that vary according to the required purposes. For example, simple mathematical models or those that consider the degradation kinetics of the substrate are highlighted. The main objective of these models is to establish the applicability of a certain substrate in the AD process and to be able to make decisions in this regard (Baquerizo C. et al., 2016). To predict the methane yield, there is the Buswell equation, which is based on the composition of the raw material. With stoichiometry and knowing the composition of the substrate, it is possible to determine the amount of methane. The Buswell equation (Equation 1) represents the transformation of organic matter toward the main products of anaerobic digestion (methane, carbon dioxide, and ammonia). This model does not consider cell synthesis, and the methane yield is given by Equation 2 (Baquerizo C. et al., 2016):

$$C_aH_bO_cN_d + \left(\frac{4a-b-2c+3d}{4}\right)H_2O \rightarrow \left(\frac{4a+b-2c-3d}{8}\right)CH_4 + \left(\frac{4a-b+2c+3d}{8}\right)CO_2 + dNH_3 \quad (1)$$

$$Y_{CH_4} = \frac{(4a+b-2c-3d) \times 22.4}{(12a+b+16c+14d) \times 8} \quad (2)$$

where a, b, c, and d represent the stoichiometric coefficients of the raw material susceptible to degradation.

The Anaerobic Digestion Model 1 (ADM1) was the first biogas production prediction model and is currently the most widely used and studied. It was developed by the International Water Association (IWA) in 2002 (Batstone et al., 2002). This model involves the degradation kinetics of carbohydrates, lipids, and proteins, and considers the hydrolysis of these molecules toward components such as sugars and amino acids, among others. In total, the model includes five stages (see Figure 11.7), with their respective products. The parameters used can be the feed concentration, substrate flow, temperature, pH, and the inhibition factor that can affect the process (Satpathy et al., 2013).

A more rigorous model found for AD simulation is the model developed by Rajendran and colleagues (2014). The process simulation model (PSM) for anaerobic digestion consists of two stages: (a) hydrolysis based on the extent of the reaction, being one of the limitations of the process; and (b) acidogenesis, acetogenesis, and methanogenesis reactions, based on kinetic information. The parameters of this model were taken from previous studies and models. Figure 11.8 shows the scheme of the model. The hydrolysis stage includes carbohydrates (cellulose and hemicellulose), proteins, and fats, and the other stages involve different organic acids and glycerol, among other components (Rajendran et al., 2014).

Multiple process simulators are available to model chemical or biochemical systems. Aspen Plus stands out as a tool that applies rigorous and comprehensive thermodynamic methods. In this sense, Aspen Plus is a powerful tool for modeling anaerobic digestion. The previously mentioned models have been applied in this software to simulate biogas production processes (Daza Serna et al., 2016; Martínez-Ruano et al., 2019, 2018; Ortiz-Sanchez et al., 2020a).

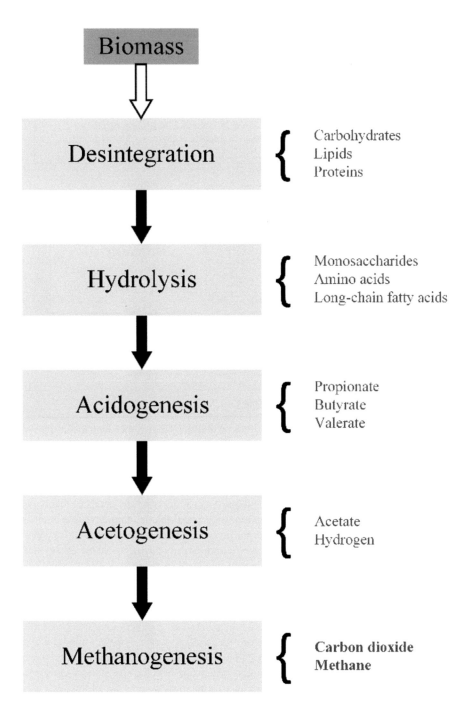

FIGURE 11.7 Stages involve in ADM1 biogas prediction model. (From Batstone et al., 2002.)

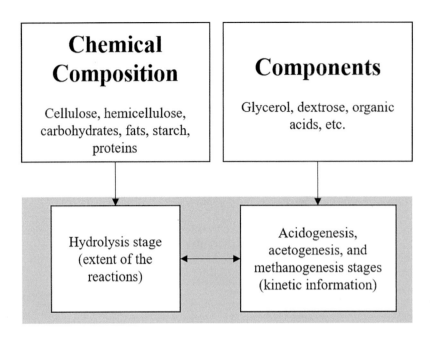

FIGURE 11.8 Process simulation model scheme. (Adapted from Rajendran et al. 2014.)

11.4 Biogas Integrated in Biorefineries

Biorefineries have been defined by several authors involving terms such as biomass, value-added products, energy vectors, and processes. For instance, biorefineries are complex systems where biomass is integrally processed or fractionated to obtain more than one product, including bioenergy, biofuels, chemicals, and high value–added compounds that only can be extracted from biobased sources (Aristizábal et al., 2016). Murthy (2019) has defined a biorefinery as a facility or cluster of facilities for processing biobased feedstocks into valuable products addressing the needs of diverse markets for fuels, feed, plastics, and other commodity chemicals in a sustainable manner. These authors highlight the potential of biomass as a source of different products. Most of the biorefineries reported in the open literature involve the production of bioenergy in the form of directly usable energy (i.e., heat and power) and energy vectors (e.g., syngas, hydrogen, methane, and biogas). The production of energy vectors in a biorefinery system is considered a crucial stage, since these products can supply a share of the utilities of the biomass upgrading process. Thus, the production of biogas into a biorefinery system can offer energy savings, if this process is correctly implemented.

Biogas as an energy vector can be used as fuel in generation, cogeneration, or trigeneration systems. Nevertheless, biogas must be upgraded before use, since the presence of water vapor and hydrogen sulfide can damage internal combustion engines (ICEs). Generation systems are only focused on the production of power using a gas engine. Cogeneration systems involve the simultaneous production of heat and power through the use of a gas turbine. Finally, trigeneration systems are addressed to produce heat, cold, and power in the same process. Most of the biogas plants implemented in the European Union (EU) are addressed to obtain both heat and power in cogeneration systems. Substrates used in these plants include lignocellulosic biomass, agricultural residues, and sludge from wastewater treatment plants. The stand-alone production of biogas seems to be an economically feasible process. Even so, most biogas production facilities have financial support through taxes.

The implementation of biogas in biorefineries does not overcome the conceptual design stage, since most biomass upgrading plants require the fractionation of the biomass components. In this way, the potential of different biomass components should be taken into consideration. For instance, the potential biogas production from starch, cellulose, hemicellulose, lignin, fats, and oils has been

TABLE 11.5

Biorefineries Reported in the Open Literature Involving the Production of Biogas

Raw Material	Biorefinery Products	Biogas Source	Biogas Use	Ref.
Food waste	Polyhydroxyalkanoates (PHA) Biogas	Food waste Activated sludge	Product	(Moretto et al., 2020)
Coffee-cut stems	Bioethanol Biogas	Exhausted solid from enzymatic hydrolysis	Product	(Aristizábal-Marulanda et al., 2020)
Marigold	Polyphenolic compounds Biogas	Stems and leaves	Product	(Poveda-Giraldo and Cardona, 2020)
Orange peel waste	Essential oil Pectin Biogas	Remaining solid fraction after pectin extraction	Product	(Ortiz-Sanchez et al., 2020b)
Municipal solid waste	Single cell protein Biomethane Biofertilizer	Municipal solid waste	Upgrading to biomethane	(Khoshnevisan et al., 2020)
Sugarcane	Bioethanol Electricity	Molasses	Electricity	(Fonseca et al., 2020)

reported in the open literature to set the maximum production yield basis. The biorefineries reported in the literature choose the C5 and lignin platforms to produce biogas. Meanwhile, the C6 platform is used to produce other value-added products such as organic acids, alcohols, and food additives. Other options related to the implementation of biogas in biorefineries involves the production of methane and products derived from this platform. Table 11.5 presents the biogas production integrated into several biorefineries reported in the open literature.

Biogas is a versatile product to be implemented into a biorefinery. This product can be obtained using several raw materials. Moreover, this product can be upgraded in a portfolio of products, including electricity and biomethane. Three different biorefineries related to marigold residues, wood chips, and plantain peel are explained to observe the effect of the biogas production on the technical and economic performance of the biorefinery.

11.4.1 Process Description

11.4.1.1 Biorefinery of Marigold Residues and Wood Chips

The biorefinery design for the processing of EC and MLS was proposed by Poveda (2019), as shown in Figure 11.9a,b. Both biorefineries are focused on the production of biomethane, vanillin, and vanillic acid. Therefore, pretreatment, reaction, and separation technologies are identical, varying in the sizing to process 58.9 tons/day of wood chips and 33.9 tons/day of calendula waste. Both biorefinery schemes assumed that the raw material was dried by solar radiation up to a moisture content of approximately 10%. Afterwards, they were ground to a particle size of less than 5 mm, and pretreated with 8% NaOH at 130°C. After alkali pretreatment, the solid (cellulose-rich) and liquid (black liquor) fractions were separated by filtration. The pH of the solid fraction was neutralized prior to anaerobic digestion at 37°C. The biogas was pressurized at 10 bar and purified through a high-pressure scrubbing column at 20°C, as reported by Cozma and colleagues (2014). Biomethane was obtained from the top of the column at a volume composition of 97.7%. Meanwhile, the liquor fraction was concentrated and oxidized at 120°C in a high-pressure reactor of 10 bar. Vanillin and vanillic acid were separated through a packed adsorption column using a non-polar resin and water according to the methodology presented by Gomes and colleagues (2018). Each phenolic compound was purified in an extraction column considering benzene as a solvent. Thereafter, vanillin and vanillic acid were produced in a composition of 95.5% after crystallization.

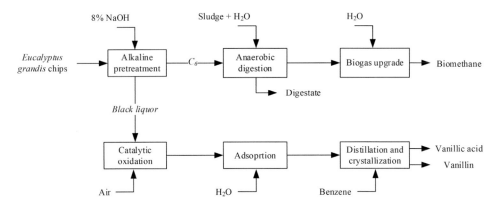

FIGURE 11.9A Block diagram of biorefinery involving biogas production: wood chips biorefinery.

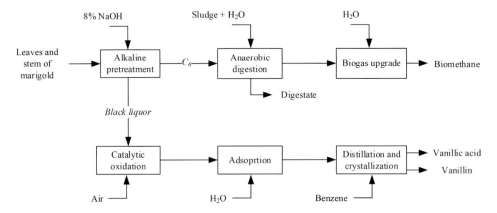

FIGURE 11.9B Block diagram of biorefinery involving biogas production: marigold biorefinery.

11.4.1.2 Biorefinery of Plantain Peel

Piedrahita-Rodriguez (2020) have proposed the scheme of biorefinery to produce biogas and ethanol from 288 tons/day of PP (Figure 11.10). The raw material was dried by a convective process until a moisture content of 10%. Subsequently, it was milled and pretreated with 5% H_2SO_4, based on the research reported by Carrasco and Roy (1992). The solid fraction was hydrolyzed by enzymes at 50°C and fermented with *Saccharomyces cerevisiae* at 32°C (Kadam et al., 2004; Moncada et al., 2016). Thereafter, ethanol was separated by distillations columns, and purified by adsorption columns packed with molecular sieves. In contrast, the liquid fraction, rich in xylose, was mixed with the remnant solid after enzymatic hydrolysis. The pH was neutralized with lime before fermentation with anaerobic bacteria. For the anaerobic digestion, the Buswell model was used to estimate the methane and carbon dioxide production (Baquerizo C. et al., 2016). Biogas, at volumetric concentrations of 50%–60%, was stored for sale. In contrast, the digestate was solar-dried until moisture removal up to 85% and used as a marketable byproduct.

11.4.2 Comparison of Biorefinery Results

The proposed biorefineries can be compared considering technical and economic aspects. Technical aspects involve analyzing the mass indicators such as product yield and process mass intensity (PMI). In addition, the economic analysis involves the calculation of cash flow indicators such as net present value (NPV). The results of the estimated mass indicators of the biorefineries are presented in Table 11.6.

The biogas yield of the proposed biorefineries allows to elucidate the potential of biogas production in biorefinery systems. Indeed, the production of vanillin and vanillic acid is the core of the proposed

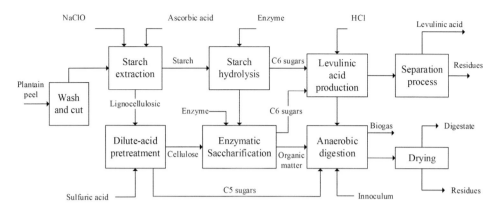

FIGURE 11.10 Block diagram of the plantain peel–based biorefinery.

TABLE 11.6
Mass Indicators of the Proposed Biorefineries

Product	Marigold Biorefinery		Wood Biorefinery		Plantain Peel Biorefinery	
	Yield (%)	PMI	Yield (%)	PMI	Yield (%)	PMI
Biogas	N.A.	54.22	N.A.	37.89	0.0025	72.23
Biomethane	10.9		2.27		N.A.	
Digestate	28.8		68.1		19.90	
Vanillin	0.48		0.47		N.A.	
Vanillic acid	0.11		0.21		N.A.	
Levulinic acid	N.A.		N.A.		0.005	

N.A.: not applicable.

processes. These products are derived from the lignin fraction of the raw materials (i.e., marigold and woodchips). Nevertheless, a high share of the raw materials is wasted, since the cellulosic fraction is not a source of the desired products. Thus, biogas appears to be a feasible option to valorize the C6 fraction of the raw materials, since the production of other value-added products using this fraction is not always feasible and possible. Reasons to avoid the production of other value-added products using the C6 fraction can be related to: (a) the stakeholders of the process not having an established market to commercialize other products; (b) the project not having more funds to implement a new processing line; (c) implementation of a new processing line could decrease the environmental performance of the process; or (d) the production of an energy vector could help to reduce operational expenditures of the process. Therefore, the production of biogas and biomethane was proposed. The implementation of this processing line (C6 → biogas → biomethane) allows to valorization of a high mass flow of waste to an energy vector to be easily commercialized (i.e., biogas). If this product is not considered a source of income in the entire process, biogas can be used directly as an energy source or upgraded to produce more energy.

The yield of biomethane in the marigold-based and wood chips–based biorefineries (i.e., 10.9% and 2.27%, respectively) is higher than the yield of the main products in both processes. This behavior is desired in a biorefinery system due to the biomethane production being justified. High volumes of an energy vector can produce more energy to supply the process requirements. Thus, the production of biogas and biomethane seems to be a good alternative. Nevertheless, the yields of the marigold and woodchips biorefineries allow elucidating the following statement: the biogas production in a biorefinery system is feasible from a technical perspective as long as the raw material to be upgraded does not have inhibitors, as well as the amount of biogas produced being sufficiently higher to supply more than

50% of the energy demand of the process. This statement is correct due to low volumes of biogas with low energy content (i.e., methane concentrations lower than 50%) being unable to decrease the entire biorefinery's operational expenditures. On the contrary, the production of biogas without accomplishing the previous statement increases capital expenditures and decreases the biorefinery's economic performance. In the same way, the biogas upgrading to biomethane, since the low amount of biomethane produced could not validate the capital investment of the process. In the case of the plantain peel–based biorefinery, the yields of both products are lower. These low yields demonstrate the low suitability of the raw material for production of the desired products. Thus, the process is not technically feasible. Meanwhile, the relatively low yields of the marigold-based and wood chips–based biorefineries could be compensated by the high value of the products.

Mass yields are important indicators to analyze the amount of product(s) obtained using any biomass source as raw material. Most of the biomass upgrading processes have relatively low yields, since the conversion of biomass components to any specific product could have some limitations related to biological systems (e.g., fermentation) or availability of the desired component to be upgraded (e.g., lignin). Even so, low yields suppose using higher amounts of raw materials to reach economic feasibility in most cases. The other important yield of the proposed biorefineries is related to the digestate. It can be used as a fertilizer since the nitrogen–phosphorous–potassium ratio of this stream is useful for some crops. As shown in Table 11.6, the digestate yield is higher than the biogas/biomethane yield in all biorefineries.

Another important indicator to establish the technical feasibility of a process or biorefinery is the process mass intensity (PMI) (Budzinski et al., 2019). This indicator relates to the amount of raw materials to produce 1 kg of products. The values of the PMI in the proposed biorefineries are presented in Table 11.6. A high value of the PMI indicator makes the biorefinery less feasible from a technical perspective. In this way, the plantain peel biorefinery is the least feasible process. Meanwhile, the woodchips biorefinery is the most feasible among the three studied processes. The PMI index is subjected to process conditions and technologies. Thus, the proposed raw materials could be upgraded to other products. A PMI value to be used as a reference in biorefinery systems has not been established in the open literature, since low processes have been implemented at the industrial scale. Nevertheless, these values are higher than the typical PMI values recommended to chemical industries (i.e., 2–6) dedicated to produce bulk chemicals even though these values are higher than those recommended to produce fine chemicals (i.e., 6–30). In this sense, the proposed biorefineries could be improved through a mass integration or a technological change.

Other indicators associated with green chemistry, such as the mass of hazardous materials input, toxic release intensity, and specific solid waste, can be calculated. These indicators would improve in all cases if the production of biogas is included in the biorefinery. The production of biogas can increase the technical performance of the process, considering the mass and energy balances. As previously stated, biogas production decreases the amount of organic waste released to the environment (either liquid or solid wastes). Nevertheless, the use of waste streams should be studied, since the composition of these streams could include inhibitor components (e.g., furfural) (Ortiz-Sanchez et al., 2020b).

The economic assessment is another important study required to recognize the feasibility of the proposed biorefineries. This analysis gives preliminary results to elucidate the economic feasibility of the biorefinery. The base case of the proposed biorefineries is given in Table 11.7. The cost distribution of the proposed biorefineries is presented in Figure 11.9. The capital depreciation is the essential factor in the cost distribution. This value is superior due to the high amount of equipment involved in the biorefineries. A high depreciation value implies a high total capital investment in the process. In this way, the wood chips biorefinery has the highest value of the capital investments. Meanwhile, the plantain peel biorefinery has lower total capital investment due to lower production capacity. The raw material costs are also a representative cost in biorefineries. Most of the biorefineries reported in the open literature have a high share of raw materials costs. This high cost of raw materials is attributed to the number of reagents used in the process. As in the case of the capital depreciation costs, the wood chips biorefinery's raw material costs are higher. Nevertheless, this cost is attributed to the high amount of raw materials used in the process. The cost distribution presented in Figure 11.11 is similar to other shares reported in the open literature (Martínez-Ruano et al., 2018).

TABLE 11.7

Mass Flow Used to Perform the Simulations of the Proposed Biorefineries

Biorefinery	Mass Flow of Raw Materials (kg/h)
Marigold	1,412
Wood chips	2,450
Plantain peel	12

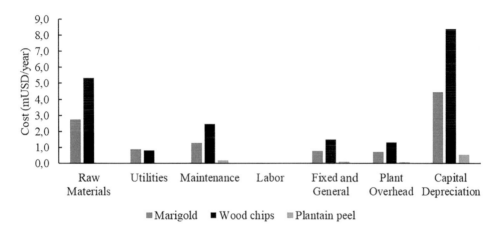

FIGURE 11.11 Cost distribution of the three proposed biorefineries.

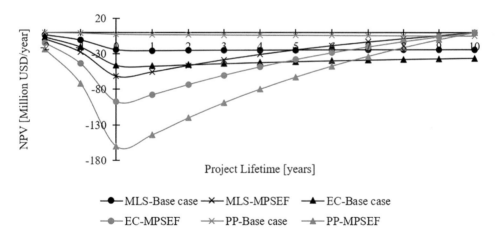

FIGURE 11.12 Scale analysis of the proposed biorefineries.

The proposed biorefineries do not have a positive economic performance since the net present value (NPV) was not positive. For this reason, a scale analysis is necessary. Thus, the minimum processing scale for economic feasibility (MPSEF) is a scale such that NPV = 0. This scale is the minimum required to reach economic feasibility. Figure 11.12 depicts the base case and the minimum processing scale of each biorefinery. Higher values of the MPSEF gives positive values of NPV. Thus, the process is feasible from an economic point of view. Nevertheless, the scale analysis should be restricted to avoid finding an unfeasible scale from a technical perspective (i.e., scales higher than national production). If biogas is commercialized as a product of the biorefinery, the scale to reach economic feasibility can decrease. Even so, a comparison between the savings of the process when biogas is used to supply a share of the energy demand of the process and the profit of the biorefinery if biogas is commercialized must be carried out. In this way, biogas has the potential to increase the economic performance of a process. Thus, this energy vector should be the first option to be implemented in a biorefinery system if organic residues are present.

11.5 Conclusion

This chapter demonstrates the potential for the production of biogas from lignocellulosic waste. Through the BMP assays of PP, MLS, and EC, it was possible to understand the factors that affected the performance of AD. It was concluded that the possible parameters that affected the rate of biogas production were related to the toxicity of VFA and ammonia concentrations. Likewise, stable alkaline compounds and sodium ions might negatively affect process yields. Although alkaline pretreatment allows the lignin fraction removal and increases the substrate porosity, prolonged times at high temperatures lead to removing hemicellulose oligosaccharides. Therefore, lower biodegradation yields could be expected since the carbon source is decreased. Additionally, the recalcitrance, rigidity, and crystallinity of the remaining material hinder the performance of AD. Therefore, higher biogas yields for PP than EC were demonstrated. Regarding the analysis of the biorefineries, biogas production helps to improve the development of medium- and high-scale processes by considering both technical and economic aspects. However, this is achieved when there is no investment availability to other value-added products, the biogas generated has a specific market, or the biogas can supply 50% of the energy demand of the process.

Acknowledgments

The authors express their gratitude for the financial support given by the Fondo de Ciencia, Tecnología e Innovación del Sistema General de Regalías, Fondo Nacional de Financiamiento para la Ciencia, la Tecnología y la Innovación "Francisco José de Caldas," Ministerio de Ciencia, Tecnología e Innovación-MINCIENCIAS, Programa Colombia BIO, Gobernación de Boyacá, through FP44842–298–2018 contract, Vicerrectoría de Investigación, Ciencia y Tecnología-Universidad de Boyacá. This chapter is the result of the research work developed through the program: Programa de investigación Reconstrucción del Tejido Social en Zonas de Pos-Conflicto en Colombia SIGP Code: 57579 within the research project "Competencias Empresariales y de Innovación Para el Desarrollo Económico y La Inclusión Productiva de Las Regiones Afectadas Por el Conflicto Colombiano" SIGP Code: 58907. Funded within the framework of Colombia Científica, Contract No FP44842-213-2018. Finally, this work was supported by the project entitled "Driving the Development of Biosurfactants through Their Systematic Life Cycle" from Minciencias contract number 80740-903-2020.

REFERENCES

Abraham, A., Mathew, A.K., Park, H., Choi, O., Sindhu, R., Parameswaran, B., Pandey, A., Park, J.H., Sang, B.I., 2020. Pretreatment strategies for enhanced biogas production from lignocellulosic biomass. *Bioresour. Technol.* 301, 122725. https://doi.org/10.1016/j.biortech.2019.122725

Angelidaki, I., Ahring, B.K., 1993. Thermophilic anaerobic digestion of livestock waste: The effect of ammonia. *Appl. Microbiol. Biotechnol.* 38, 560–564.

Angelidaki, I., Ahring, B.K., 2000. Methods for increasing the biogas potential from the recalcitrant organic matter contained in manure. *Water Sci. Technol.* 41, 189–194. https://doi.org/10.2166/wst.2000.0071

Angelidaki, I., Petersen, S.P., Ahring, B.K., 1990. Effects of lipids on thermophilic anaerobic digestion and reduction of lipid inhibition upon addition of bentonite. *Appl. Microbiol. Biotechnol.* 33, 469–472. https://doi.org/10.1007/BF00176668

Aristizábal, V., Moncada, J., Cardona, C.A., 2016. Design strategies for sustainable biorefineries. *Biochem. Eng. J.* 116, 122–134. https://doi.org/10.1016/j.bej.2016.06.009

Aristizábal-Marulanda, V., Solarte-Toro, J.C., Cardona, C.A., 2020. Study of biorefineries based on experimental data: Production of bioethanol, biogas, syngas, and electricity using coffee-cut stems as raw material. *Environ. Sci. Pollut. Res.* https://doi.org/10.1007/s11356-020-09804-y

Azman, S., Khadem, A.F., Van Lier, J.B., Zeeman, G., Plugge, C.M., 2015. Presence and role of anaerobic hydrolytic microbes in conversion of lignocellulosic biomass for biogas production. *Crit. Rev. Environ. Sci. Technol.* 45, 2523–2564. https://doi.org/10.1080/10643389.2015.1053727

Baquerizo C., R.J., Pagés D., J., Pereda R., I., 2016. El modelo de Buswell. Aplicación y comparación. Principales factores que influyen en su aplicación. *Rev. virtual pro* 168, 1–23.

Batstone, D.J., Keller, J., Angelidaki, I., Kalyuzhnyi, S. V, Pavlostathis, S.G., Rozzi, A., 2002. The IWA Anaerobic Digestion Model No 1 (ADM1). *Water Sci. Technol.* 45, 65–73.

Budzinski, K., Blewis, M., Dahlin, P., D'Aquila, D., Esparza, J., Gavin, J., Ho, S.V., Hutchens, C., Kahn, D., Koenig, S.G., Kottmeier, R., Millard, J., Snyder, M., Stanard, B., Sun, L., 2019. Introduction of a process mass intensity metric for biologics. *N. Biotechnol.* 49, 37–42. https://doi.org/10.1016/j.nbt.2018.07.005

Cárdenas Cleves, L.M., Parra Orobio, B.A., Torres Lozada, P., Vásquez Franco, C.H., 2016. Perspectivas del ensayo de Potencial Bioquímico de Metano—PBM para el control del proceso de digestión anaerobia de residuos. *Rev. ION* 29, 95–108. https://doi.org/10.18273/revion.v29n1-2016008

Carrasco, F., Roy, C., 1992. Kinetic study of dilute-acid prehydrolysis of xylan-containing biomass. *Wood Sci. Technol.* 26, 189–208. https://doi.org/10.1007/BF00224292

Chanakya, H.N., Ganguli, N.K., Anand, V., Jagadish, K.S., 1995. Performance characteristics of a solid-phase biogas fermentor. *Energy Sustain. Dev.* 1, 43–46. https://doi.org/10.1016/S0973-0826(08)60100-3

Chynoweth, D.P., Owens, J.M., Legrand, R., 2001. Renewable methane from anaerobic digestion of biomass. *Renew. Energy* 22, 1–8. https://doi.org/10.1016/S0960-1481(00)00019-7

Corro, G., Pal, U., Bañuelos, F., Rosas, M., 2013. Generation of biogas from coffee-pulp and cow-dung co-digestion: Infrared studies of postcombustion emissions. *Energy Convers. Manag.* 74, 471–481. https://doi.org/10.1016/j.enconman.2013.07.017

Cozma, P., Wukovits, W., Mámáligá, I., Friedl, A., Gavrilescu, M., 2014. Modeling and simulation of high pressure water scrubbing technology applied for biogas upgrading. *Clean Technol. Environ. Policy* 17. https://doi.org/10.1007/s10098-014-0787-7

Dasgupta, N., Ranjan, S., Saha, P., Jain, R., Malhotra, S., Arabi Mohamed Saleh, M.A., 2012. Antibacterial activity of leaf extract of Mexican marigold (*Tagetes erecta*) against different gram positive and gram negative bacterial strains. *J. Pharm. Res.* 5, 4201–4203.

Daza Serna, L.V., Solarte Toro, J.C., Serna Loaiza, S., Chacón Perez, Y., Cardona Alzate, C.A., 2016. Agricultural waste management through energy producing biorefineries: The Colombian case. *Waste Biomass Valorization* 7, 789–798. https://doi.org/10.1007/s12649-016-9576-3

Deressa, L., Libsu, S., Chavan, R.B., Manaye, D., Dabassa, A., 2015. Production of biogas from fruit and vegetable wastes mixed with different wastes. *Environ. Ecol. Res.* 3, 65–71. https://doi.org/10.13189/eer.2015.030303

Deublein, D., Steinhauser, A., 2010. *Biogas from Waste and Renewable Resources: An Introduction* (2nd Revised and Expanded Edition). Weinheim: Wiley.

Efstratiou, E., Hussain, A.I., Nigam, P.S., Moore, J.E., Ayub, M.A., Rao, J.R., 2012. Antimicrobial activity of *Calendula officinalis* petal extracts against fungi, as well as Gram-negative and Gram-positive clinical pathogens. *Complement. Ther. Clin. Pract.* 18, 173–176. https://doi.org/10.1016/j.ctcp.2012.02.003

EurObserv'ER consortium, 2019. *The State of Renewable Energies in Europe*. Available at: www.eurobserv-er.org.

Filer, J., Ding, H.H., Chang, S., 2019. Biochemical Methane Potential (BMP) assay method for anaerobic digestion research. *Water* 11, 921. https://doi.org/10.3390/w11050921

Fonseca, G.C., Costa, C.B.B., Cruz, A.J.G., 2020. Economic analysis of a second-generation ethanol and electricity biorefinery using superstructural optimization. *Energy* 204, 117988. https://doi.org/10.1016/j.energy.2020.117988

Fujishima, S., Miyahara, T., Noike, T., 2000. Effect of moisture content on anaerobic digestion of dewatered sludge: Ammonia inhibition to carbohydrate removal and methane production. *Water Sci. Technol.* 41, 119–127. https://doi.org/10.2166/wst.2000.0063

Gielen, D., Boshell, F., Saygin, D., Bazilian, M.D., Wagner, N., Gorini, R., 2019. The role of renewable energy in the global energy transformation. *Energy Strateg. Rev.* 24, 38–50. https://doi.org/10.1016/j.esr.2019.01.006

Gomes, E.D., Mota, M.I., Rodrigues, A.E., 2018. Fractionation of acids, ketones and aldehydes from alkaline lignin oxidation solution with SP700 resin. *Sep. Purif. Technol.* 194, 256–264. https://doi.org/10.1016/j.seppur.2017.11.050

Gupta, V.K., Tuohy, M.G., 2014. Progress in physical and chemical pretreatment of lignocellulosic biomass. In Gupta, V. (Ed.), *Biofuel Technologies: Recent Developments* (pp. 53–96). Berlin and Heidelberg: Springer. https://doi.org/10.1007/978-3-642-34519-7

Hilkiah Igoni, A., Ayotamuno, M.J., Eze, C.L., Ogaji, S.O.T., Probert, S.D., 2008. Designs of anaerobic digesters for producing biogas from municipal solid-waste. *Appl. Energy* 85, 430–438. https://doi.org/10.1016/j.apenergy.2007.07.013

Hwu, C.S., Donlon, B., Lettinga, G., 1996. Comparative toxicity of long-chain fatty acid to anaerobic sludges from various origins. *Water Sci. Technol.* 34, 351–358. https://doi.org/10.1016/0273-1223(96)00665-8

Jin, G., Bierma, T., Walker, P.M., 2014. Low-heat, mild alkaline pretreatment of switchgrass for anaerobic digestion. *J. Environ. Sci. Heal.—Part A Toxic/Hazardous Subst. Environ. Eng.* 49, 565–574. https://doi.org/10.1080/10934529.2014.859453

Joseph Igwe, N., 2014. Production of biogas from plantain peels and swine droppings. *IOSR J. Pharm. Biol. Sci.* 9, 50–60. https://doi.org/10.9790/3008-09555060

Kadam, K.L., Rydholm, E.C., McMillan, J.D., 2004. Development and validation of a kinetic model for enzymatic saccharification of lignocellulosic biomass. *Biotechnol. Prog.* 20, 698–705. https://doi.org/10.1021/bp034316x

Kadam, P.C., Boone, D.R., 1996. Influence of pH on ammonia accumulation and toxicity in halophilic, methylotrophic methanogens. *Appl. Environ. Microbiol.* 62, 4486–4492. https://doi.org/10.1128/aem.62.12.4486-4492.1996

Khoshnevisan, B., Tabatabaei, M., Tsapekos, P., Rafiee, S., Aghbashlo, M., Lindeneg, S., Angelidaki, I., 2020. Environmental life cycle assessment of different biorefinery platforms valorizing municipal solid waste to bioenergy, microbial protein, lactic and succinic acid. *Renew. Sustain. Energy Rev.* 117, 109493. https://doi.org/10.1016/j.rser.2019.109493

Koster, I.W., Lettinga, G., 1984. The influence of ammonium-nitrogen on the specific activity of pelletized methanogenic sludge. *Agric. Wastes* 9, 205–216. https://doi.org/10.1016/0141-4607(84)90080-5

Kulkarni, M.B., Ghanegaonkar, P.M., 2019. Biogas generation from floral waste using different techniques. *Glob. J. Environ. Sci. Manag.* 5, 17–30. https://doi.org/10.22034/gjesm.2019.01.02

Kumar, P., Barrett, D.M., Delwiche, M.J., Stroeve, P., 2009. Methods for pretreatment of lignocellulosic biomass for efficient hydrolysis and biofuel production. *Ind. Eng. Chem. Res.* 48, 3713–3729. https://doi.org/10.1021/ie801542g

Lane, A.G., 1983. Anaerobic digestion of spent coffee grounds. *Biomass* 3, 247–268.

Lauka, D., Pastare, L., Blumberga, D., Romagnoli, F., 2015. Preliminary analysis of anaerobic digestion process using cerathophyllum demersum and low carbon content additives: A batch test study. *Energy Procedia* 72, 142–147. https://doi.org/10.1016/j.egypro.2015.06.020

Lehto, J., Alén, R., 2013. Alkaline pre-treatment of hardwood chips prior to delignification. *J. Wood Chem. Technol.* 33, 77–91. https://doi.org/10.1080/02773813.2012.748077

Lelieveld, J., Klingmüller, K., Pozzer, A., Burnett, R.T., Haines, A., Ramanathan, V., 2019. Effects of fossil fuel and total anthropogenic emission removal on public health and climate. *Proc. Natl. Acad. Sci. U. S. A.* 116, 7192–7197. https://doi.org/10.1073/pnas.1819989116

Lemmer, A., Merkle, W., Baer, K., Graf, F., 2017. Effects of high-pressure anaerobic digestion up to 30 bar on pH-value, production kinetics and specific methane yield. *Energy* 138, 659–667. https://doi.org/10.1016/j.energy.2017.07.095

Li, C., Liu, G., Nges, I.A., Deng, L., Nistor, M., Liu, J., 2016. Fresh banana pseudo-stems as a tropical lignocellulosic feedstock for methane production. *Energy. Sustain. Soc.* 6, 27. https://doi.org/10.1186/s13705-016-0093-9

Li, J., Kumar Jha, A., He, J., Ban, Q., Chang, S., Wang, P., 2011. Assessment of the effects of dry anaerobic co-digestion of cow dung with waste water sludge on biogas yield and biodegradability. *Int. J. Phys. Sci.* 6, 3723–3732. https://doi.org/10.5897/IJPS11.753

Li, R., Tan, W., Zhao, X., Dang, Q., Song, Q., Xi, B., Zhang, X., 2019. Evaluation on the methane production potential of wood waste pretreated with NaOH and co-digested with pig manure. *Catalysts* 9, 539. https://doi.org/10.3390/catal9060539

Liu, T., Sun, L., Müller, B., Schnürer, A., 2017. Importance of inoculum source and initial community structure for biogas production from agricultural substrates. *Bioresour. Technol.* 245, 768–777. https://doi.org/10.1016/j.biortech.2017.08.213

Makinde, O., Odokuma, L., 2015. Comparative study of the biogas potential of plantain and yam peels. *Br. J. Appl. Sci. Technol.* 9, 354–359. https://doi.org/10.9734/bjast/2015/18135

Mama, C.N., Grace, A.C., Chukwu, C.I., Hephzibah, M.C., 2019. Evaluation of biogas production from the digestion of swine dung, plantain peel and fluted pumpkin stem. Int. *J. Civil Mech. Energy Sci.* 5, 14–23. https://doi.org/10.22161/ijcmes.5.2.3

Martínez-Ruano, J.A., Caballero-Galván, A.S., Restrepo-Serna, D.L., Cardona, C.A., 2018. Techno-economic and environmental assessment of biogas production from banana peel (*Musa paradisiaca*) in a biorefinery concept. *Environ. Sci. Pollut. Res*. 25, 35971–35980. https://doi.org/10.1007/s11356-018-1848-y

Martínez-Ruano, J.A., Restrepo-Serna, D.L., Carmona-Garcia, E., Giraldo, J.A.P., Aroca, G., Cardona, C.A., 2019. Effect of co-digestion of milk-whey and potato stem on heat and power generation using biogas as an energy vector: Techno-economic assessment. *Appl. Energy* 241, 504–518. https://doi.org/10.1016/j.apenergy.2019.03.005

Mes, T.Z.D. de, Stams, A., Reith, J.H., Zeeman, G., 2003. Methane production by anaerobic digestion of wastewater and solid wastes. In *Bio-Methane & Bio-Hydrogen: Status and Perspectives of Biological Methane and Hydrogen Production* (pp. 58–102). Petten: Dutch Biological Hydrogen Foundation

Mirahmadi, K., Kabir, M.M., Jeihanipour, A., Karimi, K., Taherzadeh, M.J., 2010. Alkaline pretreatment of spruce and birch to improve bioethanol and biogas production. *BioResources* 5, 928–938.

Møller, H.B., Sommer, S.G., Ahring, B.K., 2004. Methane productivity of manure, straw and solid fractions of manure. *Biomass Bioenergy* 26, 485–495. https://doi.org/10.1016/j.biombioe.2003.08.008

Moncada, J., Cardona, C.A., Higuita, J.C., Vélez, J.J., López-Suarez, F.E., 2016. Wood residue (*Pinus patula* bark) as an alternative feedstock for producing ethanol and furfural in Colombia: Experimental, techno-economic and environmental assessments. *Chem. Eng. Sci*. 140, 309–318. https://doi.org/10.1016/j.ces.2015.10.027

Moretto, G., Russo, I., Bolzonella, D., Pavan, P., Majone, M., Valentino, F., 2020. An urban biorefinery for food waste and biological sludge conversion into polyhydroxyalkanoates and biogas. *Water Res*. 170, 115371. https://doi.org/10.1016/j.watres.2019.115371

Murthy, G., 2019. Systems analysis frameworks for biorefineries. In *Biofuels: Alternative Feedstocks and Conversion Processes for the Production of Liquid and Gaseous Biofuels* (pp. 77–92). London: Academic Press.

Nakamura, Y., Mtui, G., 2003. Anaerobic fermentation of woody biomass treated by various methods. *Biotechnol. Bioprocess Eng*. 8, 179–182.

Ofoefule, A.U., Uzodinma, E.O., Onukwuli, O.D., 2009. Comparative study of the effect of different pretreatment methods on biogas yield from water Hyacinth (*Eichhornia crassipes*). *Int. J. Phys. Sci*. 4, 535–539. https://doi.org/10.5897/IJPS.9000125

Oi, S., Tanaka, T., Yamamoto, T., 1981. Methane fermentation of coffee grounds and some factors to improve the fermentation. *Agric. Biol. Chem*. 45, 871–878. https://doi.org/10.1080/00021369.1981.10864634

Ortiz-Sanchez, Mariana, Solarte-Toro, J.-C., González-Aguirre, J.-A., Peltonen, K.E., Richard, P., Cardona-Alzate, C.A., 2020a. Pre-feasibility analysis of the production of mucic acid from Orange Peel Waste under the biorefinery concept. *Biochem. Eng. J*. 161, 107680. https://doi.org/10.1016/j.bej.2020.107680

Ortiz-Sanchez, M., Solarte-Toro, J.C., Orrego-Alzate, C.E., Acosta-Medina, C.D., Cardona-Alzate, C.A., 2020b. Integral use of orange peel waste through the biorefinery concept: An experimental, technical, energy, and economic assessment. *Biomass Convers. Biorefinery* 1–15. https://doi.org/10.1007/s13399-020-00627-y

Parawira, W., Murto, M., Zvauya, R., Mattiasson, B., 2004. Anaerobic batch digestion of solid potato waste alone and in combination with sugar beet leaves. *Renew. Energy* 29, 1811–1823. https://doi.org/10.1016/j.renene.2004.02.005

Pengyu, D., Lianhua, L., Feng, Z., Xiaoying, K., Yongming, S., Yi, Z., 2017. Comparison of dry and wet milling pretreatment methods for improving the anaerobic digestion performance of the Pennisetum hybrid. *RSC Adv*. 7, 12610–12619. https://doi.org/10.1039/c6ra27822a

Piedrahita-Rodriguez, S. 2020. *Design of Biorefineries with Multiple Raw Materials for the Use of Agroindustrial Waste in Post-Conflict Zones in Colombia* (Master thesis). Universidad Nacional de Colombia sede Manizales, Colombia.

Pisutpaisal, N., Boonyawanich, S., Saowaluck, H., 2014. Feasibility of biomethane production from banana peel. *Energy Procedia* 50, 782–788. https://doi.org/10.1016/j.egypro.2014.06.096

Poveda G., J.A., 2019. *Biorefineries Based on Catalytic Conversion of Biomass for the Production of Phenolic Compounds from Eucalyptus grandis and Calendula officinalis Residues* (Master thesis). Universidad Nacional de Colombia sede Manizales, Colombia.

Poveda-Giraldo, J.A., Cardona, C.A., 2020. Biorefinery potential of *Eucalyptus grandis* to produce phenolic compounds and biogas. *Can. J. For. Res*. 1–49. https://doi.org/10.1139/cjfr-2020-0201

Rajendran, K., Kankanala, H.R., Lundin, M., Taherzadeh, M.J., 2014. A novel process simulation model (PSM) for anaerobic digestion using Aspen Plus. *Bioresour. Technol.* 168, 7–13. https://doi.org/10.1016/j.biortech.2014.01.051

Rajput, A.A., Sheikh, Z., 2019. Effect of inoculum type and organic loading on biogas production of sunflower meal and wheat straw. *Sustain. Environ. Res.* 1, 1–10. https://doi.org/10.1186/s42834-019-0003-x

Ranjitha, J., Vijayalakshmi, S., Vijaya kumar, P., Nitin Ralph, P., 2014. Production production of bio-gas from flowers and vegetable wastes using anaerobic digestionof bio-gas from flowers and vegetable wastes using anaerobic digestion. *Int. J. Res. Eng. Technol.* 3, 279–283. https://doi.org/10.15623/ijret.2014.0308044

Raposo, F., De La Rubia, M.A., Fernández-Cegrí, V., Borja, R., 2012. Anaerobic digestion of solid organic substrates in batch mode: An overview relating to methane yields and experimental procedures. *Renew. Sustain. Energy Rev.* 16, 861–877. https://doi.org/10.1016/j.rser.2011.09.008

Rocamora, I., Wagland, S.T., Villa, R., Simpson, E.W., Fernández, O., Bajón-Fernández, Y., 2020. Dry anaerobic digestion of organic waste: A review of operational parameters and their impact on process performance. *Bioresour. Technol.* https://doi.org/10.1016/j.biortech.2019.122681

Salehian, P., Karimi, K., Zilouei, H., Jeihanipour, A., 2013. Improvement of biogas production from pine wood by alkali pretreatment. *Fuel* 106, 484–489. https://doi.org/10.1016/j.fuel.2012.12.092

Salminen, E., Rintala, J., 2002. Anaerobic digestion of organic solid poultry slaughterhouse waste—A review. *Bioresour. Technol.* 83, 13–26. https://doi.org/10.1016/S0960-8524(01)00199-7

Sarker, S., 2020. Exploring biogas potential data of cattle manure and olive cake to gain insight into farm and commercial scale production. *Data Br.* 32, 106045. https://doi.org/10.1016/j.dib.2020.106045

Satpathy, P., Steinigeweg, S., Uhlenhut, F., Siefert, E., 2013. Application of Anaerobic Digestion Model 1 (ADM1) for prediction of biogas production. *Int. J. Sci. Eng. Res.* 4, 86–89.

Scarlat, N., Dallemand, J.F., Fahl, F., 2018. Biogas: Developments and perspectives in Europe. *Renew. Energy* 129, 457–472. https://doi.org/10.1016/j.renene.2018.03.006

Shafiei, M., Karimi, K., Zilouei, H., Taherzadeh, M.J., 2014. Enhanced ethanol and biogas production from pinewood by NMMO pretreatment and detailed biomass analysis. *Biomed Res. Int.* 2014, 1–10. https://doi.org/10.1155/2014/469378

Solarte Toro, J.C., Mariscal Moreno, J.P., Aristizábal Zuluaga, B.H., 2017. Evaluación de la digestión y co-digestión anaerobia de residuos de comida y de poda en bioreactores a escala laboratorio. *Rev. ION* 30, 105–116. https://doi.org/10.18273/revion.v30n1-2017008

Sun, Y.-M., Huang, X.-M., Kang, Y.-H., 2018. Effects of different material total solid on biogas production characteristics. *E3S Web Conf.* 38, 02006. https://doi.org/10.1051/e3sconf/20183802006

Swapnavahini, K., Srinivas, T., Kumar, P.L., Kumari, M.S., Lakshmi, T., 2010. Feasibility study of anaerobic digestion of *Ocimum sanctum* leaf waste generated from *Sanctum sanctorum*. *BioResources* 5, 372–388. https://doi.org/10.15376/biores.5.1.389-396

Wainaina, S., Lukitawesa, Kumar Awasthi, M., Taherzadeh, M.J., 2019. Bioengineering of anaerobic digestion for volatile fatty acids, hydrogen or methane production: A critical review. *Bioengineered* 10, 437–458. https://doi.org/10.1080/21655979.2019.1673937

Weiland, P., 2010. Biogas production: Current state and perspectives. *Appl. Microbiol. Biotechnol.* 85, 849–860. https://doi.org/10.1007/s00253-009-2246-7

Wiesenthal, T., Mourelatou, A., Petersen, J.E., Taylor, P., 2006. *How Much Bioenergy Can Europe Produce without Harming the Environment?* EEA Report No. 7.

Zheng, W., Phoungthong, K., Lü, F., Shao, L.-M., He, P.-J., 2013. Evaluation of a classification method for biodegradable solid wastes using anaerobic degradation parameters. *Waste Manag.* 33, 2632–2640. https://doi.org/10.1016/j.wasman.2013.08.015

12 Plant-Based Biofuel and Sustainability Issues

Lakshmi Gopakumar

CONTENTS

12.1	Introduction	269
12.2	Need for Other Sources of Biofuels	270
12.3	Plant-Based Biofuels	270
	12.3.1 First-Generation Plant-Based Biofuels	271
	12.3.2 Second-Generation Plant-Based Biofuels	271
	12.3.3 Third-Generation Biofuels	272
	12.3.4 Fourth-Generation Biofuels	272
12.4	An Overview of Biofuel Production	272
12.5	Biochemical and Thermochemical Conversion for Biofuel Production	272
12.6	Production of Bioethanol and Biodiesel	274
	12.6.1 Production of Bioethanol	274
	12.6.2 Production of Biodiesel	274
	12.6.2.1 *Jatropha curcas*	274
	12.6.2.2 *Pongamia pinnata*	274
	12.6.2.3 Grasses	275
12.7	Advantages and Disadvantages of Plant-Based Biofuels: A Comparison	275
	12.7.1 Advantages of Plant-Based Biofuels	275
	12.7.2 Disadvantages of Plant-Based Biofuels	276
12.8	Sustainability Issues Related to Plant-Based Biofuels	276
	12.8.1 Loss of Biodiversity	277
	12.8.2 Competition with Food Crops	277
	12.8.3 Overexploitation of Natural Resources	277
	12.8.4 Socioeconomic-Related Issues	277
12.9	Advantages of Using Microalgae for Biofuel Production	278
12.10	Use of Nanotechnology for Production of Biofuels	278
	12.10.1 Use of Nanoparticles for Lipid Extraction in Microalgae	279
	12.10.2 Use of Nanocatalysts for Biodiesel Production	279
	12.10.3 Use of Nanoparticles for Enzyme Immobilization	280
12.11	Scope for Improvement of Plant-Based Biofuel Scenarios	280
12.12	Conclusion	281
Acknowledgment		281
References		281

12.1 Introduction

Energy is the one of the most important of the factors which influence and control the socioeconomic growth of the world. Increased energy demand and reduced energy availability is a global issue faced by developing and developed countries (Waqas et al. 2018). The energy shortage is mainly due to limited energy resources and the need for advanced and efficient technologies for energy production, coupled with political and economic reasons. Hence, to assist global development, the wider use of energy-based systems help the world to improve economically and foster human living standards (Walker et al. 2016; Arshad and Ahmed 2015). The global consumption of energy is increasing day by day. The major share of today's energy demand is met by non-renewable resources like coal, mineral oils, and other fossil fuels. Fossil fuels play an important role in meeting energy demands (Pfenninger and Keirstead 2015), contributing to about 80.3% of fuel consumption (Escobar et al. 2009). The World Energy Outlook, 2007 estimates that energy from fossil fuels will meet only about 84% of the energy demand in 2030 (Shafiee and Topal 2009). The extensive use and increased exploitation of fossil fuels result in serious depletion of natural resources. In the near future, depletion of energy sources of animal origin will be a major threat faced by the world in relation to energy accessibility. Apart from this, the high amount of carbon dioxide released from fossil fuels contributes to global warming. To address the increased future demand of fuel energy, there arises the need for plant-based energy sources of various categories. This might also challenge the global community to think about alternative renewable, eco-friendly, clean, and cost-effective energy resources like bioenergy. Therefore, sustainable biofuel production becomes an urgent need to ensure economic development with reduced environmental pollution (Oh et al. 2018; Shields-Menard et al. 2018; Saravanan et al. 2018). To address these issues, biofuel becomes a viable solution as it offers reduced environmental pollution, increased socioeconomic benefits (van Eijck et al. 2014; Creutzig et al. 2015), and controlling the depletion of fuel reservoirs (Smith 2013).

Biofuels are mostly fuels produced from living sources, and they offer a number of advantages over conventional fuels (Table 12.1). While considering plant-based biofuels as an alternative fuel source, there also underlies the problem of natural resource depletion, but trees are a renewable source of energy and hence can be considered as a better option compared to animal-based fuel sources in terms of availability and renewability.

12.2 Need for Other Sources of Biofuels

The need for alternative fuel sources has become very critical while considering the exhaustibility of resources, overexploitation of forests, and increasing population, coupled with the increased standard of living in developing countries. Hence, more focus is needed to be concentrated on other fuel sources, whose production can be improved to satisfy the industrial needs suited to the increased demand. The attractiveness of biofuels lies in the fact that it has many advantages compared to the conventional biofuels.

12.3 Plant-Based Biofuels

Biofuels are fuels produced from biological feedstock (Arshad et al. 2017), which usually offers reduced emissions of greenhouse gases. Plant-based biofuels are the biofuels produced from plant biomass. Hence,

TABLE 12.1

Conventional Fuels and Biofuels: A Comparison

Characteristics	Conventional Fuels	Biofuels
Resource sustainability	Exhaustible	Non-exhaustible (biomass being the main source)
Accessibility of raw materials	Not easily accessible (except coal in Indonesia, Germany, Australia, etc.)	Easily accessible
Production	Factories with good infrastructure are required at all the stages	Factories are required only at the extraction stage; the cultivation and collection of biomass can be done at the micro level by farmers
Pollution prevention	Increases pollution emissions	Reduces pollution (biofuel plants fix atmospheric carbon in their biomass)
Contribution to climate change	Aggravates climate change by contributing to pollution	Reduces climate change by fixing atmospheric carbon

Source: Gopakumar (2020).

TABLE 12.2
Classification of Biofuels Based on Generations

Generation of Biofuels	Source
First-generation biofuels	Edible oil seeds
Second-generation biofuels	Non-edible crops
Third-generation biofuels	Algae/microorganisms
Fourth-generation biofuels	Uses genetically engineered organisms for biofuel production

plant-based biofuels have same properties of biofuels—and in some cases, certain plant-based biofuels have certain advantages over the others. Based on the source of origin, plant-based biofuels are mainly grouped as plant-based primary biofuels and plant-based secondary biofuels. Plant-based primary biofuels includes biofuels from plant/crop residue (Enagi et al. 2018), while plant-based secondary biofuels include fuels produced from plant biomass and microbes (Sekoai et al. 2019). The biofuels are separated into first-, second-, third-, and fourth-generation biofuels, described briefly in Table 12.2. More details on the various generations of biofuels, with examples, are explained in the remainder of this section.

12.3.1 First-Generation Plant-Based Biofuels

First-generation biofuels are produced using on fermentation of starch from edible crops (Sirajunnisa and Surendhiran 2016).

The important crops used for the production of first-generation biofuels are corn, sugarcane, soya bean, wheat, sugar beets, rapeseed, peanuts, and a number of other food crops. Many first-generation biofuels are dependent on subsidies and are not cost-competitive with fossil fuels such as mineral oil. Most of the first-generation biofuels produce only limited greenhouse gas emission savings.

12.3.2 Second-Generation Plant-Based Biofuels

Second-generation biofuels are produced from biomass residues of different crops/plants (Leong et al. 2018). For the last two decades, cellulosic or lignocellulose biomass (Sirajunnisa and Surendhiran 2016) has been used and considered as cheap, renewable resource, eco-friendly and without posing any food security threat (Hong et al. 2014). The production of second-generation biofuels from plant sources is through a different method which includes many steps.

The second-generation biofuels can be categorized into different headings (Table 12.3). From the table, it can be seen that lignocellulose is the common type of feedstock used for commonly known biofuels like synthetic biofuels, bioethanol, and biodiesel.

TABLE 12.3
Classification of Second-Generation Biofuels

Biofuel Group	Name of Biofuel	Feedstock	Process of Production
Bioethanol	Cellulosic ethanol	Lignocellulose	Advanced hydrolysis and fermentation
Synthetic biofuels	Biomass to liquids Fischer-Tropsch biodiesel Synthetic diesel Biomethanol Heavier alcohols Dimethyl ether P series (ethanol + MTHF)	Lignocellulose	Gasification and synthesis
Biodiesel	NexBTL H Bio Green pyrolysis diesel	Vegetable oils, animal fat, lignocellulose	Hydrotreatment
Methane	Biosynthetic natural gas (SNG)	Lignocellulose	Gasification and synthesis
Biohydrogen	Hydrogen	Lignocellulose	Gasification and synthesis or biological process

12.3.3 Third-Generation Biofuels

Third-generation biofuels are the biofuels produced from microalgae (Alaswad et al. 2015; Leong et al. 2018). The potential for production of biofuel from algae is due to the fact that algae produce oil that can be easily converted into gasoline or biodiesel. The feedstock for algal biofuels can involve cyanobacteria, microalgae, and macroalgae. These are produced from lipids and carbohydrates produced in algae (Sirajunnisa and Surendhiran 2016). The advantages of third-generation biofuels over the others is that: (a) they don't cause conflict with food security; and (b) there is no conflict of fuel production on natural resources like land and water. Third-generation biofuels are considered superior due to use of atmospheric carbon dioxide and increased biomass yield per acre compared to first- and second-generation biofuels. Also, the algal culture can be done under variable range of environmental conditions (Khetkorn et al. 2017), and hence can be undertaken globally based on climatic preferences. These also do not cause conflict with food security (Ahmad et al. 2011).

12.3.4 Fourth-Generation Biofuels

Fourth-generation biofuels are produced from genetically engineered plants or animals. In the case of fourth-generation biofuels, biomass crops can act as carbon capturing machines. These crops can absorb carbon dioxide from the atmosphere and lock it in their biomass. This biomass is converted into fuel and gas using the techniques for production of second-generation biofuels. In this method, the carbon dioxide is captured and the greenhouse gas is geosequestered, stored in oil and gas fields, unminable coal seams, or saline aquifers. This can be locked up for thousands of years. The fuels and gases formed as a result of this method is renewable and carbon negative.

12.4 An Overview of Biofuel Production

Biofuel production includes various chemical process like gasification, pyrolysis, and torrefaction. The first step for the production of plant-based biofuels is gasification. In gasification, carbon-based materials are converted into carbon monoxide, and hydrogen. The process occurs with limited supply of oxygen, and produces synthesis gas (syngas) which produces energy/heat. The common feedstock used for the process involves wood, black liquor, brown liquor, and other feedstock. The second step is pyrolysis. Pyrolysis is the process carried out in the absence of oxygen and in the presence of an inert gas like halogen. This facilitates the conversion of the fuel into two products, namely tar and char. The common feedstock generally used for the production of bio-oil is wood and other energy crops. The third chemical process is called torrefaction, a process similar to pyrolysis. This occurs at a lower temperature. Torrefaction converts biomass into an easily storable and transportable form.

The common methods used for biofuel production are outlined in Figure 12.1. The production process for various biofuels implies that the production is not a low-cost method. Steps like extraction, hydrolysis and gasification are the steps which require a variety of chemicals. Additionally, for the production of bioethanol, fermentation is an important process which requires pretreatment and enzyme hydrolysis.

12.5 Biochemical and Thermochemical Conversion for Biofuel Production

There are mainly two routes for the production of biofuels, namely thermochemical production and biochemical production. In thermochemical processes, the biomass is treated using varying levels of

Plant-Based Biofuels and Sustainability Issues

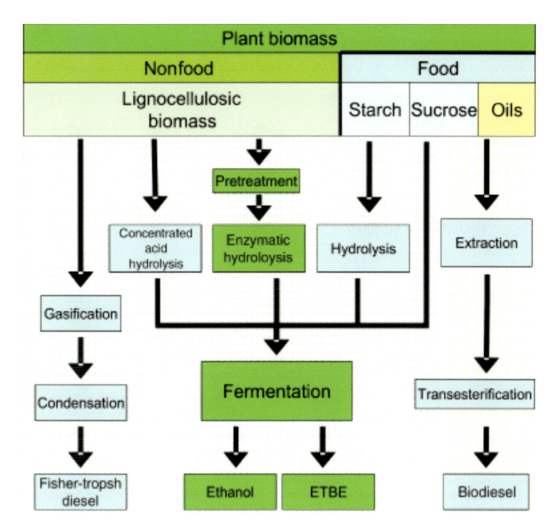

FIGURE 12.1 Production of biofuels from plant biomass. (From Gomez et al. 2008.)

oxygen. Heating in low concentrations of oxygen can result in the formation of hydrogen and other gases which can be liquified and converted into fuels. On the other hand, biochemical treatment involves the conversion of biomass and sugars that can be converted into alcohols. In most cases, ethanol or butanol are produced. Saccharification is an important step in biochemical processing, which includes the conversion of biomass to sugars. The major problem during biochemical processing of plant biofuel feedstocks is the saccharification process. Plant cells have a lignocellulosic cell wall, which is resistant to enzymatic digestion. Hence, it is difficult to release the sugars present in plant cell walls for the production of biofuels. A pretreatment is needed to release the carbohydrates present in plants. Pretreatment can be done using acid digestion or enzymatic processes, but pretreatment using acids is very expensive and pretreatment using enzymes may not be very efficient due to the resistance of the cell walls to various enzymes (Gomez et al. 2008). These methods are mostly used for the production of biofuel from second-generation biofuel feedstock. For the production of biofuels from landfill gas and municipal waste, fermentation for production of ethanol is done with the help of genetically modified bacteria.

12.6 Production of Bioethanol and Biodiesel

Bioethanol and biodiesel are two important plant-based biofuels the production of these two and their details are discussed in what follows.

12.6.1 Production of Bioethanol

Bioethanol is produced by fermentation of sugar. In large-scale production of bioethanol, the steps include fermentation, distillation, dehydration, and denaturing (an optional process). Some crops need saccharification prior to fermentation. Certain crops need saccharification/hydrolysis of carbohydrates (cellulose/starch) into sugars. Starch is converted into sugars by enzymes.

12.6.2 Production of Biodiesel

Biodiesel is commonly produced by transesterification. Transesterification (alcoholysis) is the displacement of an alcohol from an ester by another alcohol in a process similar to hydrolysis. The transesterification reaction is represented by the general equation represented in Figure 12.2, which is the key reaction for biodiesel production.

Transesterification using alkali-catalysis gives high levels of conversion of triglycerides to their corresponding methyl esters in short reaction times. Enzymatic transesterification enables the reuse of the byproduct glycerol in the biodiesel manufacturing process (Fukuda et al. 2001).

There are a number of plants used for the production of second-generation biofuels. The important plants used for biofuel production which have been used as a source of second-generation fuels are the following.

12.6.2.1 Jatropha curcas

Jatropha curcas, a shrub belonging to family Euphorbeaceae, has been extensively popularized as a source of biodiesel. The plant, which is a second-generation biofuel source, has many properties which were highlighted during the time of its popularization as a biofuel source like being farmer-friendly (based on yields, low inputs, and low initial investments) and provision of ecosystem services like soil erosion control, reclamation of soils, use for live fencing, fodder, etc. (Openshaw 2000; Achten et al. 2010). But it was proved that theoretical success was not followed by success at field level. Hence, in many countries, economic production of biofuel from *Jatropha* still seems an unachievable target.

12.6.2.2 Pongamia pinnata

Pongamia pinnata is a leguminous tree which can be used as a source of second-generation biofuels. It is a fast-growing evergreen tree reaching up to 40 meters in height. Like *Jatropha*, this plant was not used extensively for biofuels and hence proved to be a failure.

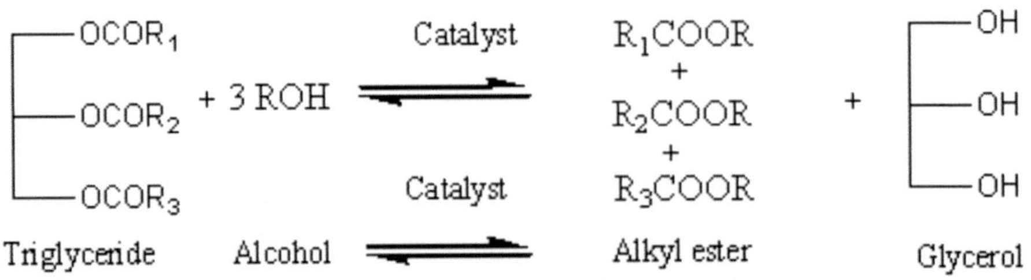

$$\begin{matrix} \text{OCOR}_1 \\ \text{OCOR}_2 \\ \text{OCOR}_3 \end{matrix} + 3\,\text{ROH} \overset{\text{Catalyst}}{\underset{\text{Catalyst}}{\rightleftharpoons}} \begin{matrix} R_1\text{COOR} \\ + \\ R_2\text{COOR} \\ + \\ R_3\text{COOR} \end{matrix} + \begin{matrix} \text{OH} \\ \text{OH} \\ \text{OH} \end{matrix}$$

Triglyceride   Alcohol   Alkyl ester   Glycerol

FIGURE 12.2 Representation of transesterification of triglyceride with alcohol. (From www.pcra.org/English/general/biodiesel.htm.)

12.6.2.3 Grasses

There are different types of grasses—like switchgrass, Indian grass, and miscanthus—which can be used as a source of biofuels. The selection of a particular type of grass depends on the climate and location. For example, in Southeast Asia, miscanthus can be used, while in United States, switchgrass can be used. While considering grasses as the source of biofuels, the following are the advantages:

- Grasses, being perennials, need very less initial investment for planting.
- Grasses are fast growing, which enables several harvests per year.
- Grasses can grow well in marginal lands.
- These plants require very little amounts of fertilizer.
- Grasses have high percentages of energy yield.

Even though grasses have many advantages, the disadvantages of the same cannot be ignored. The disadvantages of grasses involve:

- Grasses cannot be used for biodiesel production.
- Extensive processing is required to convert grasses into ethanol.
- The time period to reach harvest density is long (several years).
- Weeds can create competition with grasses, which in turn can affect grass growth; hence, additional investment will be needed.
- Grasses need moist soil, and are not suited to arid lands.

So, it is proven that grasses cannot offer a viable solution as a future biofuel source. Hence, third-generation plant-based biofuels were introduced.

12.7 Advantages and Disadvantages of Plant-Based Biofuels: A Comparison

While considering the different generations of biofuels, it can be seen that they have many advanatages and disadvanatages. These are discussed in the following section.

12.7.1 Advantages of Plant-Based Biofuels

Biofuels are considered superior to conventional petroleum-based fuels considering their eco-friendliness and production processes. The drawbacks of first-generation biofuels—like increasing food prices, negative impact on food security (Aro 2016), and their contribution to monoculture and deforestation (Altieri 2009)—gave way to development of second-generation biofuels, which generally uses less water and other inputs compared to first-generation biofuels (Carriquiry et al. 2010). The advantage of second-generation biofuels compared to the first-generation fuels is that they will not cause food vs. fuel conflict, unlike first-generation biofuels. By analyzing the various benefits of biofuels over conventional fuels, the following general advantages can be noted.

- Biofuels offer minimal environmental pollution. As there are reduced carbon monoxide emissions, particulate matter, unburned hydrocarbons, and sulfates compared to petroleum diesel fuel, they offer much less negative impact to environment.
- Compared to conventional fuels, biodiesel has 10% built-in oxygen. The result is complete combustion of the fuel.
- The cetane number for biodiesel is usually higher than conventional fuels.
- The extraction procedure and economic inputs needed for the production of biofuels are smaller when compared to fuels derived from petroleum.

- When biodiesel is used as transport fuel, the vehicles do not require engine modification since they are commonly used as blends in different proportions with petrol or diesel.
- Biofuels like biodiesel can provide excellent lubricity to the fuel injection system; hence, lubricating additives need not be added (Gonsalves 2006).
- Since they have low volatility when compared to petroleum, they are safe to handle and can be produced locally.
- Biofuel plantations give scope for better use of wastelands, thereby providing an additional income and employment to a large sector of people.
- Economic improvement: biofuels can be produced locally, which will facilitate the creation of new jobs on a regional basis.
- Improving energy security: biofuels are produced locally, which increase energy self-sufficiency and reduces a nation's dependence on imported energy.

Although these advantages of biofuels seem very promising, biofuels are not free from disadvantages. The important disadvantages include competition with food crops and destruction of natural habitats for biofuel production. When the biofuels are plant based, there is an increased concern over the disadvantages than advantages, as plant protection seems to be an unavoidable part of environmental conservation. The following section examines the disadvantages of plant-based biofuels.

12.7.2 Disadvantages of Plant-Based Biofuels

The main issues associated with first-generation biofuel are high cost of feedstocks, replacement of food crops with bioenergy crops, food scarcity, increase in food prices (Hong et al. 2014; Sirajunnisa and Surendhiran 2016), and agriculture-related problems land like erosion, water contamination, and ecotoxicity (Singh et al. 2011; John et al. 2011).

As crops yielding first-generation biofuels are edible, the costs involved in obtaining the feedstock are very high. The more profits which could be obtained from biofuels may force the people to replace food crops with biofuel plants, leading to "food versus fuel" conflict. This can cause an increase in food price, adding to global food scarcity and increased famine. Additionally, certain agriculture-related problems can also arise. Water contamination due to use of pesticides and fertilizers can happen, which can in turn lead to increased water pollution and toxicity. As fuel production becomes more profit targeted, natural habitats like forest areas will be destroyed for the establishment of large plantations. Another important aspect is the competition of biofuel crops with food crops. According to the International Water Management Institute, about 70 percent of freshwater withdrawn worldwide is used for agricultural purposes (Scheierling and Treguer 2018). Therefore, water scarcity will be a problem emerging with increased biofuel crop cultivation. Biofuels, like biodiesel, cannot be transported in pipelines and can cause inner fuel lines of older vehicles to lose their long-lasting qualities by creating problems like clogging. It attracts moisture when compared with petroleum diesel, and can cause fuel freezing in cold weather. Second-generation biofuels can be considered as a better option compared to first-generation biofuels. But the high costs involved in hydrolysis and low yield of fuels affects the production efficiency of second-generation biofuels (Fu et al. 2010; John et al. 2011). The drawbacks of first- and second-generation biofuels like food security threat, increased agricultural inputs, and social challenges (Sirajunnisa and Surendhiran 2016; Ahmed and Sarkar 2018) forced researchers to evolve new cost-effective and eco-friendly technology from new feedstock for biofuel production.

12.8 Sustainability Issues Related to Plant-Based Biofuels

It is true that plant-based biofuels like second-generation biofuels offer a good scope for production of biofuels from alternative plant sources, but there are still many underlying problems related to the use and implementation of plant-based biofuels. These include environmental related problems and socioeconomic related problems. Brief overviews of these problems are described.

12.8.1 Loss of Biodiversity

The extensive use of land for biofuel cultivation will lead to biodiversity loss. This will happen, as vegetation needs to be cleared for cultivation of biofuels. The removal of certain types of vegetation can lead to loss of the related organisms, causing a reduction of biodiversity. This can include a number of plant species and animals which are beneficial to local ecosystems.

12.8.2 Competition with Food Crops

The preference of biofuel crops over food crops can result in competition of biofuel crops with food crops. As people find biofuel crops more profitable compared to cultivation of food crops, there is a chance for the removal of food crops to increase cultivation of biofuel crops. This can create scarcity of food production and increase the scarcity of food products, which can eventually lead to increased famine.

12.8.3 Overexploitation of Natural Resources

The cultivation of biofuel crops involve use of natural resources like soil and water. Agriculture is an energy-intensive process which requires enough water supply, fertile soil, etc. Hence, the cultivation of biofuel plants can lead to exploitation of natural resources. The need for more biomass for biofuel production can cause increased input of water into fields, which will increase the water scarcity.

12.8.4 Socioeconomic-Related Issues

Apart from the previously discussed issues, there are other socioeconomic-related problems while considering the cultivation of biofuel crops. The socioeconomic issues mostly concentrate in relation to the economic upliftment of people with low economic development and the economic viability of biofuel production. An example of these two issues can be seen from the failure of *Jatropha* plantations. Cultivation of *Jatropha*, which was put forward for the use of marginal lands and biofuel production—which mainly focused on economic upliftment of farmers, along with socioeconomic development through increased returns—failed mostly due to low yields from the crops.

Currently, the price of biofuel products is relatively higher than fossil fuels (Arshad et al. 2017). Feedstock is highly significant for commercial biofuel production, and its availability is dependent on climate, soil, geographical location, soil conditions, agricultural practices (Arshad et al. 2017), yield (Tabatabaei et al. 2015), and oil contents. Considering the plant feedstock, first-generation biofuel sources (edible oil crops) are not a better option compared to the second- and third-generation feedstock. With second-generation feedstock, the main problems include insufficient yields and increased need for inputs and natural resources like water, along with demand for large area for cultivation. Considering these options, third-generation biofuels can be a better option in terms of use of natural resources. These can be considered viable compared to first- and second-generation biofuels because algae require much less area for growth. They can be cultivated in ponds, and require no irrigation or natural resources like soil. The technique can be globally used based on algae available on a regional basis.

Recent advances in the field of biotechnology have enabled scientists/researchers to develop alternative or modern techniques (Nizami and Rehan 2018) to produce biofuels for sustainable energy production systems. Sustainable biofuel energy systems offer to tackle various environmental issues, along with renewable energy production (Rai and Da Silva 2017). However, sustainable biofuel production is dependent on a number of factors including biomass/feedstock pretreatment, process parameters and optimization, reactor designs, product quality and yields, capital costs, public acceptance, and market availability for various biofuels (Nizami and Rehan 2018).

Biofuels like biogas, biodiesel, and biohydrogen are synthesized through different processes using cheap and variable feedstock and considered eco-friendly (Santoro et al. 2017). It is estimated that 90%

of biofuel comprises mainly biogas, biodiesel, and bioethanol (Demirbas and Demirbas 2011). Biodiesel is the most prominent eco-friendly biofuel produced mainly from non-edible oils feedstock (Kirubakaran and Selvan 2017). Biodiesel technology is one of the most growing technologies, with a growth rate of 7.3% per annum and industry worth of 54.8 billion USD by 2025 (Sekoai et al. 2019). Biogas is another widely accepted sustainable biofuel, synthesized through anaerobic digestion by microorganisms (Bundhoo and Mohee 2016) and comprises CH_4, CO_2, and H_2S (Sekoai et al. 2019).

Biogas is produced by anaerobic digestion of organic matter by microbes. Biogas plants produce methane and carbon dioxide from plant biomass of various origin, ranging from organic household waste to industrial waste/bioenergy plants. Apart from that, feedstocks containing sugars, starch, and lignocellulosic biomass are used for commercial bioethanol production through fermentation (Arshad et al. 2017).

Certain issues are there which are associated with all types of biofuels and reflect the need for optimization of new technologies to increase biofuel production efficiency. But compared to first- and second-generation biofuels, third-generation biofuels are beneficial when considering the less number of disadvantages.

12.9 Advantages of Using Microalgae for Biofuel Production

Microalgae can be used as a major source of biofuel production due to their features like fast growth rates and high lipid contents (Sharma et al. 2011). They can be addressed as microalgal biorefineries (Li et al. 2015). Microalgae have high amounts of carbohydrates and fermentable sugars (Nguyen and Van Hanh 2012), and hence can be used for biodiesel and biohydrogen production (Ghirardi et al. 2000; Metzger and Largeau 2005). In addition to biofuels, other economically important byproducts like nutrients, pharmaceuticals, and bioplastics can be obtained from algae. The production of a variety of products from algae is due to the high amounts of lipids, nucleic acids, and proteins and polysaccharides present in the organism compared to higher plant species (Moody et al. 2014; Kim et al. 2016). The following are the various process involved in microalgal biorefinery: cultivation, harvesting, lipid extraction, and conversion (Seo et al. 2017). The algal biorefinery can have many technology limitations which can be overcome by nanoparticle engineering (Lee et al. 2015; Wang et al. 2015). The application of nanoparticle engineering can improve biomass production, lipid extraction, output biodiesel production, and cost effectiveness of the whole process (Seo et al. 2017). The advancement of technology has enabled the application of many modern methods like biotechnology for biofuel production. One of the most sought-out technologies is nanotechnology for the production of biofuels.

12.10 Use of Nanotechnology for Production of Biofuels

Advances in biotechnology give way to use the technology for cost-effective biofuel industry (Sekhon Randhawa and Rahman 2014). This method uses a combination of green chemistry and engineering (Ramsurn and Gupta 2013).

The various processes which use nanotechnology in biofuel production technologies are anaerobic digestion, gasification, hydrogenation, pyrolysis, and transesterification. This will facilitate the production of biogas, fatty esters, and renewable hydrocarbons. The nanotechnology technique uses functional catalysts (Trindade 2011), which have superior features compared to conventional catalysts. The common nanomaterials used include metal oxide nanocatalysts (Liu et al. 2007; Verziu et al. 2008; Gardy et al. 2017), mesoporous nanocatalysts (Yahya et al. 2016), and carbon-based nanocatalysts (Dehkhoda et al. 2010; Guan et al. 2017; Mahto et al. 2018). Nanoparticles can be also used as fuel additives.

The incorporation of nanotechnology will help to increase the biofuel efficiency because of the increased reactivity, smaller size, stability, catalytic activity, and increased absorption capacity (Huang et al. 2010).

Nanoparticle-aided cultivation can be used to improve photosynthetic cell growth of the algae and induction of intracellular lipid accumulation without causing stress kill to the cells. Additional problems like contamination with other microalgae can also be avoided. The main advantages of nanoparticles for use in algal biofuels are as follows:

- Nanoparticles can increase photosynthetic cell growth and induce intracellular accumulation of lipids in algae. This is done without killing cells under stressed conditions (Lee et al. 2015).
- Indirect application of nanoparticles includes the placement of localized surface plasmon resonances (LSPR) in the outer side of closed photobioreactors for the adsorption and scattering of light at specific wavelengths (Pattarkine and Pattarkine 2012).
- There is an increase in light uptake by microalgae with the application of silver nanoparticles (Torkamani et al. 2010) and gold nanoparticles either as single or in combination around photobioreactors (Eroglu et al. 2013).
- The incorporation of nanoparticles into culture mediums for enhancing microalgal cultivation is another application of nanoparticles. Iron nanoparticles, magnesium sulfate nanoparticles, silica nanoparticles, and titanium dioxide nanoparticles can be used for this purpose.
- Microalgal harvesting is one of the major problems encountered in algal culture for biofuel production. The problems in harvesting occur due to small size of the algae, high dispersity of cells, and high concentration of algal culture on the upper part (Pienkos and Darzins 2009; Wang et al. 2015). Magnetic nanoparticles have high and fast harvesting efficiency, can be automated, offer low contamination (Borlido et al. 2013), and can be efficiently used for algal harvesting. Multi-functional nanoparticles offer integrated use of microalgae harvesting and post-harvesting stages like disruption of cells, extraction of lipids, and conversion of oils (Seo et al. 2017).

12.10.1 Use of Nanoparticles for Lipid Extraction in Microalgae

Microalgae usually contain rigid cell walls. Hence, chemicals need to be used for the pretreatment process (Lee et al. 2015; Kim et al. 2016). The usual methods used include solvent extraction and mechanical methods (Kumar et al. 2015), but these methods have certain drawbacks like high cost, increased energy consumption, reduced efficiency, reduced quality, and stability of the lipid extracted from the algae (Seo et al. 2017). Hence, alternate technologies were considered for lipid extraction. Nanoparticle-aided extraction of lipids has proven to be more efficient for this purpose. The nanoparticles used for lipid extraction have magnetic, dielectric, and enzymatic properties. To increase lipid extraction using nanoparticles, engineered nanoparticles are used as flocculants for microalgae harvesting.

12.10.2 Use of Nanocatalysts for Biodiesel Production

The use of nanocatalysts for biofuel production is an important application of nanoparticles in biofuel production. The use of nanoparticles in microalgal biorefinery helps to increase harvesting performance, lipid yield, conversion, and selectivity for green diesel. Recent nanotechnology uses enzyme-functional nanoparticles to boost lipid extraction by disrupting the rigid cell walls of microalgae. Enzyme-functionalized nanoparticles are used for the pretreatment process and increase lipid extraction efficiency, but have specific requirements of pH, temperature, and incubation time.

Nanocatalysts used for biofuel production are of two types: (a) homogeneous catalysts; and (b) heterogeneous catalysts. Homogenous catalysts are used for esterification and transesterification reactions. But there are certain problems like neutralization of wastewater, recycling problems related with catalysts, and expensive equipment (Lam and Lee 2012; Chiang et al. 2015). Hence, heterogeneous nanocatalysts are considered important due to their recyclability and cost-effective nature (Carrero et al. 2011; Lam and

Lee 2012). Another advantage is that their catalytic properties can be controlled by altering their physical characteristics.

12.10.3 Use of Nanoparticles for Enzyme Immobilization

Biofuels are produced on large scale with enzymes, but there are some limitations on the use of enzymes like enzyme inactivation by solvents, high enzyme costs, and problems scaling up (Watanabe et al. 2000). Immobilization of enzymes is considered a potential way to reduce biofuel system costs; separation, reusability, and stability at extreme conditions (Ansari and Husain 2012; Hwang and Gu 2013; Verma et al. 2013) make them compatible with improved product quality (Puri et al. 2013). These nanomaterials are considered as an alternative of conventional material for enzymes immobilization, owing to higher enzyme loading, high biocatalytic potential, and larger surface area (Gupta et al. 2011; Hwang and Gu 2013).

Nanofibers can be considered as a good alternative, as they are easy to handle with flexibility in reactor designing (Nair et al. 2007), and they have high durability and easy separable properties. Using nanofibers can facilitate the recovery of non-magnetic nanomaterials, in addition to controlling the dispersion.

12.11 Scope for Improvement of Plant-Based Biofuel Scenarios

From the information gathered related to biofuels of different generations, it is well understood that major steps need to be adopted for production of plant-based biofuels. Following are the general points to be considered as recommendations for improvement (Table 12.4).

TABLE 12.4

Steps to Increase the Sustainability of Plant-Based Biofuels

Points	Steps to Be Followed
1 Legality	Biofuel production has to follow all applicable laws and regulations
2 Planning, monitoring, and continuous improvement	Biofuel operations need to be planned, implemented, and continuously improved to ensure sustainability through an open, transparent, and consultative Environmental and Social Impact Assessment (ESIA) and an economic viability analysis
3 Greenhouse gas emissions	Biofuels contribute to climate change mitigation by reducing greenhouse gas emissions compared to fossil fuels
4 Human and labor rights	Biofuel production should not violate human rights, and has to promote decent work and the well-being of workers
5 Rural and social development	In regions of poverty, biofuel production has to contribute to the socioeconomic development of local, rural, and indigenous people
6 Local food security	Biofuel production has to improve food security in food-insecure regions
7 Conservation	Biofuel production should avoid negative impacts on biodiversity, ecosystems, and high conservation value areas
8 Soil	Biofuel production should implement practices to maintain soil health and reverse soil degradation
9 Water	Biofuel production should maintain/enhance the quality and quantity of surface and ground water resources, and respect prior formal or customary water rights
10 Air	Air pollution from biofuel production has to be minimized along the supply chain
11 Use of technology, inputs, and management of waste	The use of technologies in biofuel production shall seek to maximize production efficiency and social and environmental performance, and minimize the risk of damages to the environment and people
12 Land rights	Biofuel production should respect land rights and land use rights

12.12 Conclusion

The use of plant-based biofuels can be advantageous to improve energy security in the future. Compared to first- and second-generation biofuels, third-generation biofuels can be a better option and have many advantages to meet the future energy demands; hence, the future development of plant-based biofuels can be focused on extraction from algae based on new technologies to improve production efficiency like biotechnology and nanotechnology.

Acknowledgment

The author thanks the School of Environmental Studies, Cochin University of Science and Technology (CUSAT), Kerala, India for providing the facilities needed for completion of this chapter. Thanks to the Centre for Sustainable Technologies (CST), Indian Institute of Science (IISc) and Prof. N. H. Ravindranath who was my mentor on *Jatropha*-related field research in South India.

REFERENCES

Achten, W.M., Maes, W., Aerts, R., Verchot, L., Trabucco, A., Mathijs, E., . . . Muys, B., 2010. *Jatropha*: from global hype to local opportunity. *J. Arid Environ.* 74(1), 164–165.

Ahmad, A.L., Mat Yasin, N.H., Derek, C.J.C., Lim, J.K., 2011. Microalgae as a sustainable energy source for biodiesel production: a review. *Renew. Sust. Energ. Rev.* 15, 584–593.

Ahmed, W., Sarkar, B., 2018. Impact of carbon emissions in a sustainable supply chain management for a second generation biofuel. *J. Clean Prod.* 186, 807–820.

Alaswad, A., Dassisti, M., Prescott, T., Olabi, A., 2015. Technologies and developments of third generation biofuel production. *Renew. Sust. Energ. Rev.* 51, 1446–1460.

Altieri, M.A., 2009. The ecological impacts of large-scale agrofuel monoculture production systems in the Americas. *Bull. Sci. Technol. Soc.* 29(3), 236–244.

Ansari, S.A., Husain, Q., 2012. Potential applications of enzymes immobilized on/in nano materials: A review. *Biotechnol. Adv.* 30(3), 512–523.

Aro, E.M., 2016. From first generation biofuels to advanced solar biofuels. *Ambio*, 45(1), 24–31.

Arshad, M., Ahmed, S., 2015. Cogeneration through bagasse: a renewable strategy to meet the future energy needs. *Renew. Sust. Energ. Rev.* 54, 732–737.

Arshad, M., Zia, M.A., Shah, F.A., Ahmad, M., 2017. An overview of biofuel. In: Arshad, M. (ed) *Perspectives of water usage for biofuels production*. Cham: Springer International Publishing.

Borlido, L., Moura, L., Azevedo, A.M., Roque, A.C., Aires-Barros, M.R., Farinha, J.P.S., 2013. Stimuli-responsive magnetic nanoparticles for monoclonal antibody purification. *Biotechnol. J.* 8(6), 709–717.

Bundhoo, M.A.Z., Mohee, R., 2016. Inhibition of dark fermentative bio-hydrogen production: a review. *Int. J. Hydrog. Energ.* 41, 6713–6733.

Carrero, A., Vicente, G., Rodríguez, R., Linares, M., Del Peso, G.L., 2011. Hierarchical zeolites as catalysts for biodiesel production from *Nannochloropsis microalga* oil. *Catal. Today* 167(1), 148–153.

Carriquiry, M.A., Du, X., Timilsina, G.R., 2010. *Second-generation biofuels* (57p). Washington, DC: World Bank.

Chiang, Y.D., Dutta, S., Chen, C.T., Huang, Y.T., Lin, K.S., Wu, J.C., Yamauchi, Y., Suzuki, N., Wu, K.C.W., 2015. Functionalized Fe3O4@ silica core–shell nanoparticles as microalgae harvester and catalyst for biodiesel production. *Chem. Sus. Chem.* 8(5), 789–794.

Creutzig, F., Ravindranath, N.H., Berndes, G., Bolwig, S., Bright, R., Cherubini, F., . . . Fargione, J., 2015. Bioenergy and climate change mitigation: an assessment. *GCB Bioenerg.* 7, 916–944.

Dehkhoda, A.M., West, A.H., Ellis, N., 2010. Biochar based solid acid catalyst for biodiesel production. *Appl. Catal. A Gen.* 382(2), 197–204.

Demirbas, A., Demirbas, M.F., 2011. Importance of algae oil as a source of biodiesel. *Energy Convers. Manag.* 52(1), 163–170.

Enagi, I.I. Al-Attab, K.A., Zainal, Z.A., 2018. Liquid biofuels utilization for gas turbines: a review. *Renew. Sust. Energ. Rev.* 90, 43–55.

Eroglu, E., Tiwari, P.M., Waffo, A.B., Miller, M.E., Vig, K., Dennis, V.A., Singh, S.R., 2013. A nonviral pHEMA+ chitosan nanosphere-mediated high-efficiency gene delivery system. *Int. J. Nanomed.* 8, 1403.

Escobar, J.C., Lora, E.S., Venturini, O.J., Yanez, E.E., et al., 2009. Biofuels: environment, technology and food security. *Renew. Sust. Energ. Rev.* 13, 1275–1287.

Fu, C.C., Hung, T.C., Chen, J.Y., Su, C.H., Wu, W.T., 2010. Hydrolysis of microalgae cell walls for production of reducing sugar and lipid extraction. *Bioresour. Technol.* 101, 8750–8754.

Fukuda, H., Kondo, A., Noda, H., 2001. Biodiesel fuel production by transesterification of oils. *J. Biosci. Bioeng.* 92(5), 405–416.

Gardy, J., Hassanpour, A., Lai, X., Ahmed, M. H., Rehan, M., 2017. Biodiesel production from used cooking oil using a novel surface functionalized TiO_2 nano-catalyst. *Appl. Catal. B Environ.* 207, 297–310.

Ghirardi, M.L., Zhang, L., Lee, J.W., Flynn, T., Seibert, M., Greenbaum, E., 2000. Microalgae: a green source of renewable H2. *Trends Biotechnol.* 18, 506–511.

Gomez, L.D., Steele-King, C.G., McQueen-Mason, S.J., 2008. Sustainable liquid biofuels from biomass: the writing's on the walls. *New Phytol.* 178(3), 473–485.

Gonsalves, J.B., 2006. *An assessment of the biofuels industry in India*. Geneva: UNCTAD.

Gopakumar, L., 2020. *Jatropha* cultivation in South India—policy implications. In *Sustainability and Law* (pp. 453–472). Cham: Springer.Lee, Y.C., Lee, K., Oh, Y.K., 2015b. Recent nanoparticle engineering advances in microalgal cultivation and harvesting processes of biodiesel production: A review. *Bioresour. Technol.* 184, 63–72.

Lee, Y. C., Lee, K., Oh, Y. K., 2015. Recent nanoparticle engineering advances in microalgal cultivation and harvesting processes of biodiesel production: a review. *Bioresour technol* 184, 63–72.

Leong, W.H., Lim, J.W., Lam, M.K., Uemura, Y., Ho, Y.C., 2018. Third generation biofuels: a nutritional perspective in enhancing microbial lipid production. *Renew. Sust. Energ. Rev.* 91, 950–961.

Li, J., Liu, Y., Cheng, J.J., Mos, M., Daroch, M., 2015. Biological potential of microalgae in China for biorefinery-based production of biofuels and high value compounds. *New Biotechnol.* 32(6), 588–596.

Liu, Y., Li, J., Qiu, X., Burda, C., 2007. Bactericidal activity of nitrogen-doped metal oxide nanocatalysts and the influence of bacterial extracellular polymeric substances (EPS). *J. Photochem. Photobiol. A. Chem.* 190(1), 94–100.

Mahto, M.K., Samanta, D., Konar, S., Kalita, H., Pathak, A., 2018. N, S doped carbon dots—Plasmonic Au nanocomposites for visible-light photocatalytic reduction of nitroaromatics. *J. Mater. Res.* 33(23), 3906–3916.

Metzger, P., Largeau, C., 2005. *Botryococcus braunii*: a rich source for hydrocarbons and related ether lipids. *Appl. Microbiol. Biotechnol.* 6(25), 486–496.

Moody, J.W., McGinty, C.M., Quinn, J.C., 2014. Global evaluation of biofuel potential from microalgae. *Proc. Natl. Acad. Sci. U.S.A.* 111(23), 8691–8696.

Nair, S., Kim, J., Crawford, B., Kim, S.H., 2007. Improving biocatalytic activity of enzyme-loaded nanofibers by dispersing entangled nanofiber structure. *Biomacromolecules* 8(4), 1266–1270.

Nguyen, T.H.M., Van Hanh, V., 2012. Bioethanol production from marine algae biomass: Prospect and troubles. *J. Vietnamese Environ.* 3(1), 25–29.

Nizami, A.S., Rehan, M., 2018. Towards nanotechnology-based biofuel industry. *Biofuel Res. J.* 18, 798–799. https://doi.org/10.18331/BRJ2018.5.2.2

Oh, Y.K., Hwang, K.R., Kim, C., Kim, J.R., Lee, J.S., 2018. Recent developments and key barriers, to advanced biofuels: a short review. *Bioresour. Technol.* 257, 320–333.

Openshaw, K., 2000. A review of *Jatropha curcas*: an oil plant of unfulfilled promise. *Biomass Bioenergy* 19(1), 1–15.

Pattarkine, M. V., Pattarkine, V. M., 2012. Nanotechnology for algal biofuels. In *The science of algal fuels* (pp. 147–163). Dordrecht: Springer.

Pfenninger, S., Keirstead, J., 2015. Renewables, nuclear, or fossil fuels? Scenarios for Great Britain's power system considering costs, emissions and energy security. *Appl. Energ.* 152, 83–93.

Pienkos, P.T., Darzins, A.L., 2009. The promise and challenges of microalgal-derived biofuels. *Biofuel Bioprod. Bior.* 3(4), 431–440.

Puri, M., Barrow, C.J., Verma, M.L., 2013. Enzyme immobilization on nanomaterials for biofuel production. *Trends Biotechnol.* 31(4), 215–216.

Rai, M., Da Silva, S.S. (Eds.). (2017). *Nanotechnology for bioenergy and biofuel production*. Cham: Springer International Publishing.

Ramsurn, H., Gupta, R. B., 2013. Nanotechnology in solar and biofuels. *ACS Sustain. Chem. Eng.* 1(7), 779–797.

Santoro, C., Arbizzani, C., Erable, B., 2017. Ieropoulos I. Microbial fuel cells: from fundamentals to applications. A review. *J. Power Source* 356, 225–244.

Saravanan, A.P., Mathimani, T., Deviram, G., Rajendran, K., Pugazhendhi, A., 2018. Biofuel policy in India: a review of policy barriers in sustainable marketing of biofuel. *J. Clean Prod.* 193, 734–747.

Scheierling S.M., Treguer, D.O., 2018. *Beyond crop per drop assessing agricultural water productivity and efficiency in a maturing water economy.* Washington, DC: World Bank.

Sekoai, P.T., Oumaa, C.P.M., du Preeza, S.P., Modishaa, P., Engelbrechta, N., Bessarabova, D.G., Ghimireb, A., 2019. Application of nanoparticles in biofuels: an overview. *Fuel*, 237, 380–397.

Sekhon Randhawa, K.K., & Rahman, P.K., 2014. Rhamnolipid biosurfactants—past, present, and future scenario of global market. *Front. Microbiol.* 5, 454.

Seo, J.Y., Kim, M.G., Lee, K., Lee, Y.C., Na, J.G., Jeon, S.G., Park, S.B., Oh, Y.K., 2017. Multifunctional nanoparticle applications to microalgal biorefinery. In *Nanotechnology for bioenergy and biofuel production* (pp. 59–87). Cham: Springer.

Shafiee, S., Topal, E., 2009. When will fossil fuel reserves be diminished? *Energy Policy*, 37(1), 181–189.

Sharma, Y.C., Singh, B., Korstad, J., 2011. A critical review on recent methods used for economically viable and eco-friendly development of microalgae as a potential feedstock for synthesis of biodiesel. *Green Chem.* 13(11), 2993–3006.

Shields-Menard, S.A., Amirsadeghi, M., French, W.T., Boopathy, R., 2018. A review on microbial lipids as a potential biofuel. *Bioresour. Technol.* 259, 451–460.

Singh, A., Nigam, P.S., Murphy, J.D., 2011. Renewable fuels from algae: an answer to debatable land based fuels. *Bioresour. Technol.* 102, 10–16.

Sirajunnisa, A.R., Surendhiran, D., 2016. Algae—A quintessential and positive resource of bioethanol production: a comprehensive review. *Renew. Sust. Energ. Rev.* 66, 248–267.

Smith, K., 2013. *Biofuels, air pollution, and health: A global review.* New York: Springer Science & Business Media.

Tabatabaei, M., Karimi, K., Sárvári Horváth, I., Kumar, R., 2015. Recent trends in biodiesel production. *Biofuel Res. J.* 2(3), 258–267.

Torkamani, S., Wani, S.N., Tang, Y.J., Sureshkumar, R., 2010. Plasmon-enhanced microalgal growth in mini-photobioreactors. *Appl. Phys. Lett.* 97(4), 043703.

Trindade, A. C., Canejo, J. P., Pinto, L. F. V., Patrício, P., Brogueira, P., Teixeira, P. I. C., Godinho, M. H., 2011. Wrinkling labyrinth patterns on elastomeric janus particles. *Macromolecules* 44(7), 2220–2228.

van Eijck, J., Romijn, H., Smeets, E., Bailis, R., Rooijakkers, M., Hooijkaas, N., et al., 2014. Comparative analysis of key socio-economic and environmental impacts of smallholder and plantation based *Jatropha* biofuel production systems in Tanzania. *Biomass Bioenerg.* 61, 25–45.

Verma, M.L., Chaudhary, R., Tsuzuki, T., Barrow, C.J., Puri, M., 2013. Immobilization of β-glucosidase on a magnetic nanoparticle improves thermostability: Application in cellobiose hydrolysis. *Biores. Technol.* 135, 2–6.

Verziu, M., Cojocaru, B., Hu, J., Richards, R., Ciuculescu, C., Filip, P., Parvulescu, V.I., 2008. Sunflower and rapeseed oil transesterification to biodiesel over different nanocrystalline MgO catalysts. *Green Chem.* 10(4), 373–381.

Walker, G., Simcock, N., Day, R., 2016. Necessary energy uses and a minimum standard of living in the United Kingdom: energy justice or escalating expectations? *Energ. Res. Soc. Sci.* 18, 129–138.

Wang, Y.J., Dong, H., Lyu, G.M., Zhang, H.Y., Ke, J., Kang, Teng, J.L., Sun, L.D., Si, R., Zhang, J., Liu, Y.J., Zhang, Y.W., Huang, Y.H., Yan, C.H., 2015. Engineering the defect state and reducibility of ceria based nanoparticles for improved anti-oxidation performance. *Nanoscale*, 7(33), 13981–13990.

Wang, S.K., Stiles, A.R., Guo, C., & Liu, C.Z., 2015. Harvesting microalgae by magnetic separation: A review. *Algal Res.*, 9, 178–185.

Waqas, M., Aburiazaiza, A.S., Minadad, R., Rehan, M., Barakat, M.A., Nizami, A.S., 2018. Development of biochar as fuel and catalyst in energy recovery technologies. *J. Clean. Prod.* 188, 477–488.

Watanabe, Y., Shimada, Y., Sugihara, A., Noda, H., Fukuda, H., Tominaga, Y., 2000. Continuous production of biodiesel fuel from vegetable oil using immobilized *Candida antarctica* lipase. *J. Am. Oil Chem.' Soc.* 77(4), 355–360.

Yahya, N.Y., Ngadi, N., Jusoh, M., Halim, N.A.A., 2016. Characterization and parametric study of mesoporous calcium titanate catalyst for transesterification of waste cooking oil into biodiesel. *Energy Convers. Manag.* 129, 275–283.

13 Microbial-Based Biofuel Production: A Green Sustainable Approach

J. Ranjitha, Shreya Subedi, M. Anand, R. Shobana, S. Vijayalakshmi, and Bhaskar Das

CONTENTS

- 13.1 Introduction 285
  - 13.1.1 Biofuel from Microalgae 287
  - 13.1.2 Biodiesel from Microalgae 287
  - 13.1.3 Biohydrogen from Microalgae 289
  - 13.1.4 Biomethane from Microalgae 289
- 13.2 Biofuel Production from Fungi 290
- 13.3 Biofuel Production from Bacteria 291
- 13.4 Conclusion 293
- References 294

13.1 Introduction

Over the past decade, the temperature of the earth has been drastically varying, causing ozone depletion, melting of ice sheets, rise in sea level, and ocean acidification, resulting in a main threat to civilization. Similarly, the increasing use of energy led to a dependency on non-renewable energy sources like oil, natural gas, and coal which will soon be exhausted. The significant factor which is responsible for causing global warming or climate change is fossil fuel; when fossil fuels are burned, they release greenhouse gases into the atmosphere (Edenhofer et al., 2011). To reduce carbon dioxide emission and the depletion of crude oil reserves, biofuels are an attractive source of energy and a proven fuel. Biofuels can be used as a substitute for fossil fuels because they can minimize carbon dioxide emission, and microbial biomass is one of the better sources of bioenergy (Ayoub and Abdullah, 2012). They are broadly classified into first, second and third generation biofuels. The most common type of biofuel in use is biodiesel and bioethanol, which falls under the category of first-generation biofuels. Most of the biodiesel are produced from the waste of cooking oil, vegetable oil, plants, seeds, vegetable oil, animal fats, etc. (Demirbas, 2009). The main benefit of using biodiesel is that it produces a negligible amount of CO_2 gas when burned (Panahi et al., 2019). Bioethanol can be produced from sugar-based crops, such as fruits, sugar beet, sweet sorghum, and sugarcane using fermentation process (Mohr and Raman, 2013). In the fermentation process, yeast can be converted sugar molecules into ethanol as a main product (Robak and Balcerek, 2018). In the current situation, more than 90% of the bioethanol is produced from starch- and sugar-based feedstock. Most of the countries like the United States and Brazil produce over 15.7 billion gallons of ethanol by primarily using corn (or) sugarcane (Patil et al., 2008). Even though first-generation biofuel can be easily processed, the major drawback is that the fuel feedstock is also used for feed and food purposes, which leads to the most controversial debate of food vs. fuel (Helwani et al., 2009).

Second-generation biofuels are mainly produced using lignocellulose material of plant biomass, agriculture residues such as straws and bagasse, and waste products obtained from waste cooking oil. They do not compete with the food vs. fuel debate in the production process and also arable land is not required. Second-generation liquid biofuels are generally produced by biological or thermochemical processing of the feedstock (Patil et al., 2008). The drawbacks with second-generation biofuel are that the manufacturing technology is comparatively immature, production cost needs to be reduced, and there should be an increase in the efficiency of production. Its production requires more material and energy. Considerably, these feedstocks are not practically and economically feasible to produce energy, because they lack the sourcing material and conversion rates are low (Leong et al., 2018). Alternative sources of fuels regardless the first or second generation have major setbacks. The crops (corn, maize, rapeseed, and sugarcane) used for the production of the fuel have caused strain in the production of food globally, excessive use of lands, destruction of world's forests, and water shortage (Leong et al., 2018). In the middle of world energy demand, trying to find a viable alternative to the first and second generation is a major concern, a fuel which will not compete with food, thus third-generation biofuels derived from microbes have been considered to be the most sustainable alternative fuel (Behera et al., 2014; Raja et al., 2008). Much research has been conducted for the production of biofuel from various species of microbes which shows that they can be used as a substrate as they can synthesize and store a large number of fatty acids in their biomass (Agu et al., 2019).

The major source of third-generation biofuels is oleaginous species of microorganisms such as yeast, bacteria, fungi, and microalgae, as shown in Figure 13.1. Oleaginous microorganisms are defined as microbial with the content of microbial lipid excess of 20% (Fontana, 2010). The extent of lipid

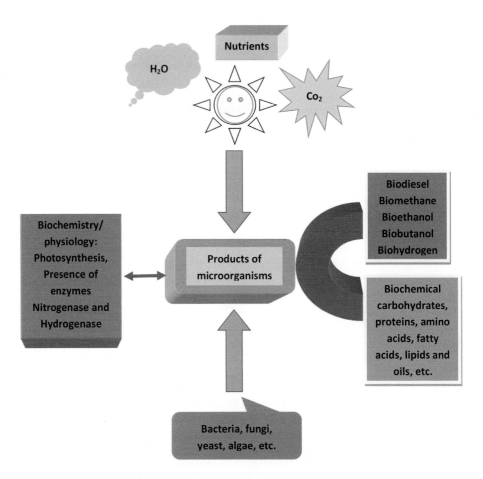

FIGURE 13.1 Microbial-based biofuel production.

accumulation is based upon the genetic constitution of the organism; the maximum attainable lipid contents vary among species (Smeets et al., 2014). These sources do not conflict with food sources and are renewable. Certain types of algae can produce 50% oil and 50% biomass, which can be converted into fuel products; it can become one of the world most valuable, renewable, and sustainable sources of fuel. Third-generation biofuels can reduce the direct or indirect dependence on fossil fuels for sustainability (Behera et al., 2015).

13.1.1 Biofuel from Microalgae

Biofuel production from microalgae appears a potential alternative for first- and second-generation biofuels. Since the production of biofuels from microalgae is a new finding, there are a few issues such as designing an economical and sustainable method for oil production and designing an engineered metabolic pathway to control the reactions of lipid synthesis to produce algal cells with desired lipid content. Even so, the use of third-generation biofuel is more advantageous than the first- and second-generation biofuels, as microalgae can retain higher growth tendency and biomass productivity, which increases the production of the fuel. Besides, the residual microalgae biomass can further be converted into other products such as biomethane, bio-oil, bioethanol, biohydrogen, etc., through successions of biorefinery processes (Banerjee et al., 2002). To enhance biofuel production, fourth-generation biofuel uses genetically modified algae. Even though it is a great alternative, the potential environmental and health-related risks of the modified algae are still not determined (Melis, 2002) because for fourth-generation biofuel, synthetic biology is used to intensify the lipid extraction from algae. The barriers for fourth-generation biofuels are the open-pond system, which is an economical solution for large-scale cultivation of microalgae; however, the concern regarding the health and environmental risk of cultivation, production of genetically modified algae, and disposal of their residues. As genetically modified algae are minute in size, they can easily invade the ecosystem and be a major concern for ecological change. The research for the fourth-generation biofuel is at the primary stage to consider it as a sustainable source for alternative fuel.

Microalgae uses sunlight, CO_2, and inorganic nutrients (nitrogen, phosphorus, vitamins, and silica) to grow, and in a controlled environment, algae can multiply itself rapidly. In the presence of sunlight, microalgae produces energy; they use the excessive amount of CO_2 for their production, thus resulting in a carbon-neutral emission. Microalgae are capable of fixing CO_2 using different sources like atmosphere, discharge gases, and soluble carbonates (Giselrd et al., 2008). They are generally more like sunlight-driven cell factories that efficiently convert solar energy into biomass and also double their lipid content 80% by weight of dry biomass (Spolaore et al., 2006). Generally, all types of microalgae can grow in risky surroundings like high salinity and temperature, and are flexible to cultivate at large scale (Hu et al., 2008). The genetically modified microalgae strains are enhanced in lipid content. The percentage yield of biodiesel production from microalgae is always greater than other oleaginous seeds (Hsieh and Wu, 2001, 2009). The scientific reports clearly demonstrated that microalgae produce higher oil yield percentage than other plant oils. Microalgae can produce 100,000 kg oil per hectare per year compared with castor, coconut, palm, and sunflower yield up to 5,950, 2,689, 1,413 and 9,521 per hectare per year, respectively (Hu et al., 2008; Deng et al., 2009; Tonon et al., 2002; Xin et al., 2008). Microalgal harvesting, lipid extraction, and biodiesel are clearly shown in Figure 13.2.

13.1.2 Biodiesel from Microalgae

In heterotrophic medium, the microalgae *Chlorella protothecoids* was grown by the acidic transesterification method and produced biodiesel in high quality, as reported by Xu and colleagues (2006). The production of biodiesel from this work is done through integrated method. The result of a novel process in this study for the biodiesel production was cost-effective, efficient, and feasible. In the same way, the biomass production and accumulation of lipid enhanced by the aeration of CO_2 in *Nannochloropsis oculata* for lipid yield and long-term biomass was done by semi-continuous culture (Chiu et al., 2009). The *Chlorella vulgaris* microalgae contain high lipid content of 42% and the lipid productivity is 148 mg/L/d. It was

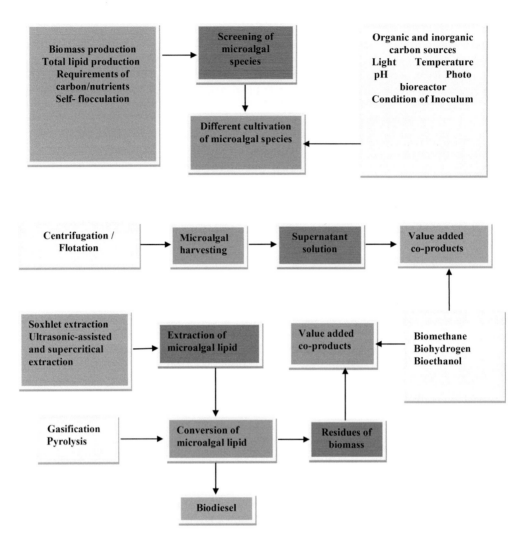

FIGURE 13.2 Flow diagram of microalgal-based biodiesel production.

cultivated in artificial water waste through a semi-continuous cultivation system. It is cost-effective when compared with the petroleum at 63.97 USD/container, along with possible credit for treatment of wastewater (Feng et al., 2011). In the case of *Nannochloropsis oculata*, lipid content was enhanced by the addition of nitrogen salts at various concentration at 22–25°C, and the results showed that the lipid content was doubled from 7.95% to 14.93%. Similarly, in the case of *C. vulgaris* from 5.98% to 16.45%, correspondingly (Converti et al., 2009). The both relative intensity of fluorescence and of neutral lipids on marine algae *C. vulgaris* for the effect of growth and accumulation of lipids were investigated and reported that lipids and the extracted lipid content in cultures, the Fe^{3+} supplement were thrice to sevenfold (1.2×10^{-5} mol l^{-1}), appropriately in other cultures (Liu et al., 2008). To assess the lipid production and biomass, the *Chlorella* sp. were cultivated by fed-batch and semi-continuous culture modes Hsieh and Wu (2009). By this method, the maximum productivity of lipid in semi-continuously culture was higher, when compared with fed-batch method. With the help of CO_2 and flue gas, the high biomass and lipid production microalgae were cultivated. The result shows *Scenedesmus* sp. is highly biomass productive for CO_2 mitigating, because of its C-fixation ability (Yoo et al., 2010).

13.1.3 Biohydrogen from Microalgae

Likewise, hydrogen can be produced from microalgae and depends on strain-specific ability which is mainly based the enzymes nature, as well as environmental conditions. Recently, genetic engineering has played a vital role in hydrogen production using targeted engineering of certain enzymes like hydrogenases and nitrogenases to enhance the hydrogen production via metabolic pathways (Khetkorn et al., 2013). Certain strains of microalgae have been developed using genetic engineering approach to improve the efficiency of hydrogen production (Baebprasert et al., 2011; Nyberg et al., 2015). Several processes are there for the production of hydrogen, such as steam reforming, electrolysis, photolysis, etc. (Holladay et al., 2009). In the steam reforming process, hydrogen is produced from fossil fuels at a very high temperature. In this process, scientists developed several catalysts to enhance hydrogen production. Electrolysis is used to cleave water molecules into O_2 and H_2. These two methods produce carbon monoxide and carbon dioxide as byproducts, which is not eco-friendly. There are several methods are available for H_2 production, but photobiological hydrogen production is efficient, eco-friendly, and less energy-intensive as compared to electrochemical processes (Khetkorn et al., 2013; Nyberg et al., 2015). Microalgae are photosynthetic microorganisms which are suitable for photobiological production of hydrogen (Nagarajan et al., 2017). Extensive literature survey clearly states that *Scenedesmus, Synechocystis, Tetraspora, Botryococcus, Chlamydomonas, Chlorococcum, Chlorella, Anabaena, Nostoc*, etc., were able to produce hydrogenase enzyme for hydrogen production (Eroglu et al., 2011). Generally, algae can grow with simple nutrient supplementation, with the ability to fix CO_2 from the atmosphere as a carbon source and sunlight as a source of energy to produce hydrogen. A biological process of hydrogen production has several merits over the conventional methods.

13.1.4 Biomethane from Microalgae

Microalgae can be used as a feedstock for the production of biogas production and it is proved to be efficient, realistic and cost-effective (Ward et al., 2014; Saratale et al., 2018; Zamalloa et al., 2012). The lignocellulosic materials were used for biomethane production, since the biodegradability of lignocellulosic biomass is slow for the complex structure and nature of biomass (Zabed et al., 2017) Similarly, lignin is highly challenging to hydrolysis and lethal for some microorganisms in anaerobic digestion, which results in a decrease in the biodegradation process (Passos et al., 2018; Neves et al., 2006). The methane yield of various microalgae biomass as shown in Table 13.1 and compared with other lignocellulosic biomass—specifically like rice straw, wheat straw, corn residue and sugarcane waste—which were considerably slighter than microalgae strains (Chandra et al., 2012). Extensive literature survey revealed that microalgae can produce higher amounts of methane than other biomass and also methane yield can vary depending on the habitats of microalgae (Song et al., 2015).

Frigon et al. (2013) revealed that biomethane yield varied between 298 ± 83 mL g^{-1} VS in marine algae and 329 ± 43 mL g^{-1} VS in freshwater algae. The methane from microalgae have been found to be energy significant than microalgae-based biodiesel production (Jones and Mayfield, 2012; Collet et al., 2011). On the other side, ability of using lipid-extracted macroalagal biomass from biodiesel production can propose an integrated concept by incorporating biodiesel production and biogas from the same biomass. The extracted lipid from substantial amounts of proteins and carbohydrates can be biodegraded into biogas during anaerobic digestion (Chisti, 2008; Sialve et al., 2008). The upgradation of biomethane is one of the major pertaining issues for future bioenergy industries (Wang et al., 2017), because of operational costs and environmental contexts (Table 13.1) (Xia and Murphy, 2016). Microalgae are capable of growing in the recycled biogas slurry, which can offer opportunities for biogas upgrading and processing of the biogas slurry in cleaner way (Zhao et al., 2015; Bahr et al., 2013) because carbon dioxide and other nutrients in the slurry can be used by microalgae (Holm-Nielsen et al., 2009).

TABLE 13.1

Biomethane Production Using Different Microalgal Strains

Si.no	Microorganism	Biofuel	Biofuel Yield (g l−1)	Reference
1	*Yarrowiali polytica*	Fatty acids	55	Beopoulos et al., 2009
2	*Saccharomyces cerevisiae*	Isoprenoid-based biofuel	40	Westfall et al., 2012
3	*Synechococcus* sp.	Limonene	0.04	Davies et al., 2014
4	*Trichoderma reesei*	Bioethanol	10	Huang et al., 2014
5	*Zymomonas mobilis*	Bioethanol	–	Kremer et al., 2015
6	*Clostridium acetobutylicum*	Biobutanol	3	Lütke-Eversloh and Bahl, 2011
7	*Synechococcus elongates*	1,3-Propanediol	0.28	Hirokawa et al., 2016
8	*Cryptococcus vishniaccii*	Lipids	7.8	Deeba et al., 2016
9	*Caldicellulosiruptor* sp.	Bioethanol	0.70	Chung et al., 2014
10	*Saccharomyces cerevisiae*	Fatty acids	0.38	Yu et al., 2016
11	*Zymomonas mobilis*	2,3-Butanediol	10	Yang et al., 2016
12	*Escherichia coli*	Bioethanol	25	Romero-García et al., 2016
13	*Clostridium thermocellum*	Bioisobutanol	5.4	Lin et al., 2015
14	*Escherichia coli*	Biobutanol	30	Shen et al., 2011
15	*Pseudomonas putida*	Biobutanol	0.05	Nielsen et al., 2009

13.2 Biofuel Production from Fungi

Fungi are non-photosynthetic organisms which are well known for their growth with short lifecycles, their lack of a need for light energy, easy scalability, and the ability to use a wide range of carbon sources, viz. lignocellulosic biomass, agroindustrial residues, and wastewater, and also fungal lipid productivity is independent of seasonal variations (Elreesh and Haleem, 2013; Subhash and Mohan, 2014). Most of the fungi are investigated for the production of lipids such as docosahexenoic acid (DHA), gamma-linolenic acid (GLA), eicosapentaenoic acid (EPA), and arachidonic acid (ARA), which are used for various applications, and there are a few reports recently on biodiesel production from fungal oil (Liang et al., 2013). In comparison to the microalgae, the growth of microorganisms such as yeast and fungi can be carried out in conventional microbial bioreactors, which will improve the biomass yield and will reduce the cost of biomass and oil production (Vicente et al., 2004). Consequently, yeasts are unicellular microorganisms which are used for various industrial applications. Among 600 described species of yeast, 30 species have been found to accumulate sufficient quantities of lipid, and on average, these yeasts can accumulate more than 30% of lipids from cell dry weight—but some of them accumulate up to 70% of neutral lipids under nitrogen or phosphorus starvation (Yuzbasheva et al., 2014). Oleaginous yeasts store lipids mainly in the form of triacylglycerols (TAG) and diacylglycerols (DAG) in intracellular lipid bodies (Kitcha and Cheirsilp, 2011). *Mucor circinelloides* contains high lipid content suitable as a feedstock for biodiesel production (Vicente et al., 2004). The performance of the lipid extraction process was studied for three different solvent systems (chloroform: methanol, chloroform: methanol: water, and n-hexane) and biodiesel was produced by both direct transesterification and indirect transesterification methods (Khot et al., 2012). The filamentous fungus *Aspergillus* sp. was studied for biofuel production, when grown using Sabouraud dextrose broth (SDB) and corncob waste liquor (CWL) as substrates in which SDB medium showed high biomass and lipid productivity. Both direct and indirect transesterification methods were carried out; direct transesterification method showed higher biodiesel productivity. Results showed that the direct process produced fatty acid methyl esters (FAMEs) with higher purities (>99% for all catalysts) than those from the two-step process (91.4%–98.0%).

Papanikolaou and colleagues (2004) cultured *Mortierella isabellina* in nitrogen-limited media, which showed notable biomass production (35.9 g/L) and high glucose uptake in culture media. In nitrogen depletion conditions, the lipid accumulation in fungal mycelia was increased up to 60% in dry biomass with high fungal oil production of 18.1 g/L. Similarly, lipid production was optimized using different carbon sources in the filamentous fungus *Mortierella isabellina* (Ruan et al., 2012). Li and colleagues (2012) isolated and identified 32 strains of oleaginous microbes from various habitats (wetlands, lawn, hot springs, alpine permafrost, and saline-alkali soil) in the Tibetan Plateau. Carbon use analysis indicated that the filamentous fungi such as *Backusella ctenidia*, *Fusarium* sp. and *Gibberella fujikuroi* have the strongest capability to use xylose and carboxymethyl cellulose (CMC) as a carbon source and enhanced their lipid productivity. These fungal strains were used for biodiesel production (Neema and Kumari, 2013). The phenomenon of intraradical and extraradical mycelia of fungi *Scutellospora calospora* and *Glomus intraradices*, when colonizing *Plantago lanceolate*, were observed. The arbuscular mycorrhizal fungal phospholipid accumulation was studied in two fungal species, *Scutellospora calospora* and *Glomus intraradices*, when colonizing *Plantago lanceolata*. No vesicles formation was found in *S. calospora*, and it also accumulated more neutral lipids in extraradical than in intraradical mycelium, but vesicles formation was occurred in *G. intraradices*; as a result, more neutral lipids in intraradical than in extraradical (Aarle and Olsson, 2003). *Rodotorula*110 was screened from 176 yeasts isolated from 34 soil samples used for biodiesel production. The lipid content and lipid productivity was found as 15.29 g/L and 58.2% in optimized conditions, respectively (Enshaeieh et al., 2013). The fungal species *Mucorrouxii* MTCC386 contains high lipid content of 32.10 ± 1.50% and is used for biodiesel production (Somashekar et al., 2001). The lipid-producing capability of 16 different filamentous fungal species were studied for developing biodiesel using its lipid. The results showed that the fungus *Cunninghamella japonica* showed high lipid content of 43%. Development of low-cost medium for growing the fungal species was also studied using ammonium nitrate as a nitrogen source. A notable increase in biomass and lipid percentage was observed. The fatty acid profile showed 50% of total fatty acid was dominated by oleic acid. Iodine index of the lipid fraction and heat of combustion were 86.61 and 37.13 MJ/kg, respectively (Sergeeva et al., 2008). Elreesh and Haleem (2013) screened DGB1 which showed high lipid content (40%) from 30 fungal isolates, and based on the morphological and molecular analysis, the isolated DGB1 was identified as a fungi. Optimization was carried out to improve the lipid productivity, and it was found that the lipid content reached a highest percentage yield. The produced fungal oil was analyzed using GC-MS spectral analysis and reported that 50% of the fatty acids was heptadecanoic acid. The filamentous fungi contains large amounts of palmitic, oleic, and linoleic acids, which are important fatty acids for biodiesel production. Wahlen and colleagues (2012) studied the biodiesel production from the lipid of the yeast *Cryptococcus curvatus* and investigated the engine performance using the produced biodiesel. The study concluded that the properties like heating value, viscosity, density, and cetane index were found to be the same as the biodiesel produced from soybean biodiesel. This proved that the oil from the yeast can be potentially used for biodiesel production.

Miah and colleagues (2004) increased biogas production during thermophilic digestion (65°C) of sewage sludge caused by the protease activity of a *Geobacillus* sp. strain. Similarly, Bagiand colleagues (2007) worked on mesophilic *Enterobacter cloacae* and thermophilic *Caldicellulosyruptor saccharolyticus* strains during anaerobic digestion of wastewater sludge, pig manure, and dried plant biomass of artichoke, and achieved a remarkable increase of biogas production (160%). This increase was explained by the improved H_2 level, and the strains showed significant hydrogen-producing bacteria Ravikumar, K., J. Dakshayini and S.T. Girisha 2012). *Trichoderma viride* has shown promising results on improving the subsequent anaerobic digestibility in biogas reactors, as have the possibility of *Anaeromyces* and *Piromyces* strains to mix into biogas-producing anaerobic sludge bacterial communities, to improve degradation of substrate polysaccharides, and similarly, to enhance methane production. The results were obtained during the bioaugmentation of swine manure–fed biogas reactors with different strains of anaerobic fungi. Amendment with fungal biomass led to 4%–22% higher gas yields and up to 2.5% higher methane concentration (Wagner et al., 2013). A recent study showed that bioaugmentation with anaerobic fungi did not increase the overall methane yield, but that it speeds up initial gas production and thus may help to reduce retention time (Fliegerová et al., 2012).

13.3 Biofuel Production from Bacteria

Bacteria are prokaryotic microorganisms which produces fewer lipid bodies, and so they are not satisfactory for biodiesel production, but it was found that some bacteria can produce biodiesel under stressed or special environmental conditions. It has been reported that only a few bacterial species—belonging to *Nocardia, Rhodococcus, Streptomyces, Gordonia*, and *Mycobacterium*—can accumulate TAG (triglycerides), from which *Rhodococcus* sp. and *Gordonia* sp. contain 80% of oil under optimized conditions (Liang and Jiang, 2013). Lipid productivity of two bacterial species, *Gordonia* sp. DG and *Rhodococcusopacus* PD630, were studied using sodium gluconate as carbon source. In the beginning of stationary phase, the lipid content was noted 80% and 72% for *Rhodococcus opacus* PD630 and *Gordonia* sp., respectively. Optimization of lipid content was carried out using various carbon sources, and maximum oil content of 93% and 96% for *Rhodococcus opacus* PD630 and *Gordonia* sp., respectively, were observed when sugarcane molasses was used. When orange waste was used as carbon source, high TAG content of 88.9 and 57.8 mg/L was acquired by *Rhodococcus opacus* PD630 and *Gordonia* sp., respectively. The culture medium was optimized statistically and achieved high lipid content using agrowastes (Gouda et al., 2008). Kalscheuer et al., (2006) genetically modified the bacteria *Escherichia coli* to increase the fatty acid ethyl esters (FAEEs). By heterologous expression in *E. coli* of the *Zymomonas mobilis* pyruvate decarboxylase and alcohol dehydrogenase and the unspecific acyltransferase from *Acinetobacter baylyi* strain ADP1, the increase in FAEEs was obtained. *Escherichia coli* were genetically modified to produce high fatty acids content by inserting different genetic changes into the *E. coli* genome. The genetically modified *E. coli* strain was cultured in defined media fermentation conditions, under fed-batch condition to enhance the lipid productivity (Lu et al., 2008). Among the isolated bacterial associates, *Bacillus subtilis* (RRL-8) from *Aurora globostellata* and *Pseudomonas* spp. (RRL-28) from *Heteronemaerecta* were observed to produce high lipid content of 16.9% and 31.7%, respectively. It was also observed that increase in C:N ratio raised the lipid productivity to 33.4% and 42.7% in *Bacillus subtilis* (RRL-8) and *Pseudomonas* spp. (RRL-28), respectively (Patnayak and Sree, 2005). Similarly, in another study, *Rhodococcus opacus* bacteria were studied for the biodiesel production using their lipid. The biodiesel was successfully produced and its engine performance was investigated using a two-cylinder diesel engine. Exhaust emission analysis revealed that there was muchless CO_2 and hydrocarbon emission than petroleum diesel when biodiesel from the bacteria was used (Wahlen et al., 2012). Compared with other microorganisms, the growth rate of bacteria is very high in very simple media, but most of the bacteria can accumulate lipids under special stressed conditions. It was known that nearly all bacteria can produce complex lipids and only a few are found to produce TAG during their stationary phase. Gene regulation mechanisms in fatty acid synthesis of bacteria are clearly studied, and hence, it is easy to genetically modify the bacteria to increase their lipid productivity (Thevenieau and Nicaud, 2013). Subsequently, cyanobacteria are prokaryotic organisms that combine the advantages of both eukaryotic algae, as a photosynthetic microorganism, and *E. coli*, as a tractable and naturally transformable host. CO_2 is converted into energetic compounds through photosynthetic reactions. They have a superiority in the production of biodiesel with high growth rate (only needing 12–24 hours to reach huge biomass), easy culture method, and genetically modifiable biological feature. The lipid produced is present in the thylakoid membranes. This constant cultivation of these microorganisms allows the yearlong cultivation process, leading to much higher oil yield. They are able to survive in high saline water with light as the only energy source. They also grow in a varied temperatures, but they thrive in warmer temperature (Meng et al., 2009). Cyanobacteria area strong candidate for biodiesel production; they don't need arable land, and hence they fall out of the food vs. fuel debate (Quintana et al., 2011). Wastewater rich in nitrogen and phosphorous can be used for sustainable biofuels production. They use excessive amounts of carbon dioxide for their growth, which makes them act as a carbon neutralizer. The growth rate is comparatively higher than the crops used for first-generation biofuel. They accumulate large amounts of lipids, as they contain generous amounts of polysaccharides and can be converted into ethanol. Biohydrogen is produced by cyanobacteria by using the process of biophotolysis of water or by photo-fermentation of organic substrates from photosynthetic bacteria (Kapdan and Kargi, 2006). *Cyanothece* sp. ATCC 51142, a cyanobacteria, has been capable to generate high levels of hydrogen under aerobic conditions (Bandyopadhyay et al., 2010). Cultivation of cyanobacteria is relatively simple and inexpensive compared to other organisms. The main merit of cyanobacteria in the production of ethanol is ferment naturally without the addition of yeast cultures than the fermentation of other traditional energy crops. Heyer and Krumbein

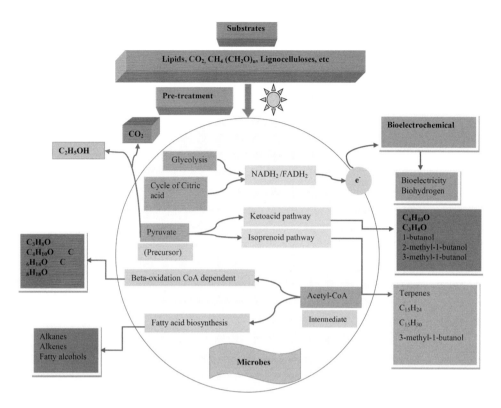

FIGURE 13.3 General metabolic pathways of product formation using microbes.

(1991) screened some cyanobacterial strains to check their ability of fermentation and fermented products. Of the 37 strains studied, it was found that 16 strains were able to produce ethanol as one of the fermentation products, while significant quantities of ethanol were produced in two *Oscillatoria* strains. To increase production of ethanol, genetically modified cyanobacteria are to be used. Even though cyanobacteria allow for a low-cost, low-maintenance, simple production method for biofuels, there are certain challenges that need to be overcome in order to achieve an efficient and sustainable production of biofuel. Various cyanobacterial strains can be used for the production of ethanol; butanol; alkanes; alkenes; biohydrogen; 2, 3-butanediol; 1, 3-propanediol; etc., using different experimental processes, as shown in Figure 13.3.

13.4 Conclusion

The harmful environmental consequences brought on by the excessive use of fossil fuels have caused concerns about petroleum supplies and spurred the search for renewable biofuels. To reduce present biodiesel cost, and also to improve the quality, it can be replaced by conventional fuel. As an alternative to fossil fuel, first- and second-generation biofuels have been in use. Due to the rising concerns about the food vs. fuel debate, researchers have moved in their research toward the cultivation of biofuels from oleaginous microorganisms. Microalgae, cyanobacteria, fungi, yeast, and bacteria are important and promising alternative feedstocks for biofuel production. Producing low-cost microbial fuel from oleaginous microorganisms can be genetically modified to yield more oil. Biotechnology improvement is important for increasing biofuel production, and microorganisms containing high lipid content and/or strains of engineered organisms will potentially become sustainable ways in the future.

REFERENCES

Aarle, I.M.V. and P.A. Olsson (2003). Fungal lipid accumulation and development of mycelial structures by two arbuscular mycorrhizal fungi, *Applied and Environmental Microbiology*, Vol. 69, No. 11, pp. 6762–6767.

Agu, C.V., V. Ujor and T.C. Ezeji (2019). Metabolic engineering of *Clostridium beijerinckii* to improve glycerol metabolism and furfural tolerance, *Biotechnology for Biofuels*, Vol. 12, No. 50, pp. 1–20.

Ayoub, M. and A.Z. Abdullah (2012). Critical review on the current scenario and significance of crude glycerol resulting from biodiesel industry towards more sustainable renewable energy industry, *Renewable and Sustainable Energy Reviews*, Vol. 16, pp. 2671–2686.

Baebprasert, W., S. Jantaro, W. Khetkorn, P. Lindblad and A. Incharoensakdi (2011). Increased H2 production in the cyanobacterium *Synechocystis* sp. strain PCC 6803 by redirecting the electron supply via genetic engineering of the nitrate assimilation pathway, *Metabolic Engineering*, Vol. 13, pp. 610–616.

Bagi, Z., N. Ács, B. Bálint, L. Horváth, K. Dobó, K.R. Perei, G. Rákhely and K.L. Kovács (2007). Biotechnological intensification of biogas production, *Applied Microbiology and Biotechnology*, Vol. 76, No. 2, pp. 473–482. https://doi.org/10.1007/s00253-007-1009-6

Bahr, M., I. Díaz, A. Dominguez, A. Gonzalez Sanchez and R. (2013). Muneoz microalgal-biotechnology as a platform for an integral biogas upgrading and nutrient removal from anaerobic effluents, *Environmental Science & Technology*, Vol. 48, pp. 573–581.

Bandyopadhyay, A., J. Stöckel, H. Min, L.A. Sherman and H.B. Pakrasi (2010). High rates of photobiological H_2 production by a cyanobacterium under aerobic conditions, *Nature Communications*, Vol. 1, No. 139.

Banerjee, A., R. Sharma, Y. Chisti and U.C. Banerjee (2002). *Botryococcus braunii*: A renewable source of hydrocarbons and other chemicals, *Critical Reviews in Biotechnology*, Vol. 22, pp. 245–279.

Behera, S., R. Singh, R. Arora, N.K. Sharma, M. Shukla and S. Kumar (2014). Scope of algae as third generation biofuels, *Frontiers in Bioengineering and Biotechnology*, Vol. 2, p. 90.

Behera, S., R. Singh, R. Arora, N. K. Sharma, M. Shukla, and S. Kumar. (2015). Scope of algae as third generation biofuels, *Front. Bioeng. Biotechnol.* doi: 10.3389/fbioe.2014.00090.

Beopoulos, A., J. Cescut, R. Haddouche, J.-L. Uribelarrea, C. Molina-Jouve and J.-M. Nicaud (2009). *Yarrowia lipolytica* as a model for bio-oil production. *Progress in Lipid Research*, Vol. 48, 375–387.

Chandra, R., H. Takeuchi and T. Hasegawa (2012). Methane production from lignocellulosic agricultural crop wastes: A review in context to second generation of biofuel production, *Renewable and Sustainable Energy Reviews*, Vol. 16, pp. 1462–1476.

Chisti, Y. (2008). Biodiesel from microalgae beats bioethanol, *Trends in Biotechnology*, Vol. 26, pp. 126–131.

Chiu, S.Y., C.Y. Kao, M.T. Tsai, S.C. Ong, C.H. Chen and C.S. Lin (2009). Lipid accumulation and CO_2 utilization of *Nannochloropsis oculata* in response to CO_2 aeration, *Bioresource Technology*, Vol. 100, No. 2, pp. 833–838.

Chung, D., M. Cha, A.M. Guss and J. Westpheling (2014). Direct conversion of plant biomass to ethanol by engineered *Caldicellulosiruptor bescii*, *Proceedings of the National Academy of Sciences of the United States of America*, Vol. 111, pp. 8931–8936.

Collet, P., A. Helias, L. Lardon, M. Ras, R.-A. Goy and J.-P. Steyer (2011). Life-cycle assessment of microalgae culture coupled to biogas production, *Bioresource Technology*, Vol. 102, pp. 207–214.

Converti, A., A.A. Casazza, E.Y. Ortiz, P. Perego and M. Del Borghi (2009). Effect of temperature and nitrogen concentration on the growth and lipid content of *Nannochloropsis oculata* and *Chlorella vulgaris* for biodiesel production, *Chemical Engineering and Processing: Process Intensification*, Vol. 48, pp. 1146–1151.

Davies, F.K., V.H. Work, A.S. Beliaev and M.C. Posewitz (2014). Engineering limonene and bisabolene production in wild type and a glycogen-deficient mutant of *Synechococcus* sp. PCC 7002, *Frontiers in Bio engineering and Biotechnology*, Vol. 2, p. 21.

Deeba, F., V. Pruthi and Y.S. Negi (2016). Converting paper mill sludge into neutral lipids by oleaginous yeast *Cryptococcus vishniaccii* for biodiesel production, *Bioresource Technology*, Vol. 213, pp. 96–102.

Demirbas, A. (2009). Biofuels securing the planet's future energy needs, *Energy Conversion and Management*, Vol. 50, pp. 2239–2249.

Deng, X., Y. Li and X. Fee (2009). Microalgae: A promising feedstock for biodiesel, *African Journal of Microbiology Research*, Vol. 3, pp. 1008–1014.

Edenhofer, O., R. Pichs-Madruga, Y. Sokona, K. Seyboth, P. Matschoss, S. Kadner and C. von Stechow (2011). *Renewable energy sources and climate change mitigation*, Cambridge: Cambridge University Press. https://doi.org/10.1017/CBO9781139151153.

Elreesh, G.A. and D.A.E. Haleem (2013). An effective lipid-producing fungal sp. strain DGB1 and its use for biodiesel production, *African Journal of Biotechnology*, Vol. 12, No. 34, pp. 5347–5353.

Enshaeieh, M., A. Abdoli, I. Nahvi and M. Madani (2013). Selection and optimization of single cell oil production from *Rodotorula* 110 using environmental waste as substrate, *Journal of Cell and Molecular Research*, Vol. 4, No. 2, pp. 68–75.

Eroglu, E., S. Okada and A. Melis (2011). Hydrocarbon productivities in different *Botryococcus* strains: Comparative methods in product quantification, *Journal of Applied Phycology*, Vol. 23, 763–775.

Feng, Y., C. Li and D. Zhang (2011). Lipid production of *Chlorella vulgaris* cultured in artificial wastewater medium, *Bioresource Technology*, Vol. 102, No. 1, pp. 101–105.

Fliegerová, K., J. Procházka, J. Mrázek, Z. Novotná, L. Štrosová and M. Dohányos (2012). Biogas and rumen fungi. In: Litonjua, R. and Cvetkovski, I. (eds). *Biogas: Production, consumption and applications*, Prague: Nova Science Pub Inc, pp. 161–180.

Fontana, G. (2010). Genetically modified micro-organisms: The EU regulatory framework and the new Directive 2009/41/EC on the contained use, *Chemical Engineering Transactions*, Vol. 20, pp. 1–6.

Frigon, J.-C., F. Matteau-Lebrun, R. HamaniAbdou, P.J. McGinn, S.J.B. O'Leary and S.R. Guiot (2013). Screening microalgae strains for their productivity in methane following anaerobic digestion, *Applied Energy*, Vol. 108, pp. 100–107.

Giselrød, H.R., V. Patil and K. Tran (2008). Towards sustainable production of biofuels from microalgae, *International Journal of Molecular Sciences*, Vol. 9, p. 1188e95.

Gouda, M.K., S.H. Omar and L.M. Aouad (2008). Single cell oil production by *Gordonia* sp. DG using agro-industrial wastes, *World Journal of Microbiology Biotechnology*, Vol. 24, pp. 1703–1711.

Helwani, Z., M.R. Othman, N. Aziz, J. Kim and W.J.N. Fernando (2009). Solid heterogeneous catalysts for transesterification of triglycerides with methanol: A review. *Applied Catalysis A: General*, Vol. 363, pp. 1–10.

Heyer, H. and W.E. Krumbein (1991). Excretion of fermentation products in dark and anaerobically incubated cyanobacteria, *Archives of Microbiology*, Vol. 155, pp. 284–287.

Hirokawa Y., Y. Maki, T. Tatsuke and T. Hanai (2016). Cyanobacterial production of 1, 3-propanediol directly from carbon dioxide using a synthetic metabolic pathway. *Metabolic Engineering*, Vol. 34, pp. 97–103.

Holladay, J.D., J. Hu, D.L. King and Y. Wang (2009). An overview of hydrogen production technologies, *Catalysis Today*, Vol. 139, No. 4, pp. 244–260.

Holm-Nielsen, J.B., T. Al Seadi and P. Oleskowicz-Popiel (2009). The future of anaerobic digestion and biogas utilization, *Bioresource Technology*, Vol. 100, pp. 5478–5484.

Hsieh, C.H. and W.T. Wu (2001). Cultivation of microalgae for oil production with a cultivation strategy of urea limitation, *Bioresource Technology*, Vol. 100, No. 17, pp. 3921–3926.

Hsieh, C.H. and W.T. Wu (2009). Cultivation of microalgae for oil production with a cultivation strategy of urea limitation, *Bioresource Technology*, Vol. 100, pp. 3921–3926.

Hu, Q., M. Sommerfeld, E. Jarvis, M. Ghirardi, M. Posewitz, M. Seibert and A. Darzins (2008). Microalgal triacyglycerols as feedstocks for biofuel production: Perspectives and advances, *Plant Journal*, Vol. 54, pp. 621–639.

Huang J., D. Chen, Y. Wei, Q. Wang, Z. Li, and Y. Chen, et al. (2014). Direct ethanol production from lignocellulosic sugars and sugarcane bagasse by a recombinant *Trichoderma reesei* strain HJ48, *Scientific World Journal*, Vol. 2014, Article ID 798683.

Jones, C.S. and S.P. Mayfield (2012) Alga's biofuels: Versatility for the future of bioenergy, *Current Opinion in Biotechnology*, Vol. 23, pp. 346–351.

Kalscheuer, R., T. Stolting and A. Steinbuche (2006). Microdiesel: *Escherichia coli* engineered for fuel production, *Microbiology*, Vol. 152, pp. 2529–2536.

Kapdan, I.K. and F. Kargi (2006). Bio-hydrogen production from waste materials, *Enzyme and Microbial Technology*, Vol. 38, pp. 569–582.

Khetkorn, W., N. Khanna, A. Incharoensakdi and P. Lindblad (2013). Metabolic and genetic engineering of cyanobacteria for enhanced hydrogen production, *Biofuels*, Vol. 4, pp. 535–561.

Khot, M., S. Kamat, S. Zinjarde, A. Pant, B. Chopade and A. Ravi Kumar (2012). Single cell oil of *Oleaginous fungi* from the tropical mangrove wetlands as a potential feedstock for biodiesel, *Microbial Cell Factories*, Vol. 11, No. 71, pp. 1–13.

Kitcha, S. and B. Cheirsilp (2011). Screening of oleaginous yeasts and optimization for lipid production using crude glycerol as a carbon source, *Energy Procedia*, Vol. 9, pp. 274–282.

Kremer, T.A., B. LaSarre, A.L. Posto and J.B. McKinlay (2015). N2 gas is an effective fertilizer for bioethanol production by *Zymomonas mobilis*, *Proceedings of the National Academy of Sciences of the United States of America*, Vol. 112, pp. 2222–2226.

Leong, W.-H., J.-W. Lim, M.-K. Lam, Y. Uemura and Y.-C.J.R. Ho (2018). Third generation biofuels: A nutritional perspective in enhancing microbial lipid production, *Renewable and Sustainable Energy Reviews*, Vol. 91, pp. 950–961.

Li, S.L., Q. Lin, X.R. Li, H. Xu, Y.X. Yang, D.R. Qiao and Y. Cao (2012). Biodiversity of the oleaginous microorganisms in Tibetan plateau, *Brazilian Journal of Microbiology*, Vol. 43, No. 2, pp. 627–634.

Liang, M.H. and J.G. Jiang (2013). Advancing oleaginous microorganisms to produce lipid via metabolic engineering technology, *Progress in Lipid Research*, Vol. 52, No. 4, pp. 395–408.

Lin, P.P., L. Mi, A.H. Morioka, K.M. Yoshino, S. Konishi and S.C. Xu (2015). Consolidated bioprocessing of cellulose to isobutanol using *Clostridium thermocellum*. *Metabolic Engineering*, Vol. 31, pp. 44–52.

Liu, Z.Y., G.C. Wang and B.C. Zhou (2008). Effect of iron on growth and lipid accumulation in *Chlorella vulgaris*, *Bioresource Technology*, Vol. 99, No. 11, pp. 4717–4722.

Lu, X., H. Vora and C. Khosla (2008). Overproduction of free fatty acids in *E. coli*: Implications for biodiesel production, *Metabolic Engineering*, Vol. 10, No. 6, pp. 333–339.

Lütke-Eversloh T. and H. Bahl (2011). Metabolic engineering of *Clostridium acetobutylicum*: Recent advances to improve butanol production, *Current Opinion in Biotechnology*, Vol. 22, pp. 634–647.

Melis, A. (2002). Green alga hydrogen production: Progress, challenges and prospects, *International Journal of Hydrogen Energy*, Vol. 27, pp. 1217–1228.

Meng, J., X. Yang, L. Xu, Q. Zhang and M. Nie Xian (2009). Biodiesel production from oleaginous microorganisms, *Renewable Energy*, Vol. 34, pp. 1–5.

Miah, M.S., C. Tada and S. Sawayama (2004). Enhancement of biogas production from sewage sludge with the addition of *Geobacillus* sp. strain AT1 culture, *Japanese Journal of Water Treatment Biology*, Vol. 40, No. 3, pp. 97–104. https://doi.org/10.2521/jswtb.40.97

Mohr, A. and S. Raman (2013). Lessons from first generation biofuels and implications for the sustainability appraisal of second generation biofuels, *Energy Policy*, Vol. 63, No. 100, pp. 114–122.

Nagarajan, D., D.-J. Lee, A. Kondo and J.-S. Chang (2017). Recent insights into biohydrogen production by microalgae—From biophotolysis to dark fermentation, *Bioresource Technology*, Vol. 227, pp. 373–387.

Neema, P.M. and A. Kumari (2013). Isolation of lipid producing yeast and fungi from secondary sewage sludge and soil, *Australian Journal of Basic and Applied Sciences*, Vol. 7, No. 9, pp. 283–288.

Neves, L., R. Oliveira and M.M. Alves (2006). Anaerobic co-digestion of coffee waste and sewage sludge, *Waste Management*, Vol. 26, pp. 176–181.

Nielsen, D.R., E. Leonard, S.H. Yoon, H.C. Tseng, C. Yuan and K.L.J. Prather (2009). Engineering alternative butanol production platforms in heterologous bacteria, *Metabolic Engineering*, Vol. 11, pp. 262–273.

Nyberg, M., T. Heidorn and P. Lindblad (2015). Hydrogen production by the engineered cyanobacterial strain Nostoc PCC 7120 ΔhupW examined in a flat panel photobioreactor, *Journal of Biotechnology*, Vol. 215, pp. 35–43.

Panahi, H.K.S., M. Dehhaghi, J.E. Kinder and T.C. Ezeji (2019). A review on green liquid fuels for the transportation sector: A prospect of microbial solutions to climate change, *Biofuel Research Journal*, Vol. 23, pp. 995–1024.

Papanikolaou, S., M. Komaitis and G. Aggelis (2004). Single cell oil (SCO) production by *Mortierella isabellina* grown on high-sugar content media, *Bioresource Technology*, Vol. 95, No. 3, pp. 287–291.

Passos, F., P.H.M. Cordeiro, B.E.L. Baeta, S.F. de Aquino and S.I. Perez-Elvira (2018). Anaerobic co-digestion of coffee husks and microalgal biomass after thermal hydrolysis, *Bioresource Technology*, Vol. 253, pp. 49–54.

Patil, V., K.-Q. Tran and H.R. Giselrød (2008). Towards sustainable production of biofuels from microalgae, *International Journal of Molecular Sciences*, Vol. 9, No. 7, pp. 1188–1195.

Patnayak, S. and A. Sree (2005). Screening of bacterial associates of marine sponges for single cell oil and PUFA, *Letters in Applied Microbiology*, Vol. 40, No. 5, pp. 358–363.

Quintana, N., F. Van der Kooy, M.D. Van de Rhee, G.P. Voshol and R. Verpoorte (2011). Renewable energy from Cyanobacteria: energy production optimization by metabolic pathway engineering, *Applied Microbiology and Biotechnology*, Vol. 91, pp. 471–490.

Raja, R., S. Hemaiswarya, N. Ashok Kumar, S. Sridhar and R. Rengasamy (2008). A perspective on the biotechnological potential of microalgae, *Critical Reviews in Microbiology*, Vol. 34, p. 77.

Ravikumar, K., J. Dakshayini and S.T. Girisha (2012). Biodiesel production from *Oleaginous fungi*, *International Journal of Life Sciences*, Vol. 6, No. 1, pp. 43–49.

Robak, K. and M. Balcerek (2018). Review of second generation bioethanol production from residual biomass, *Food Technology and Biotechnology*, Vol. 56, No. 2, pp. 174–187.

Romero-García, J.M., C. Martínez-Pati-o, E. Ruiz, I. Romero and E. Castro (2016). Ethanol production from olive stone hydrolysates by xylose fermenting microorganisms, *Bioethanol*, Vol. 2, pp. 51–65.

Ruan, Z., M. Zanotti, X. Wang, C. Ducey and Y. Liu (2012). Evaluation of lipid accumulation from lignocellulosic sugars by *Mortierella isabellina* for biodiesel production, *Bioresource Technology*, Vol. 110, pp. 198–205.

Saratale, R.G., G. Kumar, R. Banu, A. Xia, S. Periyasamy and G.D. Saratale (2018). A critical review on anaerobic digestion of microalgae and macroalgae and co-digestion of biomass for enhanced methane generation, *Bioresource Technology*, Vol. 262, pp. 319–332.

Sergeeva, E.Y., L.A. Galanina, D.A. Andrianova and E.P. Feofilova (2008). Lipids of filamentous fungi as a material for producing biodiesel fuel, *Applied Biochemistry and Microbiology*, Vol. 44, No. 5, pp. 523–527.

Shen, C.R., E.I. Lan, Y. Dekishima, A. Baez, K.M. Cho, J.C. Liao (2011). Driving forces enable high-titer anaerobic 1-butanol synthesis in *Escherichia coli*. *Applied and Environmental Microbiology*, Vol. 77, pp. 2905–2915.

Sialve, B., N. Bernet and O. Bernard (2009). Anaerobic digestion of microalgae as a necessary step to make microalgal biodiesel sustainable, *Biotechnology Advances*, Vol. 27, pp. 409–416.

Smeets, E., A. Tabeau, S. Van Berkum, J. Moorad, H. Van Meijl and G. Woltjer (2014). The impact of the rebound effect of the use of first generation biofuels in the EU on greenhouse gas emissions: A critical review, *Renewable and Sustainable Energy Reviews*, Vol. 38, pp. 393–403.

Somashekar, D., G. Venkateshwaran, C. Srividya, Krishnanand, K. Sambaiahand and B.R. Lokesh (2001). Efficacy of extraction methods for lipid and fatty acid composition from fungal cultures, *World Journal of Microbiology and Biotechnology*, Vol. 17, No. 3, pp. 317–320.

Song, M., H.D. Pham, J. Seon and H.C. Woo (2015). Overview of anaerobic digestion process for biofuels production from marine macroalgae: A developmental perspective on brown algae, *Korean Journal of Chemical Engineering*, Vol. 32, pp. 567–575.

Spolaore, P., C. Joannis-Cassan, E. Duran and A. Isambet (2006). Commercial applications of microalgae, *Journal of Bioscience and Bioengineering*, Vol. 101, No. 2, pp. 87–96.

Subhash, G.V. and S.V. Mohan (2014). Lipid accumulation for biodiesel production by oleaginous fungus *Aspergillus awamori*: Influence of critical factors, *Fuel*, Vol. 116, pp. 509–515.

Thevenieau, F. and J.M. Nicaud (2013). Microorganisms as sources of oils, *Oil Seeds and Fat, Crops and Lipids*, Vol. 20, No. 6, pp. 1–8.

Tonon, T., D. Harvey, T.R. Larson and I.A. Graham (2002). Long chain polyunsaturated fatty acid production and partitioning to triacylglycerols in four microalgae, *Phytochemistry*, Vol. 61, pp. 15–24.

Vicente, G., M. Martinez and J. Aracil (2004). Integrated biodiesel production: A comparison of different homogeneous catalysts systems, *Bioresource Technology*, Vol. 92, No. 3, pp. 297–305.

Patil, V., K-Q. Tran and H.R. Giselrød (2008). Towards sustainable production of biofuels from microalgae, *International Journal of Molecular Science*, Vol. 9, No. 7, pp. 1188–1195.

Wagner, A.O., T. Schwarzenauer and P. Illmer (2013). Improvement of methane generation capacity by aerobic pre-treatment of organic waste with a cellulolytic *Trichoderma viride* culture, *Journal of Environmental Management*, Vol. 129, pp. 357–360.

Wahlen, B.D., M.R. Morgan, A.T. McCurdy, R.M. Willis, M.D. Morgan, D.J. Dye, B.Bugbee, B.D. Wood and L.C. Seefeldt (2012). Biodiesel from microalgae, yeast, and bacteria: Engine performance and exhaust emissions, *Energy Fuels*, Vol. 27, No. 1, pp. 220–228.

Wang, X., K. Bao, W. Cao, Y. Zhao and C.W. Hu (2017). Screening of microalgae for integral biogas slurry nutrient removal and biogas upgrading by different microalgae cultivation technology, *Scientific Reports*, Vol. 7, p. 5426.

Ward, A.J., D.M. Lewis and F.B. Green (2014). Anaerobic digestion of algae biomass: A review, *Algal Research*, Vol. 5, pp. 204–214.

Westfall, P.J., D.J. Pitera, J.R. Lenihan, D. Eng, F.X. Woolard, R. Regentin. et al. (2012). Production of amorphadiene in yeast, and its conversion to dihydroartemisinic acid, a precursor to the antimalarial agent artemisinin. *Proc. Natl. Acad. Sci. U.S.A.* 109, E111–E118. doi: 10.1073/pnas.1110740109.

Xia, A. and J.D. Murphy (2016). Microalgal cultivation in treating liquid digestate from biogas systems, *Trends in Biotechnology*, Vol. 34, pp. 264–275.

Xin, M., J. Yang, X. Xu, L. Zhang, Q. Nie and M. Xian (2008). Biodiesel production from oleaginous microorganisms, *Renewable Energy*, Vol. 34, No. 1, pp. 1–5.

Xu H., X. Miao and Q. Wu (2006). High quality biodiesel production from a microalga *Chlorella protothecoides* by heterotrophic growth in fermenters, *Journal of Biotechnology*, Vol. 126, pp. 499–507.

Yang S., A. Mohagheghi, M.A. Franden, Y.C. Chou, X. Chen, N. Dowe, et al. (2016). Metabolic engineering of *Zymomonas mobilis* for 2, 3-butanediol production from lignocellulosic biomass sugars, *Biotechnology for Biofuels*, Vol. 9, p. 189.

Yoo, C., S.Y. Jun, J.Y. Lee, C.Y. Ahn and H.M. Oh (2010). Selection of microalgae for lipid production under high levels carbon dioxide, *Bioresource Technology*, Vol. 101, No. 1, pp. 71–74.

Yu A.Q., N.K.P. Juwono, J.L. Foo, S.S.J. Leong and M.W. Chang (2016). Metabolic engineering of *Saccharomyces cerevisiae* for the overproduction of short branched-chain fatty acids, *Metabolic Engineering*, Vol. 34, pp. 36–43.

Yuzbasheva, E.Y., T.V. Yuzbashev, E.B. Mostova, N.I. Perkovskaya and S.P. Sineokii (2014). Microbial synthesis of biodiesel and its prospects, *Applied Biochemistry and Microbiology*, Vol. 50, No. 9, pp. 789–801.

Zabed, H., J.N. Sahu and A. Suely (2017). Bioethanol production from lignocellulosic biomass: an overview of pretreatment, hydrolysis, and fermentation. In: Mondal, P. and Daiai, A.K. (eds.) *Sustainable utilization of natural resources*. Boca Raton, FL: CRC Press, pp. 145–186.

Zamalloa, C., N. Boon and W. Verstraete (2012). Anaerobic digestibility of *Scenedesmus obliquus* and *Phaeodactylum tricornutum* under mesophilic and thermophilic conditions, *Applied Energy*, Vol. 92, pp. 733–738.

Zhao, Y., S. Sun, C. Hu, H. Zhang, J. Xu and L. Ping (2015). Performance of three microalgal strains in biogas slurry purification and biogas upgrade in response to various mixed light-emitting diode light wavelengths, *Bioresource Technology*, Vol. 187, pp. 338–345.

14 Biodiesel Production from Mimusops elengi Seed Oil through Means of Co-Solvent-Based Transesterification Using an Ionic Liquid Catalyst

Gokul Raghavendra Srinivasan, Shalini Palani, Mamoona Munir,
Muhammad Saeed, and Ranjitha Jambulingam

CONTENTS

14.1 Introduction 299
14.2 Materials and Methodology 302
    14.2.1 Collection and Preparation of Sample 302
    14.2.2 Proximate Analysis on *M. elengi* Seeds 302
    14.2.3 Mechanical Extraction of MEO 303
    14.2.4 Refining of MEO 303
    14.2.5 Characterization of MEO and MeMEE 303
    14.2.6 Methanol- and Ethanol-Based Transesterification of MEO 304
    14.2.7 Co-Solvent-Based Transesterification of MEO 304
    14.2.8 Characterization of [D-Vaemim]Cl Catalyst 305
    14.2.9 Fuel Properties of MEO Biodiesel 306
14.3 Results and Discussions 306
    14.3.1 Characterization of MEO and MeMEE 306
    14.3.2 Optimization of Methanol and Ethanol Transesterification
            (Using Statistical Optimization) 307
    14.3.3 Optimization of Methanol-Ethanol-Based Co-Solvent Transesterification
            (Using OFAT Method) 309
            14.3.3.1 Effect of Methanol-to-Ethanol Ratio on MeMEE Yield 309
            14.3.3.2 Effect of Catalyst Concentration on MeMEE Yield 311
            14.3.3.3 Effect of Reaction Temperature on MeMEE Yield 312
            14.3.3.4 Effect of Reaction Time on MeMEE Yield 312
    14.3.4 Fuel Properties of Diesel and MEO Biodiesel 313
14.4 Conclusion 315
References 316

14.1 Introduction

Overconsumption and exploitation of existing fossil fuels have raised concerns about ensured energy supply for future generations, besides the threats like environmental pollution, greenhouse gases, and global warming caused by them. Serving as an immediate yet an effective solution, renewable biofuels—especially biodiesel—satisfies both the everlasting demand and also projects sustainability (Fadhil et al., 2020; Munir et al., 2019). Technically, biodiesel is a long carbon chain fatty acid molecule bonded covalently with an alcohol moiety at its carboxylic end, and tends to be high oxidized fuel with good cetane number and decent calorific value, which exhibits superior engine performance and reduced emissions

(Srinivasan and Jambulingam, 2018). In general, these fatty acids (FAs) are available in various natural sources in form of triglycerides and occur in different types of oils and fats. Most commonly preferred sources rich in these triglycerides include non-edible seeds (Munir et al., 2019), discarded fleshing and meat processing wastes (Srinivasan et al., 2020), and greases and waste cooking oils (Issariyakul et al., 2007).

Among these, oil extracted from non-edible seeds are deemed as highly preferred feedstocks for biodiesel, accounting for the easy availability of raw materials, simplicity in extracting and refining, easy handling, and fewest concerns about crystallization. In fact, many non-edible seeds have been identified as effective raw material for biodiesel production, yet seeds with good lipid potential and easy availability are always preferred over the rarely occurring plants. One such easily available raw material available in abundance is seeds of the *Mimusops elengi* (*M. elengi*) tree (Krupakaran et al., 2018); and indeed, it has been proven to be the widely celebrated feedstock due to its multi-speciality features. To begin with, *M. elengi* trees are floral species commonly occurring widely across Indian soil and hold their applications in numerous pharmaceutical and chemical industries. Preciously, each part of this plant possesses value-added property and serves a vital role in medical and allied applications. Looking into its classification in the Plantae (plant) kingdom, *M. elengi* belongs to the order of Ericales and family of Sapotaceae under the genus *Mimusops*, with species name as *M. elengi* L. this tree occurs most commonly all over the southern part of India and in the evergreen forests of the Andaman Islands. These trees are easily identified based on their leafy head and short erect trunk with smooth and scaly bark of gray color, and their leaves in elliptical shape with their tapered heads and acuminated bases. Besides that, bud ovoid shaped flowers in these trees produce fragrance and fertilizes in berry-shaped fruit bearing yellow color when ripe and high oil potential seeds (Gami et al., 2012).

As mentioned earlier, different parts of this tree can be used for various medical and allied purposes, and it has been widely used in the field of ayurvedic treatments (Bharat and Parabia, 2010). For instance, the gray-colored scaly bark and fragrant flowers taste sweet and can be used for treatments of teeth and gum diseases and biliousness in view of their antiscorbutic, alexipharmic, cardiotonic, astringent, and antispasmodic properties, while its flowers are also used to treat headaches, nasal disorders, liver problems, and asthma, and are also used as stimulant laxatives (Bhuyan et al., 2004). The leaves of the *M. elengi* tree possess very good antipyretic and analgesic properties (Sakshi et al., 2011). Also, the fruits of this tree provide effective solutions for aphrodisiac, bowel-astringent, diuretic, and gonorrhea purposes, while their seeds are used as healing agent for headaches and nasal irritations, and restore loose teeth, apart from treating the ailments mentioned earlier (Shanmugam et al., 2011).

Though these seeds possess greater medical importance, only a small quantity of them are being used effectively, leaving behind a large quantity of these seeds as unused biomass. Eventually, these seeds showcase a good amount of oil content, which can be rendered and processed into useful biofuel for energy applications (Deepanraj et al., 2015). Looking into the lipid chemistry of *M. elengi* seeds, capric acid, lauric acid, myristic acid, palmitic acid (16.71%), stearic acid (17.23%), oleic acid (53.48%), linoleic acid (16.71%), and arachidic acid were identified as their dominant FAs; with traces of β- and γ-sitosterol in the *M. elengi* oil (Gami et al., 2012). Ensuring their presence, these seeds are identified as a high-potential raw material for extracting a high quality of oil with desired amounts of FAs. This oil can be used for engine applications in their blended or processed forms. In fact, the FAs in this oil perform better in compression ignition engines in their ester forms due to their reduced density and viscosity, and increased cetane numbers and calorific value.

In general, conversion of these FAs into their ester form is done through an effective technique called transesterification, which involves in reacting the extracted oil (triglycerides) from these seeds with any short chained alcohols (especially methanol or ethanol) in the presence of suitable catalyst, which tends to enhance the overall rate of the reaction (Van Gerpen, 2005). Indeed, many researches have been focused globally on improvising the overall reaction rate and reduced the energy consumption with minimal resource use during the production of biodiesel, in order to face competitive fossil fuels in the commercial market. For example, Hsiao and colleagues (2018) recorded highest conversion rate of 97.1% upon transesterification of cooking oil in a high-speed homogenizer using sodium methoxide as a homogeneous catalyst (Hsiao et al., 2018). Likewise, even ultrasonics and microwaves are used for assisting transesterification of oils into biodiesel, and have been found to produce very high promising yields up to

98% under reduced reaction time (Varghese et al., 2018). Surprisingly, the resulting biodiesel properties were well bound within the permitted range and were found to be readily appropriate for engine applications (Inayat et al., 2018).

Apart from that, novel homogeneous and heterogeneous catalysts have proven to boost the overall yield and reaction rate during the biodiesel production, simultaneously preventing saponification of the feedstock (Varghese et al., 2018; Srinivasan et al., 2018). One such catalyst is ionic liquid (IL), which showcases superior catalytic activities when compared to other known catalysts. Some of their superior qualities include: its existence in liquid phase at room temperatures, non-volatility as a result of sufficient ion attraction, lower viscosity, enhanced catalytic activities, and promised effectiveness even after multiple uses (enhanced recyclability) (Liang et al., 2010; Umar et al., 2020). In addition, these IL catalysts avoided the formation of soaps during transesterification in the event of high free fatty acids (FFA) feedstocks (Srinivasan et al., 2020).

Supporting this, Srinivasan and colleagues (2020) transesterified leather fleshing and meat processing wastes–based animal fat using ethanol as primary solvent and D-valine amido ethyl methyl imidazolium chloride as IL catalyst, and reported a maximum biodiesel yield of 97.36%. The optimum reaction parameters were as follows: molar ratio 1:6 (WAF:ethanol), catalyst concentration 10% wt. of D-valine amido ethyl methyl imidazolium chloride ([D-Vaemim]Cl) catalyst, reaction temperature 75°C, and reaction time 90 minutes. Though this study focused on the influence of fatty acids during the biodiesel production and engine characteristics, this study strongly concluded that these IL catalysts were found to be effective up to ten cycles with promised efficiency (Srinivasan et al., 2020).

Looking away from this, the quality of biodiesel always depends on its fuel properties, and must be always superior, for effective engine applications. However, it is impossible to always produce high-quality biodiesel, owing to the physico-chemical properties of the oil used as raw feedstock. Yet, these properties can be enhanced by simply blending any organic solvent as an oxidant or additive; simply blending with another superior quality biodiesel (Shrivastava et al., 2020). Though promising, these techniques fail during stages of commercial production, considering the overall energy demand and capital investment involved. Eventually, these complications can be encountered by introducing a co-solvent during the transesterification reaction, which enhances the overall reaction kinetics through reducing the reaction time and produces new subspecies of ester molecules, which tends to enhance the fuel properties of resultant biodiesel (Jambulingam et al., 2020).

Among the currently known state-of-the-art techniques in biodiesel production, co-solvent–based transesterification has been found to be more effective and promising on account of the following advantages: (a) influence of co-solvent was well reflected in the resultant biodiesel and their fuel properties; (b) increased the miscible rate between oil and solvents, thus reducing the mass transfer resistance (Sakthivel et al., 2013); and (c) enhanced reaction rate due to reduced reaction time. In general, transesterification of triglycerides involves the cleavage of FA moieties from the glyceride spine to recombine with alkyl ion from the solvent through nucleophilic substitution to form alkyl acid fatty acids esters (Trejo-Zárraga et al., 2018), however, addition of co-solvent tends to act as an effective ester exchange agent, thereby improving the rate of nucleophilic substitution. Also, the product of this technique eased down complications during the biodiesel-glycerol phase separation and remained inert against soap formation (Sakthivel et al., 2013).

Advantages of this technique on different types of feedstocks—especially on non-edible seeds, vegetable, and waste cooking oils—have been well illustrated based on numerous studies carried out by different researchers. For instance, Alhassan and colleagues (2014) observed the influence of different co-solvents in deciding the overall reaction parameters for the transesterification of cotton seed oil; by maintaining methanol as the primary solvent whereas solvents like acetone, dichlorobenzene and diethyl ether were used as co-solvents. Upon identifying acetone as the most ideal co-solvent, the optimal reaction parameters for transesterification of the cotton oilseed were identified as follows: molar ratio 1:6, catalyst concentration 0.75% w/w, reaction temperature 55°C, and reaction time 10 minutes. Interestingly, the use of co-solvent reduced the overall reaction duration by 60%, in addition to promising 90% of yield with the first ten minutes of the reaction (Alhassan et al., 2014).

Following this, Luu et al. (2014) attempted the transesterification on waste cooking oil using acetone as co-solvent for the following reaction parameters: molar ratio 1:5, co-solvent volume 20% acetone wt., catalyst concentration 1 wt percent potassium hydroxide, reaction temperature 40°C, and time 30 minutes. Indeed, addition of acetone as a co-solvent reduced the overall reaction time, thereby increasing the reaction rate, as co-solvent helped in overcoming the mass transfer barrier between the solvent and oil. Upon looking in the concentration, post refining, the concentration of residual methanol and acetone were found to be 95 ppm and 247 ppm, respectively (Luu et al., 2014).

In like manner, Julianto and Nurlestari (2018) carried out the acetone-based transesterification on waste cooking oil using methanol as primary solvent, and attempted in determining the ideal acetone-to-methanol molar ratio for the effective biodiesel production by maintaining the catalyst concentration as 1% (w/w) KOH for a reaction duration of 15 minutes at room temperature. Here, the oil-to-methanol molar ratio was maintained at 1:12, while the acetone-to-methanol ratio was varied as 1:4, 1:2, and 1:1, and they concluded 1:4 as the most optimum acetone-to-methanol ratio, which yielded a maximum yield of 99.93% (Julianto and Nurlestari, 2018).

Jambulingam and colleagues (2020) carried out methanol-ethanol–based transesterification on waste beef tallow to study the role of ethanol in improvising the reaction rate and fuel properties of resultant biodiesel. Upon adding the ethanol with methanol as co-solvent in ratio of 3:3, the optimized reaction parameters were found to be as follows (a) oil to alcohol molar ratio, 1:6; (b) catalyst concentration, 0.55% KOH; (c) reaction temperature, 70°C; and (d) reaction time, 35 minutes—and reported a maximum yield of 97.2% ± 1.08%. Looking further, this study strongly concluded that addition of ethanol increased the overall reaction kinetics by helping the tallow and solvent overcome mass transfer barrier, and introduced tallow ethyl esters into the resultant biodiesel, which enhanced the overall fuel properties of the latter remarkably. Also, this study projected that ethanol failed to react effectively with unsaturated fatty acids, thus leading to reduced concentration of unsaturated fatty acid ethyl ester (Jambulingam et al., 2020).

Indeed, these studies tend to prove that co-solvent–assisted transesterification is an effective technique for producing high-quality biodiesel under reduced reaction time and enhanced reaction kinetics. This technique can especially be used for unsaturated feedstocks, considering very minimal requirements for resources. Since *Mimusops elengi* oil (MEO) exhibits increased unsaturation and long-chain FAs in its triglyceride molecules, it can be chosen as an ideal feedstock for biodiesel production. Contradicting the poor performance of unsaturated biodiesel, this co-solvent technique can be attempted on this MEO, aiming at producing high-quality biodiesel. Considering this, the present study aims to carry out co-solvent transesterification, with methanol as primary solvent and ethanol as co-solvent, on *M. elengi* oil extracted from seed kernels. To understand better, the study also focuses on process optimization and the fuel properties of the resultant MEO biodiesel (MeMEE). To enhance the reaction rate further, novel IL catalyst [D-Vaemim]Cl has been used for the study purposes.

14.2 Materials and Methodology

14.2.1 Collection and Preparation of Sample

The dried fruits of *M. elengi* were collected from in and around the campus and nursery of VIT University (Vellore Institute of Technology), Vellore, Tamil Nadu, India. The collected fruits were scrapped to remove the dried outer covering and the seeds were washed with water to remove any remaining flesh. Since the kernels of *M. elengi* seeds were contained in semi-hard casing, these kernels were broken free from the casings, either using a rubber mallet or stomper tool depending on the quantity available. After cracking, the kernels were separated from the casing wastes and were preserved in an airtight container. For the purpose of this study, approximately 500 g of *M. elengi* kernels were heated in a hot air oven at 110°C for around five hours to remove moisture content from it.

14.2.2 Proximate Analysis on *M. elengi* Seeds

As per description by Association of Official Analytical Chemists (AOAC, 2005), the preserved kernel sample was evaluated for its proximate composition to determine the moisture, lipid, protein, and ash content. To begin with, moisture content in the kernel samples were determined using a hot air oven as per AOAC standards (AOAC, 2005) using a material test chamber M720, by drying the samples until the constant weight was achieved. The moisture content in the kernel sample was calculated using Equation 1 (Ooi et al., 2012):

$$Moisture\ content\ (in\ \%) = \left(1 - \frac{Weight_{dry\ sample}}{Weight_{wet\ sample}}\right) * 100 \qquad (1)$$

Continuing on, the lipid content in the dried kernel samples was estimated by means of the Soxtec method by using Soxtec 2050 automated analyzer as described in the work by Noureddini and Byun (2010). For this purpose, petroleum ether was taken as extraction solvent and the overall lipid content in the dried kernel samples was computed using Equation 2 (Ooi et al., 2012):

$$Lipid\ Content\ (in\ \%) = \left(\frac{Weight_{(extraction\ cup + residue)} - Weight_{extraction\ cup}}{Weight_{sample}}\right) * 100 \qquad (2)$$

Next, the protein content in the kernel samples was decided based on the total nitrogen content calculated using a Kjeltec 2200 Auto Distillation Unit as per the method defined by Ng and colleagues (2008). For the purpose of nitrogen-to-protein conversion, a factor of 4.4 was considered and the total protein content was calculated from the nitrogen content determined using this technique. Finally, the total ash content of the kernel sample was determined using dry ashing method (AOAC, 2005), which involves with the incineration of these samples at 550°C in a furnace. After incineration, the residual inorganic materials were cooled and weighed for the ash content by comparing with initial weight of the sample. To ensure increased accuracy in the results, the experiments were carried out in triplicates and presented in terms of their mean ± standard deviation (Fadhil et al., 2015).

14.2.3 Mechanical Extraction of MEO

MEO was extracted from the seed kernels through the means of mechanical extraction using a mini screw press. This equipment was equipped with suitable base plate, feed hopper, and oil discharge chute, and was capable of handling biomass up to 1 kg for extraction; the dried seed kernels were loaded into the screw press cumulatively and the press was operated manually by rotating the attached crank. The force exerted due to the spinning of worm shaft, along with the axial pressure caused by the volumetric compression of the screw press, extracted the oil from kernels effectively (Bhuiya et al., 2020). The extraction process was carried out for 20 minutes and the yield of oil (in wt.%) was calculated using Equation 3 (Bhuiya et al., 2020):

$$Yield\ of\ oil\ (in\ \%) = \frac{weight\ of\ oil\ extracted\ (in\ g)}{weight\ of\ kernel\ (in\ g)} * 100 \qquad (3)$$

14.2.4 Refining of MEO

Following extraction, MEO was refined to remove any solid sediment and dissolved chemicals present before proceeding to its transesterification. To begin with, the extracted oil was filtered to remove any solid sediment from it, and it was subjected to degumming to remove phospholipids carried away during extraction. Accordingly, the filtered MEO were continuously heated at 60°C, and 1% of orthophosphoric acid was added to it under continuous stirring for first ten minutes, with prolonged heating for another ten minutes. After completion of degumming, residual phospholipids were settled down at bottom as dark brown residual lecithin.

14.2.5 Characterization of MEO and MeMEE

The FAs in the triglycerides of refined MEO were characterized using gas chromatographic (GC) quantification, with their presence identified based on the peak obtained for their corresponding retention time and area under the peak signifying their availability. The refined MEO was prepared for GC

TABLE 14.1

Technical Specifications of Gas Chromatograph–Mass Spectrometer for Characterization Purposes

Gas Chromatograph		Mass Spectrometer	
Equipment	Agilent 6890 chromatograph	**Equipment**	JOEL GC mate II bench top
Injector Liner	direct/2 mm	**Type**	Double focusing magnetic sector MS
Column	15 m all tech ec-5 (25 μm ID, 0.25 μm thickness)	**Operation Mode**	Electron ionization (EI) mode
Split Ratio	10:01	**Software**	TSS-2001
Oven Temperature	35°C/2 min	**Resolving Power**	1000 (20% height definition)
Ramp	20°C/min @ 300°C for five min	**Scanning Feature**	25–700 m/z @ 0.3 s/scan
Helium Carrier Gas	2 ml/min (constant flow mode)	**Interscan Delay**	0.2 s

Source: Jambulingam et al. (2020).

characterization by refluxing it with methanol, along with 1%–2% of concentrated sulfuric acid (concentrated H_2SO_4), as explained by Christie (1993). In the similar manner, the MeMEE was quantified to characterize the methyl and ethyl esters of their corresponding fatty acids, and was injected directly into the equipment without any preparations. Table 14.1 furnishes the technical details and specifications of GC–MS equipment used for the characterization study.

14.2.6 Methanol- and Ethanol-Based Transesterification of MEO

In general practice, any transesterification must be always carried out under optimized reaction conditions in order to discourage saponification and enhance the conversion yields. Accordingly, the ideal range for the reaction parameters of co-solvent–based transesterification was decided based on the optimized reaction parameters of mono-solvent (methanol- and ethanol-based) transesterification. For this purpose, the reaction parameters of mono-solvent transesterification reactions were optimized statistically using the software Design-Expert v11, with help of response surface methodology (RSM) technique under Box–Behnken design (BBD) by maintaining molar ratio (A), catalyst concentration (B), reaction temperature (C), and reaction time (D) as four independent input variables, while biodiesel yields were taken as output variable. In total, a set of 29 experimental runs were carried out to identify the output variable, which was later used in statistical regression for developing quadratic equations and correlating two parameters using quadratic response surface. Table 14.2 tabulates the coded factors, along with their range for the reaction parameters. Also, the necessary factors for developing the quadratic equation were decided based on the ANOVA table and their significance test; and were substituted in the following Equation 4 (Munir et al., 2019; Srinivasan et al., 2020):

$$y = b_o + \sum_{i=1}^{k} b_i X_i + \sum_{i=1}^{k} b_{ii} X_i^2 + \sum_{i=1}^{k} b_{ij} X_i X_j + e \quad (4)$$

where y = biodiesel yield; b_o, b_i, b_{ii}, b_{ij} = coefficient of variables, (I = 1, 2, 3 ... k).

14.2.7 Co-Solvent-Based Transesterification of MEO

Upon identifying the ideal range using statistical optimization, the optimal reaction parameters for methanol-ethanol–based co-solvent transesterification like molar ratio, methanol-to-ethanol ratio, catalyst concentration, reaction temperature, and time were decided using the one-factor-at-a-time (OFAT) method. For the purpose of optimization, each parameter was varied within its permissible range while other parameters were maintained constant. Schematic representation of the transesterification setup used for biodiesel production is illustrated in Figure 14.1. To ensure increased accuracy in the results, the experiments were carried out in triplicates and presented in terms of their mean ± standard deviation.

TABLE 14.2

Experimental Design Data, Coded Factor, Range of Reaction Parameters for Methanol- and Ethanol-Based Transesterification

			Methanol-Based Transesterification		Ethanol-Based Transesterification	
\multicolumn{7}{c}{Experimental Design: Box–Behnken Design (BBD)}
\multicolumn{7}{c}{Total Number of Independent Input Variables: 4}
\multicolumn{7}{c}{Total Number of Independent Output Variables: 1}
\multicolumn{7}{c}{Total Number of Experimental Runs: 29}
Reaction Parameter	**Units**	**Coded Factor**	**Minimum**	**Maximum**	**Minimum**	**Maximum**
Molar Ratio	–	A	3	9	3	9
Catalyst Concentration	%	B	4	8	4	10
Reaction Temperature	°C	C	55	65	70	80
Reaction Time	minutes	D	60	90	30	60

Source: Jambulingam et al. (2020).

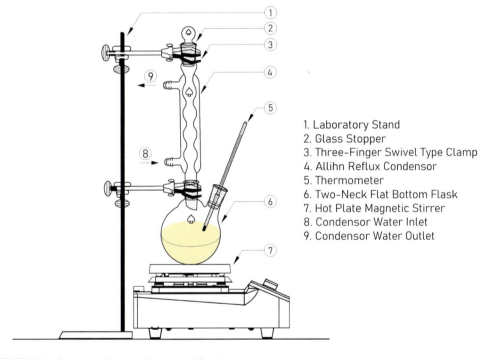

FIGURE 14.1 Schematic diagram of transesterification setup.

1. Laboratory Stand
2. Glass Stopper
3. Three-Finger Swivel Type Clamp
4. Allihn Reflux Condenser
5. Thermometer
6. Two-Neck Flat Bottom Flask
7. Hot Plate Magnetic Stirrer
8. Condenser Water Inlet
9. Condenser Water Outlet

14.2.8 Characterization of [D-Vaemim]Cl Catalyst

The synthesis of [D-Vaemim]Cl catalyst involves refluxing of aqueous solution of 1-ethyl-3-methylimidazolium chloride with Boc-D-valine amide in a round-bottomed flask using the calculated molality, as instructed by Pal and colleagues (2015). Here, D-valine amido ethyl methyl imidazolium moiety and chloride ion were identified as the serving cation and active center of the synthesized catalyst, respectively. To ensure purity and various molecular bond activities, the synthesized IL catalyst was characterized using ^{1}H-NMR, and ^{13}C-NMR, and FT-IR spectra, respectively. Structure of synthesized [D-Vaemim]Cl catalyst was confirmed through ^{1}H-NMR and ^{13}C-NMR spectra and was found to be as follows: ^{1}H-NMR (DMSO-d6): 3.7 (s, 3H), 7.3 (d, 1H), 6.9 (d, 1H), 8.1 (s, 1H), 5.3 (t, 2H), 2.6 (t, 2H), 7.9 (s, 1H), 4.2 (d, 1H), 1.9 (m, 1H), 0.9 (d, 6H), 10.5 (s, 1H). ^{13}C-NMR: (DMSO-d6): 36.59, 128.33, 128.95, 140.67, 54.47, 23.05, 177.03, 184.37, 65.98, and 20.98. Also, FT-IR spectra signified the secondary amine (N-H) and carbonyl (>C = O) stretching based on the distinct peaks obtained at 3393.69 cm^{-1} and 1693.47 cm^{-1}, respectively (Srinivasan et al., 2020).

14.2.9 Fuel Properties of MEO Biodiesel

Fuel properties were evaluated for *M. elengi* methyl ethyl ester (MeMEE), *M. elengi* methyl ester (MeME), *M. elengi* ethyl ester (MeEE), and diesel (taken as reference sample) as per ASTM D 6751 standards. First, density of test samples were measured as per ASTM D1298 using a standard hydrometer (BS718 M50SP, ± 0.0006 g/ml @ 15°C), which helps in working out the quantity of fuel consumed during engine application. Meanwhile, kinematic viscosity, which decides the degree of atomization of fuel, was calculated for the test samples using calibrated glass-viscosity as per ASTM D445 method. Again, ASTM D613 was used to calculate the cetane numbers of these test samples, which were later on used to signify their ignition delay time during combustion inside the cylinder. Also, a bomb calorimeter was used to determine the calorific value of the test samples as per ASTM D240, and was used to define the net energy content present in them. Similarly, test samples were ensured for their safety and handling by determining their flash point (lowest temperature which produce vapors and ignites upon introducing flame) as per ASTM D93-16 using a Pensky–Martens closed-cup apparatus. Properties like cloud point and pour point, which denotes the cloudy appearances and loss of flow characteristics due to cold conditions were determined as per ASTM D2500 and ASTM D7346-15, respectively. Acid values in the test samples were used to denote the free fatty acids present in them and were calculated using ASTM D664 method. Finally, the test samples were evaluated for their elemental composition to calculate the distribution of carbon (C), hydrogen (H), oxygen (O) and sulfur (S). To ensure increased accuracy in the results, the experiments were carried out in triplicates and presented in terms of their mean ± standard deviation.

14.3 Results and Discussions

14.3.1 Characterization of MEO and MeMEE

M. elengi seeds were easily available throughout the study area and their surroundings, since these trees were found in plenty and were regarded as a native tree in that area; and in general, weighed 2–5 grams. Looking into their proximate elemental composition, the overall lipid content in the dried seed kernel of *M. elengi* was measured as 32.87% ± 1.36%, and was found to be fairly resourceful for extracting oil which is later on used in the biodiesel production process. In addition, the seed kernels were also made up of high moisture content (45.32%), followed by protein content (19.64%). Upon using mechanical extraction, the maximum amount of oil that can be extracted using the mini screw press was found to be 24.65% ± 1.68%, thus producing a rendering efficiency of 75%; which was found to be 21.04% less than solvent extraction technique. Moreover, the FFA content in the extracted MEO was estimated to 1.78% ± 0.2%, and was found to be lying very well within the permissible range, thus ruling out the necessity for any pretreatments (esterification or glycerolysis).

Based on GC characterization of extracted MEO, the concentration of overall saturated FAs was calculated as 33.02% and was contributed by stearic acid (15.19%), palmitic acid (11.29%), and myristic acid (6.54%). On the other hand, concentration of overall unsaturated FAs was estimated to be 66.98%, contributed by linoleic acid (36.93%), oleic acid (28.29%), linolenic acid (1.34%), and erucic acid (0.42%). Moreover, presence of long-chain FAs in the triglycerides of MEO demanded higher molar ratio and reaction temperature (Joram et al., 2018), whereas the unsaturated FAs (linoleic acid, oleic acid, linolenic acid) led to an increased rate of oxidation and hydrolysis due to the unsaturated double bond in their FA moiety.

Likewise, GC characterization of MeMEE reported methyl and ethyl esters of linoleic acid (24.27%; 11.76%), oleic acid (17.86%; 10.1%), stearic acid (8.23%; 7.61%), and palmitic acid (7.13%; 5.01%) as their dominant fatty acid methyl and ethyl esters. Again, the concentration of overall saturated FAEs was calculated as 34.71% and was contributed by the methyl and ethyl esters of stearic acid (8.23%; 7.61%), palmitic acid (7.13%; 5.01%) and myristic acid (3.76%; 2.97%). Meanwhile, concentration of overall unsaturated FAEs was estimated as 65.29%; and was contributed by the methyl and ethyl esters of Linoleic acid (24.27%; 11.76%), oleic acid (17.86%; 10.1%), linolenic acid (0.61%; 0.34%), and erucic acid (0.21%; 0.14%). Upon comparison, concentration of MeME was found to be 63.64% greater than MeEE, whereas concentration of unsaturated MeMEEs was 88.1% greater than saturated MeMEEs. The reduced concentration of FAEEs was explained by the greater reactivity of methoxide ions when compared to ethoxide ions (Issariyakul et al., 2007), while reduced concentration of unsaturated fatty acid ethyl esters (FAEEs) was a result of reduced interaction between the ethyl ions and unsaturated FA moiety (Jambulingam et al., 2020).

14.3.2 Optimization of Methanol and Ethanol Transesterification (Using Statistical Optimization)

Based on the molar ratio–catalyst concentration plots (Figure 14.2 and Figure 14.4), yield of MEO biodiesel was found to be increasing up to a molar ratio of 1:6, followed by reduction in yield for increasing ratios. This was explained by the stoichiometry in the reaction mixture that ensured sufficient availability of alcohol up to 1:6, beyond which excess alcohol forced the resultant products into ester-glycerol recombination. Looking into the response of biodiesel yield for varying catalyst concentrations, the yield increased up to 6% and 7% for methanol- and ethanol-based transesterification systems, respectively, and remained constant thereon, regardless of any superior reaction parameters. Again, looking into temperature–time plot (Figure 14.3 and Figure 14.5), maximum biodiesel

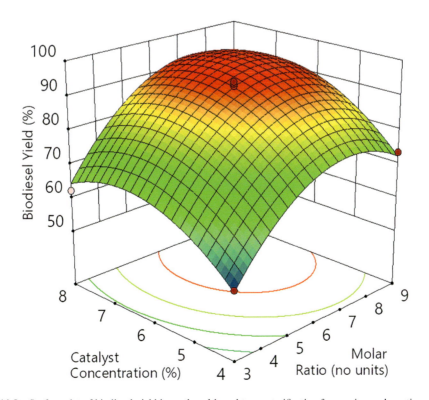

FIGURE 14.2 Surface plot of biodiesel yield in methanol-based transesterification for varying molar ratio and catalyst concentration.

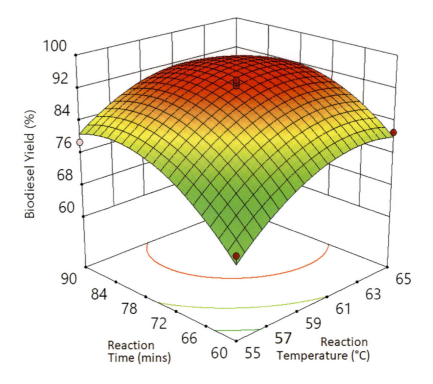

FIGURE 14.3 Surface plot of biodiesel yield in methanol-based transesterification for varying reaction temperature and time.

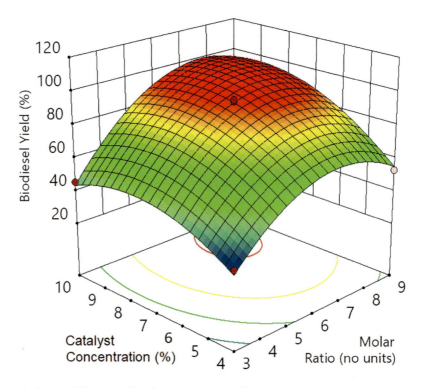

FIGURE 14.4 Surface plot of biodiesel yield in ethanol-based transesterification for varying molar ratio and catalyst concentration.

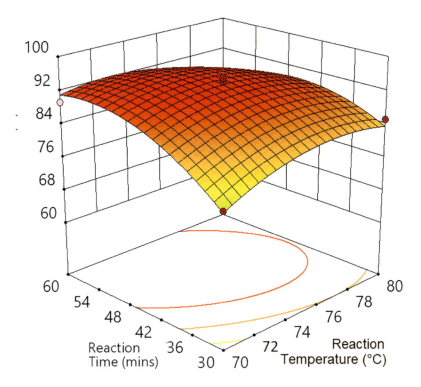

FIGURE 14.5 Surface plot of biodiesel yield in ethanol-based transesterification for varying reaction temperature and time.

yields were reported at reaction temperatures of 60°C (in the case of methanol) and 75°C (in the case of ethanol), and was on account of homogeneity in the phase mixture, which allowed the maximum interaction between oil and alcohol. However, high reaction temperatures led to evaporation of solvent in the reaction system, which simply resulted in slightly reduced yields. Also, reaction time played a significant role in producing higher biodiesel yields, which were recorded at 75 minutes (in the case of methanol) and 45 minutes (in the case of ethanol). Further, elapsed reaction time simply reverted the reaction and reduced the yield as it led back to monoglyceride formation. Here, MEO responded well for molar ratio of 1:6 instead of 1:3 in view of long-chain FAs in their triglyceride molecules, which demanded excess alcohol for their conversion into FAEs, while prolonged reaction durations were deemed to be sufficient enough to overcome mass transfer barriers and induce bond cleavage to FA moieties and reconfiguration into their respective ester molecules. Based on statistical optimization, most optimal reaction parameters for transesterification of MEO using methanol and ethanol are as follows: methanol-based transesterification (molar ratio 1:6, catalyst concentration 6% [D-Vaemim]Cl, reaction temperature 60°C and reaction time 75 minutes) and ethanol-based transesterification (molar ratio 1:6, catalyst concentration 7% [D-Vaemim]Cl, reaction temperature 75°C and reaction time 45 minutes). Ultimately, the highest biodiesel yields were noted in a range of 92%–95% upon using these optimized sets of reaction parameters, which were also used for deciding the range for reaction parameters of co-solvent transesterification to be determined using OFAT method.

14.3.3 Optimization of Methanol-Ethanol-Based Co-Solvent Transesterification (Using OFAT Method)

14.3.3.1 *Effect of Methanol-to-Ethanol Ratio on MeMEE Yield*

In common practice, feedstocks with long-chain FAs demand higher molar ratio like 1:6 or higher (oil/fat:alcohol) and coincided well with the optimized molar ratio (1:6) of MEO upon transesterification using

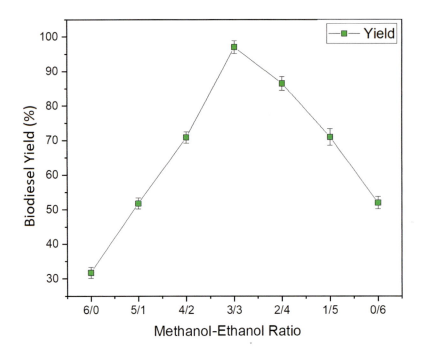

FIGURE 14.6 Yield of biodiesel for varying methanol-to-ethanol ratio (molar ratio 1:6; catalyst concentration 6.5% [D-Vaemim]Cl; reaction temperature 70°C; reaction time 20 minutes).

mono-solvents (methanol and ethanol). Hence, this ratio was maintained constant throughout the study, while the methanol-to-ethanol molar ratio was varied for different proportions by blending ethanol with methanol in the proportions of 6:0 (methanol only), 5:1, 4:2, 3:3, 2:4, 1:5, and 0:6 (ethanol only). From Figure 14.6, MeMEE yield reported its steady rise until equivalent proportion of methanol-ethanol (3:3), beyond which, it reduced gradually as concentration of ethanol increased. Looking into this, reduced reaction time failed to produce higher yields for methyl-based systems in spite of favorable reaction conditions; meanwhile, low reaction temperature and time failed the ethyl-based system. Thus, 3:3 is regarded as the most optimum methanol-to-ethanol molar ratio found effective for transesterification of MEO; and was slightly closer to the methanol-to-ethanol molar ratio (4:2) optimized by Ma and colleagues (2016) for co-solvent transesterification of waste cooking oil. Moreover, both molar ratio and optimal co-solvent–to-solvent ratio are decided entirely based on the composition of FAs present in the feedstock and alcohol used (Encinar et al., 2010).

Eventually, this equivalent methanol/ethanol molar ratio (3:3) reported a maximum MeMEE yield of 97.85 ± 1.25% and was comparatively, 5.51% and 3.35% greater than methanol- and ethanol-based transesterification, respectively. Looking into their cause, ethanol acted as an effective ester exchange agent and its low polarity enhanced the solubility of MEO (as oil is non-polar) into the reaction mixture, which also reduced the mass transfer barrier, thereby leading to formation of esters rapidly. Eventually, the emulsifying nature of ethanol upon interacting with MEO was reduced due to methanol, which also reduced the stearic effect of former. Alarmingly, slightly excess methanol must be added to the reaction, since the reaction temperature is maintained slightly higher than its boiling point. Here, extra care must be given upon adding co-solvent to the reaction system, since the excess volume simply diluted the primary solvent, which reduced their activity, and the overall biodiesel yield (Luu et al., 2014).

The amount of co-solvent required for the transesterification of MEO for varying methanol-ethanol molar ratio was calculated using the following equation 5 (Jambulingam et al., 2020):

$$V_{alchol} = \frac{V_{sample} * m * \rho_{TG} * \{(a * M_{methanol}) + (b * M_{ethanol})\}}{\left[92.17 - 3 + \left[3\left(\sum_{i=1}^{n} M_{FA} * x_i\right) - 17\right]\right] * \{(a * \rho_{methanol}) + (b * \rho_{ethanol})\}} \quad (5)$$

14.3.3.2 Effect of Catalyst Concentration on MeMEE Yield

[D-Vaemim]Cl was identified as effective IL catalyst for production of MeMEE by the means of co-solvent–based transesterification on MEO, and is widely known for its non-volatility, lower viscosity, enhanced catalytic activities, and promised effectiveness, even after multiple uses (enhanced recyclability) (Liang et al., 2010; Umar et al., 2020). Accordingly, the highest MeMEE yield was reported as 97.8% ± 1.41% (Figure 14.7), for a catalyst concentration of 6.5% of MEO used; and it remained constant beyond this threshold value, regardless of increasing the catalyst concentration. Moreover, constant stirring of the reaction mixture enhanced the molecular interaction between the catalytic active centers and triglyceride molecules, which improvised the yield. Also, these catalysts used the traces of excess alcohol to undergo ester-glycerol recombination, thereby favoring the monoglyceride formation, which also reduced the yield slightly. Interestingly, this novel IL catalyst was found to be effective for ten cycles and retained nearly the same yield for all these cycles. For this purpose, the IL catalyst was separated through fractional distillation and was evaluated for its purity using FT-IR spectra before being using for another cycle.

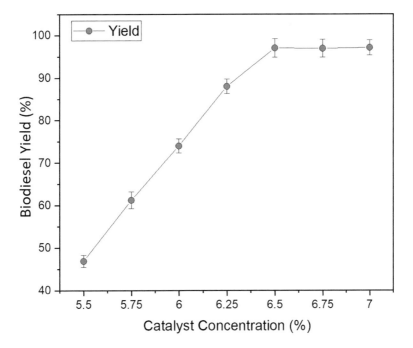

FIGURE 14.7 Yield of biodiesel for varying catalyst concentrations (molar ratio 1:6; methanol-to-ethanol ratio 3:3; reaction temperature 70°C; reaction time 20 minutes).

14.3.3.3 Effect of Reaction Temperature on MeMEE Yield

As explained earlier, temperatures for transesterification reaction must be always maintained closer to the boiling point of the used solvent (methanol: 64.7°C, ethanol: 78.37°C), and higher temperatures simply reduced the yield. Considering this, the temperature range for co-solvent transesterification was entirely decided based on these boiling points and reaction temperatures optimized earlier for mono-solvent–based reactions. Based on temperature plot (Figure 14.8), MeMEE yield rose to its peak value up to a maximum temperature of 70°C and decreased gradually beyond it. Surprisingly, adding ethanol with methanol in equivalent proportion (3:3) simply increased the boiling point of methanol by 8.19%, which made possible for the latter to work effectively even beyond its normal boiling point (by 16.67%) (Aldrich and Querfeld, 1931). High temperatures enhanced the activation energy of triglyceride and alcohol molecules, which increased the overall rate of reaction; however, low and elevated temperatures failed to produce desirable yields, as lower temperatures deprived the sufficient energy for overcoming the energy barrier while elevated temperatures simply evaporated the solvents from the system. However, unsaturated feedstocks required low reaction temperatures (45°C–65°C); high reaction temperatures were encouraged for MEO on account of its long chain FAs (Fadhil and Ahmed, 2018).

14.3.3.4 Effect of Reaction Time on MeMEE Yield

For the co-solvent transesterification of MEO, the optimal reaction time was found to be 20 minutes (from Figure 14.9); which was 73.33% and 55.56% less than the time taken for transesterification of MEO using methanol- and ethanol-based solvents, respectively. The addition of ethanol simply increased the solubility of MEO with reaction mixture (Ma et al., 2016), which helped the reactants in attaining the phase equilibrium in reduced reaction duration, thus completing 85% of the reaction within the first 12 minutes of the reaction time and remaining during another eight minutes of the reaction time with no change in yield beyond the recorded time (Jambulingam et al., 2020). Here, MeMEE yield

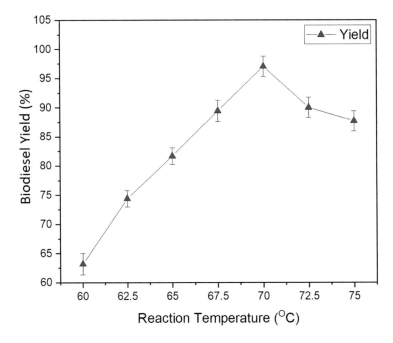

FIGURE 14.8 Yield of biodiesel for varying reaction temperatures (molar ratio 1:6; methanol-to-ethanol ratio 3:3; catalyst concentration 6.5% [D-Vaemim]Cl; reaction time 20 minutes).

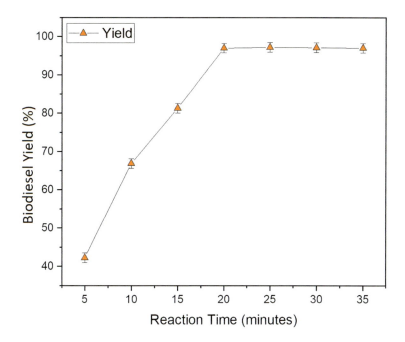

FIGURE 14.9 Yield of biodiesel for varying reaction times (molar ratio 1:6; methanol-to-ethanol ratio 3:3; catalyst concentration 6.5% [D-Vaemim]Cl; reaction temperature 70°C).

increased steadily with elapsing reaction time due to the molecular interaction (collision and adsorption) between the triglycerides and alcohol molecules (reactants) until the equilibrium was attained between them. Contrary to this, poor yields were reported during the initiation of the reaction due to inter- and intramolecular forces of attraction between the reactants, which induced mass transfer resistance and required sufficient reaction time to overcome this incomplete immiscibility. To be highlighted, reaction time is predominantly decided based on the chemical nature of the co-solvent, and is often found to be much less upon using unsaturated feedstocks for biodiesel production (Ma et al., 2016; Fadhil et al., 2015).

Using OFAT method, optimal reaction parameters for co-solvent transesterification of the MEO were found to be as follows: molar ratio 1:6 (oil to alcohol), methanol-to-ethanol ratio 3/3, catalyst concentration 6.5% [D-Vaemim]Cl, reaction temperature 70°C, reaction time 20 minutes. Maximum MeMEE yield achieved upon using these reaction parameters was calculated as 97.85% ± 1.25%, and remained significantly higher than MeME and MeEE biodiesel.

14.3.4 Fuel Properties of Diesel and MEO Biodiesel

Table 14.3 tabulates the fuel (physical, chemical, and thermal) properties of MeME, MeEE, MeMEE, and neat diesel evaluated as per ASTM D6751 standards. Looking at their fuel properties, density of MeMEE was found to be 0.48% lower than MeME, and 6.08% and 0.12% higher than neat diesel and MeEE, respectively, on account of the long-chain FAs present in the ester molecules. Minimal variation in the density of MeME and MeEE signified the contribution of fatty acid moieties in deciding the overall density of biodiesel instead of their alcohol moieties. On the other hand, kinematic viscosity (KV) of MeMEE was measured as 4.21% ± 0.02 mm²/s using ASTM D445 and was marginally greater (1.2%) than MeME. In spite of slightly increased KV resulting in poor atomization, MeMEE exhibited superior engine performance in view of steady combustion due to this increased KV.

TABLE 14.3
Fuel Properties of *M. elengi* Oil Methyl Ester (MeME), *M. elengi* Oil Ethyl Ester (MeEE), *M. elengi* Oil Methyl Ethyl Ester (MeMEE), and Diesel

Properties	Units	Standards	*M. elengi* Oil Methyl Ester (MeME)	*M. elengi* Oil Ethyl Ester (MeEE)	*M. elengi* Oil Methyl Ethyl Ester (MeMEE)	Diesel	Permissible Range
Physical Properties							
Density	kg/m^3	D1298	842 ± 0.4	837 ± 0.45	838 ± 0.4	790	<900
Specific Gravity	–	D1298	0.842	0.837	0.838	0.79	<0.9
Kinematic Viscosity	mm^2/s	D445	4.16 ± 0.03	4.24 ± 0.02	4.21 ± 0.02	3.6	1.90–6.0
Thermal Properties							
Flash Point	°C	D93-16	135 ± 1.1	147 ± 1.2	142 ± 0.95	45	130 min
Cloud Point	°C	D2500	3 ± 0.15	1 ± 0.15	1.5 ± 0.15	0	−3 to 12
Pour Point	°C	D7346-15	−6 ± 0.3	−9 ± 0.3	−8 ± 0.25	−13	−15 to 10
Cetane Number	–	D613	52 ± 0.5	56 ± 0.47	55 ± 0.5	47	47 min
Calorific Value	MJ/kg	D240	36.17 ± 0.15	36.97 ± 0.17	36.74 ± 0.19	42.5	35–43
Chemical Properties							
Molecular Weight	g/mol	–	289.45	303.3	294.7	200	–
Acid Value	mg KOH/g oil	D664	0.21 ± 0.02	0.18 ± 0.02	0.18 ± 0.02	0.36	0.8
FFA Content	%	D664	0.11 ± 0.01	0.1 ± 0.01	0.1 ± 0.01	0.17	0.25
Carbon Content	Wt.%	D5291	76.28 ± 0.4	76.47 ± 0.35	76.51 ± 0.4	87.28	–
Hydrogen Content	Wt.%	D5291	12.41 ± 0.1	12.64 ± 0.1	12.19 ± 0.1	12.72	–
Oxygen Content	Wt.%	D5291	11.31 ± 0.1	10.89 ± 0.1	11.3 ± 0.1	0	–

Following this, the cetane number (CN) of MeMEE was found to be greater than that of MeME by 5.77% due to the increased CN of individual FAEEs. In fact, this higher cetane number was on account of their long-chain FAs and oxygen content, and was in contrast with diesel, which reported reduced CN due to short carbon chain length and increased aromatic and chained hydrocarbon content (Wadumesthrige et al., 2008). Slightly increased CN helped in igniting the fuel earlier during combustion inside the engine cylinder, thus providing sufficient time for complete combustion of high viscous biodiesel and liberating high NOx emissions. Likewise, calorific value (CV) of MEMEE was measured as 36.74 ± 0.19 MJ/Kg, and was slightly reduced due to the presence of unsaturated FAEs in it. However, CV of MeMEE was found to be 1.58% greater than calorific value of MeME, but 13.55% lesser than the CV of neat diesel. Similarly, the flash point of MeMEE was 5.19% higher than that of MeME and was accounted by the ethyl esters in it. As a result, the presence of ethyl esters in MeMEE reduced its cloud point by 50% relative to MeME and its pour point to −8°C ± 0.25°C, thus maintaining its flow characteristics even at low temperatures.

The FFA content of ester samples was observed to be 0.11% (on average) and was explained by the conversion of FAs into their corresponding FAEs due to enhanced methanol's reactivity and ethanol's solvent activity. From elemental composition, the molecular formula of the resulting MeMEE was formulated as $C_{19}H_{36}O_2$, with a molecular weight significantly greater than the molecular weight of MeME by 1.81%. The average carbon content remained the same in both MeMEE and MeME, considering the incorporation of ethanol into the reaction mixture.

It is evident from these findings that the addition of ethanol as co-solvent greatly improvised the overall reaction rate and yield, and that the fuel properties of the resulting biodiesel (MeMEE) compared to that of biodiesel (MeME) developed using transesterification based on mono-solvent (methanol). Specifically, this technique has recorded a substantial rise in reaction temperature and a decrease in reaction time, effectively leading to an increased biodiesel yield. In view of presence of MeEE in the resultant biodiesel, the overall fuel properties of MeMEE significantly improved as compared to MeME.

14.4 Conclusion

Thus, co-solvent–based transesterification of *Mimusops elengi* oil was carried out using methanol as primary solvent and ethanol as co-solvent, in presence of D-valine amido ethyl methyl imidazolium chloride ([D-Vaemim]Cl) as homogeneous catalyst. In addition, the production process and its reaction parameters were optimized for effective yields, while fuel properties were evaluated as per ASTM standards. Following are the major conclusions deduced from this present study:

1. Maximum oil content available in the seed kernels of *M. elengi* was estimated to be 32.87% ± 1.36%, while the maximum renderable oil contents were found to be as follows: 31.22% ± 0.78% (Soxhlet's extraction technique) and 24.65% ± 1.68% (mechanical extraction technique using mini screw press).
2. Looking into mono-solvent transesterification reactions, maximum biodiesel yields were reported as 92.74% ± 1.32% and 94.68% ± 1.18% for methanol- and ethanol-based reaction systems, respectively.
3. Likewise, maximum yield of MeMEE produced using co-solvent–based transesterification was measured as 97.85% ± 1.25%. This was explained by ethanol's role as an active ester exchange agent and its low polarity, which reduced the overall reaction time by 73.33% and increased the overall biodiesel yield by 5.51% as compared to MeME. Interestingly, this technique reduced the mass transfer barrier between the oil and solvent, and use of IL catalyst completely eliminated the chances of saponification of the oil.
4. From GC characterization data, MeMEE reported 62.07% and 37.93% of methyl and ethyl esters, respectively. Reduced distribution of MeEE than MeME was explained by the involvement of ethyl ions in reacting with saturated FAs than unsaturated FAs. However, presence of MeEE had significant effect on improving the overall quality of the resultant biodiesel.

5. Eventually, the fuel properties of MeMEE reported significant rise in its molecular weight, kinematic viscosity, cetane number, and calorific value as compared to MeME. Also, properties related to thermal behavior like flash point, fire point, cloud points, and pour point improvised remarkably and were, ultimately, contributed by the MeEE present.
6. The resulting MeMEE showed decreased density by 0.34% and increased kinematic viscosity (2.89%), cetane number (1.41%), and calorific value (1.49%) compared to MeME. Thermal properties such as flash point, cloud point, and pour point were also improved by adding this. The ethyl esters present in them gradually led to superior properties in MeMEE.

Thus, it can be strongly concluded that co-solvent transesterification serves as a simple yet effective technique for producing biodiesel with superior fuel qualities using minimal resources. Also, this technique reduces the demand for single solvent, thereby easing the pressure focused on its supply chain. Owing to superior fuel properties, these hybrid biodiesels can supply renewable power for sustainable growth into the future.

REFERENCES

Aldrich, E.W. and Querfeld, D.W., 1931. Freezing and boiling points of the ternary system ethanol-methanol-water. *Industrial & Engineering Chemistry*, 23(6), pp. 708–711.

Alhassan, Y., Kumar, N., Bugaje, I.M., Pali, H.S. and Kathkar, P., 2014. Co-solvents transesterification of cotton seed oil into biodiesel: Effects of reaction conditions on quality of fatty acids methyl esters. *Energy Conversion and Management*, 84, pp. 640–648.

Association of Official Analytical Chemists (AOAC), 2005. *Official methods of analysis of AOAC international* (18th ed.). Gaithersburg, MD: AOAC International.

Bharat, G. and Parabia, M.H., 2010. Pharmacognostic evaluation of bark and seeds of *Mimusops elengi* L. *International Journal of Pharmacy and Pharmaceutical Sciences*, 2(4), pp. 110–113.

Bhuiya, M.M.K., Rasul, M., Khan, M., Ashwath, N. and Mofijur, M., 2020. Comparison of oil extraction between screw press and solvent (n-hexane) extraction technique from beauty leaf (*Calophyllum inophyllum* L.) feedstock. *Industrial Crops and Products*, 144, p. 112024.

Bhuyan, R., Saikia, C.N. and Das, K.K., 2004. Extraction and identification of colour components from the barks of *Mimusops elengi* and *Terminalia arjuna* and evaluation of their dyeing characteristics on wool. *IJFTR*, 29(4), pp. 470–476.

Christie, W.W., 1993. Preparation of ester derivatives of fatty acids for chromatographic analysis. *Advances in Lipid Methodology*, 2(69), p. e111.

Deepanraj, A., Gokul, R., Raja, S., Vijayalakshmi, S. and Ranjitha, J., 2015. Facile acid-catalysed biodiesel production from the seeds of *Mimusops elengi*. *International Journal of Applied Engineering Research*, 10(2), pp. 2106–2109.

Encinar, J.M., Gonzalez, J.F., Pardal, H. and Martinez, G.M., 2010. Transesterification of rapeseed oil with methanol in the presence of various co-solvents. In: *Proceedings Venice 2010, third international symposium on energy from biomass and waste, CISA, Environmental Sanitary Engineering Centre, Venice, Italy* (pp. 8–11).

Fadhil, A.B. and Ahmed, A.I., 2018. Production of mixed methyl/ethyl esters from waste fish oil through transesterification with mixed methanol/ethanol system. *Chemical Engineering Communications*, 205(9), pp. 1157–1166.

Fadhil, A.B., Al-Tikrity, E.T. and Albadree, M.A., 2015. Transesterification of a novel feedstock, *Cyprinus carpio* fish oil: Influence of co-solvent and characterization of biodiesel. *Fuel*, 162, pp. 215–223.

Fadhil, A.B., Saleh, L.A. and Altamer, D.H., 2020. Production of biodiesel from non-edible oil, wild mustard (*Brassica Juncea* L.) seed oil through cleaner routes. *Energy Sources, Part A: Recovery, Utilization, and Environmental Effects*, 42(15), pp. 1831–1843.

Gami, B., Pathak, S. and Parabia, M., 2012. Ethnobotanical, phytochemical and pharmacological review of *Mimusops elengi* Linn. *Asian Pacific Journal of Tropical Biomedicine*, 2(9), pp. 743–748.

Hsiao, M.C., Kuo, J.Y., Hsieh, P.H. and Hou, S.S., 2018. Improving biodiesel conversions from blends of high- and low-acid-value waste cooking oils using sodium methoxide as a catalyst based on a high speed homogenizer. *Energies*, 11(9), p. 2298.

Inayat, A., Ghani, C., Jamil, F., Alobaidli, A.S., Bawazir, H.M. and Ali, N.A., 2018. Biodiesel production from date seeds via microwave assisted technique. In: *2018 5th international conference on renewable energy: generation and applications (ICREGA), Al Ain, United Arab Emirates, IEEE, USA* (pp. 61–63).

Issariyakul, T., Kulkarni, M.G., Dalai, A.K. and Bakhshi, N.N., 2007. Production of biodiesel from waste fryer grease using mixed methanol/ethanol system. *Fuel Processing Technology*, 88(5), pp. 429–436.

Jambulingam, R., Srinivasan, G.R., Palani, S., Munir, M., Saeed, M. and Mohanam, A., 2020. Process optimization of biodiesel production from waste beef tallow using ethanol as co-solvent. *SN Applied Sciences*, 2(8), pp. 1–18.

Joram, A., Sharma, R. and Sharma, A.K., 2018. Synthesis, spectral and thermo-gravimetric analysis of novel macromolecular organo-copper surfactants. *The Open Chemistry Journal*, 5(1).

Julianto, T.S. and Nurlestari, R., 2018. The effect of acetone amount ratio as co-solvent to methanol in transesterification reaction of waste cooking oil. In: *IOP conference series: Materials science and engineering, The 12th Joint Conference on Chemistry 19–20 September 2017, Indonesia* (vol. 349, no. 1, p. 012063).

Krupakaran, R.L., Hariprasad, T. and Gopalakrishna, A., 2018. Impact of various blends of *Mimusops elengi* methyl esters on performance and emission characteristics of a diesel engine. *International Journal of Green Energy*, 15(7), pp. 415–426.

Liang, J.H., Ren, X.Q., Wang, J.T. and Li, Z.J., 2010. Preparation of biodiesel by transesterification from cottonseed oil using the basic dication ionic liquids as catalysts. *Journal of Fuel Chemistry and Technology*, 38(3), pp. 275–280.

Luu, P.D., Takenaka, N., Van Luu, B., Pham, L.N., Imamura, K. and Maeda, Y., 2014. Co-solvent method produce biodiesel form waste cooking oil with small pilot plant. *Energy Procedia*, 61, pp. 2822–2832.

Ma, Y., Wang, Q., Zheng, L., Gao, Z., Wang, Q. and Ma, Y., 2016. Mixed methanol/ethanol on transesterification of waste cooking oil using Mg/Al hydrotalcite catalyst. *Energy*, 107, pp. 523–531.

Munir, M., Saeed, M., Ahmad, M., Waseem, A., Sultana, S., Zafar, M. and Srinivasan, G.R., 2019. Optimization of novel *Lepidium perfoliatum* Linn. Biodiesel using zirconium-modified montmorillonite clay catalyst. *Energy Sources, Part A: Recovery, Utilization, and Environmental Effects*, pp. 1–16.

Ng, E.C., Dunford, N.T. and Chenault, K., 2008. Chemical characteristics and volatile profile of genetically modified peanut cultivars. *Journal of Bioscience and Bioengineering*, 106(4), pp. 350–356.

Noureddini, H. and Byun, J., 2010. Dilute-acid pretreatment of distillers' grains and corn fiber. *Bioresource Technology*, 101(3), pp. 1060–1067.

Ooi, D.J., Iqbal, S. and Ismail, M., 2012. Proximate composition, nutritional attributes and mineral composition of *Peperomia pellucida* L. (Ketumpangan Air) grown in Malaysia. *Molecules*, 17(9), pp. 11139–11145.

Pal, A., Kumar, H., Maan, R., Sharma, H.K. and Sharma, S., 2015. Solute–solvent interactions of glycine, l-alanine, and lvaline in aqueous 1-methyl imidazolium chloride ionic liquid solutions in the temperature interval (288.15 to 308.15) K. *Journal of Chemical Thermodynamics*, 91, pp. 146–155.

Sakshi, S., Vineet, G., Rajiv, G. and Shubhini, A.S., 2011. Analgesic and antipyretic activity of *Mimusops elengi* L.(bakul) leaves. *Pharmacologyonline*, 3, pp. 1–6.

Sakthivel, S., Halder, S. and Gupta, P.D., 2013. Influence of co-solvent on the production of biodiesel in batch and continuous process. *International Journal of Green Energy*, 10(8), pp. 876–884.

Shanmugam, S., Annadurai, M. and Rajendran, K., 2011. Ethnomedicinal plants used to cure diarrhoea and dysentery in Pachalur hills of Dindigul district in Tamil Nadu, Southern India. *Journal of Applied Pharmaceutical Science*, 1(8), p. 94.

Shrivastava, P., Salam, S., Verma, T.N. and Samuel, O.D., 2020. Experimental and empirical analysis of an IC engine operating with ternary blends of diesel, karanja and roselle biodiesel. *Fuel*, 262, p. 116608.

Srinivasan, G.R. and Jambulingam, R., 2018. Comprehensive study on biodiesel produced from waste animal fats: A review. *Journal of Environmental Science and Technology*, 11(3), pp. 157–166.

Srinivasan, G.R., Shankar, V., Chandra Sekharan, S., Munir, M., Balakrishnan, D., Mohanam, A. and Jambulingam, R., 2020. Influence of fatty acid composition on process optimization and characteristics assessment of biodiesel produced from waste animal fat. *Energy Sources, Part A: Recovery, Utilization, and Environmental Effects*, pp. 1–19.

Trejo-Zárraga, F., de Jesús Hernández-Loyo, F., Chavarría-Hernández, J.C. and Sotelo-Boyás, R., 2018. Kinetics of transesterification processes for biodiesel production. In: *Biofuels: State of development* (pp. 149–179). London: IntechOpen Publishers. https://doi.org/10.5772/intechopen.75927

Umar, A., Munir, M., Riaz, M.A., Murtaza, M., Sultana, R., Srinivasan, G.R., Firdous, A. and Saeed, M., 2020. Properties and green applications based review on highly efficient deep eutectic solvents. *Egyptian Journal of Chemistry*, 63(1), pp. 59–69.

Van Gerpen, J., 2005. Biodiesel processing and production. *Fuel Processing Technology*, 86(10), pp. 1097–1107.

Varghese, R., Henry, J.P. and Irudayaraj, J., 2018. Ultrasonication-assisted transesterification for biodiesel production by using heterogeneous ZnO nanocatalyst. *Environmental Progress & Sustainable Energy*, 37(3), pp. 1176–1182.

Wadumesthrige, K., Smith, J.C., Wilson, J.R., Salley, S.O. and Ng, K.S., 2008. Investigation of the parameters affecting the cetane number of biodiesel. *Journal of the American Oil Chemists' Society*, 85(11), pp. 1073–1081.

15 Conventional Breeding Methods for the Genetic Improvement of Jatropha curcas L.: A Biodiesel Plant

Nitish Kumar

CONTENTS

15.1 Introduction 319
15.2 Overview of Diversity Studies..320
15.3 Domestication...322
    15.3.1 Breeding Targets for *J. curcas* Crop Improvement 323
15.4 Characterization of Breeding Systems in *Jatropha curcas* 323
    15.4.1 Genetic Diversity between Populations..324
    15.4.2 Crop Improvement..325
    15.4.3 Mass Selection and Recurrent Selection 325
    15.4.4 Mutation and Heterosis...325
    15.4.5 Interspecific Hybridization 326
    15.4.6 Toxic and Non-Toxic *Jatropha*: A Potential Key Trait in Breeding Programs............326
    15.4.7 Molecular Breeding 327
15.5 Conclusion 328
References...328

15.1 Introduction

Jatropha curcas (family Euphorbiaceae) has assumed paramount importance as a potential biodiesel crop in more than 50 countries. It is a plant with several attributes, multiple uses, and considerable potential (Heller 1996; Openshaw 2000). The major limitation with the currently used planting material is the narrow genetic base, low productivity, and vulnerability to a wide array of biotic and abiotic stresses. Genetic variability and divergence studies in seed traits and oil content revealed variability for these traits (Kaushik et al. 2007). However, the evaluation was carried out with germplasm collected from trees of different regions, different aged plants (3–20 years), and plants propagated through seeds or vegetative cuttings. Comparison of yield-contributing traits based on such material results in erroneous conclusions about the superiority of the identified clone, as it is influenced by the mode of propagation and climatic conditions. Results of provenance trials indicated strong genotype and environment interactions for several quantitative traits (Heller 1996). Genetic variability assessment of *J. curcas* germplasm using molecular markers indicated modest levels of interaccessional variability (Basha and Sujatha 2007). Hence, there is an immediate need to widen the genetic base of *J. curcas*. Among the various crop breeding approaches, interspecific hybridization is an immediate option for genetic enhancement of *J. curcas* (Sujatha 2006).

    The genus *Jatropha* is morphologically diverse, with 160–175 Old World and New World woody species comprised of trees, shrubs, rhizomatous subshrubs, tuberous perennial herbs, geophytes, and facultative annuals, which are distributed chiefly in the tropical and subtropical regions of the Americas, Africa, and India (Dehgan 1984). Dehgan and Webster (1979) recognized two subgenera (*Curcas*, *Jatropha*), ten sections, and ten subsections. The subgenus *Curcas* comprises all Mexican, one Costa

Rican, two African, and one Indian species, while the subgenus *Jatropha* includes all South American, African (except two), Antillean, all Indian (except one), and two North American species (Dehgan 1984). Of the 175 reported species, only nine species are available in India. *J. villosa* and its allied species are of Indian origin. With the exception of *J. curcas*, all other species existing in India belong to the subgenus *Jatropha*.

Several *Jatropha* species are cultivated for their ornamental leaves and flowers, while some are grown in the tropics for their economic uses. Variation for fatty acid profiles, photoperiod insensitivity, flowering, and fruiting pattern has been reported in different *Jatropha* species (Banerji et al. 1985; Sujatha 1996). Screening of *Jatropha* species against foliage feeders, which attack other Euphorbiaceous members, revealed varying levels of resistance within the *Jatropha* species, with *J. integerrima* conferring maximum resistance in terms of larval mortality, feeding cessation, and with or without pupation (Lakshminarayana and Sujatha 2001). *Jatrophas* are rich sources of hydrocarbons, and *J. multifida* possesses higher oil content (50%) as compared to *J. curcas* (23%–38%) (Banerji et al. 1985; Sujatha 1996). Determination of the energy values of the oils indicated much higher energy content for *J. gossypifolia* (42.2 MJ/kg), *J. glandulifera* (47.2 MJ/kg), and *J. multifida* (57.1 MJ/kg) than for *J. curcas* (39.8–41.8 MJ/kg) (Banerji et al. 1985; Jones and Miller 1991). *Jatropha multifida*, *J. podagrica*, *J. integerrima*, and *J. gossypifolia* are well known and cultivated throughout the tropics as ornamental plants. *J. gossypifolia*, a facultative annual, has heavy fruit bearing ability and thrives well in saline soils. The species *J. integerrima*, *J. multifida*, and *J. podagrica* are drought hardy and have continuous bearing, unlike *J. curcas*, which has between two and four flowering flushes, depending on the agroecological conditions. There is immense scope for transfer of beneficial traits from other *Jatropha* species to *J. curcas*, such as heavy bearing, photoperiod insensitivity, improved fuel characteristics, high oil content, desired oil quality, plant architecture, earliness, reduced toxicity of endosperm proteins, and wider adaptability (Sujatha 2006).

McVaugh (1945), Wilbur (1954), and Dehgan and Webster (1979) regarded *J. curcas* as the most primitive member of the genus because of its ability to interbreed with species from both subgenera, palmately lobed leaves, arborecsent growth habits, and occasional hermaphrodite flowers. Inter- and intrasectional hybrids can be produced with *J. curcas* as the maternal parent, as the barriers to sexual crossability are weak (Dehgan 1984; Sujatha 1996). Dehgan (1984) attempted interspecific hybridization of 20 species in eight of the ten sections and established the phylogenetic significance of interspecific hybridization in *Jatropha*. The study was confined to identification of crossability barriers and morphological characterization of the F1 hybrids. The species that could be crossed unilaterally with *J. curcas* as an ovule parent include *J. macrorhiza*, *J. capensis*, *J. cathartica*, *J. multifida*, *J. podagrica*, *J. cordata*, and *J. cinerea* (Dehgan 1984). Reciprocal crosses are possible with *J. integerrima*, and interspecific hybrids have been developed between *J. curcas* and *J. integerrima* (Rupert et al. 1970; Dehgan 1984; Sujatha and Prabakaran 2003). One to two backcrosses of the F1 hybrids to *J. curcas* resulted in transgressive segregants exhibiting variation for fruit and seed characters (Sujatha and Prabakaran 2003).

The main bottlenecks in cultivation of *J. curcas* are the lack of improved hybrid/variety, locally available inferior planting material, fewer female flowers, early maturity, resistance to lodging, resistance to pests and disease, reduced plant height, and high natural ramification of branches. Therefore, success of genetic improvement of *J. curcas* depends on the collection of large number of germplasm from diverse agroecological regions and the existence of genetic variability for desired traits in the collected germplasm. The present chapter was an attempt to review the recent advances made in the assessment of diversity and genetic improvement of *J. curcas* as a potential biodiesel crop, which facilitates the breeder to use the available literature for further improvement of *J. curcas* for oil and yield in various agroecological regions.

15.2 Overview of Diversity Studies

The systematic work on germplasm exploration, characterization, use, and documentation has been at a nascent stage. Whatever provincial variability is recorded, it is due to genotype and environmental interaction. Priority should be given to assess intra- and interaccessional variability in the available germplasm, selection of pure lines, and then multiplication. The existence of natural hybrid complexes has been observed between

J. curcas-canascens in Mexico (Dehgan and Webster 1978), *J. integerrima—hastata* complex in Cuba and the West Indies (Pax 1910), and *J. curcas-gossypifolia* (*J. tanjorensis*) in India (Prabhakaran and Sujatha 1999). It was reported that highest interspecific genetic divergence (0.419) was found between *J. glandulifera* and *J multifida*. The least interspecific genetic divergence (0.085) was found between *J. gossipifolia* and *J. tanjorensis*. Sun and colleagues (2008) reported that the existence of low variation in microsatellite simple sequence repeat (SSR) markers within populations of *Jatropha*, even in its natural distribution (Mexico). Studies based on genetic markers uncovered only modest levels of diversity in India, indicating that the gene pool applied at a large scale may rest on a fairly fragile genetic foundation (Ranade et al. 2008). Tatikonda and colleagues (2009) studied the diversity of 48 accessions from India based on amplified fragment-length polymorphism (AFLP) markers and found 680 polymorphic fragments, which provided discriminative power for the classification of germplasm accessions into five major clusters. Ganesh-Ram and colleagues (2008) studied five accessions of *Jatropha* and seven *Jatropha* species, and found that highest genetic similarity coefficient (0.85) was measured between TNMC 1 and TNMC 6. Cluster analysis indicated that three distinct clusters—one comprising all the accessions of *J. curcas*, while the second cluster included six species viz *J. ramanadensis*, *J. gossipifolia*, *J. podagrica*, *J. tanjorensis*, *J. villosa* and *J. integerrima*. Then *J. gladulifera* formed the third cluster. The latter species has genetic distinctness and wider geographical distribution in India compared to other seven species studied. Makkar and colleagues (1997), reported large variations in contents of crude protein, crude fat, neutral detergent fiber and ash on 18 different provenances of *Jatropha* from the countries in West and East Africa, the Americas and Asia. In China, Li Kun and colleagues (2007) studied variation in *J. curcas* in different regions and reported that maximum seed weight (698.9 g) was recorded from Liuku of Lujiang river and that minimum seed weight of 500.7 g was recorded from extensive heat regions of the Yuanmou Basin. The oil content of seed ranged from 53.3 (Yuanmou Basin) to 64.25% (Taoyuan of Yongsheng County, which is considered as hot region). Heller (1996) conducted multi-location field trials in 13 provenances in 1987 and 1988 in two countries of the Sahel region of Africa, Senegal and Cape Verde. Significant differences in the vegetative trait were recorded, except leaf shape, among the various provenances at all locations. A total of 1855 candidate plus trees were identified by the National Vegetable Oil Development Board (NOVOD). Beside this, around 5,000 accessions were collected through a network of various interinstitutional research initiatives with an oil content ranging from 26%–42.7% (Punia 2007). The production of quality planting material (30%–40% oil content with 3–5 tons seed yield/ha) under micro mission was undertaken by the Department of Biotechnology, Government of India. Kumar and Sharma (2008) recorded that female flower/inflorescence showed maximum variation, while male flowers/inflorescence showed minimum variation. This may be due to intracellular and extracellular activity at different development stages. They also reported that strong correlation existed between plant height and branch length, number of branches, and collar diameter, which helps in the selection of superior genotypes. Wide variation in 100 seed weight (57–79 g) and oil content (30%–37%) for accession was collected from Andhra Pradesh, India (Rao et al. 2008). Kaushik and colleagues (2007) reported that accessions from Uttaranchal recorded a high percentage (73%) of high-yielding plants. Kaushik and colleagues (2007) explored the variability in the accessions collected from the state of Haryana in India, and found wide variation in 100 seed weight (49–69 g) and oil content (28%–39%). Wani and colleagues (2006) recorded variation in Indian accessions for oil content (27.8%–38.4%) and 100 seed weight (44–77 g). On-farm trials conducted involving 103 accessions over a period of five years at the Regional Agricultural Research Station (RARS), Tirupati revealed that mean seed yield ranged from 39–1,312 g per plant, with an oil content ranging from 20%–30%. Similar trails were conducted at the International Crops Research Institute for the Semi-Arid Tropics (ICRISAT), Patancheru, and it was found that 100 seed weight ranged from 49.2–77.2 g, with an oil content ranging from 27.8%–38.4%. Phenotypic variation in plant height (34–225 cm), number of primary branches (1–14), plant spread (14–161), number of clusters per plant (1–20) and oil content (21.5%–39.8%) from accessions collected from the states of Chhattisgarh and Andhra Pradesh was reported by Sunil and colleagues (2008). Based on analysis using DIVA-GIS software, the same authors reported that the Prakasam district of Andhra Pradesh had higher CV value for oil content (29%–36%). The richness in oil content using rarefaction method of DIVA-GIS showed that the Ranga Reddy district was found to be a potential area for germplasm with high oil content. Kumar and Sharma (2008) studied the intraspecific variation for various morphological traits and found the following ranges: plant height (140.95–175.30 cm), collar diameter (4.70–6.27 cm), number of branches

(4.75–7.93), and branch length (102.06–132.96 cm). Rao and colleagues (2008) observed four clusters with phylogeographic patterns of genetic diversity among 32 high-yielding candidate plus trees of *J. curcas* for seed traits. Gohil and Pandya (2008) analyzed diversity based on phenotypic traits of nine *Jatropha* genotypes and suggested that for varietal improvement, hybridization among the genotypes of divergent clusters (clusters III–V) may be done in order to obtain better results in terms of variability and diversity. Kaushik and colleagues (2007) subjected 24 diverse accessions to non-hierarchical Euclidian cluster analysis for seed traits and found that crossing between accessions of clusters IV and VI will yield wide spectra of variability in subsequent generations. To increase genetic species diversity and add new alleles, they recommended interspecific cross-pollination between *J. curcas* and other *Jatropha* species to develop new hybrids with higher yield potential and resistance to diseases. Among all the crosses, the cross between *J. curcas* and *J. integerrima* produced successful hybrids with more seed set, while the other crosses failed to produce seeds due to existence of cross ability barriers, as reported by Parthiban and colleagues (2009), who attempted crosses between *J. curcas* and other *Jatropha* species, and identified 27 distinct hybrid progeny clones. Clones such as FC RI HC 3 (55.26%), FCRI HC 15 (48.50%), and FCRI HC 13 (37.01%) exhibited superiority in terms of oil content. Other progenies, such as FCRI 22 (357.48 g), FCRI HC 21 (328.07 g), FCRI HC 10 (325.01 g), FCRI HC 18 (305.43 g), FCRI HC 12 (255 g), FCRI HC 20 (252.26 g), and FCRI HC 27 (250 g), recorded maximum seed yield and early flowering at nine months after planting. These clones can be promoted and used effectively as biofuel crop. Gujarat Agricultural University released the first variety of *J. curcas* (SDAUJ I, Chatrapati) for commercial cultivation in India. The Regional Agricultural Research Station (RARS), Tirupati of Acharya N. G. Ranga Agricultural University, Hyderabad, Andhra Pradesh, India has released variety of *J. curcas* for cultivation in the Rain Shadow Districts of Andhra Pradesh.

15.3 Domestication

In the tropics, there are number of woody and non-woody perennial species that have provided many products of ethnomedicine and other daily needs. Such useful plants are virtually undomesticated. Domestication of crop plants has been a continuous process, and as as result, today's major food crops are the products of years of artificial selection and breeding. The domestication of tree species is a dynamic process from background socioeconomic studies, the collection of germplasm, genetic selection, and improvement to the integration of domesticated species in land use. Domestication is an ongoing process in which genetic and cultivation improvements are continuously refined. In genetic terms, domestication is accelerated and human-induced evolution. Domestication, however, is not only about selection and breeding. In nature, the forces of the environment naturally select the fittest trees in the population. In artificial selection, we choose those trees that offer the best combination of adaptability, growth, quality and quantity of products, and disease resistance. It integrates the four key processes of the identification, production, management, and adoption of tree resources.

Farmers' access to good-performing and well-adapted *Jatropha* accessions with a wide genetic base seems be the most logical domestication strategy for small-scale farmer systems. A wide genetic base increases the sustainability of production. Other requirements for implementation of this domestication strategy are as follows:

Researchers and farmers should disseminate their agronomic findings on land suitability, agronomy, and integrated pest management, for example.

Farmers will need to team up with other production chain partners to have a guaranteed offset of their production. Without market access, farmers should not embark on joint domestication programs on *Jatropha*. The next step will be to establish domestication and breeding programs with farmers, researchers, extension workers, and, preferably, private enterprises aiming at small-scale farming and participatory on-farm involvement, as carried out for fruit trees in western and southern Africa.

A well-designed domestication program will include an exit strategy for all actors, both the farmers and their buyers. This exit strategy provides a route toward independence of all project actors after the project implementation period.

Part of a well-designed program will be training in seed collection practices, in order to prevent narrowing of the genetic base of subsequent *Jatropha* generations, and post-harvest handling to ensure viability.

Without such basic practices, selection for favorable traits—such as production, oil content, and/or seed size—will not yield benefits. Selection may, depending on the heritability of the selected traits, give an initial positive response due to selection of superior genotypes, but this positive effect could be lost in subsequent generations, due to a narrowing of the genetic base (Lengkeek 2007).

15.3.1 Breeding Targets for *J. curcas* Crop Improvement

A number of traits such as seed yield, oil content, seed toxicity, female-to-male flower ratio, increased branching, early flowering, synchronous maturity, and adaptation to biotic and abiotic stresses are considered relevant for commercial exploitation of the crop. Improvement in seed yield could be achieved by increasing the ratio of female flowers, number of branches, number of days from fruiting to maturity, stem diameter, total leaf area, and modifying plant architecture (Abdelgadir et al. 2009). Increasing the oil content can be achieved by altering the expression levels of genes involved in fatty acid and triacylglycerol synthesis. Further, developing non-toxic varieties by blocking the expression of curcin and phorbol esters is important to make the crop environment friendly. Curcin and other toxic compounds are known to be toxic to both human handlers and the environment in the long term. Increased seed yield was achieved conventionally by increasing the ratio of lateral branches (Abdelgadir et al. 2009) and by application of paclobutrazol, a growth regulator (Ghosh et al. 2011). Knowledge of pest resistance and drought resistance in wild *Jatropha* germplasm can be of great potential in improvement programs (Divakara et al. 2010; Shanker and Dhyani 2006). Further, the UN Food and Agriculture Organization (FAO) in its released review (Brittaine and Lutaladio 2010) highlighted several pro–poor breeding objectives, which will benefit small *Jatropha* growers, especially the small and marginal farmers. Needing immediate attention are: improving drought resistance and productivity under abiotic stress conditions, enhanced resistance to pests and diseases, altered plant architecture to allow intercropping and mechanization, enabling easier shelling and seed crushing, increased adaptability and productivity under low to medium soil nutrient conditions, and plants with inedible leaves to ensure utility as a livestock hedge.

15.4 Characterization of Breeding Systems in *Jatropha curcas*

Inbreeding depression in tree species is the process by which self- or related matings lead to homozygosity and the accumulation of deleterious mutations (Dawson et al. 2009; Boshier et al. 2000; Lowe et al. 2005). Inbreeding depression reduces individual fitness, survival, and growth variables (Nielsen et al. 2007), and raises the possibilities of population and/or species extinction (Hansson and Westerberg 2002; Charlesworth and Charlesworth 1987; Reed and Frankham 2003). The negative effects of inbreeding in trees are well documented and include embryo abortion, limited fruit set, reduced overall seed yield, and lower germination rate (Dawson et al. 2009). Furthermore, selfed or inbred progeny can suffer from lower seedling vigor and poor growth form, and end up being less productive when they reach maturity (Hardner and Potts 1997; Gigord et al. 1998; Wu et al. 1998; Koelewijn et al. 1999; Stacy 2001). Inbreeding depression is worsened by the large variations in fecundity often observed in tree species (El-Kassaby and Cook 1994; El-Kassaby et al. 1989). This phenomenon, in which a small number of trees contribute disproportionately to the seed crop, can result in the effective population size of a tree (N_e)—the size of an "idealized" population that would have the same genetic properties as that observed for a real population—being much lower than the census size, and lower than that required to maintain heterozygosity and productivity (Wright 1931; Lengkeek et al. 2005b). Ne is also lowered if the reproductive connectivity between trees in a landscape is weak. Connectivity depends on the density and evenness of distribution of sexually mature individuals in the landscape and, if a species relies on animal pollinators and/or seed dispersers, on the presence of these agents to facilitate gene flow (Nason and

Hamrick 1997). As explained previously, *Jatropha* can set seed after both insect and self-pollination. However, self-pollinated fruits are lighter in general (Abdelgadir et al. 2008) and aborted before maturation in 25% of cases (Raju and Ezradanam 2002). This could be due to early acting inbreeding depression (Husband and Schemske 1996), and thus may reflect a high natural outcrossing rate. Chang-Wei and colleagues (2007) suggested that *Jatropha* is not only able to reproduce via selfing, but also through apomixis (without sexual reproduction). However, the issue with contrasting traits apparently present in the breeding system seems not to be fully clarified. Preliminary studies indicate very low variation in microsatellite simple sequence repeat (SSR) markers within populations of *Jatropha*, even in its natural distribution (Mexico) (Sun et al. 2008). Pamidimarri et al. applied SSR, AFLP, and random amplification of polymorphic DNA markers to discriminate between two Mexican accessions of *Jatropha* (one toxic and one non-toxic). Although they could discriminate between the accessions, they found no variation between individuals within each accession. This could be an indication of a population structure with a high level of homozygosity, as well. It is important to confirm or disconfirm the hypothesis of non-panmictic breeding, because understanding the breeding pattern is central for design of domestication strategies. Breeding, large-scale mass propagation, and distribution across landscapes will obviously be much easier if the species is reproducing by natural selfing without inbreeding depression or, especially, if it reproduces by apomixis (Poehlman and Sleper 1995).

15.4.1 Genetic Diversity between Populations

The genome size of *Jatropha* is fairly small (2C DNA content of 0.850 ± 0.006 pg, or C DNA content of 0.416×10^9 base pairs) compared with other members of Euphorbiaceae (Silvertown and Charlesworth 2001; Henning et al. 2007; Carvalho et al. 2008). The level of genetic diversity and genetic differentiation in *Jatropha* populations deserves special attention, due to its introduction history as an exotic species in many countries. In such a situation, plant populations may result in a complex genetic history, including several potential genetic bottlenecks (Kjær and Siegismund 1996; Lengkeek et al. 2005b). Given the successive introductions of *Jatropha* and its ability of clonal mass propagation within a short time, it is possible that all African and/or Asian populations result from a narrow germplasm origin (Henning et al. 2007; Lengkeek 2007). Genetic bottlenecks resulting from such founder effects are also known in other important (agroforestry) crops, most notably coffee and banana (Fernie et al. 2006; Heslop-Harrison and Schwarzacher 2007; López-Gartner et al. 2009). "One-off" introductions are of particular concern if they are already of narrow genetic base, and/or represent low quality or poorly adapted material. Many tropical trees in farm landscapes also demonstrate both extremely low densities and highly aggregated distributions, which—even if long-distance pollen transfer is possible—will reduce effective population sizes and promote inbreeding (Dawson et al. 2009). Data on other tree species that propagate vegetatively show comparable concerns of genetic erosion (Lengkeek 2003). Species that are reproduced vegetatively are vulnerable to clone losses, unless new clones are introduced or old clones redistributed. Generally, a certain number of individual clones respond more successfully to propagation, and simple mathematical simulation models show that, after some generations, only a few clones may dominate an area (Heslop-Harrison and Schwarzacher 2007). Recent studies based on genetic markers uncovered surprisingly low levels of genetic diversity in *Jatropha* landraces from China (Sun et al. 2008) and only modest levels of diversity in India (Ranade et al. 2008; Basha and Sujatha 2007), indicating that the gene pool applied at a large scale may rest on a fairly fragile genetic foundation. Tatikonda and colleagues (2009) studied the diversity of 48 accessions from India based on AFLP markers and found 680 polymorphic fragments, which provided discriminative power for the classification of germplasm accessions into five major clusters. There are limited published data characterizing the genetic variation in the African gene pool. Basha and colleagues (2009) included accessions from Egypt and Uganda, which in general clustered closer to the Asian landraces compared with the Mexican accessions, although the emerging pattern was not completely clear. Regarding the level of genetic variation within African populations, preliminary results based on SSR markers and AFLP markers indicate surprisingly low levels of genetic diversity in landraces from Mali, Kenya, and Tanzania (Nielsen 2009). Correlations of growth and oil production with DNA polymorphism have not yet been investigated. Species with naturally high levels of inbreeding ("selfers") are expected to show less inbreeding depression (Waller 1993), but at present, very little is known about

Jatropha's breeding system (see flower morphology as discussed earlier in this chapter). Both the small population sizes during introduction history and the losses of introduced genotypes due to imbalanced clonal propagation by farmers may have led to purging of recessive, deleterious alleles. This could have counteracted inbreeding depression (Glémin 2003). Owing to the lack of knowledge, it is indeed possible that the *Jatropha* landraces have reduced growth, due to imbedded inbreeding depression. Given the low genomic diversity in *Jatropha* landraces, we believe that "smart" outcrossing between superior Asian individuals with new introductions from the Americas should be performed. Such crosses should release any inbreeding depression and thereby increase vigor and fruit production if the genetic diversity of American landraces is effectively larger. The introduction of genetic variability can be performed by intraspecific and/or interspecific crossing. Actually, interspecific crossing experiments have successfully been carried out between *Jatropha curcas* L. and *Jatropha gossypifolia* L., *Jatropha glandulifera* Roxb., *Jatropha integerrima* Jacq., *Jatropha multifida* L., *Jatropha villosa* (Forssk.) Müll. Arg., and *Jatropha maheshwarii* Subram. & M.P. Nayar (Basha and Sujatha 2009). F1, F2, and F3 hybrids, as well as various back-crossings, are available and should be evaluated with respect to seed production, oil yield, and performance when grown under different climatic conditions (Parthiban et al. 2009). The results from such studies will provide important insight during the next few years. Random amplification of polymorphic DNA markers and AFLP markers are available for interspecific hybrids identification (Pamidiamarri et al. 2009).

15.4.2 Crop Improvement

A sound breeding program depends upon the availability of genetic variability for desired trait. Collection, characterization, and evaluation of germplasm for oil and yield and agromorphological traits are in a nascent stage. The major activity of genetic improvement is selection and breeding. As *J. curcas* is often a cross-pollinated crop, exploitation of genetic variation may be carried out through mass selection, recurrent selection, mutation breeding, heterosis breeding, and interspecific hybridization.

15.4.3 Mass Selection and Recurrent Selection

Individual superior plants are selected based on phenotypic performance, and bulk seed is used to produce the next-generation crop for genetic improvement. To gain genetic improvement for desired traits, there must be positive offspring-parent regression, which depends on degree of environmental effects in the parental population. Mishra (2009) devised the paired comparison method for selecting plus phenotypes of *J. curcas* with emphasis on seed and oil yield, which can overcome the problem of inbreeding depression by controlling pollen source and environment effects, and reduced population size. Evaluation trials in *J. curcas* to study its degree of variability undertaken by Montes and colleagues (2014) involving 225 lines collected from Asia, Africa, and Latin America revealed low genetic variability in African and Indian accessions and high genetic variability in Guatemalan and Latin American lines. This confirmed the studies on evaluation of *J. curcas* for phenotypic and genotypic by Basha and Sujatha (2007). However, wild *J. curcas* is available in tropical Africa, South Asia, and the Americas, as reported by Heller (1996).

In *J. curcas*, recurrent selection is advantageous to overcome the deficiencies of mass selection. Heterosis breeding and improving specific combining ability aids in isolation superior inbred from the population and subjected to recurrent selection for further use in the development of hybrid and synthetic varieties. This method increases the frequency of desirable genes within a population while maintaining variability for continued selection. Development of open-pollinated varieties using mass selection and recurrent selection methods is under testing in India. After obtaining the required data on seed yield, oil content, oil quality, and resistance to disease and insects, the best performing genotypes will be released as new varieties of *Jatropha* by adopting the standard procedure, as reported by Punia (2007).

15.4.4 Mutation and Heterosis

Mutation breeding in tree crops is preferred due to demerits of conventional breeding such as time consumption, unpredictable results, long juvenile phase, high heterozygosity, and fear of loss of unique genotype. Mutation breeding work was carried out in Thailand using fast neutrons and isolated dwarf or early flowering mutants from the M3 generation, but the potential productivity of these variants under intensive cultivation conditions was not proven. Dwimahyani and Ishak (2004) used induced mutations in *J. curcas* for improvement of agronomic characters with irradiation dose of 10 Gy and identified mutant plants with early maturity, 100 seeds weight (30% over control) and better branch growth. In India, mutation breeding using chemical and physical mutagens has been initiated to create genetic variation for various traits, and developed mutants are being characterized using DNA markers (Punia 2007). Mutation studies undertaken at the National Botanical Research Institute (NBRI), Lucknow, India has led to induction of cotyledonary 514 variabilities in *J. curcas* (Pandey and Datta 1995). The mutants themselves may not be suitable for direct release, but they do provide the necessary alleles for developing superior cultivars with desirable traits.

Heterosis in tree species is evident in many hybrids; perhaps the best examples are genera Eucalyptus and Populus. *J. curcas* is often cross-pollinated; heterosis can be exploited by using inbred lines as parent for production of hybrid variety. Improvement of seed yield and oil can be achieved by selection of superior germplasm and release as cultivar. However, little work has been done in *J. curcas* for exploitation of heterosis. Paramathma and colleagues (2006) explained the interspecific hybridization using *J. curcas* as the female parent and *J. integerrima* as the male parent with a wide range of variation for vegetative, flowering, and fruiting characters in F1 hybrids.

15.4.5 Interspecific Hybridization

The main breeding objective in *J. curcas* would be seed and oil yield per unit area, which depends on greater number of pistillate flowers per inflorescence, number of capsule per shrub, 1,000 seed weight, oil content of seeds, and plants per hectare. Studies on phylogenetic significance of interspecific hybridization in *Jatropha* by Dehgan (1984) confined to identification of crossability barriers and morphological characterization revealed that all F1 hybrids, except *J. curcas* × *J. multifida*, were more vigorous than the parental species. The species that could be crossed unilaterally with *J. curcas* as female parent include *J. macrorhiza*, *J. capensis*, *J. cathartica*, *J. multifida*, *J. podagrica*, *J. cordata*, and *J. cinerea*. Interspecific hybridization has immense scope for improving the genetic architecture and agronomic attributes of *J. curcas* Sujatha (2006). Basha and Sujatha (2009) produced artificial hybrids between *J. curcas* and all *Jatropha* species used in the study with the exception of *J. podagrica* without any crossability barriers. Evaluation of backcross interspecific derivatives of cross involving *J. curcas* and *J. integerrima* indicate scope for pre-breeding and genetic enhancement of *J. curcas* through interspecific hybridization (Sujatha and Prabhakaran 2003; Parthiban et al. 2009). Parthiban and colleagues (2009) made crosses between *J. curcas* and other species. Cross between *J. curcas* and *J. integerrima* was successful, as it evolved hybrids with more seed set, and other hybrids failed to produce seeds due to existence of crossability barriers.

15.4.6 Toxic and Non-Toxic *Jatropha*: A Potential Key Trait in Breeding Programs

An important aspect of the domestication of *Jatropha* is the toxicity of the plant (Pamidiamarri et al. 2009). Consumption of *Jatropha* seeds may result in various symptoms, including vomiting and diarrhea, and has proven lethal in animal experiments (Abduaguye et al. 1986; Becker and Makkar 1998; Chimbari and Shiff 2008). The most problematic toxic components in *Jatropha* are probably a number of phorbol esters that, in general, are found in high concentrations in the seeds (Makkar et al. 1997; Adolf et al. 1984; Rakshit et al. 2008). Phorbol esters are compounds known to cause severe biological effects including inflammation and tumor promotion (Haas et al. 2002; Goel et al. 2007). Removal of the phorbol esters during processing is possible (Makkar et al. 2009), but this is not an easily deployed process and the presence of possibly toxic phorbol ester degradation products after treatment cannot be ruled out (Rakshit et al. 2008). From the user's perspective, the toxic phorbol esters provide a potential health risk for workers and limit the use of the protein-rich press cake that otherwise would be suitable for animal diets (Becker and Makkar 1998; Goel et al. 2007). Still, phorbol esters may protect the plants against pests, and will therefore be important when testing if phorbol ester–free

Jatropha plants are more susceptible to damage from pests. To our knowledge, no such experiments on the insect resistance of non-toxic versus toxic plants exist, and it is therefore advisable to include such studies in any breeding program that aims at removing the toxic phorbol esters from the phenotype. The fate of degradation of phorbol esters in soils is also an important aspect, because residuals from *Jatropha* oil production (e.g., the press cake) are often applied as fertilizers (Achten et al. 2008). Although it is claimed that the phorbol esters of the press cake decompose completely within six days after application to the soil (Rug et al. 1997), it is still uncertain whether crops fertilized by the *Jatropha* press cake (e.g., horticulture and cereals) can take up phorbol esters during that period (Achten et al. 2008). For these reasons (e.g., human health, soil and seed cake use), breeding for non-toxic *Jatropha* provides interesting prospects. In this context, it is highly interesting that plants from some provenances in Mexico contain very low or non-detectable levels of phorbol esters (Basha et al. 2009; Makkar et al. 1997; Martinez-Herrera et al. 2006; Makkar et al. 1998a, 1998b, 2008). The presence of naturally occurring plants with low levels of phorbol esters is very interesting in a domestication context, because it makes it likely that plant material without phorbol esters can easily be developed without the use of advanced molecular breeding or transgenic modification. However, introducing non-toxic material may raise new complications, as non-toxic and toxic *Jatropha* are morphologically alike. Additionally, close proximity of toxic and non-toxic *Jatropha* can trigger unexpected traits through cross-pollination. *Jatropha* seeds also contain a trypsin inhibitor, lectins, and phytate that are antinutritional factors in relation to a potential use of the press cake for animal feed (Makkar et al. 1997). The trypsin inhibitor and lectins can be removed by heat treatment, and phytate is neutralized by the addition of microbial phytase (Makkar et al. 1998a, 1998b).

15.4.7 Molecular Breeding

As toxicity in terms of phorbol esters seems to be expressed qualitatively, it may be regulated by only one or a few genes. The inheritance of toxicity is not settled, but Sujatha et al. (2005) suggest that the phenotype of the mother tree is passed on to the seed (i.e., non-toxic mothers give non-toxic seed and toxic mothers give toxic seeds, independent of the phenotype of the father). Maternal inheritance (of chloroplast-specific markers) was also observed in natural and artificial intraspecific hybrids (Basha and Sujatha 2009). This pattern of inheritance could be explained in different ways, including hypotheses of apomixes or involvement of suppressor genes. Potentially, the non-toxic phenotype may be driven by a single suppressor gene that inhibits the production of all phorbol esters when present. Established F1 hybrids between toxic and non-toxic *Jatropha* and first-generation backcrossing could be used to detect the genetic mechanism behind toxicity. By localizing the locus (or loci) responsible for (non) toxicity in the respective parent with molecular markers, it will be possible to use the markers for future use in breeding. Additionally, established clonal field trials can serve as material for association studies based on cosegregation between genotypes (e.g., clones with high oil content and Indian Andhra Pradesh accessions (Sunil et al. 2009) and molecular markers as, for instance, microsatellites (SSRs), AFLPs, and/or inter-SSR markers. Still, more segregation studies are needed to resolve the mode of heritance of toxicity and its molecular background, as such understanding will guide the design of multiple breeding programs. The application of genetic markers (e.g., polymorphic microsatellites) also makes it possible to carry out earlier selections, and thus reduce the time between the recurrent selections and increase the genetic gains per year. Finally, the use of the markers will also help to increase breeding efficiency. The ability to reveal genetic markers associated with certain traits depends on the size of the material sufficient polymorphic markers and precise estimates of genetic values. Transgenic approaches may also prove important in the future (Gressel et al. 2008). *Jatropha* can be transformed, based on *Agrobacterium tumefaciens* infection (Li et al. 2008). The approach can target removal of toxicity and antinutrient components, but work has also been carried out on the biosynthesis of fatty oils (Li et al. 2008). If the improved value of byproducts as livestock feed is a primary target for breeding, reduction of antinutrient factors can potentially be an interesting goal for molecular breeding, as suggested in other crops (Bohn et al. 2008).

15.5 Conclusion

Jatropha curcas is known for its wide spread distribution and adaptability under varying eco-geographic conditions. However, there are several bottlenecks in the commercial exploitation of *J. curcas* as a biodiesel plant, due lack of improved varieties for seed and oil yield, thus making cultivation of *J. curcas* as risky enterprise. Apart from agronomic, socioeconomic, and institutional constraints, planned crop improvement programs are lacking globally. Hence, *J. curcas* can be improved through assessment of variation in wild sources and selection of superior/elite genotypes and application of mutation, alien gene transfer through interspecific hybridization, and biotechnological interventions to bring change in the desired traits. Enhancement of productivity can achieved by evolving hybrids/varieties bearing a greater number of pistillate flowers by exploiting heterosis, thus increasing seed and oil yield per unit area so as to make it profitable venture.

REFERENCES

Abdelgadir HA, Johnson SD, Van Staden J. Approaches to improve seed production of *Jatropha curcas* L. *S. Afr. J. Bot.* 74(2), 359 (2008).

Abdelgadir HA, Johnson SD, Van Staden J. Pollinator effectiveness, breeding system, and tests for inbreeding depression in the biofuel seed crop, *Jatropha curcas*. *J. Horticult. Sci. Biotechnol.* 84, 319–324 (2009).

Abduaguye I, Sannusi A, Alafiyatayo RA, Bhusnurmath SR. Acute toxicity studies with *Jatropha curcas* L. *Human Toxicol.* 5(4), 269–274 (1986).

Achten WMJ, Verchot L, Franken YJ et al. *Jatropha* bio-diesel production and use. *Biomass Bioenergy* 32(12), 1063–1084 (2008).

Adolf W, Opferkuch HJ, Hecker E. Irritant phorbol derivatives from four *Jatropha* species. *Phytochemistry* 23(1), 129–132 (1984).

Banerji R, Chowdhury AR, Misra G, Sudarsanam G, Verma SC, Srivastava GS. *Jatropha* seed oils for energy. *Biomass* 8, 277–282 (1985).

Basha SD, Francis G, Makkar HPS, Becker K, Sujatha M. A comparative study of biochemical traits and olecular markers for assessment of genetic relationships between *Jatropha curcas* L. germplasm from different countries. *Plant Sci.* 176(6), 812–823 (2009).

Basha SD, Sujatha M. Genetic analysis of *Jatropha* species and interspecific hybrids of *Jatropha curcas* using nuclear and organelle specific markers. *Euphytica* 168, 197–214 (2009).

Basha SD, Sujatha M. Inter and intrapopulation variability of *Jatropha curcas* (L.) characterized by RAPD and ISSR markersand development of population-specific SCAR markers. *Euphytica* 156(3), 375–386 (2007).

Becker K, Makkar HPS. Toxic effects of phorbolesters in carp (*Cyprinus carpio* L.). *Vet. Hum. Toxicol.* 40(2), 82–86 (1998).

Bohn L, Meyer AS, Rasmussen SK. Phytate: impact on environment and human nutrition. A challenge for molecular breeding. *J. Zhejiang Univ. Sci. B* 9(3), 165–191 (2008).

Boshier DH. Mating systems. In: *Forest Conservation Genetics: Principles and Practice*. Young AG, Boshier DH, Boyle TJ (Eds). Wallingford, UK: CSIRO Publishing and CABI Publishing, 63–79 (2000).

Brittaine R, Lutaladio N. *Jatropha: A Smallholder Bioenergy Crop – The Potential for Pro-Poor Development*. Rome, Italy: Food and Agriculture Organization of the United Nations (2010). Accessed August 25, 2011.

Carvalho CR, Clarindo WR, Praca MM, Araujo FS, Carels N. Genome size, base composition and karyotype of *Jatropha curcas* L., an important biofuel plant. *Plant Sci.* 174(6), 613–317 (2008).

Chang-Wei L, Kun L, You C, Yongyu S. Floral display and breeding system of *Jatropha curcas* L. *For. Stud. China* 9(2), 114–119 (2007).

Charlesworth D, Charlesworth B. Inbreeding depression and its evolutionary consequences. *Annu. Rev. Ecol. Syst.* 18, 237–268 (1987).

Chimbari MJ, Shiff CJ. A laboratory assessment of the potential molluscicidal potency of *Jatropha curcas* aqueous extracts. *Afr. J. Aquat.* Sci. 33(3), 269–273 (2008).

Dawson IK, Lengkeek A, Weber JC, Jamnadass R. Managing genetic variation in tropical trees: linking knowledge with action in agroforestry ecosystems for improved conservation and enhanced livelihoods. *Biodivers. Conserv.* 18(4), 969–986 (2009).

Dehgan B, Webster GL. Morphology and infrageneric relationships of the genus *Jatropha* (*Euphorbiaceae*). *Univ. Calif. Publ. Bot.* 74, 1–73 (1979).

Dehgan B, Webster GL. Three new species of *Jatropha* from western Mexico. *Madrono* 25, 30–39 (1978).

Dehgan B. Phylogenetic significance of interspecific hybridization in *Jatropha* (*Euphorbiaceae*). *Syst. Bot.* 9, 467–478 (1984).

Divakara BN, Upadhyaya HD, Wani SP, Gowda C, Laxmipathi L. Biology and genetic improvement of *Jatropha curcas* L.: a review. *Appl. Energy* 87, 732–742 (2010).

Dwimahyani I, Ishak. Induced mutation on Jatropha (*Jatropha curcas* L.) for improvement of agronomic characters variability. *Atom. Indonesia J.* 30, 53–60 (2004).

El-Kassaby YA, Cook C. Female reproductive energy and reproductive success in a Douglas-Fir seed orchard and its impact on genetic diversity. *Silvae Genet.* 43(4), 243–246 (1994).

El-Kassaby YA, Fashler AMK, Crown M. Variation in fruitfulness in a Douglas-Fir seed orchard and its effect on crop-management decisions. *Silvae Genet.* 38(3–4), 113–121 (1989).

Fernie AR, Tadmor Y, Zamir D. Natural genetic variation for improving crop quality. *Curr. Opin. Plant Biol.* 9(2), 196–202 (2006).

Ganesh-Ram S, Parthiban KT, Senthil- Kumar R, Thiruvengadam V, Paramathma M. Genetic diversity among *Jatropha* species as revealed by RAPD markers. *Genet. Resour. Crop. Evol.* 55, 803–809 (2008).

Ghosh A, Chikara J, Chaudhary DR. Diminution of economic yield as affected by pruning and chemical manipulation of *Jatropha curcas* L. *Biomass Bioenergy* 35, 1021–1029 (2011).

Gigord L, Lavinage C, Shykoff JA. Partial self-incompatibility and inbreeding depression in a native tree species of La Reunion (Indian Ocean). *Oecologia* 117(3), 342–352 (1998).

Glémin S. How are deleterious mutations purged? Drift versus non-random mating. *Evolution* 57(12), 2678–2687 (2003).

Goel G, Makkar HPS, Francis G, Becker K. Phorbol esters: structure, biological activity, and toxicity in animals. *Int. J. Toxicol.* 26(4), 279–288 (2007).

Gohil RH, Pandya JB. Genetic diversity assessment in physic nut (*Jatropha curcas* L.). *Int. J. Plant Prod.* 2(4), 321–326 (2008).

Gressel J. Transgenics are imperative for biofuel crops. *Plant Sci.* 174(3), 246–263 (2008).

Haas W, Sterk H, Mittelbach M. Novel 12-deoxy-16-hydroxyphorbol diesters isolated from the seed oil of *Jatropha curcas*. *J. Nat. Prod.* 65(10), 1434–1440 (2002).

Hansson B, Westerberg L. On the correlation between heterozygosity and fitness in natural populations. *Mol. Ecol.* 11(12), 2467–2474 (2002).

Hardner CM, Potts BM. Post-dispersal selection following mixed mating in *Eucalyptus regnans*. *Evolution* 51(1), 103–111 (1997).

Heller J. *Physic Nut Jatropha curcas L. Promoting the Conservation and Use of Underutilized and Neglected Crops*. Rome, Italy: Institute of Plant Genetic and Crop Plant Research, International Plant Genetic Resource Institute (1996).

Henning RK. *Jatropha curcas* L. In: *Plant Resources of the Tropical Africa* (Volume 14). van der Vossen HAM, Mkamilo GS (Eds). Wageningen, The Netherlands: Vegetable Oils PROTA Fondation, 116–122 (2007).

Heslop-Harrison JS, Schwarzacher T. Domestication genomics and the future for banana. *Ann. Bot.* 100(5), 1073–1084 (2007).

Husband B, Schemske DW. Evolution of the magnitude and timing of inbreeding depression in plants. *Evolution* 50(1), 54–70 (1996).

Jones N, Miller JH. *Jatropha curcas*. A multipurpose species for problematic sites. *Land Resour. Ser.* 1, 1–12 (1991).

Kaushik N, Kumar K, Kumar S, Kaushik N, Roy S. Genetic variability and divergence studies in seed traits and oil content of Jatropha (*Jatropha curcas* L.) accessions. *Biomass Bioenergy* 31, 497–502 (2007).

Kjær ED, Siegismund H. Allozyme diversity of two Tanzanian and two Nicaraguan landraces of Teak (*Tectona grandis* Lf). *Forest Genet.* 3(1), 45–52 (1996).

Koelewijn HP, Koshi V, Savolainen O. Magnitude and timing of inbreeding depression in Scots pine (*Pinus sylvestris* L.). *Evolution* 53(33), 758–768 (1999).

Kumar A, Sharma S. An evaluation of multipurpose oil seed crop for industrial uses (*Jatropha curcas* L.): a review. *Ind. Crops Prod.* 28, 1–10 (2008).

Kun L, Wen-yun Y, Li L, Chun-hua Z, Yong-zhong C, Yong-yu S. Distribution and development strategy for *Jatropha curcas* L. in Yunnan Province, Southwest China. *Forest. Stud. China* 9(2), 120–126 (2007).

Lakshminarayana M, Sujatha M. Screening of Jatropha species against major defoliators of castor (*Ricinus communis* L.). *J. Oilseeds Res.* 18, 228–230 (2001).

Lengkeek AG. *Diversity Makes a Difference: Farmers Managing Inter- and Intra-Specific Tree Species Diversity in Meru Kenya*. The Netherlands: Wageningen University (2003).

Lengkeek AG. The *Jatropha curcas* agroforestry strategy of Mali Biocarburant SA. In: *Proceedings of the FACT Seminar on Jatropha curcas L. Agronomy and Genetics*. The Netherlands: Wageningen University, 6 (2007).

Lengkeek AG, Kindt R, van der Maesen LJG, Simons AJ, van Oijen DCC. Tree density and germplasm source in agroforestry ecosystems in Meru, Mount Kenya. *Genet. Resour. Crop Ev.* 52(6), 709–721 (2005b).

Li MR, Li HQ, Jiang HW, Pan XP, Wu GJ. Establishment of an Agrobacteriuim-mediated cotyledon disc transformation method for *Jatropha curcas*. *Plant Cell Tiss. Org.* 92(2), 173–181 (2008).

López-Gartner G, Cortina H, McCouch SR, Del Pilar Moncada M. Analysis of genetic structure in a sample of coffee (Coffea arabica L.) using fluorescent SSR markers. *Tree Genet. Genomes* 5(3), 435–446 (2009).

Lowe AJ, Boshier D, Ward M, Bacles CFE, Navarro C. Genetic resource impacts of habitat loss and degradation: reconciling empirical evidence and predicted theory for neotropical trees. *Heredity* 95(4), 255–273 (2005).

Makkar HPS, Aderibigbe AO, Becker K. Comparative evaluation of non-toxic and toxic varieties of *Jatropha curcas* for chemical composition, digestibility, protein degradability and toxic factors. *Food Chem.* 62(2), 207–215 (1998a).

Makkar HPS, Becker K, Schmook B. Edible provenances of *Jatropha curcas* from Quintana Roo state of Mexico and effect of roasting on antinutrient and toxic factors in seeds. *Plant Food Hum. Nutr.* 52(1), 31–36 (1998b).

Makkar HPS, Becker K, Sporer F, Wink M. Studies on nutritive potential and toxic constituents of different provenances of *Jatropha curcas*. *J. Agr. Food Chem.* 45(8), 3152–3157 (1997).

Makkar HPS, Francis G, Becker K. Protein concentrate from *Jatropha curcas* screw-pressed seed cake and toxic and antinutritional factors in protein concentrate. *J. Sci. Food Agr.* 88(9), 1542–1548 (2008).

Makkar HPS, Maes J, Greyt WD, Becker K. Removal and degradation of phorbol esters during pre-treatment and transesterification of *Jatropha curcas* oil. *J. Am. Oil Chem. Soc.* 86(2), 173–181 (2009).

Martinez-Herrera J, Siddhuraju P, Francis G, Davila-Ortiz G, Becker K. Chemical composition, toxic/antimetabolic constituents, and effects of different treatments on their levels, in four provenances of *Jatropha curcas* L. from Mexico. *Food Chem.* 96(1), 80–89 (2006).

McVaugh R. The genus Jatropha in America: principal intergeneric groups. *Bull. Torrey Bot. Club* 72, 271–294 (1945).

Mishra DK. Selection of candidate plus phenotypes of *Jatropha curcas* L. using method of paired comparisons. *Biomass Bioenergy* 33, 542–545 (2009).

Montes JM, Technow F, Martin M, Becker K. Genetic diversity in *Jatropha curcas* L. assessed with SSR and SNP markers. *Diversity* 6, 551–566 (2014).

Nason JD, Hamrick JL. Reproductive and genetic consequences of forest fragmentation two case studies of neotropical canopy trees. *J. Hered.* 88(4), 264–276 (1997).

Nielsen F. *Jatropha curcas* oil production for local development in Mozambique. *Afr. Crop Sci. Conf. Proc.*, 9, 71–75 (2009).

Nielsen LR, Siegismund HR, Hansen T. Inbreeding depression in the partially self-incompatible endemic plant species Scalesia affinis (*Asteraceae*) from Galápagos Islands. *Evol. Ecol.* 21(1), 1–12 (2007).

Openshaw K. A review of *Jatropha curcas*: an oil plant of unfulfilled promise. *Biomass Bioenergy* 19, 1–15 (2000).

Pamidiamarri DVNS, Singh S, Mastan SG, Patel J, Reddy MP. Molecular characterization and identification of markers for toxic and non-toxic varieties of *Jatropha curcas* L. using RAPD, AFLP and SSR markers. *Mol. Biol. Rep.* 36(6), 1357–1364 (2009).

Pandey RK, Datta SK. Gamma ray induced cotyledonary variabilities in *Jatropha curcas* L. *J. Nuclear Agricult. Biol.* 24, 62–66 (1995).

Paramathma M, Reeja S, Parthiban KT, Malarvizhi D. Development of interspecific hybrids in Jatropha. In: *Proceedings of the Biodiesel Conference Toward Energy Independence-Focus on Jatropha, June 9–10, Rashtrapati Bhawan, Hyderabad, India*. Singh B, Swaminathan R, Ponraj V (Eds). 136–142 (2006).

Parthiban KT, Senthil Kumar R, Thiyagarajan P, Subbulakshmi V, Vennila S, Govinda Rao M. Hybrid progenies in Jatropha—a new development. *Curr. Sci. India* 96(6), 815–823 (2009).

Pax F. Euphorbiaceae—Jatropheae. In: *Das Pflanzenreich IV 147*. Engler A (Ed). Leipzig: Verlag von Wilhelm Engelmann, 42 (1910).

Poehlman JM, Sleper DA. *Breeding Field Crops* (Fourth Edition). Ames, IA: Iowa State University Press (1995).

Prabakaran AJ, Sujatha M. *Jatropha tanjorensis* Ellis and *Saroja*, a natural interspecific hybrid occurring in Tamil Nadu, India. *Genet. Res. Crop Evolut.* 46, 213–218 (1999).

Punia MS. Current status of research and development on Jatropha (*Jatropha curcas*) for sustainable biofuel production in India. In: *USDA Global Conference on Agricultural Biofuels: Research and Economics, August 20–22, 2007, Minneopolis, Minnesota* (2007).

Raju AJS, Ezradanam V. Pollination ecology and fruiting behavior in a monoecious species, *Jatropha curcas* L. (*Euphorbiaceae*). *Curr. Sci. India* 83(11), 1395–1398 (2002).

Rakshit KD, Darukeshwara J, Raj KR, Narasimhamurthy K, Saibaba P, Bhagya S. Toxicity studies of detoxified Jatropha meal (*Jatropha curcas*) in rats. *Food Chem. Toxicol.* 46(12), 3621–3625 (2008).

Ranade SA, Srivastava AP, Rana TS, Srivastava J, Tuli R. Easy assessment of diversity in *Jatropha curcas* L. plants using two single-primer amplification reaction (SPAR) methods. *Biomass Bioenergy* 32(6), 533–540 (2008).

Rao GR, Korwar GR, Shanker AK, Ramakrishna YS. Genetic associations, variability and diversity in seed characters, growth, reproductive phenology and yield in *Jatropha curcas* (L.) accessions. *Trees Struct. Funct.* 22(5), 697–709 (2008).

Reed DH, Frankham R. Correlation between fitness and genetic diversity. *Conserv. Biol.* 17(1), 230–237 (2003).

Rug M, Sporer F, Wink M, Liu SY, Henning R, Ruppel A. Molluscicidal properties of *J. curcas* against vector snails of the human parasites *Schistosoma mansoni* and *S. japonicum*. In: *Biofuels and Industrial Products from Jatropha curcas—Proceedings from the Symposium "Jatropha 97"*. Gübitz GM, Mittelbach M, Trabi M (Eds). Managua, Nicaragua, 227–232 (1997).

Rupert EA, Dehgan B, Webster GL. Experimental studies of relationships in the genus *Jatropha*. L. *J. curcas J. integerrima*. *Bull. Torrey Bot. Club* 97, 321–325 (1970).

Shanker C, Dhyani SK. Insect pests of *Jatropha curcas* L. and the potential for their management. *Curr. Sci.* 91(2), 162–163 (2006).

Silvertown J, Charlesworth D. The evolution of plant life history: breeding systems. In: *Introduction to Plant Population Biology*. Silvertown J, Charlesworth D (Eds). Oxford, UK: Blackwell Publishing, 271–303 (2001).

Stacy EA. Cross-fertility in two tropical tree species. Evidence of inbreeding depression within populations and genetic divergence among populations. *Am. J. Bot.* 88(6), 1041–1051 (2001).

Sujatha M. *Genetic and Tissue Culture Studies in Castor (Ricinus communis L.) and Related Genera*. Ph.D. Dissertation, Osmania University, Hyderabad (1996).

Sujatha M. Genetic improvement of *Jatropha curcas* L.: possibilities and prospects. *Indian J. Agrofor.* 8, 58–65 (2006).

Sujatha M, Makkar HPS, Becker K. Shoot bud proliferation from axillary nodes and leaf sections of non-toxic *Jatropha curcas* L. *Plant Growth Regul.* 47(1), 83–90 (2005).

Sujatha M, Prabakaran AJ. New ornamental Jatropha through interspecific hybridization. *Genet. Resource Crop Evolut.* 50, 75–82 (2003).

Sun Q-B, Li L-F, Li Y, Wu G-J, Ge X-J. SSR and AFLP markers reveal low genetic diversity in the biofuel plant *Jatropha curcas* in China. *Crop Sci.* 48(5), 1865–1871 (2008).

Sunil N, Sivaraj N, Abraham D et al. Analysis of diversity and distribution of *Jatropha curcas* L. germplasm using geographic information system (DIVA-GIS). *Genet. Resour. Crop Ev.* 56(1), 115–119 (2009).

Sunil N, Varaprasad KS, Sivaraj N, Kumar TS, Abraham B, Prasad RBN. Assessing *Jatropha curcas* L. Germplasm in situ—a case study. *Biomass Bioenergy* 32, 198–202 (2008).

Tatikonda L, Wani SW, Kannan S et al. AFLP-based molecular characterization of an elite germplasm collection of *Jatropha curcas* L., a biofuel plant. *Plant Sci.* 176(4), 505–513 (2009).

Waller DM. The statistics and dynamics of mating system evolution. In: *The Natural History of Inbreeding and Outbreeding*. Thornhill N (Ed.), Chicago, IL: University of Chicago Press, 97–117 (1993).

Wani SP, Osman M, D'silva E, Sreedevi TK. Improved livelihoods and environmental protection through biodiesel plantations in Asia. *Asian Biotechnol. Develop. Rev.* 8(2), 11–29 (2006).

Wilbur RL. A synopsis of Jatropha, subsection *Eucurcas*, with the description of two new species from Mexico. *J. Elisha Mitch. Sci. Soc.* 70, 92–101 (1954).

Wright S. Evolution in Mendelian populations. *Genetics* 16, 97–159 (1931).

Wu HK, Matheson AC, Spencer D. Inbreeding in *Pinus radiata*. I. The effects of inbreeding on growth, survival and variance. *Theor. Appl. Genet.* 97(8), 1256–1268 (1998).